High Performance Computing in Science and Engineering '17

T0135279

and Prediction

Wolfgang E. Nagel • Dietmar H. Kröner •
Michael M. Resch

Editors

High Performance Computing in Science and Engineering '17

Transactions of the High Performance
Computing Center, Stuttgart (HLRS) 2017

 Springer

Editors
Wolfgang E. Nagel
Zentrum für Informationsdienste
und Hochleistungsrechnen (ZIH)
Technische Universität Dresden
Dresden, Germany

Dietmar H. Kröner
Abteilung für Angewandte Mathematik
Universität Freiburg
Freiburg, Germany

Michael M. Resch
Höchstleistungsrechenzentrum
Stuttgart (HLRS)
Universität Stuttgart
Stuttgart, Germany

Front cover figure: Turbulent structures in the draft tube of a hydraulic propeller turbine operated in part load condition. The turbulent flow structures are colored with the eddy viscosity ratio. A high viscosity ratio (red) indicates regions with high turbulence. Details can be found in "Numerical Analysis of a Propeller Turbine Operated in Part Load Conditions" by Bernd Junginger and Stefan Riedelbauch, Institute of Fluid Mechanics and Hydraulic Machinery, University of Stuttgart, Germany on page 355ff.

ISBN 978-3-319-88595-7   ISBN 978-3-319-68394-2   (eBook)
https://doi.org/10.1007/978-3-319-68394-2

Library of Congress Control Number: 2017964382

Mathematics Subject Classification (2010): 65Cxx, 65C99, 68U20

© Springer International Publishing AG 2018
Softcover reprint of the hardcover 1st edition 2018 978-3-319-68393-5

Printed on acid-free paper

This Springer imprint is published by Springer Nature
The registered company is Springer International Publishing AG
The registered company address is: Gewerbestrasse 11, 6330 Cham, Switzerland

Contents

Part I
Physics

Peter Nielaba

In this section, six physics projects are described, which achieved important scientific results in 2016/2017 by using Hazel Hen at the HLRS and ForHLR of the Steinbuch Center.

Fascinating new results are being presented in the following pages on soft matter systems (simulations of fibrinogen properties and of the behaviour of blood proteins at nanoparticles), astrophysical systems (simulations of galaxy formation and of photoionization and radiative charge transfer), many body quantum systems (simulations of trapped ultracold quantum systems) and high energy physics systems (simulations of partial pressures of mesons and baryons with different strangeness quantum numbers and of quantum anomalies, topologies and particle production in strongly correlated gauge theories).

Studies of the soft matter systems have focused on the properties of fibrinogen and on the behaviour of blood proteins at nanoparticles.

T. Schäfer, J. Zhou, F. Schmid and G. Settanni from the University of Mainz investigated in their project *Flexadfg* properties of fibrinogen as well as the behaviour of blood proteins in the presence of the polymer poly(ethylene-glycol) by atomistic Molecular Dynamics simulations. Their simulations reveal important interaction details and permit the construction of simplified models.

The studies of the astrophysical systems have focused on galaxy formation and on photoionization and radiative charge transfer of several particles.

V. Springel, A. Pillepich, R. Weinberger, R. Pakmor, L. Hernquist, D. Nelson, S. Genel, M. Vogelsberger, F. Marinacci, J. Naiman, and P. Torrey from Heidelberg (V.S., A.P., R.W., R.P.), Cambridge, USA (L.H., J.N., M.V., F.M., P.T.), New York

P. Nielaba (✉)
Fachbereich Physik, Universität Konstanz, 78457 Konstanz, Germany
e-mail: peter.nielaba@uni-konstanz.de

(S.G.) and Garching (D.N.), present in their project *GCS-ILLU* results from a new generation of hydrodynamical simulations ("IllustrisTNG" project, AREPO code), including new black hole physics, and using more accurate techniques, an enlarged dynamical range, and system sizes with side lengths of up to 300 megaparsec. The authors report on early results on matter and galaxy clustering and find very good agreement with observation data for the two-point galaxy correlation function. According to their investigations, the effect of baryonic physics has to be taken into account in cosmological precision studies.

B.M. McLaughlin, C.P. Ballance, M.S. Pindzola, P.C. Stancil, J.F. Babb, S. Schippers and A. Müller from the Universities of Belfast (B.M.M., C.P.B.), Auburn (M.S.P.), Georgia (P.C.S.), Cambridge USA (J.F.B.), and Giessen (S.S., A.M.) investigated in their project *PAMOP* atomic, molecular and optical collisions on petaflop machines, relevant for astrophysical applications, for magnetically-confined fusion and plasma modeling, and as theoretical support for experiments and satellite observations. The Schrödinger and Dirac equations have been solved with the R-matrix or R-matrix with pseudo-states approach, and the time dependent close-coupling method has been used. Various systems and phenomena have been investigated, ranging from the photoionization of chlorine ions, of cobalt ions and of tungsten ions, and of He and H_2, to the collision scenario of SiO and H_2, and to the radiative charge transfer in systems of carbon atoms and He ions, and the radiative association in systems of carbon atoms and H ions.

In addition to the projects discussed above, interesting general relativistic properties of neutron stars and mergers have been investigated in project *HypeBBH* by using Hazel Hen.

In the last granting period, quantum mechanical properties of high energy physics systems have been investigated as well as the quantum many body dynamics of trapped bosonic systems.

P. Alba, R. Bellwied, S. Borsanyi, Z. Fodor, J. Günther, S.D. Katz , V. Mantovani Sarti, J. Noronha-Hostler, P. Parotto, A. Pasztor, I. Portillo Vazquez, and C. Ratti from the Universities of Frankfurt (P.A.), Houston (R.B., J.N.-H., P.P., I.P.V., C.R.), Wuppertal (S.B., Z.F., J.G., A.P.), Budapest (Z.F., S.D.K.), Torino (V.M.S.), and the FZ Jülich (Z.F.) computed in their project *GCS-POSR* the partial pressures of mesons and baryons with different strangeness quantum numbers by lattice simulations in the confined phase of quantum chromodynamics. A comparison of the effect of different hadronic spectra on thermodynamic observables in the Hadron Resonance Gas model with results from lattice quantum chromodynamics simulations reveals that additional states, e.g. from the quark model or the particle data group booklet, are are required in the Hadron Resonance Gas model for a better agreement.

N. Mueller, J. Berges, O. Garcia, and N. Tanji from the University of Heidelberg investigate in the project *RTGT* quantum anomalies, topologies and particle production in strongly correlated gauge theories by real-time lattice simulations in the Glasma model, using computer time at the ForHLR. The project has achieved many interesting results for various high energy physics phenomena, ranging from out-of-equilibrium transient anomalous charge production from coherent color fields, to the anomalous transport in ultra-relativistic heavy ion collisions, and to the anomalous dynamical refringence in quantum electrodynamics experiments beyond the Schwinger limits.

O.E. Alon, R. Beinke, C. Bruder, L.S. Cederbaum, S. Klaiman, A.U.J. Lode, K. Sakman, M. Theisen, M.C. Tsatsos, S.E. Weiner, and A.I. Streltsov from the Universities of Haifa (O.E.A.), Heidelberg (R.B., L.S.C., S.K., M.T., A.I.S.), Basel (C.B., A.U.J.L.), Wien (K.S.), Sao Paulo (M.C.T.), and Berkeley (A.S.) studied in their project *MCTDHB* trapped ultracold atomic systems by their method termed multiconfigurational time-dependent Hartree for bosons (MCTDHB). The principal investigators investigated the interplay of light and ultra-cold matter on fragmented superradiance, the effect of angular momentum on fragmented vortices, the phenomena accompanying a splitting of a two-dimensional Bose-Einstein-Condensate, many-body excitations in finite condensates in double-well and triple-well traps, and the differences of the Bose-Einstein-Condensate wave function in mean-field and many-body computations.

Blood Proteins and Their Interactions with Nanoparticles Investigated Using Molecular Dynamics Simulations

Timo Schäfer, Jiajia Zhou, Friederike Schmid, and Giovanni Settanni

Abstract Blood proteins play a fundamental role in determining the response of the organism to the injection of drugs or, more in general, of therapeutic preparations in the blood stream. Some of these proteins are responsible for mediating immune response and coagulation. Nanoparticles, which are being intensely investigated as possible drug nanocarriers, heavily interact with blood proteins and their ultimate fate is determined by these interactions. Here we report the results of molecular dynamics simulations of several blood proteins aimed to determining their possible behavior at the nanoparticle surface. On one hand we investigated the behavior of fibrinogen, a glycoprotein, which polymerizes into fibrin during coagulation. On the other hand we investigated the behavior of several blood proteins in the presence of the polymer poly (ethylene-glycol), often used as nanoparticle coating to reduce unspecific interactions with the surrounding environment.

1 Introduction

Our research activity, centered on studying the behavior of blood proteins at the interface with (nano)materials has followed two main directions: on one hand, after revealing the flexibility of Fibrinogen [1], a central protein of the coagulation cascade, and its adsorption properties on model surfaces [2], we have studied with enhanced sampling techniques the full extent of its flexibility, paving the way to effective simulations of its adsorption process on realistic surface; on the other hand we have studied the interactions of proteins with poly(ethylene-glycol), a polymer which is very frequently used to improve the properties of therapeutic preparations including nanoparticles. The report is thus subdivided in two sections.

T. Schäfer • F. Schmid • G. Settanni (✉)
Institut für Physik, Johannes Gutenberg University, Mainz, Germany
e-mail: settanni@uni-mainz.de

J. Zhou
Institute of Physics, Johannes-Gutenberg University, Mainz, Germany

© Springer International Publishing AG 2018
W.E. Nagel et al. (eds.), *High Performance Computing in Science and Engineering '17*, https://doi.org/10.1007/978-3-319-68394-2_1

5

2   Metadynamics of the Fibrinogen Protomer

Fibrinogen (Fg) is a multi-chain serum glycoprotein consisting of a dimer of identical protomers, each made up of three different peptide chains Aα, Bβ and γ and two carbohydrates (Fig. 1). Fg has a molecular weight of 340 kDa and an elongated structure. Each protomer consists of the globular D- and E-domains, connected via a coiled-coil region [3]. At the dimerization interface in the E-domain, the three chains of either protomer are connected covalently via disulphide bridges. Each protomer is diglycosylated at the D-domain and the coiled-coil region. Parts of ends of the three peptide chains are unstructured and flexible.

Fg is a major element in blood clot formation [4]. In the coagulation cascade, thrombin transforms Fg into fibrin, one of the main components of blood clots. Fibrinolysis, the cleavage of the fibrin mesh by the enzyme plasmin, controls the dissolution of the blood clot. In this process, plasmin cleaves fibrin at specific plasmin cleavage sites, some of which are located at the coiled-coil region between the D- and E-domains.

In our previous investigations of Fg dynamics [1] and adsorption behavior [2], carried out using classical atomistic molecular dynamics simulations, the presence of consistent bending motions at a hinge centered on the coiled-coil region of each protomer was observed. It was shown that these motions could represent a conserved mechanism for the exposure of early plasmin cleavage sites.

While Fg underwent significant bending motions in our molecular dynamics simulations under physiological conditions, considerable simulation time would be necessary to fully sample them due to their large timescale and variability. To speed up their sampling and reduce the computational burden, we have employed metadynamics [5], an enhanced sampling method which biases the system towards unexplored regions of the conformational space. In these simulations, the projections of the motion along its principal components (obtained via principal component analysis (PCA) [6] of our previous simulations of Fg [1]) were adopted as collective variables (CVs) that capture the main characteristics of Fg motion. In a metadynamics simulation, an external biasing potential acting on the CVs is steadily built up to force the system and explore new regions of CV space.

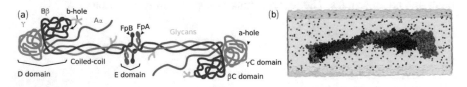

Fig. 1 Fibrinogen Structure and protomer system: (**a**) Sketch of the Fibrinogen Dimer. (**b**) Van der Waals representation of the protomer in solution, water molecules omitted for clarity, sodium and chloride ions in blue and green, respectively

2.1 Methods

The crystal structure of human Fg (PDB ID: 3GHG) [3] was used for the initial atomic coordinates of the protein. Due to the flexibility of the αC region and the N-terminal regions of all three chains, they do not reach a consistent stable state in the crystal, leaving their coordinates unresolved. Only the resolved carbohydrate group attached at residue β364 was included. One protomer (with the addition of only a small segment of the second protomer to complete the E-domain) was used as the minimal system that captures all relevant aspects of Fg structure.

The protomer was solvated in a rectangular simulation box with explicit TIP3P water [7] and a physiological concentration of ions (0.15 Mol [NaCl]) using VMD [8]. The initial size of 10.14 nm $\times$ 30.33 nm $\times$ 10.14 nm of the periodic unit cell of the system ensured a minimal padding of 1.5 nm, with a total number of atoms of 316,862. Preparation of the system was performed with GROMACS 4.6.5 [9], using an energy minimization followed by an equilibration in the Isothermal-isobaric ensemble at a temperature of 310 K and a pressure of 1 atm for 150 ps. A stochastic velocity rescaling thermostat [10] with a time constant of 0.1 ps and the Parrinello-Rahman barostat [11] with a time constant of 1 ps were used. The CHARMM36 force field with CMAP correction [12] was employed. Cutoffs of 1.2 nm for van der Waals and short-range electrostatic interactions were used, with treatment of long-range electrostatic interactions via Particle-Mesh-Ewald with a grid spacing of 0.1 nm. The first 50 ps of equilibration were performed using a time step of 1 fs, for everything else, a time step of 2 fs was used. All covalent bonds involving hydrogen were constrained in their length.

The metadynamics bias potential V is built up over the course of a simulation by periodically adding (i.e., at regular intervals τ) to the potential energy of the system a Gaussian function in CV space centered at the current position of the system s_t. For the purpose of performing metadynamics of Fg, version 5.0.4 of GROMACS [13] patched with PLUMED 2.2 [14] was used.

Results of principal component analysis (PCA) [6] performed on the Cα atoms of equilibrium trajectories of the Fg protomer in solution [1] were used to characterize relevant degrees of freedom. The first two PCs, describing bending motions at the hinge region, were used as CVs for the metadynamics. A multiple walker implementation of metadynamics with 10 parallel simulations sharing the same bias potential was used to speed up the sampling of PC space. Gaussian hills with a fixed height of $w = 0.7437 \frac{kJ}{mol}$, corresponding to a thermal energy of 0.3 $k_B T$ were deposited by each walker every $\tau = 4$ ps. The Gaussian width of the hills was set to $\delta = 100$ nm^2 for both PCs.

Due to the significant computational cost of calculating the PCs, a multiple time stepping (MTS) integrator of order 10 was applied to the bias potential [15]. To ensure sufficient accuracy of the simulation, the effective energy drift [15] introduced by the MTS integrator was monitored. The effective energy of the bias potential describes the generalized detailed balance violation of the system, taking into account the inherent non-equilibrium character of metadynamics.

GROMACS is a molecular dynamics package optimized for performance and scaling. PLUMED provides the implementation of the metadynamics scheme and interfaces with the GROMACS code. In this way, PLUMED extracts coordinates and forces from GROMACS at runtime and adds the forces caused by the metadynamics bias as well as providing the communication between the individual simulations in the multiple walker scheme. For the production run of the simulations performed on Hazelhen at the High performance computing center Stuttgart, GRO-MACS and PLUMED were compiled specifically for the CRAY XC40 architecture with MPI parallelization. Good scaling up for up to thousands of cores was achieved. Each production job used 100 nodes, lasted at most 3.5 h and gathered 5 ns of trajectory distributed equally onto the ten walkers.

2.2 Results

The length of the individual trajectories of the walkers reached 100 ns, resulting in 1 μs of total simulation time. At one part of the simulation, jumps in the PCs appeared as artifacts, due to the ambiguity of the optimal alignment of instantaneous and reference coordinates. The decision was therefore made to remove this part of the simulations from all further analysis, with a total of 980 ns of trajectories remaining. For the remaining simulations the average effective energy drift of $4.8 \frac{kJ}{mol \cdot ns}$ was comparatively low, indicating multiple time stepping without significant losses in accuracy.

At convergence, the negative of the accumulated bias potential of metadynamics simulations provides an estimate of the underlying free energy surface of the system in terms of the CVs. The free energy landscape of Fg as identified by our simulations showed a single minimum centered roughly around the crystallographic conformation (Fig. 2a). The minimum was relatively deep, differing by $158 \frac{kJ}{mol}$ from

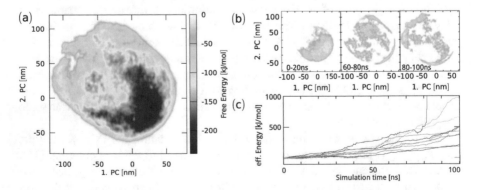

Fig. 2 Free energy surface: (**a**) Estimated free energy from full simulation. The blue dot marks the position of the crystallographic structure. (**b**) Free energy contribution of each walker from 0 to 20 ns, 60 to 80 ns and 80 to 100 ns of simulation time. (**c**) Effective energy of each walker, third walker (blue) reaches 35,000 kJ/mol after 100 ns

the average value of free energy measured across the sampled regions. In the initial part of the simulations the various walkers sampled mostly the minimum region. Later, the accumulated bias potential filled up this minimum allowing all the walkers to explore a wide area of CV space.

One pair of collective variables that has shown its use in the description of the protomer is the bending and torsion angles γ and ϕ. γ is the angle between the center of mass of E-domain, hinge and D-domain, while two different points in the E-domain together with the centers of hinge and D-domain were used to define the torsion angle ϕ. The correlation between bending and torsion at the hinge and the PCs that was noted before [1] was also present in the metadynamics simulations: the Pearson correlation coefficient r between γ and a linear combination of the first two PCs (1st–2nd–3rd) was 0.98, while r was 0.45 between ϕ and a combination of the first four PCs (-2nd/$5+3$rd$+4$th).

Figure 3a shows the overall distribution of the bending angle γ and the torsion angle ϕ as observed in the metadynamics simulations presented here. The angle distribution showed that Fg reaches heavily bent conformations with a bending angle below 30°, and that such a bending is only possible at certain torsion angles. In general, torsion angles around $\pm 180°$ or $-50°$ allowed strong bending, while values of ϕ around 70° heavily restrict it.

Analysis reveals two distinct bending modes (B1 and B2) characterized by different values of the torsion angle (Fig. 3) as well as an area where bending is heavily restricted (R). B1 and B2 conformations were clearly separated in PC-space, occupying roughly half of the accessible PC-space each. Figure 3 shows representative snapshots of the R, B1 and B2 conformations.

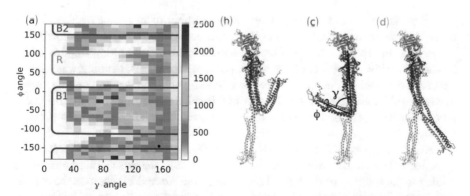

Fig. 3 Bending modes of the fibrinogen protomer: (**a**) Distribution of bending and torsion angles γ and ϕ. γ is the angle between the center of mass of the Cα atoms of the D-domain (β200–458, γ140–394), the hinge region (α99–110, β130–55, γ70–100) and the E-domain(α50–58, β82–90, γ23–31). For ϕ, the centers of mass of the Aα and γ chain or the Bβ and γ chain of the E-domain are used separately to create four reference points together with the centers of mass of the hinge region and the D-domain. The black dot marks the position of the crystallographic structure in angle space. (**b–d**) Representative snapshots showing bending in B1, B2 and R conformations

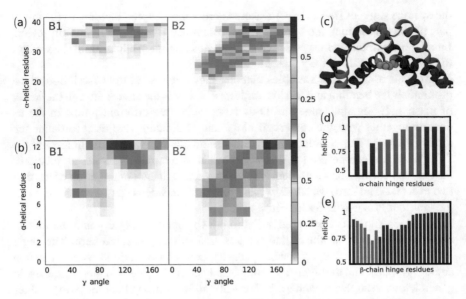

Fig. 4 α-helical content of the hinge region. (**a**) Distribution of the bending angle γ and the number of α-helical residues in the hinge region (α99–110, β130–155) of the Aα and Bβ chains for B1 and B2 conformations. (**b**) Distribution of the bending angle γ and the number of α-helical residues in the plasmin cleavage sites (α102–107, β131–136) for B1 and B2 conformations. (**c**) Bent B1 conformation, showing exposed plasmin cleavage sites. (**d–e**) Frequency of α-helical secondary structure by residue for the Aα and Bβ chain, respectively. Plasmin cleavage sites highlighted

Figure 4a shows the relation between the number of α-helical residues in the hinge and the bending angle. In bent B1 conformations (bending angle below 120°) the Aα and Bβ chains lost 2.4 of their α-helical residues in the hinge region, while it was 7.6 residues in B2 conformations. In R conformations, bending was heavily suppressed especially, never reaching low bending angles. Even at bending angles between 100° and 120° 10 residues lost their α-helical structure in the Bβ chain alone. Generally, the non α-helical residues formed hydrogen-bonded turns, bends or loops.

Overall, the simulations showed that bending occurs along two preferential directions B1 and B2, while it is heavily restricted otherwise. This behavior is related to the stability of the helical segments of the Aα and Bβ chain in the coiled-coil region. Bending in B1 conformations only required small non-helical turns in the hinge region of the Aα and Bβ chains. Bending in B2 conformations on the other hand relied on significantly larger regions of flexible, non-helical residues in the Aα and Bβ chains. In R conformations the Bβ chain heavily restricted bending, even slight bending required large losses of α-helical structure.

Apart from an impact on overall flexibility of Fg and fibrin fibers, bending in the hinge region was also relevant for the exposure of plasmin cleavage sites, as previously hypothesized [16]. The two cleavage sites at the residues α103–104 and β133–134 are part of the hinge region of the Aα and Bβ chain, respectively. Previous simulations of Fg showed the correlation between strong bending and non-helical secondary structure around the cleavage sites [1].

This result could be confirmed here for both the B1 and B2 bending modes. While the B1 bending generally required only a few non-helical residues in the Aα and Bβ chains (Fig. 4a), they were predominantly located at the plasmin cleavage sites. The frequency of α-helical secondary structure for the residues in the hinge region presented in Fig. 4d–e showed that the residues around the plasmin cleavage sites lost helicity more frequently than other parts of the hinge region. This means that bending of Fg in either B1 or B2 conformations resulted in an increased exposure of the plasmin cleavage sites to the surrounding environment.

3   Interactions Between Proteins and Poly(Ethylene-Glycol) Investigated Using Molecular Dynamics Simulations

Poly(ethylene-glycol) (PEG) is a non-toxic polymer highly soluble both in water and non-polar solvents and finds application in a broad variety of contexts, including for example, as dispersant in toothpastes, lubricant in skin creams and anti-foaming agent in the food industry. In biomedical applications, PEG, when attached to a therapeutic molecule or nanoparticle (PEGylation), confers them extra resistance to degradation, extended blood clearance time and, ultimately, improved therapeutic efficacy. These improvements are obtained by a combination of factors like increased overall volume, which prevents renal clearance, but also a reduction of unspecific interactions with the surrounding environment. This last phenomenon goes under the name of stealth effect, and results in the ability of PEG to screen the PEGylated species from the immune response of the host organism.

Although PEG helps reduce unspecific interactions with the surrounding biological milieu, it does not remove them completely. It has been shown that, also due the constant exposure to PEG-containing products, the organism can produce antibodies against PEG. In addition, many different proteins adsorb on PEG-coated nanoparticles, as soon as these enter in contact with blood [17]. A molecular understanding of these protein adsorption phenomena on PEG-coated surfaces and more in general of the interactions between PEG and proteins is still missing.

Atomistic molecular dynamics simulations can help to fill this gap. They have been successfully used to address protein adsorption on a large variety of surfaces [18, 19], as well as the behavior of specific blood proteins on model surfaces [1, 2], While accurate atomistic models and force field already exist for many metallic, metal-oxides and some polymeric surfaces [19], the surface exposed by PEGylated species poses extra challenges due to its brushy nature. These challenges have been recently addressed by simulating the proteins of interest in a mixture of water and PEG molecules [20, 21]. With this approach, which is similar in spirit to what done in computational drug screening exercises [22, 23], the possible binding modes of

PEG with the protein surface are sampled accurately and their relative importance is determined from the simulations.

Here, we report the results obtained with these simulations, which were performed mostly on Hazelhen and have been published in ref. [20] and in ref. [24] where we have also determined preferential binding coefficients of several blood proteins for PEG.

3.1 Methods

All the simulations were carried out using the program NAMD [25], specifically compiled for the Cray XC40 architecture, and the charmm27 force field [26] with the extension for PEG [27]. Tip3p [7] was used as model for the explicit treatment of water. An integration time step of 1 fs was used across the simulations. Simulations were carried out using periodic boundary conditions. 1 atm Pressure and 300 K temperature were maintained constant during the simulations using the Langevin piston algorithm and Langevin thermostat [28, 29]. A cutoff of 1.2 nm was used for the non-bonded interactions with a switch function. Long range electrostatic interactions were treated using the smooth particle mesh Ewald (PME) method [30] with a grid spacing of about 0.1 nm. To prepare the PEG-water mixture, 64 PEG molecules ($H-[O-CH_2-CH_2]_n-OH$, with n either 4 or 7) were placed on a 4 × 4 × 4 grid with 1.0 nm spacing between grid points. Then 1 ns high temperature (700 K) simulations in vacuum with damped electrostatics interactions (dielectric constant 200) were run to randomize the initial dihedral distribution of the PEG molecules. The PEG molecules were then immersed in a box of water molecules and sodium and chlorine ions were added to reach physiological concentration (0.15 M). Mixtures with different concentrations of PEG were obtained by changing the size of the water box surrounding the PEG molecules. The prepared mixtures were then equilibrated first at high temperature (373 K) for 1.0 ns and then at 300 K for 1.0 ns. Several plasma proteins have been considered: human serum albumin(HSA), bovine serum albumin(BSA), transferrin(TF) and apolipoprotein A1 (ApoA1). The initial coordinates of the proteins were taken from the Protein Data Bank(PDB) (see Table 1 for the list of PDBids). Each protein was immersed in a box filled by replicating the coordinates of the PEG-water mixtures obtained before in the three space directions and removing mixture atoms in close contact with protein atoms. The final PEG concentration in the simulation boxes is reported in Table 1. The size of the boxes was large enough to leave at least 1.0 /nm from each protein atom and the box boundary. In the case of HSA, larger box sizes with 1.5 and 2.0 nm distances between protein and box boundary were simulated to investigate the dependence of the simulation results upon box size. The total charge of the systems was further neutralized by changing an adequate number of water molecules into ions. The complete systems were then minimized using the steepest descent algorithm for 10,000 steps with harmonic restraints on the heavy atoms of the proteins. Then the systems were equilibrated at room temperature and pressure for 1.0 ns during which

Table 1 List of the simulations performed

System	Box size (Å)	N. Atoms	PEG length	[PEG] (g/ml)	Simulation time (ns)
HSA+lipids	103	114,000	4	0.13	4×105
HSA	98.8	100,881	4	0.08	4×100
	98.2	99,301	4	0.11	4×100
	108.6	134,134	4	0.12	4×125
	118.2	172,541	4	0.12	5×100
BSA	103.6	116,489	4	0.11	4×100
Transferrin	108	132,009	4	0.11	5×200
	108.7	134,663	4	0.07	5×200
	108.4	133,303	7	0.07	5×200
	108.8	134,372	7	0.04	5×200
ApoA1	192	739,653	4	0.13	$4 \times \approx 100$

the harmonic restraints were gradually removed and for 1.0 ns without restraints. Finally production runs were started with 4 or 5 replicas for each system. Length of the runs ranged from 100 to 200 ns (see Table 1). The simulations have been run using NAMD an MPI-based molecular dynamics simulation engine which has been specifically compiled for the Cray XC40 architecture of Hazelhen. The large number of atoms included in the simulations (10^5–10^6) assure good scaling up to thousands cores. Typical job sizes of 100 nodes have been used for the simulations. Typical job length of 30 min–3 h were submitted. Depending on system size this has allowed to collect from 1 to 8 ns/h/100nodes.

The direct PEG-protein interactions along the simulations were measured using the NAMD pair-interaction utility. The time series along the trajectories of the number of PEG and water heavy atoms found within a distance d of the protein or of each amino acid type was determined using the "pbwithin" selection command of VMD. The ratio of these numbers was compared to the ratio between all the PEG and water heavy atoms in the simulation box (bulk ratio). A PEG/water ratio larger than bulk for an amino acid implies the presence of an effective attractive interaction between PEG and the amino acid. On the other hand, amino acids with a PEG/Water ratio smaller than bulk exert and effective repulsion for PEG. For the residue specific PEG/Water ratio calculations we used $d = 5.0$ Å. The preferential binding coefficient measures the excess of PEG found in the local environment of the protein, i.e., in its vicinity (Fig. 5) and was defined as:

$$\Gamma = [PEG]_L - [H_2O]_L \frac{[PEG]_B}{[H_2O]_B} \tag{1}$$

where $[\cdot]_L$ is the concentration of PEG or water in the vicinity of the protein and $[\cdot]_B$ the concentration in the bulk, that is, away from the protein. The local environment of the protein is defined as a shell of thickness d around the protein. Γ converges for $d \geq 9.0$ Å, which is the value we used for these calculations.

Fig. 5 The concentration of
PEG in the local environment
of the protein and in the bulk
solution are used for the
measure of the preferential
binding coefficient Eq. (1)

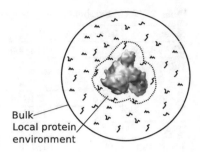

Bulk
Local protein
environment

3.2   Results and Discussion

Four different plasma proteins have been simulated immersed in a PEG/water
mixture, as reported in Sect. 3.1. Three of them (HSA, BSA, TF) are relatively
abundant in plasma but not on the surface of PEGylated nanoparticles and one
(ApoA1) is abundant on the nanoparticle [17]. The simulations were performed
at different PEG concentrations to verify a possible dependence of the results on
this parameter. Additional simulations explored the role of PEG length and box
size. Each simulation was repeated 4–5 times to improve the statistical accuracy of
the results. Only short PEG molecules were considered (4 or 7 monomers), which
diffuse fast and allow for a thorough exploration of the possible binding modes
with the protein. This insures that the observed quantities converge in the accessible
simulation times (100–200 ns). It is worth noting that, since the persistence length
(3.8 Å) of PEG is comparable with the distance between monomers, longer PEG
chains are expected to explore the same binding modes as the shorter chains
simulated here. The simulations show that the protein structures are not affected
by the presence of PEG. The average root mean square deviation with respect to
the starting structure or, in the case of ApoA1 to a control run in pure water, is in
agreement with native state fluctuations of proteins of similar size (3–5 Å for HSA,
BSA and TF).

   PEG molecules in the simulations reach and leave the surface of the protein
continuously, in a rather dynamical equilibrium. However, some regions of the
surface are more frequently populated by PEG (Fig. 6).

   We, then, analyzed in more detail the possible origin of the observed uneven
distribution of PEG on the protein surface. We hypothesized that each amino-acid
type may have a different affinity for PEG molecules. To test the hypothesis, we
measured the average ratio between the number of PEG and water heavy atoms
(PEG/Water ratio) in a 5.0 Å shell surrounding each amino acid (Fig. 7). The data
show that some amino acids are more likely to have a concentration of PEG higher
than bulk, while others have a smaller concentration. In other words, some amino
acids exert an effective attraction for PEG while other exert a repulsion. The PEG-
amino acid interaction pattern look also very similar across the different simulations.
Indeed, we verified that the residue-specific PEG/Water ratio, normalized to the bulk
value in each simulation, are highly correlated across the simulations, at least for

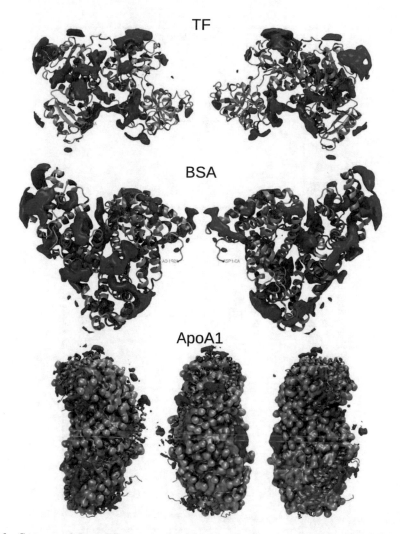

Fig. 6 Cartoons of three different simulated proteins (cyan) where regions with PEG density larger than twice as in the bulk have been highlighted (dark blue). Left and right and center pictures represent different views of the proteins. Negatively charged residues are highlighted in red. In the case of ApoA1, the lipids present in the apolipoprotein are also rendered as a continuous surface

the most exposed amino acids (Fig. 8). These information, obtained from similar simulations, allowed us to describe PEG-protein interactions with a simplified model based on the solvent accessibility of each amino acid on the protein surface [20].

A further analysis of the data consists in the measurements of the preferential binding coefficient of each protein for PEG. Preferential binding coefficients are derived from the Kirkwood-Buff integrals of the pair correlation functions in a

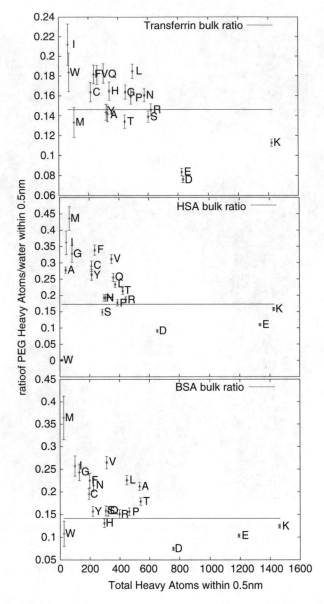

Fig. 7 Ratio between PEG heavy atoms and water molecules within 5 Å of each residue types vs sum of PEG+water (indicating the degree of exposure of each residues type). Negatively charged residues "repel" PEG

multi-component system [31]. In particular they measure the excess number of cosolvent molecules in the vicinity of each protein molecule, thus they are positive in case of a preferential binding of the cosolvent to the protein, and negative otherwise. Preferential binding coefficient are thermodynamic parameters that can

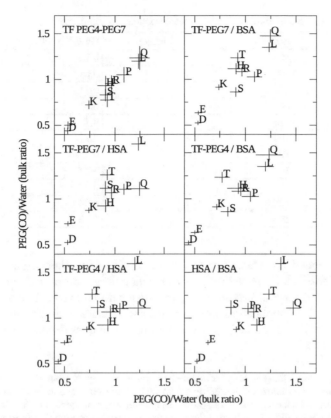

Fig. 8 Correlation plots of the observed residue specific PEG/water ratios between the various simulations. Correlation coefficients all above 0.8 indicate the substantial independence of the values from the parameters of the simulations

be accessed experimentally through high precision densitometry experiments [32], differential scanning calorimetry [33] and vapor pressure osmometry [34]. In the present case, they report the affinity of each protein for the cosolvent (PEG) *versus* the solvent (water) and can be approximated using Eq. (1) [35]. The measurements we have performed on the simulations (Table 2) show in most cases positive coefficients very close to zero, indicating a very weak preference of the analyzed proteins for PEG.

Concluding, we show how atomistic molecular dynamics simulations can be used fruitfully to assess the interaction of plasma proteins with PEG. The simulations reveal that the interaction patterns with surface amino acids are similar across the different proteins and the different simulation conditions analyzed here. This fact has allowed for building simplified models of protein interactions with PEG [20]. The simulations provide also direct measurements of the preferential binding coefficients which could be assessed experimentally to test and help improve the available force fields. The approach used in the present work is rather general and

Table 2 Preferential binding coefficients measured along the simulations

System	Box size (Å)	PEG length	[PEG] (g/ml)	Γ (n.mol.)
HSA+lipids	103	4	0.13	3.4 (1.2)
HSA	98.8	4	0.08	5.0 (0.8)
	98.2	4	0.11	7.0 (1.3)
	108.6	4	0.12	2.9 (0.9)
	118.2	4	0.12	−0.1 (1.1)
BSA	103.6	4	0.11	0.0 (1.0)
Transferrin	108	4	0.11	4.8 (0.9)
	108.7	4	0.07	7.5 (1.0)
	108.4	7	0.07	4.7 (0.7)
	108.8	7	0.04	2.9 (0.7)
ApoA1	192	4	0.13	−1.2 (1.6)

could be extended to evaluate interactions of proteins with other polymers that, like PEG, are highly soluble in waters and do not induce large conformational changes to proteins. Indeed, large efforts are being made to identify polymers that could replace PEG in biomedical applications and solve the issues related to its possible long-term toxicity and nonbiodegradability. We expect the techniques developed here to help in that direction.

Acknowledgements TS gratefully acknowledges financial support from the Graduate School Materials Science in Mainz. GS gratefully acknowledges financial support from the Max-Planck Graduate Center with the University of Mainz. We gratefully acknowledge support with computing time from the HPC facility Hazelhen at the High performance computing center Stuttgart and the HPC facility Mogon at the university of Mainz. This work was supported by the German Science Foundation within SFB 1066 project Q1.

References

1. S. Köhler, F. Schmid, G. Settanni, PLoS Comput. Biol. **11**(9), 1 (2015)
2. S. Köhler, F. Schmid, G. Settanni, Langmuir **31**(48), 13180 (2015)
3. J. Kollman, L. Pandi, M. Sawaya, M. Riley, R. Doolittle, Biochem **48**(18), 3877 (2009)
4. G. Marguerie, E. Plow, T. Edgington, J. Biol. Chem. **254**(12), 5357 (1979)
5. A. Laio, M. Parrinello, Proc. Natl. Acad. Sci. USA **99**(20), 12562 (2002)
6. A.E. García, Phys. Rev. Lett. **68**, 2696 (1992)
7. W.L. Jorgensen, J. Chandrasekhar, J.D. Madura, R.W. Impey, M.L. Klein, J. Chem. Phys. **79**(2), 926 (1983)
8. W. Humphrey, A. Dalke, K. Schulten, J. Mol. Graph. **14**, 33 (1996)
9. B. Hess, C. Kutzner, D. van der Spoel, E. Lindahl, J. Chem. Theory Comput. **4**(3), 435 (2008)
10. G. Bussi, D. Donadio, M. Parrinello, J. Chem. Phys. **126**(1), 014101 (2007)
11. M. Parrinello, A. Rahman, J. Appl. Phys. **52**(12), 7182 (1981)
12. R.B. Best, X. Zhu, J. Shim, P.E.M. Lopes, J. Mittal, M. Feig, A.D. Mackerell Jr., J. Chem. Theory Comput. **8**(9), 3257 (2012)

13. M.J. Abraham, T. Murtola, R. Schulz, S. Páll, J.C. Smith, B. Hess, E. Lindahl, SoftwareX **1-2**, 19 (2015)
14. G.A. Tribello, M. Bonomi, D. Branduardi, C. Camilloni, G. Bussi, Comput. Phys. Commun. **185**(2), 604 (2014)
15. M.J. Ferrarotti, S. Bottaro, A. Párez-Villa, G. Bussi, J. Chem. Theory Comput. **11**(1), 139 (2015)
16. R.F. Doolittle, D.M. Goldbaum, L.R. Doolittle, J. Mol. Biol. **120**(2), 311 (1978)
17. S. Schöttler, G. Becker, S. Winzen, T. Steinbach, K. Mohr, K. Landfester, V. Mailänder, F.R. Wurm, Nat. Nanotechnol. **11**, 372–377 (2016)
18. H. Heinz, H. Ramezani-Dakhel, Chem. Soc. Rev. **45**(2), 412 (2016)
19. M. Ozboyaci, D.B. Kokh, S. Corni, R.C. Wade, Q. Rev. Biophys. **49**, e4 (2016)
20. G. Settanni, J. Zhou, T. Suo, S. Schöttler, K. Landfester, F. Schmid, V. Mailänder, Nanoscale **9**(6), 2138 (2017)
21. Q. Shao, Y. He, A.D. White, S. Jiang, J. Chem. Phys. **136**(22), 225101 (2012)
22. N. Basse, J.L. Kaar, G. Settanni, A.C. Joerger, T.J. Rutherford, A.R. Fersht, Chem. Biol. **17**(1), 46 (2010)
23. J. Seco, F.J. Luque, X. Barril, J. Med. Chem. **52**(8), 2363 (2009)
24. G. Settanni, J. Zhou, F. Schmid, CSP2017 (accepted)
25. J.C. Phillips, R. Braun, W. Wang, J. Gumbart, E. Villa, C. Chipot, R.D. Skeel, L. Kale, K. Schulten, J. Comput. Chem. **26**, 1781 (2005)
26. A.D. Mackerell, M. Feig, C.L. Brooks, J. Comput. Chem. **25**(11), 1400 (2004)
27. H. Lee, R.M. Venable, A.D. Mackerell, R.W. Pastor, Biophys. J. **95**(4), 1590 (2008)
28. G.J. Martyna, D.J. Tobias, M.L. Klein, J. Chem. Phys. **101**(5), 4177 (1994)
29. S.E. Feller, Y. Zhang, R.W. Pastor, B.R. Brooks, J. Chem. Phys. **103**(11), 4613 (1995)
30. U. Essmann, L. Perera, M.L. Berkowitz, T. Darden, H. Lee, L.G. Pedersen, J. Chem. Phys. **103**(19), 8577 (1995)
31. J.G. Kirkwood, F.P. Buff, J. Chem. Phys. **19**(6), 774 (1951)
32. J.C. Lee, S.N. Timasheff, J. Biol. Chem. **256**(14), 7193 (1981)
33. N. Poklar, N. Petrovčič, M. Oblak, G. Vesnaver, Protein. Sci. **8**(4), 832 (1999)
34. E.S. Courtenay, M.W. Capp, C.F. Anderson, M.T. Record, Biochemistry **39**(15), 4455 (2000)
35. B.M. Baynes, B.L. Trout, J. Phys. Chem. B **107**(50), 14058 (2003)

Cosmic Large-Scale Structure
in the IllustrisTNG Simulations

Volker Springel, Annalisa Pillepich, Rainer Weinberger, Rüdiger Pakmor, Lars Hernquist, Dylan Nelson, Shy Genel, Mark Vogelsberger, Federico Marinacci, Jill Naiman, and Paul Torrey

Abstract We have finished two new, extremely large hydrodynamical simulations of galaxy formation that significantly advance the state of the art in cosmology. Together with accompanying dark matter only runs, we call them 'IllustrisTNG', the next generation Illustris simulations. Our largest and most ambitious calculation follows a cosmological volume 300 megaparsecs on a side and self-consistently solves the equations of magnetohydrodynamics and self-gravity coupled to the fundamental physical processes driving galaxy formation. We have employed AREPO, a sophisticated moving-mesh code developed by our team over the past 7 years and

V. Springel (✉)
Zentrum für Astronomie der Universität Heidelberg, Astronomisches Recheninstitut, Mönchhofstr. 12-14, 69120 Heidelberg, Germany

Heidelberg Institute for Theoretical Studies, Schloss-Wolfsbrunnenweg 35, 69118 Heidelberg, Germany
e-mail: volker.springel@h-its.org

A. Pillepich
Max-Planck Institute for Astronomy, Königstuhl 17, 69117 Heidelberg, Germany
e-mail: pillepich@mpia-hd.mpg.de

R. Weinberger • R. Pakmor
Heidelberg Institute for Theoretical Studies, Schloss-Wolfsbrunnenweg 35, 69118 Heidelberg, Germany
e-mail: rainer.weinberger@h-its.org; ruediger.pakmor@h-its.org

L. Hernquist • J. Naiman
Center for Astrophysics, Harvard University, 60 Garden Street, Cambridge, MA 02138, USA
e-mail: lars.hernquist@cfa.harvard.edu; jill.naiman@cfa.harvard.edu

D. Nelson
Max-Planck Institute for Astrophysics, Karl-Schwarzschild-Str. 1, 85740 Garching, Germany
e-mail: dnelson@mpa-garching.mpg.de

S. Genel
Department of Astronomy, Columbia University, 550 W. 120th St., New York, NY 10027, USA
e-mail: shygenelastro@gmail.com

M. Vogelsberger • F. Marinacci • P. Torrey
Kavli Institute for Astrophysics and Space Research, MIT, Cambridge, MA 02139, USA
e-mail: mvogelsb@mit.edu; fmarinac@mit.edu; ptorrey@mit.edu

W.E. Nagel et al. (eds.), *High Performance Computing in Science and Engineering '17*, https://doi.org/10.1007/978-3-319-68394-2_2

equipped with an improved, multi-purpose galaxy formation physics model. The simulated universe contains tens of thousands of galaxies encompassing a variety of environments, mass scales and evolutionary stages. The groundbreaking volume of TNG enables us to sample statistically significant sets of rare astrophysical objects like rich galaxy clusters, and to study galaxy formation and the spatial clustering of matter over a very large range of spatial scales. Here we report some early results on the matter and galaxy clustering found in the simulations. The two-point galaxy correlation function of our largest simulation agrees extremely well with the best available observational constraints from the Sloan Digital Sky Survey, both as a function of galaxy stellar mass and color. The predicted impact of baryonic physics on the matter power spectrum is sizeable and needs to be taken into account in precision studies of cosmology. Interestingly, this impact appears to be fairly robust to the details of the modelling of supermassive black holes, provided this reproduces the scaling properties of the intracluster medium of galaxy clusters.

1 Introduction

Observed galaxies range in mass from a few thousand to a few trillion times the mass of the sun, encompass physical sizes from a fraction to tens of kilo-parsecs, and span a variety of morphologies. Galaxies can reside in diverse environments— in isolation, or as members of rich groups and clusters. They are self-gravitating systems of stars and gas embedded in a halo of dark matter, and their distribution throughout space traces a 'cosmic web' defined by filaments, nodes, sheets, and voids of matter. The highly clustered large-scale structure of the Universe today, at mega-parsec and giga-parsec scales, arose from 13.8 billion years of evolution, starting from the nearly homogeneous distribution of matter in the early universe.

According to the current cosmological paradigm this large-scale structure emerges through the dominant presence of cold dark matter (CDM), which in turn fuels individual galaxies with cosmic gas, imparts gravitational torques and tides, and determines a bottom-up or 'hierarchical' growth, with smaller objects collapsing earlier by gravitational-instability and then later merging to form progressively more massive and rare systems. The cosmological environment and the resulting hierarchical growth govern the formation and transformation of galaxies through cosmic time, regulating their stellar content, star formation activity, gas and heavy-element composition, morphological structure, and so on. In order to gain a theoretical, ab-initio understanding of these processes and their role in shaping galaxies within the full cosmological context, simulations that account for the multi-scale physics involved are the tool of choice.

As a starting point, such calculations rely on initial conditions that are *known* and well constrained by observational data from the Cosmic Microwave Background radiation, now measured to exquisite precision by the Planck mission. The calculations then need to accurately compute the dominant physical force, namely gravity, acting upon all matter, within the accelerating expansion of the Universe.

Furthermore, (magneto)hydrodynamical processes for modelling the evolution of the gaseous component of the Universe need to be followed. Finally, one must account for all other relevant astrophysical processes: from the atomic level interactions that govern radiative cooling of a metal-enriched gas, to the formation of stars and supermassive black holes, with their subsequent expulsion of mass, metals, and 'feedback' energy, which can impact scales as large as entire galactic halos.

The mathematical equations governing these physical processes are non-linear, complex, and highly coupled. This makes their solution through numerical simulation the only practical approach, although the inherently multi-scale, multi-physics nature of the problem poses significant challenges. Nevertheless, rapid progress has been made. In the last 10 years, gravity-only simulations such as the Millennium Run [1], which model only the dark matter component, have reached an extraordinary level of size and sophistication—the largest such simulations now routinely include over one trillion resolution elements [2, 3], encompassing volumes 6–10 giga-parsecs on a side.

Simulations including gas have proven significantly more difficult. Yet, in the past few years projects such as Eagle [4], Horizon-AGN [5], and our Illustris simulation [6–8] have demonstrated that hydrodynamical simulations of structure formation at kilo-parsec spatial resolution can reasonably reproduce the basic properties and scaling relations of observed galaxies. Together with simulations of individual galaxies at even higher resolutions [9–13], these calculations have provided unparalleled insights into the processes underlying galaxy formation. Simultaneously, they have been calling for further improvements of the cosmological hydrodynamical simulations techniques.

In particular, the limited volume covered by simulations like Illustris— encompassing about 100 mega-parsecs on a side—have hindered the exploration of astrophysical processes which act on the largest scales of the Universe and precluded the study of very rare objects, like rich clusters of galaxies and the supermassive black holes sitting at their centers. Because of the enormous computational cost required to simulate the cosmological formation of very massive structures or very large cosmological volumes, at present just a handful of hydrodynamical simulations exist which are capable of accessing the realm of *statistical* samples of massive clusters [14–18]. Yet, none of them has the resolution needed to unveil the structural details of clusters, their intra-cluster plasma and the galaxies hosted therein. The latest generation of supercomputers and cosmological codes make it now possible to attain the simulated volumes necessary to study the regime of rare objects, yet simultaneously achieving mass and spatial resolutions suitable for galaxy formation applications.

Here we report on first results from our team's efforts in this direction, carried out with the computer time granted under our GCS allocation GCS-ILLU-44057 on the HazelHen supercomputer at the HLRS. We will first give a description of the large simulations we have carried out, and then briefly discuss in an exemplary fashion some first results on cosmic large-scale structure. We note, however, that the letter is only a small subset of a large number of new scientific results that we obtain from these simulations at the time of writing of this report.

2 TNG: The Next Generation Illustris Simulations

2.1 The Need to Go Beyond the State of the Art

A few years ago, our team has performed one of the first high-resolution, large-scale
hydrodynamic simulations of galaxy formation [6–8]. This simulation suite—
"Illustris"—used a novel and innovative numerical approach, employing a moving,
unstructured mesh as implemented in the AREPO code [19]. One of the major
achievements of Illustris was its ability to follow the small-scale evolution of gas
and stars within a representative portion of the Universe, yielding a population of
thousands of well-resolved elliptical and spiral galaxies. For the first time ever,
it reproduced the observed morphological mix of galaxies and its dependence on
stellar mass. It could also explain at the same time the characteristics of hydrogen on
large scales, and the metal and neutral hydrogen content of galaxies on small scales.

In practice, all of the dominant galaxy formation processes needed to simulate
galaxies over a broad mass range were identified, implemented, and tested in a full
cosmological setting. These include primordial and metal line cooling of a gas with
self-shielding from radiation; a time-varying, spatially uniform ultraviolet (UV)
background; the evolution of stars and stellar populations, their energetic feedback,
as well as mass and metal return from Supernova Type Ia, Type II and AGB stars;
the formation, growth, and merging of supermassive black holes; models for distinct
'quasar', 'radio', and radiative feedback from active galactic nuclei (AGN). The
Illustris stellar evolution and chemical enrichment network followed nine elements
simultaneously (H, He, C, N, O, Ne, Mg, Si, Fe), while the stellar feedback was
realised through stochastically driven, galactic-scale winds [20, 21].

While the Illustris model has been remarkably successful in reproducing a wide-
range of galaxy properties, particularly for L^* galaxies, some of the results in
Illustris were found to be in tension with a number of observable constraints [22].
In particular, the predicted central gas fractions of galaxy groups and clusters were
clearly too low [8]. At the same time, the colours of massive galaxies were too blue,
and their stellar masses too high when compared to observations. Similar tensions
existed with the stellar mass content as well as the sizes of the smallest resolved
galaxies [7, 23]. These important discrepancies could be traced to the Illustris
feedback model from active galactic nuclei (AGN), pointing to deficits in the physics
model for supermassive black hole accretion and its associated energy release.

2.2 New Modelling Techniques

In the last 3 years we have therefore undertaken a campaign of model development
and improvement, in order to address this key deficiency and other lingering

problems of the original Illustris simulation. The next generation (TNG) models[1] resulting from these efforts were used in our GCS-ILLU project to carry out new large science calculations that significantly advance the state of the art. The collective improvements of these new TNG models for galaxy formation physics include the following primary aspects:

- A new, kinetic AGN feedback scheme, based on black hole-driven winds [25]. This replaces our previous low accretion state model for local black hole (BH) energy injection, and is motivated by recent theoretical arguments for the inflow/outflow solutions of advection dominated accretion flows (ADAFs) in this regime. The new model demonstrates effective quenching of the cooling flows at the centers of massive halos through the thermalisation of small-scale shocks, producing red and passive galaxy populations.
- A refined galactic wind feedback model, with isotropic rather than bipolar outflow geometry, and a metallicity dependent wind energy scaling designed to capture unresolved radiative loses in metal-enriched interstellar medium (ISM) conditions [26]. Adjustments of the wind energetics and directionality effectively resolve the previous overabundance of low mass systems in the stellar mass function, while also bringing the stellar sizes of systems into much better agreement with observations.
- Updated stellar yields for the mass and metal return of stars, which take into account the most recent stellar evolution calculations. The numerical advection of chemical elements has also been improved to enforce realistic abundance constraints.
- A novel 'metal tagging' scheme to record the different stellar production sites (SNIa, SNII, and AGB stars) of heavy elements. This includes a semi-empirical prescription for neutron-star (NSNS) mergers, in order to investigate their role as astrophysical candidates for the origin of nuclear r-process elements.
- An update of the concordance cosmological model to the Planck 2015 values [27].
- For the first time ever, we included magneto-hydrodynamics in such detailed cosmological simulations of galaxy formation, allowing us to study the amplification of a primordial field through large-scale shear flows and dynamo processes [12, 28, 29]. This allows novel predictions for the field strength and topology in galaxies, and how this correlates with other properties and the large-scale environment.
- We also used for the first time a newly developed scheme for adaptive time integration that is based on a hierarchical subdivision of the Hamiltonian of the system, guaranteeing manifest momentum conservation for the gravitational forces even when individual and adaptive timesteps are used. Combined with improvements in our hydrodynamical time integration this lowers the numerical

---

[1]Previously, as in [24], we alluded to the corresponding simulations as 'Illustris++' instead of 'IllustrisTNG'. Both names refer to the same project.

truncation errors and numerical noise considerably, thereby yielding more accurate results for a given number of resolution elements.

With the compute time granted under our GCS allocation (GCS-ILLU-44057), we could demonstrate in the first project phase that our new galaxy formation model is superior to previous models. We have done so using both an array of comparatively small uniform simulations [24–26] and of cosmological realisations of individual objects ("zooms"), as well as by re-simulating the full volume of the original Illustris run with two different black hole models, using the same phases in the initial conditions such that object-by-object comparisons are possible.

2.3 Completed Simulation Set

Equipped with the significantly improved galaxy formation model and code described above, we have devised an ambitious simulation program that we realised in our GCS-ILLU-44057 project, and subsequently expanded in scope still further. The full project is now comprised of three distinct yet complementary cosmological simulations—The Next Generation (TNG) series—whose flagship runs are:

- TNG300 (large volume, $L = 300$ Mpc with 2×2500^3 particles/cells)
- TNG100 (intermediate volume, $L = 110$ Mpc with 2×1820^3 particles/cells)
- TNG50 (small volume, $L = 50$ Mpc with 2×2160^3 particles/cells)

Figure 1 gives a visual comparison of these different simulations. The TNG300 simulation is complementary to both the TNG100 (medium, $L_{box} \simeq 110$ Mpc) and

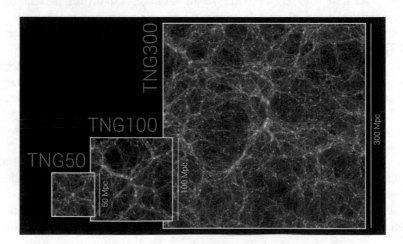

Fig. 1 Dark matter density distribution in slices through the TNG300, TNG100 and TNG50 simulations of our project, illustrating the size of the different simulation volumes. The large box TNG300 has the largest number of resolution elements and by far the largest volume of the set, which is ideal for large-scale structure studies and for sampling rich galaxy clusters. In contrast, TNG100 and TNG50 have much higher mass resolution but poorer statistics, making them a better match for detailed studies of small galaxies

TNG50 (small, $L_{box} \simeq 50\,Mpc$) volumes, but has different science objectives. It features the largest box size ($L_{box} \simeq 300\,Mpc$) and by far the largest number of resolution elements of the whole IllustrisTNG simulation project, although at a lower mass resolution than the other series.

The highest resolution implementation of the TNG300 simulation contains 2×2500^3 total resolution elements (31.25 billion), half dark matter particles and half Voronoi gas cells. In addition, 2500^3 Monte Carlo tracer particles are included. The mean baryonic gas mass resolution is $1.1 \times 10^7\,M_{\odot}$, while the DM particle mass is $5.9 \times 10^7\,M_{\odot}$. The Plummer equivalent gravitational softening of the collisionless component (dark matter and stars) is 2.95 comoving kilo-parsecs until $z = 1$, after which it is fixed to its physical value of 1.48 kilo-parsecs at redshift one. The gravitational softening of gas cells is adaptive, with a minimum of about 370 comoving parsecs. To assess resolution convergence, the flagship TNG300 run is supplemented by other lower-resolution simulations, spaced by factors of 8 in mass resolution. To isolate the effects of the galaxy formation model onto the underlying dark matter distribution, each full-baryonic run is repeated with the same initial condition phases, but including only dark matter (DM) and gravity.

TNG100 constitutes effectively a re-simulation of the original Illustris volume, and has also been finished on HazelHen in two versions with different black hole physics models, using about 18.0M cpu hours on 10,572 cores each. TNG100 has already proven that the new galaxy formation physics model is highly successful at reproducing, among others: the observed gas content of massive halos; the stellar mass function of galaxies at $z = 0$; the inferred stellar-to-halo mass relation for both massive halos as well as galaxies less massive than the Milky Way; the stellar sizes and the color distributions of simulated galaxies across a wide range of masses; the observational signatures in the X-ray and sub-mm wavelengths from groups and clusters of galaxies. Moreover, with TNG100 we have been able to test the performance of the recently implemented MHD scheme, at spatial and time scale regimes previously completely inaccessible (see Fig. 2 for an example of the magnetic field structure in a formed disk galaxy).

We have then proceeded and run the flagship TNG300 simulation on HazelHen, for which also a time extension proposal from us was granted. We consider this calculation a highlight of our GCS-ILLU-44057 project. It targets galaxy clusters and cosmic large-scale structure, and allows us to address a number of unique science goals that cannot be reached with the smaller simulations. With the chosen cosmological volume 300 mega-parsecs a side, TNG300 provides an unmatched hydrodynamical dataset at the current epoch, including five halos with total mass around $10^{15}\,M_{\odot}$, almost 300 hundred galaxy clusters exceeding $10^{14}\,M_{\odot}$, and a few thousands smaller group-scale halos, all at a novel combination of resolution and sample size.

Finally, TNG50 is meant to propel the TNG series towards uncharted terrain on the opposite end of the spectrum. It increases the mass resolution by a factor of 18 and targets, in particular, small dwarf galaxies. We have obtained computer time on HazelHen for this calculation with an independent proposal, and at the time of this writing, are still evolving this simulation on HazelHen.

Fig. 2 Magnetic field strength (color scale) and structure (fine striations indicate the local field direction) in a spiral galaxy in one of our TNG zoom simulations

We note that the completion of the TNG300 and TNG100 runs represents a significant technical achievement that was only possible thanks to the high compute performance of HazelHen and its powerful I/O subsystem. Especially the utility of the comparatively large memory per core cannot be overstated. The TNG300 run was executed on 24,000 cores, and at peak, required up to 101 TB of RAM. Given our previous experience with large MPI partitions, the combined memory requirements of the MPI library and the operating system may reach $\sim$1 GB per core. Therefore, we put a ceiling to our user space memory usage at $\simeq$4.2 GB per core (of the physical 5.33 GB/core available on Hazel Hen). We generally operated the code close to the smallest number of nodes that could fit a particular simulation given these memory constraints, thereby staying in the optimal scaling regime of the code.

The large volume of science data produced (about $\sim$130 TB per large flagship run) and the huge size of our checkpoint files represented a particular challenge. The AREPO code uses a custom parallel I/O strategy for achieving high throughput I/O by writing both checkpoint and output snapshots files (the latter in HDF5) from many different nodes into separate files in parallel. For the TNG300 primary run we have been using 250 simultaneous read/write tasks. They are spread out onto different nodes, and as individual files are completed, new tasks are launched by a built-in I/O scheduler. With this configuration, we achieved on HazelHen, on the filesystem assigned to us, a maximum I/O rate of 36 GB/s and a typical I/O rate of 26 GB/s. To write a full restart file set (simulation checkpoint) of size 36 TB this corresponds to an average time of nearly 30 min. Writing restart files twice per 24 h job we then spent $\sim$4% of our wallclock time on I/O. We stored one redundant backup of the restart file set at all times, which is rotated automatically, and includes an MD5 checksum verification to guard against potential filesystem errors or instabilities. The primary science outputs of the simulation, particle snapshots and substructure catalogs, are significantly smaller than the cumulative checkpoint data volume, and saved just 100 times over the entire simulation duration, giving a negligible increase of the overall I/O cost. The total saved output science data of the TNG300 run amounts to 138 TB. Using a multi-connection parallel transfer we moved this data with a sustained data rate of 350 MB/s from HazelHen to HITS in Heidelberg, where we carry out the science analysis on local machines. From there, we also transferred it further (albeit more slowly) to Harvard-CfA, where we keep a redundant copy. We also do a long-term archival of the data using the large-scale data facility of the state of Baden-Württemberg in Karlsruhe.

3   Early Results on Matter Clustering

The IllustrisTNG simulations allow for a detailed comparison with existing and upcoming observational measurements (for example through mock observations of the simulated data), which in turn provides a foundation for their theoretical interpretation. These include, for example, the imaging and spectroscopic characterisation of the galaxy populations and diffuse stellar light in nearby galaxy clusters like Virgo, Fornax, and Coma (SAURON, ATLAS-3D, NGVS), in local samples of massive early-type galaxies (the MASSIVE SURVEY) or in extremely massive merging systems like the Frontier Field Clusters (HST); the X-ray mapping of the outskirts of galaxy clusters with the SUZAKU telescope; and polarisation measurements with forthcoming large radio telescopes in order to measure magnetic fields in the outer regions of halos (LOFAR, SKA) and in the disks of galaxies (ASKAP). Simultaneously, TNG300 provides important theoretical guidance to maximize the scientific outcome of large future observational campaigns dedicated to constrain cosmology, in particular dark energy, via for example the abundance and spatial clustering of galaxies (EUCLID) and clusters (eROSITA, ATHENA, Planck, South Pole Telescope).

   In our collaboration we are at present busily working on all of these topics. The results will be presented in forthcoming publications over the next months. Here we briefly discuss a few early findings on large-scale structure in the IllustrisTNG simulations.

   The impact of baryonic effects on the large-scale matter power spectrum needs to be understood very accurately to take full advantage of cosmological probes for dark energy. Our simulations permit precision measurements of the clustering of matter, and in particular allow a quantification of the effects of baryons on the total matter distribution. In Fig. 3, we show measurements of the total matter power spectra in TNG300, at all three resolution levels we have run, both for the full physics calculations and the dark matter only runs. The results show the expected behaviour on large scales, which are still in the linear regime. But importantly, they can also probe deeply into the non-linear regime of matter clustering, as it occurs on smaller scales.

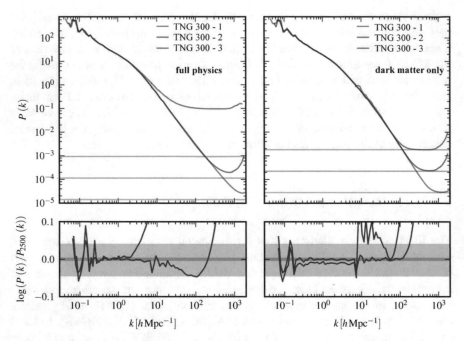

Fig. 3 Top panels show the total matter power spectrum of the full physics runs of TNG300 (left panel) and of the dark matter only runs (right panel), at $z = 0$. The thin horizontal lines show the shot noise limits. The bottom panels give the difference of the two lower resolution simulations to the highest resolution run. The light and dark grey bands in the bottom panels denote ratios of 10% and 1%, respectively

Of particular interest is the deviation between the full physics and the dark matter only simulations, which is shown for redshifts $z = 1$ and $z = 0$ in Fig. 4. There is a characteristic damping of the power on comparatively large scales, at around $k \sim 10\,h\,\mathrm{Mpc}^{-1}$, reaching up to 25%, which can be attributed to baryonic feedback processes, especially from active galactic nuclei (AGN). In the old Illustris simulation, this impact was particularly strong, whereas in IllustrisTNG it is somewhat weaker, and importantly, does not extend to quite such large scales. This is likely directly connected to the strength of the AGN feedback, which is now just right to reproduce the observed gas fractions in poor clusters of galaxies. Interestingly, the results of the Eagle simulation [30], where the same is the case, is very similar. Encouragingly, this suggests that the impact of baryonic feedback may be relatively well constrained if other observable properties in the baryonic sector, especially for galaxy clusters, are matched well.

Another view on the evolution of the large-scale structure is given in Fig. 5, where the dark matter power spectrum is shown relative to the linear theory evolution. The initial conditions contain baryonic acoustic oscillations, whose scale is an important cosmological probe for the expansion history of the universe. We also show the power spectrum as measured through two sets of different tracers (a galaxy sample

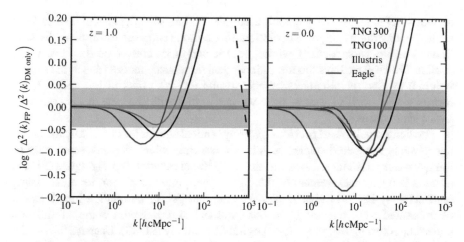

Fig. 4 The ratio between the total matter power spectrum of different full physics runs and the total power spectrum of their dark matter only companion runs. The lines turn from straight to dashed where the shot noise reaches 10% of the power for the full physics power spectrum. The light and dark grey areas denote ratios of 10% and 1%, respectively

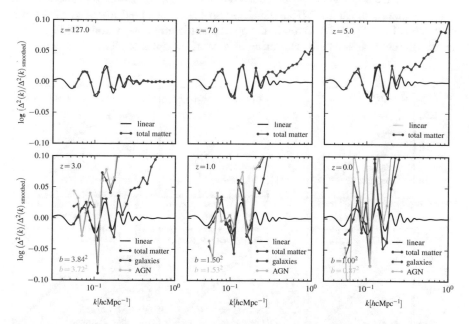

Fig. 5 Power spectra relative to the linearly evolved smoothed initial power spectrum for TNG 300 at $z = 0$. The black and red lines show the linearly evolved initial power spectrum and the total matter power spectrum, respectively. In the bottom row of panels at redshifts $z = 3$, $z = 1$ and $z = 0$ we additionally include power spectrum measurements for two different tracer populations, a galaxy and an AGN sample

and an AGN sample). They exhibit a sample- and redshift-dependent bias on large scales. With progressing time evolution, mode-mode coupling towards mildly non-linear scales occurs, partially wiping out the baryonic acoustic oscillations. While even in TNG300 we have too few independent large-scale modes (due to the limited volume) to give quantitatively precise results for this effect at low redshift, our results give a good idea how close we get to this with the newest generation of hydro simulations.

The limited volume of previous hydrodynamical simulations of galaxy formation has severely hampered studies of galaxy clustering, which depends sensitively on long-wavelength modes. However, the TNG300 simulation has enough volume to allow a faithful measurement of the galaxy two-point correlation function, nearly unaffected by box size constraints. We note that galaxy correlation functions can be measured very accurately in observational surveys, with some of the best constraints coming from the Sloan Digital Sky Survey (SDSS). Because the precise clustering of galaxies depends on the way star formation is affected in different environments, the galaxy correlation function probes a combination of cosmology and the regulation of star formation through environment and feedback effects. It thus provides a central and powerful constraint on any cosmological model.

In Fig. 6 we show the two-point correlation function of TNG300 in six different mass bins compared to the SDSS Data Release 7 at its mean redshift of $z = 0.1$ (with data taken from [31, 32]). Overall, the agreement is remarkably good. In Fig. 7, we extend the comparison to the clustering of separate samples of blue and red galaxies.

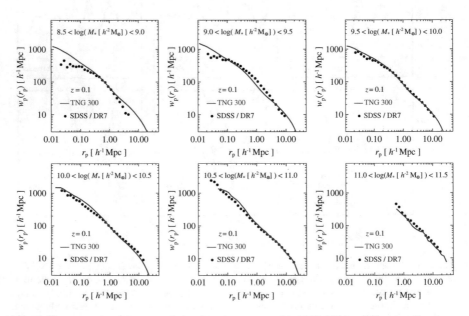

Fig. 6 The two-point galaxy correlation function measured in TNG 300 for different stellar mass ranges, compared to the Sloan Digital Sky Survey. The comparison is carried out at $z = 0.1$, close to the mean redshift of the survey data

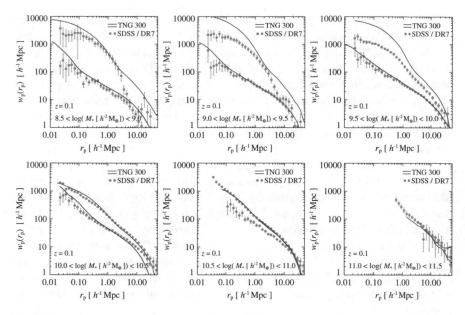

Fig. 7 Two-point galaxy correlation function split by color and stellar mass in TNG 300, compared to the Sloan Digital Sky Survey

Here the galaxies are split into red and blue galaxies based on their *g-r* colour in the Sloan bands, using the same cut as in [32]. Reassuringly, the simulation predictions are also in good agreement with the data in this sensitive test, especially for blue galaxies. The only significant discrepancy occurs for red galaxies in the mass range between 10^9 and 10^{10} $M_\odot$, where the model predictions give a somewhat too strong clustering. This suggests that galaxies of this type shut down their star formation in the simulation slightly too early, which constitutes very useful information about small deficits in the physics model implemented in TNG.

4 Conclusions

We have developed a new advanced simulation model for galaxy formation and applied it in a suite of state-of-the-art hydrodynamical simulations of cosmic structure formation which we refer to as IllustrisTNG. Our calculations set new standards in the field with respect to resolution, size, hydrodynamical accuracy and comprehensiveness of the included physical processes. For example, we for the first time accounted for magnetic fields and their amplification through small-scale dynamo processes during galaxy formation. The new simulation suite was calculated on the HazelHen supercomputer and provides a rich resource for theoretical studies in cosmology and galaxy formation.

In this report, we briefly described a few results on matter and galaxy clustering, in an exemplary fashion. We confirm that AGN feedback has a sizeable impact on the total matter power spectrum, of about 25% at scales of $k \simeq 10\,h\,\mathrm{Mpc}^{-1}$. It appears that the magnitude of this effect is relatively robust to details of the black hole modelling, provided the baryonic properties of the intracluster gas match basic cluster scaling relations. This is an encouraging result and offers the prospect to arrive at fully predictive hydrodynamic simulation that reach the nominal 1% accuracy that is required on these large scales for precision experiments that search for dark energy.

Modern galaxy surveys like SDSS have determined the galaxy two-point correlation function to high accuracy. This data is reproduced remarkably well by the predictions of the TNG300 simulation. Even when split by colour, the agreement remains generally impressive, modulo a slightly too strong clustering of red low-mass galaxies. This non-trivial result is highly encouraging and provides ample motivation to explore the simulation data in many different directions. This is currently in progress, and we look forward to present a flurry of publications that address the science enabled by the TNG simulations in the near future. These studies will be led by IllustrisTNG team members, as well as students and postdocs in our groups.

We also foresee to make all of the IllustrisTNG data publicly available approximately 1 year after our own first papers haven been published, following the successful model we had applied for our older Illustris simulation [22]. We anticipate that this will further boost the utility of the simulations for the full community, and make them a useful scientific resource for many years to come.

Acknowledgements The authors gratefully acknowledge computer time through the project GCS-ILLU on Hornet/HazelHen at HLRS. We acknowledge financial support through subproject EXAMAG of the Priority Programme 1648 'SPPEXA' of the German Science Foundation, and through the European Research Council through ERC-StG grant EXAGAL-308037, and we would like to thank the Klaus Tschira Foundation.

References

1. V. Springel, T. Di Matteo, L. Hernquist, Mon. Not. R. Astron. Soc. **361**, 776 (2005). https://doi.org/10.1111/j.1365-2966.2005.09238.x
2. J. Kim, C. Park, G. Rossi, S.M. Lee, J.R. Gott III., J. Kor. Astrono. Soc. **44**, 217 (2011). https://doi.org/10.5303/JKAS.2011.44.6.217
3. S.W. Skillman, M.S. Warren, M.J. Turk, R.H. Wechsler, D.E. Holz, P.M. Sutter (2014). arXiv e-prints 1407.2600
4. J. Schaye, R.A. Crain, R.G. Bower, M. Furlong, M. Schaller, T. Theuns, C. Dalla Vecchia, C.S. Frenk, I.G. McCarthy, J.C. Helly, A. Jenkins, Y.M. Rosas-Guevara, S.D.M. White, M. Baes, C.M. Booth, P. Camps, J.F. Navarro, Y. Qu, A. Rahmati, T. Sawala, P.A. Thomas, J. Trayford, Mon. Not. R. Astron. Soc. **446**, 521 (2015). https://doi.org/10.1093/mnras/stu2058
5. Y. Dubois, C. Pichon, C. Welker, D. Le Borgne, J. Devriendt, C. Laigle, S. Codis, D. Pogosyan, S. Arnouts, K. Benabed, E. Bertin, J. Blaizot, F. Bouchet, J.F. Cardoso, S. Colombi, V. de Lapparent, V. Desjacques, R. Gavazzi, S. Kassin, T. Kimm, H. McCracken, B. Milliard,

S. Peirani, S. Prunet, S. Rouberol, J. Silk, A. Slyz, T. Sousbie, R. Teyssier, L. Tresse, M. Treyer, D. Vibert, M. Volonteri, Mon. Not. R. Astron. Soc. **444**, 1453 (2014). https://doi.org/10.1093/mnras/stu1227

6. M. Vogelsberger, S. Genel, V. Springel, P. Torrey, D. Sijacki, D. Xu, G. Snyder, S. Bird, D. Nelson, L. Hernquist, Nature **509**, 177 (2014). https://doi.org/10.1038/nature13316

7. M. Vogelsberger, S. Genel, V. Springel, P. Torrey, D. Sijacki, D. Xu, G. Snyder, D. Nelson, L. Hernquist, Mon. Not. R. Astron. Soc. **444**, 1518 (2014). https://doi.org/10.1093/mnras/stu1536

8. S. Genel, M. Vogelsberger, V. Springel, D. Sijacki, D. Nelson, G. Snyder, V. Rodriguez-Gomez, P. Torrey, L. Hernquist, Mon. Not. R. Astron. Soc. **445**, 175 (2014). https://doi.org/10.1093/mnras/stu1654

9. J. Guedes, S. Callegari, P. Madau, L. Mayer, Astrophys. J. **742**, 76 (2011). https://doi.org/10.1088/0004-637X/742/2/76

10. G.S. Stinson, C. Brook, A.V. Macciò, J. Wadsley, T.R. Quinn, H.M.P. Couchman, Mon. Not. R. Astron. Soc. **428**, 129 (2013). https://doi.org/10.1093/mnras/sts028

11. P.F. Hopkins, D. Kereš, J. Oñorbe, C.A. Faucher-Giguère, E. Quataert, N. Murray, J.S. Bullock, Mon. Not. R. Astron. Soc. **445**, 581 (2014). https://doi.org/10.1093/mnras/stu1738

12. R.J.J. Grand, F.A. Gómez, F. Marinacci, R. Pakmor, V. Springel, D.J.R. Campbell, C.S. Frenk, A. Jenkins, S.D.M. White, Mon. Not. R. Astron. Soc. **467**, 179 (2017). https://doi.org/10.1093/mnras/stx071

13. O. Agertz, A.V. Kravtsov, Astrophys. J. **824**, 79 (2016). https://doi.org/10.3847/0004-637X/824/2/79

14. V. Biffi, F. Sembolini, M. De Petris, R. Valdarnini, G. Yepes, S. Gottlöber, Mon. Not. R. Astron. Soc. **439**, 588 (2014). http://dx.doi.org/10.1093/mnras/stu018

15. A.M.C. Le Brun, I.G. McCarthy, J. Schaye, T.J. Ponman, Mon. Not. R. Astron. Soc. **441**, 1270 (2014). http://dx.doi.org/10.1093/mnras/stu608

16. S. Planelles, S. Borgani, D. Fabjan, M. Killedar, G. Murante, G.L. Granato, C. Ragone-Figueroa, K. Dolag, Mon. Not. R. Astron. Soc. **438**, 195 (2014). http://dx.doi.org/10.1093/mnras/stt2141

17. S. Bocquet, A. Saro, K. Dolag, J.J. Mohr, Mon. Not. R. Astron. Soc. **456**, 2361 (2016). http://dx.doi.org/10.1093/mnras/stv2657

18. I.G. McCarthy, J. Schaye, S. Bird, A.M.C. Le Brun, Mon. Not. R. Astron. Soc. **465**, 2936 (2017). https://doi.org/10.1093/mnras/stw2792

19. V. Springel, Mon. Not. R. Astron. Soc. **401**, 791 (2010). https://doi.org/10.1111/j.1365-2966.2009.15715.x

20. M. Vogelsberger, S. Genel, D. Sijacki, P. Torrey, V. Springel, L. Hernquist, Mon. Not. R. Astron. Soc. **436**, 3031 (2013). https://doi.org/10.1093/mnras/stt1789

21. P. Torrey, M. Vogelsberger, S. Genel, D. Sijacki, V. Springel, L. Hernquist, Mon. Not. R. Astron. Soc. **438**, 1985 (2014). https://doi.org/10.1093/mnras/stt2295

22. D. Nelson, A. Pillepich, S. Genel, M. Vogelsberger, V. Springel, P. Torrey, V. Rodriguez-Gomez, D. Sijacki, G.F. Snyder, B. Griffen, F. Marinacci, L. Blecha, L. Sales, D. Xu, L. Hernquist, Astron. Comput. **13**, 12 (2015). https://doi.org/10.1016/j.ascom.2015.09.003

23. G.F. Snyder, P. Torrey, J.M. Lotz, S. Genel, C.K. McBride, M. Vogelsberger, A. Pillepich, D. Nelson, L.V. Sales, D. Sijacki, L. Hernquist, V. Springel, Mon. Not. R. Astron. Soc. **454**, 1886 (2015). https://doi.org/10.1093/mnras/stv2078

24. V. Springel, A. Pillepich, R. Weinberger, R. Pakmor, L. Hernquist, D. Nelson, S. Genel, M. Vogelsberger, F. Marinacci, J. Naiman, P. Torrey, in *High Performance Computing in Science and Engineering 16: Transactions of the High Performance Computing Center, Stuttgart (HLRS) 2016*, ed. by W.E. Nagel, D.H. Kröner, M.M. Resch (Springer International Publishing, Cham, 2016), pp. 5–20. https://doi.org/10.1007/978-3-319-47066-5_1

25. R. Weinberger, V. Springel, L. Hernquist, A. Pillepich, F. Marinacci, R. Pakmor, D. Nelson, S. Genel, M. Vogelsberger, J. Naiman, P. Torrey, Mon. Not. R. Astron. Soc. **465**, 3291 (2017). https://doi.org/10.1093/mnras/stw2944

26. A. Pillepich, V. Springel, D. Nelson, S. Genel, J. Naiman, R. Pakmor, L. Hernquist, P. Torrey, M. Vogelsberger, R. Weinberger, F. Marinacci (2017). arXiv e-prints 1703.02970
27. Planck Collaboration, P.A.R. Ade, N. Aghanim, C. Armitage-Caplan, M. Arnaud, M. Ashdown, F. Atrio-Barandela, J. Aumont, C. Baccigalupi, A.J. Banday et al., Astron. Astrophys. **571**, A16 (2014). https://doi.org/10.1051/0004-6361/201321591
28. R. Pakmor, V. Springel, Mon. Not. R. Astron. Soc. **432**, 176 (2013). https://doi.org/10.1093/mnras/stt428
29. R. Pakmor, F. Marinacci, V. Springel, Astrophys. J. **783**, L20 (2014). https://doi.org/10.1088/2041-8205/783/1/L20
30. W.A. Hellwing, M. Schaller, C.S. Frenk, T. Theuns, J. Schaye, R.G. Bower, R.A. Crain, Mon. Not. R. Astron. Soc. **461**, L11 (2016). https://doi.org/10.1093/mnrasl/slw081
31. Q. Guo, S. White, M. Boylan-Kolchin, G. De Lucia, G. Kauffmann, G. Lemson, C. Li, V. Springel, S. Weinmann, Mon. Not. R. Astron. Soc. **413**, 101 (2011). https://doi.org/10.1111/j.1365-2966.2010.18114.x
32. B.M.B. Henriques, S.D.M. White, P.A. Thomas, R.E. Angulo, Q. Guo, G. Lemson, W. Wang, Mon. Not. R. Astron. Soc. **469**, 2626 (2017). https://doi.org/10.1093/mnras/stx1010

PAMOP: Large-Scale Calculations Supporting Experiments and Astrophysical Applications

B.M. McLaughlin, C.P. Ballance, M.S. Pindzola, P.C. Stancil, J.F. Babb, S. Schippers, and A. Müller

Abstract Our prime computation effort is to support current and future measurements of atomic photoionization cross-sections being performed at various synchrotron radiation facilities around the globe, and computations for astrophysical applications. In our work we solve the Schrödinger or Dirac equation using the R-matrix or R-matrix with pseudo-states approach from first principles. The time dependent close-coupling (TDCC) method is also used in our work. Finally, we present cross-sections and rates determined for diatom-diatom and radiative collision processes between atoms and ions currently of great interest to astrophysics.

B.M. McLaughlin (✉) • C.P. Ballance
Centre for Theoretical Atomic, Molecular and Optical Physics (CTAMOP), School of Mathematics & Physics, Queen's University Belfast, Belfast BT7 1NN, UK
e-mail: bmclaughlin899@btinternet.com; c.ballance@qub.ac.uk

M.S. Pindzola
Department of Physics, 206 Allison Laboratory, Auburn University, Auburn, AL 36849, USA
e-mail: pindzola@physics.auburn.edu

P.C. Stancil
Department of Physics and Astronomy and the Center for Simulational Physics, University of Georgia, Athens, GA 30602-2451, USA
e-mail: stancil@physast.uga.edu

J.F. Babb
Institute for Theoretical Atomic and Molecular Physics, Harvard Smithsonian Center for Astrophysics, 60 Garden Street, Cambridge, MA 02138, USA
e-mail: jbabb@cfa.harvard.edu

S. Schippers
I. Physikalisches Institut, Justus-Liebig-Universität Giessen, 35392 Giessen, Germany
e-mail: Stefan.Schippers@physik.uni-giessen.de

A. Müller
Institut für Atom- und Molekülphysik, Justus-Liebig-Universität Giessen, 35392 Giessen, Germany
e-mail: Alfred.Mueller@iamp.physik.uni-giessen.de

© Springer International Publishing AG 2018
W.E. Nagel et al. (eds.), *High Performance Computing in Science and Engineering '17*, https://doi.org/10.1007/978-3-319-68394-2_3

37

1   Introduction

Our research efforts continue to focus on the development of computational methods to solve the Schrödinger and Dirac equations for atomic and molecular collision processes. Access to leadership-class computers such as the Cray XC40 at HLRS allows us to benchmark our theoretical solutions against dedicated collision experiments at synchrotron radiation facilities such as the Advanced Light Source (ALS), Astrid II, BESSY II, SOLEIL and PETRA III and to provide atomic and molecular data for ongoing research in laboratory and astrophysical plasma science. In order to have direct comparisons with experiment, semi-relativistic, or fully relativistic R-matrix or R-matrix with pseudo-states (RMPS) computations, involving a large number of target-coupled states are required to achieve spectroscopic accuracy. These computations could not be even attempted without access to high performance computing (HPC) resources such as those available at computational centers in Europe (HLRS) and the USA (NERSC, NICS and ORNL).

The motivation for our work is multi-fold; (a) Astrophysical Applications [27, 39, 43, 50], (b) Magnetically-confined fusion and plasma modeling, (c) Fundamental interest and (d) Support of experiment and satellite observations. For photoabsorption by heavy atomic systems [41, 42], little atomic data exists and our work provides results for new frontiers on the application of the R-matrix BREIT-PAULI and DARC parallel suite of codes. Our highly efficient R-matrix codes have evolved over the past decade and have matured to a stage now that large-scale collision calculations can be carried out in a timely manner for electron or photon impact of heavy systems where inclusion of relativistic effects are essential. These codes are widely applicable for the theoretical interpretation of present experiments being performed at leading synchrotron radiation facilities. Examples of our results are presented below in order to illustrate the predictive nature of the methods employed compared to experiment.

2   Valence Shell Photoionization Studies

For comparison with the measurements made at the ALS, state-of-the-art theoretical methods using highly correlated wavefunctions were applied that include relativistic effects. An efficient parallel version of the DARC [5, 61, 66] suite of codes continues to be developed and applied to address electron and photon interactions with atomic systems, providing for hundreds of levels and thousands of scattering channels. These codes are presently running on a variety of parallel high performance computing architectures world wide [44, 46, 47, 49]. The input wavefunctions for the DARC codes are determined by the GRASP0 code [12, 17, 22].

GRASP0 [17, 22] is used to construct a bound orbital basis set for the residual ion of the system of interest, with the Dirac-Coulomb Hamiltonian H_D, via the relation,

$$H_D = \sum_i -ic\alpha \nabla_i + (\beta - 1)c^2 - \frac{Z}{r_i} + \sum_{i<j} \frac{1}{|r_j - r_i|}, \tag{1}$$

where the electrons are labelled by i and j and the summation is taken over all electrons of the system. The matrices α and β are directly related to the Pauli spin matrices, c is the speed of light and the atomic number is Z. The relativistic orbitals are described with a large component, $\mathscr{P}_{n\ell}$ and small component $\mathscr{Q}_{n\ell}$. The residual ion target wavefunctions are appropriately defined on a radial grid for input into the relativistic Dirac Atomic R-matrix Codes (DARC) [66].

The total cross-section σ for photoionization (in Megabarns, $1\,Mb=10^{-18}\,cm^2$) by unpolarized light is obtained by integrating over all electron-ejection angles $\hat{k}$ and averaging over photon polarization [12] to give

$$\sigma = \frac{8\pi^2 \alpha a_0^2 \omega}{3(2J_i + 1)} \sum_{l,j,J} |<\Psi_f^- |M_1|\Psi_i>|^2, \tag{2}$$

where M_1 represents the dipole length operator and the equations are simplified by use of the Wigner-Eckart theorem [12]. The cross-section σ can be cast in both length (L) and velocity gauges (V) through the dipole moment operator M_1, where for the velocity gauge ω is replaced by ω^{-1}, where a_0 is the Bohr radius, α the fine structure constant, $g_i = (2J_i + 1)$ the statistical weighting of the initial state, with Ψ_i, Ψ_f^- being the initial and final state scattering wavefunctions.

2.1 Photoionization of Atomic Chlorine Ions: Cl^+

The sulphur-like ion, Cl^+ in its ground and metastable states, namely; $3s^2 3p^4\ ^3P_{2,1,0}$, and $3s^2 3p^4\ ^1D_2$, $^1S_0$, were measured recently at the ALS at Lawrence Berkeley National Laboratory using the merged beams photon-ion technique at a photon energy resolution of 15 meV in the energy range 19–28 eV [24]. These measurements are compared with large-scale Dirac Coulomb R-matrix calculations in the same energy range [40] as shown in Fig. 1. Photoionization of this sulphur-like chlorine ion is characterized by multiple Rydberg series of autoionizing resonances superimposed on a direct photoionization continuum. A wealth of resonance features observed in the experimental spectra are spectroscopically assigned and their resonance parameters tabulated and compared with the recent measurements. Metastable fractions in the parent ion beam are determined from the present study. Theoretical resonance energies and quantum defects of the prominent Rydberg series associated with fine-structure core-excitations $3s^2 3p^3 nd$, identified in the spectra as $3p \rightarrow nd$ transitions are compared with the available measurements.

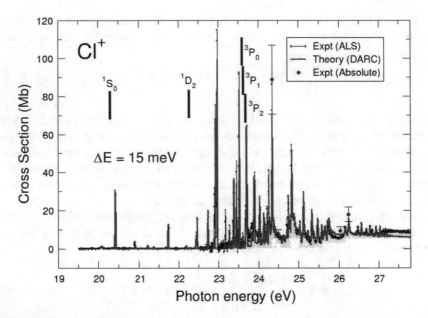

Fig. 1 Single photoionization of the sulphur-like chlorine ion as a function of the photon energy in the energy region 19–27.5 eV. The experimental results shown are from the recent high resolution measurements made at the ALS at a resolution of 15 meV FWHM [24]. The theoretical cross-section calculations are from large-scale DARC calculations [40] convoluted with a Gaussian having a profile of 15 meV FWHM. An appropriate weighting has been used for the ground and metastable states to best match the high resolution ALS measurements [24]. For further details see the recent work of McLaughlin [40]

Weaker Rydberg series $3s^2 3p^3$ ns, identified as $3p \rightarrow ns$ transitions and window resonance $3s3p^4(^4P)np$ features, due to $3s \rightarrow np$ transitions are also found in the spectra.

Excellent agreement of the theoretical photoionization cross-sections and resonance features with recent ALS measurements [24] is found [40] as illustrated in Fig. 1 and are within the ±13 meV energy uncertainty of experiment. Interloping resonances are found in the photoionization spectra from the different initial states of this sulphur-like species. The interloping resonances are seen to disrupt the regular Rydberg series pattern in the spectra [40].

2.2 Photoionization of Co$^+$ and Electron Excitation of Co^{2+} Ions

Modeling of massive stars and supernovae (SNe) plays a crucial role in understanding galaxies. From this modeling we can derive fundamental constraints on stellar evolution, mass-loss processes, mixing, and the products of nucleosynthesis.

Proper account must be taken of all important processes that populate and depopulate the levels (collisional excitation, de-excitation, ionization, recombination, photoionization, and bound-bound processes). For the analysis of Type Ia SNe and core collapse SNe (Types Ib, Ic and II) Fe group elements are particularly important. Unfortunately little data are currently available and most noticeably absent are the photoionization cross-sections for the Fe-peak elements which have high abundances in SNe. Important interactions for both photoionization (PI) and electron-impact excitation (EIE) are calculated using the relativistic DARC codes for low ionization stages of cobalt. All results are calculated up to photon energies of 45 eV and electron energies up to 20 eV. The wavefunction representation of Co III has been generated using the GRASP0 code [17, 22] by including the dominant $3d^7$, $3d^6[4s, 4p]$, $3p^43d^9$ and $3p^63d^9$ configurations, resulting in 292 fine structure levels. Electron-impact collision strengths and Maxwellian averaged effective collision strengths across a wide range of astrophysically relevant temperatures are computed for Co III. Statistically weighted PI cross-sections are presented for Co II in Fig. 2 and compared directly with existing work [20, 68, 76, 78].

The collision strength between an initial state i and a final state j can be obtained from the EIE cross-section $\sigma_{j\to i}^e$,

$$\Omega_{i\to j} = \frac{g_i k_i^2}{\pi a_0^2} \sigma_{j\to i}^e. \tag{3}$$

These collision strengths represent a detailed spectrum, complete with the autoionizing states. To present the results in a more concise format, we assume a

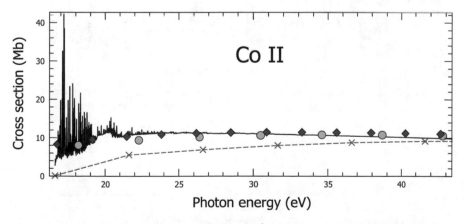

Fig. 2 Photoionization cross-section (Mb) for the Co^+ ion, as a function of photon energy (eV). The solid (black) line represents the statistically weighted 292-level DARC results [76] for all transitions from the initial ground-state term $3d^8$ ^3F to all dipole allowed final states. The solid (red diamonds) are from the work of Verner and co-workers [78], who used Dirac-Slater potentials within a central field approximation. The solid (green circles) are from the distorted wave (DW) calculations of Fontes and co-workers [20] and the (purple) crosses with the dashed line are the calculations of Reilman and Manson [68] who used the Hartree-Slater wavefunctions of Hermann and Skillman

Maxwellian distribution of the colliding electron velocities. We define the effective collision strength as a function of the electron temperature in Kelvin (K) as,

$$\Upsilon_{i\to j} = \int_0^\infty \Omega_{i\to j} \exp(-\epsilon_f/kT)\, d\left(\frac{\epsilon_f}{kT}\right) \tag{4}$$

where ϵ_f is the energy of the electron, $3800 \leq T \leq 40{,}000$ in K, and $k= 8.617\times 10^{-5}$ eV/K is Boltzmann's constant. Each $\Upsilon_{i\to j}$ is calculated for 11 electron temperatures.

We calculate the main contribution to the collision strength as defined in Eq. (3) from the partial waves up to $J = 13$ of both even and odd parity. These are obtained by considering appropriate total multiplicity and orbital angular momentum partial waves. However, in order to obtain converged results on transitions at higher energies, we explicitly calculate partial waves up to $J = 38$ and use top-up procedures to account for further contributions to the total cross-section. The effective collision strengths as a function of electron temperature were determined using Eq. (4) for the appropriate transition of interest [76].

GRASP0 [17, 22] was used to obtain A-values for transitions between the fine-structure levels. The DARC computer package has been employed to extend the problem to include photon and electron interactions. In Fig. 3 results are shown

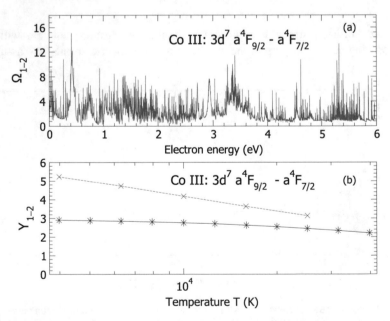

Fig. 3 Electron impact excitation of Co^{2+} ions for the $3d^7a^4F_{9/2} - a^4F_{7/2}$ transition. Panel (**a**) is the collision strength Ω, as a function of electron collision energy (eV), panel (**b**) is the corresponding effective collision strength Υ as a function of electron temperature in Kelvin (K). The solid black line with stars DARC 292-level results [76], the dashed blue line with crosses results from Storey and Sochi [75]

for for the $3d^7a^4F_{9/2} - a^4F_{7/2}$ transition in Co III, for the collision strengths (Ω) and Maxwellian averaged effective collision strength (Υ). We note that when comparisons are made with other available works one finds good agreement. The reliability of the atomic data presented has been rigorously tested through the variation of the level of sophistication of the current calculations. Great care has been taken to ensure the inclusion of important correlation and configuration-interaction in the wavefunction expansions. Furthermore, the complex resonance structures in the cross-sections (photoionization and excitation) have been accurately resolved through a series of calculations incorporating mesh sizes with finer and finer energy increments. Further details of the collision work can be found in the recent publication of Tyndall and co-workers [76].

2.3 Photoionization of Tungsten (W) Ions: W^{4+}

DARC calculations on photoionization of heavy ions carried out for Se^+ [41], Xe^+ [42], Cl^+ [40], Co^+ [76], Ar^+ [10, 15, 77], Fe^+ [18], Xe^{7+} [55], W^+ [51, 56, 57], Kr^+[25], Se^{2+} [38], and W^{2+} and W^{3+} [48] ions showed suitable agreement with high resolution ALS measurements. Large-scale DARC photoionization cross-section calculations on neutral sulfur compared to photolysis experiments, carried out in Berlin [7], and measurements performed at SOLEIL for $2p$ removal in Si^+ ions by photons [28] both showed suitable agreement. All of these cross-section calculations using either the DARC or BREIT-PAULI R-matrix codes [66] showed suitable agreement with the measurements made at the ALS, BESSY II and SOLEIL radiation facilities.

Tungsten, the element with atomic number $Z = 74$, has moved into the focus of controlled nuclear fusion research because of its unique physical and chemical properties which make it the most suitable material for the wall regions of highest particle and heat load in a fusion reactor vessel [60]. The downside of tungsten as a high-Z impurity is its extremely high potential for radiative plasma cooling. Minuscule concentrations of tungsten ions in a fusion plasma prevent ignition. Maximum tolerable relative fractions of tungsten in the plasma are of the order of 2×10^{-5} [59, 65]. By plasma-wall interactions tungsten atoms and ions are inevitably released from the surfaces of the vacuum vessel and enter the plasma. For the modeling of tungsten plasma impurities and their characteristic line emissions, detailed knowledge about collisional and spectroscopic properties is required. In order to meet some of the most important requirements a dedicated experimental project was initiated several years ago with the goal to provide cross-section data and spectroscopic information on tungsten ions exposed to collisions with electrons and photons [52]. The main topics of this project are electron-ion recombination, photoionization, and electron-impact ionization of tungsten ions. Results on the recombination of W^{18+}, W^{19+}, and W^{20+} have been published [4, 33, 34, 70, 71]. Cross-sections and rate coefficients for electron-impact ionization of W^{17+} and

W^{19+} have also been made available in the literature [11, 67]. The present work adds experimental and theoretical cross-sections for single-photon single ionization of W^{4+} to a series of photoionization studies on tungsten ions in low charge states [48, 53, 54, 56, 57].

Experimental and theoretical results are reported for single-photon single ionization of the tungsten ion W^{4+}. Absolute cross-sections have been measured employing the photon-ion merged-beams setup at the Advanced Light Source in Berkeley. Detailed photon-energy scans were performed at 200 meV bandwidth in the 40–105 eV range. Theoretical results have been obtained from a Dirac-Coulomb R-matrix approach employing basis sets of 730 levels for the photoionization of W^{4+}. Calculations were carried out for the $4f^{14}5s^2 5p^6 5d^2\ {}^3F_J$, $J{=}2$, ground level and the associated fine-structure levels with $J{=}3$ and 4 for the W^{4+} ions. In addition, cross-sections have been calculated for the metastable levels $4f^{14}5s^2 5p^6 5d^2\ {}^3P_{0,1,2}, {}^1D_2, {}^1G_4, {}^1S_0$. Very satisfying agreement of theory and experiment is found for the photoionization cross-section of W^{4+} which is remarkable given the complexity of the electronic structure of tungsten ions in low charge states. Interpretation of the resonance features found in the cross-sections was carried out using the Cowan code [16, 20, 58].

The DARC calculations for W^{q+} with $q = 0, 1, 2, 3$, and 4 show increasingly better agreement with experiments along the sequence of increasing charge states. This was expected because the physics of more highly charged ions becomes simpler in that the electron-nucleus interactions become more prominent relative to the electron-electron interactions which are difficult to treat. The comparison of the theoretical results obtained from the DARC codes and the ALS experimental cross-sections illustrated in Fig. 4 indicate excellent agreement. It would be interesting to see how well other theoretical methods can reproduce the experimental results. A statistical theory based on the concept of quantum many-body chaos has been suggested for the treatment of atomic processes involving interactions of electrons and photons with complex many-electron atoms or ions [19]. The application of this approach to recombination of tungsten ions with an open $4f$ shell such as W^{20+} [9] provided very good overall agreement with the experiment [70]. The quantum-chaos theory may turn out to be also suitable for the present problem of photoionization of W^{q+} ions in low charge states q.

Absolute experimental and theoretical cross-sections are presented for single photoionization of W^{4+}. The measurements were obtained by the merged-beams technique using synchrotron radiation and the calculations were performed in the Dirac-Coulomb R-matrix approximation. Very satisfactory agreement was obtained, as shown in Fig. 4, in fact, the best to date between measured and calculated photoionization cross-sections along the tungsten isonuclear sequence. Improvement of the theoretical description had been expected with the charge state of the ion increasing and the experimental and theoretical work up to W^{4+} demonstrate the validity of that assumption.

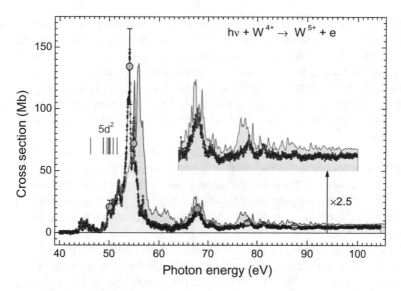

Fig. 4 Comparison of the present experimental and theoretical photoionization cross-sections for W^{4+} ions. The statistically weighted sum of the results for all levels within the $4f^{14}5s^25p^65d^2$ configuration, i.e., the configuration-averaged cross-section, obtained with the 730-level approximation was convoluted with a 200-meV FWHM Gaussian to simulate the experimental energy resolution. The cross-sections at energies beyond 64 eV were multiplied with a factor of 2.5 and displayed separately with a vertical offset of 50 Mb. The experimental energy-scan data are displayed as small circles with (orange) shading. The statistical error bars are provided as vertical black bars which can hardly be seen because they are of the size of the data points. Absolute cross-sections are shown as larger circles with (cyan) shading together with their total uncertainties. The theoretical result is represented by the (olive) solid line with light (green) shading. The ionization thresholds of all levels within the $4f^{14}5s^25p^65d^2$ ground-state configuration are given by the vertical (blue) lines. They were taken from the NIST compilation [32]. Further details can be found in the recent work of Müller and co-workers [58]

3 Multiphotoionization: He and H_2

A time-dependent close-coupling (TDCC) method [62] is used to calculate the multiphoton double ionization for the ground state of the He atom using femtosecond laser pulses. The TDCC equations are solved using standard numerical methods to obtain a discrete representation of the radial wavefunctions and all operators on a two-dimensional lattice. Total double ionization probabilities are calculated for two, three, four, and five-photon absorption in the photon energy range from 10 to 60 eV. Single and triple differential probabilities are calculated for two, three, four, and five-photon absorption at energies where the total ionization probability is near a maximum. For circular polarization the total and differential probabilities are consistently smaller compared to linear polarization as the number of photons absorbed is increased while keeping the radiation field intensity constant. For linear polarization, the total and differential probabilities vary substantially as a function of the number of photons absorbed due to the presence of more absorption pathways.

Total double ionization probabilities were calculated over the energy range from 10 to 60 eV to locate those energy ranges for which two, three, four, and five-photon absorption is the dominant process. The energy ranges are generally from 1.0 to 1.6 times the threshold energy for double ionization. Peak total double ionization probabilities were identified for the two, three, four, and five-photon absorption for both linear and circularly polarized light. At the total probabilities peak energies single and triple differential double ionization probabilities were calculated to guide experiments.

For circular polarization the total, single differential, and triple differential double ionization probabilities drop in a steady manner as the number of photons absorbed is increased. However, for linear polarization the total, single differential, and triple differential probabilities do not drop in a steady manner. Figure 5 illustrates the three-photon double ionization probabilities for linear and circular polarization in He.

For example, the five-photon double ionization probabilities are much larger than the four-photon double ionization probabilities. We hope these survey calculations [63] will stimulate experimental studies for two, three, four, and five-photon double ionization of He using femtosecond laser pulses. The time-dependent close-coupling method can be easily applied to the ground and metastable excited states of many atoms and their ions for which two electrons are found above a closed shell atomic core.

The time-dependent close-coupling method was then used to calculate the multiphoton double ionization of H_2 using circularly polarized applied laser pulses. Total double ionization probabilities are calculated for two, three, and four-photon absorption in the energy range from 10 to 50 eV. Single and triple differential

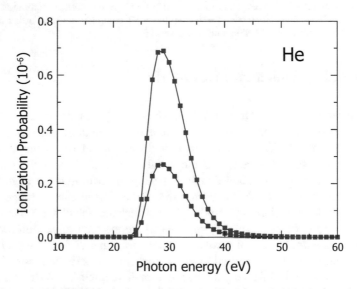

Fig. 5 Three photon, double ionization probability for He, for $L = 3$. Solid (red squares): linear polarization, and solid (blue squares): circular polarization. Further details can be found in the recent work of Pindzola and co-workers [63]

probabilities are calculated at photon energies where the total ionization probability is near a maximum. For one electron emitted along the internuclear axis, the angular distribution for the other electron is similar for two, three, and four-photon absorption. As one electron is emitted further away from the internuclear axis, the angular distribution for the other electron is similar for two and four-photon absorption, but quite different for three-photon absorption.

Multiphoton double ionization probabilities for H_2 using circularly polarized laser pulses were calculated using the TDCC method. Total double ionization probabilities were calculated over the photon energy range from 10 to 50 eV. The results are illustrated in Fig. 6. The two-photon probability has a peak at 33 eV, the three-photon probability a peak at 20 eV, and the four-photon probability a peak at 14 eV. Single and triple differential probabilities were then calculated at the peak energies for two, three, and four-photon absorption. The single differential probability at equal energy sharing has a minimum for two-photons, is flat for three-photons, and has a maximum for four photons. The triple differential probabilities remain of a similar shape for two and four-photons as the ejection of one electron is changed from 0° to 90° with respect to the internuclear axis. However, the shape for 3 photons becomes quite different as the ejection of one electron moves towards 90°.

Many issues remain to be explored in the multiphoton double ionization of H_2 processes. For example, it would be of interest to explore how the multiphoton ionization process varies as a function of the internuclear separation R. It would be preferable to examine the ionization process in models in which R is allowed to vary, but such calculations are computationally prohibitive. Instead, we hope to

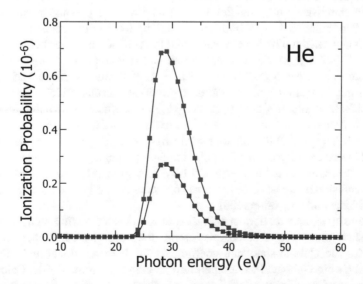

Fig. 6 Total double ionization probabilities for H_2 using circular polarization. Solid (red squares): $M = 2$, solid (blue squares): $M = 3$, and solid (green squares): $M = 4$. Here M is the symmetry of the molecule, $M = \Pi, \Delta, \Phi, \ldots$, i.e 1, 2, 3, ... Further details can be found in the recent work of Pindzola and co-workers [64]

look at multiphoton double ionization for a range of fixed R values that encompass the equilibrium separation of H_2. We hope that the present calculations [64] will stimulate experimental studies for multiphoton double ionization of H_2 using circularly polarized laser pulses. In the future we plan to apply the TDCC method to the calculation of the multiphoton ionization for other diatomic molecules.

4   Diatom–Diatom Collisions: SiO–H_2

Quantum mechanical calculations of molecular scattering are of great interest for chemical dynamics and astrophysics studies. Molecular hydrogen is the most abundant species in most interstellar environments. Collisional relaxation of rotationally or vibrationally excited molecules by H_2 impact is therefore an important process in astrophysics, astrochemistry, and in many environments where non-equilibrium kinetics plays a dominant role. In the interstellar medium (ISM), cooling processes are primarily associated with collisional thermal energy transfer between internal degrees of freedom followed by emission of radiation. The collisional data for state-to-state vibrational and rotational quenching rate coefficients are needed to accurately model the thermal balance and kinetics in the ISM.

We report full-dimensional potential energy surface (PES) and quantum mechanical close-coupling calculations for the scattering of SiO from H_2. The full-dimensional interaction potential surface was computed using the explicitly correlated coupled-cluster (CCSD(T)-F12b) method [23] available within the MOLPRO quantum chemistry suite and fitted using an invariant polynomial method (see Fig. 7). Pure rotational quenching cross-sections from initial states $v_1 = 0$, $j_1 = 1$–5 of SiO in collision with H_2 are calculated for collision energies between 1.0 and $5000 \, cm^{-1}$ (see Fig. 8). State-to-state rotational rate coefficients are calculated at temperatures between 5 and 1000 K. The rotational rate coefficients of SiO with para-H_2 are compared with previous approximate results which were obtained using SiO-He PES's or scaled from SiO-He rate coefficients. Rovibrational state-to-state and total quenching cross-sections and rate coefficients for initially excited SiO($v_1 = 1$; $j_1 = 0$ and 1) in collisions with para-H_2 ($v_2 = 0$; $j_2 = 0$) and ortho-H_2 ($v_2 = 0$; $j_2 = 1$) were calculated over the collision energy ranging from 1.0 to $5000 \, cm^{-1}$ and over the temperature between 5 and 1000 K, respectively. The application of the current collisional rate coefficients to astrophysics is briefly discussed in the recent work of Yang and co-workers [80].

State-to-state and total quenching cross-sections from SiO vibrational state $v_1 = 1$ show resonance structures at intermediate energies for both para-H_2 and ortho-H_2. The state-to-state rate coefficients for both rotational and vibrational quenching were computed for temperature ranging from 5 to 1000 K. Calculations of quenching from higher excited rotational states of SiO in $v_1 = 1$ are in progress and will be reported in the literature in due course. The current calculations together with large-scale coupled-states (CS) approximation results will be essential in

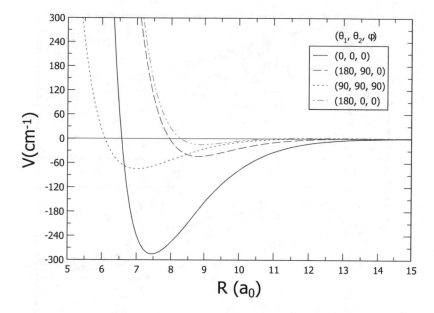

Fig. 7 The R dependence of the vibrationally-averaged (4D rigid-rotor) SiO-H$_2$ PES in ground vibrational states $v_1 = v_2 = 0$. $(\theta_1, \theta_2, \phi) = (0°, 0°, 0°)$, $(180°, 90°, 0°)$, $(90°, 90°, 90°)$, and $(180°, 0°, 0°)$. Further details can be found in Yang and co-workers [80]

the construction of a database of SiO rotational and vibrational quenching rate coefficients urgently needed for astrophysical modeling.

5 Radiative Processes in Collisions of Atoms and Ions

In certain astrophysical environments, such as, for example, interstellar clouds and protostellar winds, the particle densities are relatively low and two-body collisional processes enter into the modeling of charge transfer [72] and of molecule formation [1].

In this section we describe two studies with applications to astrophysics that relied on diatomic molecular potential energy surfaces and transition dipole moments calculated using large-scale computational methods.

5.1 Radiative Charge Transfer: Carbon Atoms with He$^+$ Ions

After the explosion of Supernova 1987A, the infrared emission of the molecule CO was detected in the ejecta. In short, explanation of the observed molecular spectra

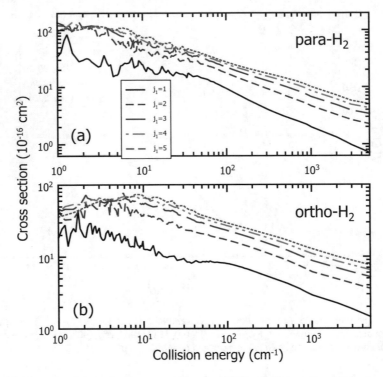

Fig. 8 Total rotational de-excitation cross-section for SiO from initial rotational states $j_1 = 1$–5 in collisions with (**a**) para-H$_2$ ($j_2=0$) and (**b**) ortho-H$_2$ ($j_2=1$). For further details see the work of Yang and co-workers [80]

led to several modeling efforts, which could explain the formation and destruction of CO in this violent environment. In particular, in their work, Lepp, Dalgarno and McCray [36] highlighted the destruction of CO through the dissociative charge transfer reaction

$$He^+ + CO \rightarrow He + C^+ + O. \tag{5}$$

Since He$^+$ is also removed by charge exchange reactions with atoms, it is necessary to have reliable estimates of the rate of He$^+$ removal in order to model the net destruction of CO. Lepp et al. [36] identified C and O, and to a lesser extent S and Si, as the most important atoms playing roles in the destruction of He$^+$. However, at the time, reliable values for the rate coefficients of the C and of the O reactions were not available and estimates were used.

Subsequently, calculations of rate coefficients of the charge transfer reactions for C and He$^+$ [29] and for O and He$^+$ were carried out [30]. Additionally, radiative charge transfer, where a photon is emitted, was considered for these species [31, 81]. In the case of O and He$^+$(^2S) it was found that the main channel for charge transfer at collisional energies corresponding to temperatures below many thousands of Kelvin is radiative charge transfer, via the initial collision channel corresponding to the $O(^3P)$ and $He^+(^2S)$ collisions. The analogous channel for radiative charge transfer of C and He^+ is $C(^3P)$ and $He^+(^2S)$, but this was not considered in [31].

We investigate radiative charge exchange collisions between a carbon atom $C(^3P)$ and a helium ion $He^+(^2S)$, in order to determine the rate coefficients for radiative charge transfer. There are a number of previous calculations of the molecular potential energy curves, but the required data for the present study are not available in the literature. Therefore detailed quantum chemistry calculations are carried out using the MRCI + Q approximation [23] that contains the Davidson correction [35]. We obtain potential energy curves and transition dipole matrix elements for doublet and quartet molecular states of the HeC^+ cation. Figure 9 illustrates the HeC^+ doublet and quartet potential energy curves involved in the radiative charge transfer process.

Our investigation finds that radiative charge exchange (RCX) is more significant than direct charge exchange (CX) as the relative collisional energy decreases in $C(^3P) + He^+$ collisions. We confirm the earlier finding of Kimura et al. [31], that radiative charge transfer via the $3^2\Pi$–$B^2\Pi$ transition is relatively negligible.

Our results show that radiative charge transfer leaving the residual ion in its ground state is the dominant mechanism, similar to the findings by Zhao and co-workers [81] for $O + He^+$. A comparison between the present result for $C + He^+$

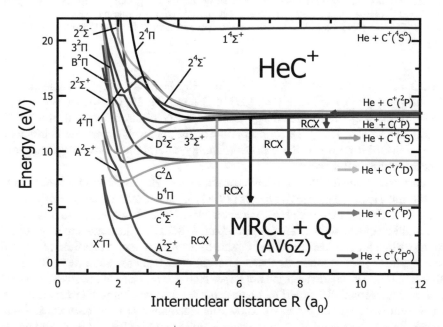

Fig. 9 Potential energies for the HeC^+ molecular ion, as a function of the internuclear distance R, corresponding to the initial channel $C(^3P)+He^+(^2S)$ and to the final channels of $C^+ + He(^1S)$. The downward pointing arrows mark the radiative charge exchange (RCX) processes for the dominant doublet and quartet transitions studied [2]

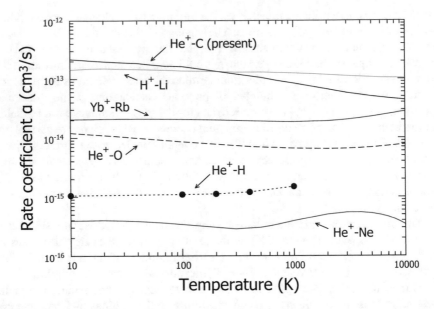

Fig. 10 A comparison of rate coefficients, $\alpha(T)$ (cm^{3}s^{-1}) as a function of temperature $T(K)$, for radiative charge transfer. In order of decreasing magnitude at 10 K, He$^+$ and C, $3^2\Pi$ to X$^2\Pi$ transition (black line) present work, H$^+$ and Li (green line) [73], Yb$^+$ + Rb (red line) [45], He$^+$ and O (dashed black line) [81], He$^+$ and H [82] (points, connected with a dotted line), and He$^+$ + Ne (blue line) [37]. Further details can be found in the recent work of Babb and McLaughlin [2]

[2], O + He$^+$[81], Li + H$^+$ [73], Rb + Yb$^+$ [45], H + He$^+$ [82] (multiplied by the factor $\frac{1}{4}$, see [73]), and Ne + He$^+$ [37] is given in Fig. 10. Compared to the other systems, the large transition energy from the 3 $^2\Pi$ electronic state to the X $^2\Pi$ electronic state of HeC$^+$ partly accounts for the relatively large rate coefficient, due to the third power of the transition energy entering the rate for spontaneous electronic dipole transitions.

Earlier calculated rate coefficients for removal of He$^+$ by C or O were found to be too small to affect the ejecta models [29, 81]. Charge exchange cross-sections and rates for collisions of Si with He$^+$ were considered recently by Satta et al. [69] using the multi-channel Landau-Zener approximation (MCLZ). The dominant mechanism is very different than that for C and O, due to the presence of a manifold of excited states (SiHe$^+$)* above the exit channel energy of Si$^+$ in its ground state. However, the calculated rate coefficients are not larger than the estimates of the 1990s. We note that a similar manifold would be present for the case of charge transfer collisions of S with He$^+$, but, as far as we know, the calculation has not been carried out. Nevertheless, the role of He$^+$ in the destruction of CO is affirmed by recent ejecta models [13, 14]. It would be interesting to revisit the models of the 1990s in order to determine if the improved charge exchange rate coefficients now available for C, O, or Si with He$^+$ modify the conclusions obtained at that time.

Recent models of ejecta chemistry now go beyond equilibrium chemistry, but, generally, still suffer from a lack of charge transfer data [26]. In addition to modeling supernovae ejecta, the present results might be applicable to modeling the boundaries of photon dominated regions (PDRs) and in x-ray dominated regions (XDRs), where the abundance of He^+ can affect the abundance of CO and C^+, C, and CO can coexist.

5.2   Radiative Association: Carbon Atoms with H^+ Ions

Radiative association is the process whereby the formation of a molecular complex is created when two systems collide, stabilizing the complex by the emission of a photon. The formation of CH^+ (methylidynium) by radiative association is a much-studied problem in theoretical astrophysics, see for example [8, 79]. However, previous studies focused on the reaction

$$C^+ + H \rightarrow CH^+ + h\nu, \tag{6}$$

where $h\nu$ represents the emitted photon, because in most astrophysical environments carbon is ionized before hydrogen. For the process (6), the molecular transitions correspond to those from the singlet A $^1\Pi$ electronic state to the singlet X $^1\Sigma^+$ electronic state. In the present study, the radiative association of C and H^+,

$$C + H^+ \rightarrow CH^+ + h\nu, \tag{7}$$

is investigated by calculating cross-sections for photon emission into bound ro-vibrational states of CH^+. The process we consider here corresponds to transitions from the vibrational continua of initial triplet states leading to bound states of CH^+ triplet states. Previously, radiative charge transfer for the triplet transitions was investigated by Stancil et al. [74], but, as far as we know, process (7) has not been quantitatively investigated. Potential energy curves and transition dipole moments are calculated using multi-reference configuration interaction (MRCI + Q) methods [23, 35] with AV6Z basis sets. Figure 11 illustrates the potential energy curves for the four triplet states of interest in the radiative processes.

The cross-sections are then evaluated using quantum-mechanical methods and rate coefficients are calculated. The rate coefficients are about 100 times larger than those for radiative association of C^+ ($^2P^o$) and H from the A $^1\Pi$ state. We also confirm that the formation of CH^+ by radiative association of C^+ ($^2P^o$) and H via the triplet c $^3\Sigma^+$ state is a minor process.

We assessed the radiative association process for the possible triplet state transitions yielding CH^+ and determined that the rate coefficients will be dominated by the cross-sections corresponding to the d $^3\Pi$ to a $^3\Pi$ electronic transitions. The corresponding calculated rates for the d $^3\Pi$–a $^3\Pi$ transitions decrease from about 6.5×10^{-15} cm^3/s at 10 K to about 6.6×10^{-16} cm^3/s at 10,000 K.

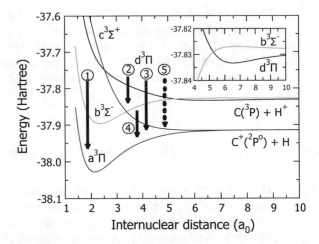

Fig. 11 Potential energy curves for the four low-lying triplet molecular states of the CH^+ molecular ion as a function of internuclear distance. The downward pointing solid arrows schematically indicate the transitions studied for radiative association from: (1) the vibrational continuum of the $b^3\Sigma^-$ state to a bound ro-vibrational level of the $a^3\Pi$ state, and similarly for, (2) $d^3\Pi$ to $b^3\Sigma^-$, (3) $d^3\Pi$ to $a^3\Pi$, and (4) $c^3\Sigma^+$ to $a^3\Pi$ transitions, while, for (5) the dotted line connecting the vibrational continuum of the $d^3\Pi$ state to the vibrational continuum of the $c^3\Sigma^+$ state indicates that radiative charge transfer is the dominant mechanism for this channel. All quantities are in atomic units. Further details can be found in the recent work of Babb and McLaughlin [3]

We ignored the effects of the spin-orbit splitting of the $C(^3P)$ atom and the consequent fine structure splittings of the $d^3\Pi$ state and $b^3\Sigma^-$ state and their rotational coupling, which may affect the validity of our results for low temperatures. In the cold interstellar medium the $C(^3P)$ atoms will be in the lowest level of the lowest term [8], which correlates with the $d^3\Pi$ state, making it likely that the main channel for the radiative association process will be via the $d^3\Pi$–$a^3\Pi$ transition. Figure 12 illustrates the rates via the $d^3\Pi$–$a^3\Pi$ transition, for radiative decay and association compared to the radiative charge transfer results of Stancil and co-workers [74]. The chance that the $C(^3P)$ atom and H^+ ion come together in the $d^3\Pi$ state, which we assumed to be $\frac{2}{3}$, is just the approach probability factor.

In most astrophysical environments where formation of CH^+ is modeled, the carbon atoms are ionized before hydrogen atoms. Therefore, the radiative association process occurs via the singlet states, process (6), and is the dominant channel, where the rate coefficients for the $A^1\Pi$–$X^1\Sigma^+$ transition are around 1.3×10^{-17} cm^3/s at 1000 K [6]. Thus, while the radiative association process via triplet states is faster than via singlet states, its relevance will depend on the relative concentration of C versus $C^+(^2P^o)$ [3].

The present results might find application to modeling an environment where neutral carbon exists in the presence of protons, such as modeling the variation of the emission of atomic carbon with the cosmic ionization rate in metal-poor galaxies [21].

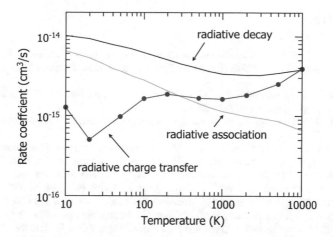

Fig. 12 For the $d^3\Pi$–$a^3\Pi$ transition, the rate coefficients as a function of temperature (K), for radiative association (solid green line) and radiative decay (solid black line) compared to the rate coefficients from Stancil and co-workers [74] for radiative charge transfer (solid blue circles with blue line). Further details can be found in the recent work of Babb and McLaughlin [3]

6 Summary

Access to leadership architectures is essential to our research work (like the Cray XC40 at HLRS and the Cray XC40 at NERSC) providing essential resources to our computational effort in atomic, molecular and optical collision processes, without which we simply could not do these calculations. The theoretical results obtained from access to and use of such high performance resources illustrates vividly the predictive nature of the methods employed and the symbiotic and synergistic relationship between theory, experiment and astrophysical applications.

Acknowledgements A. Müller acknowledges support by Deutsche Forschungsgemeinschaft under project number Mu-1068/20. B.M. McLaughlin acknowledges support from the US National Science Foundation through a grant to ITAMP at the Harvard-Smithsonian Center for Astrophysics, under the visitor's program, the hospitality of the University of Georgia at Athens during recent visits, and a visiting research fellowship (VRF) from Queen's University Belfast. M.S. Pindzola acknowledges support by NSF and NASA grants through Auburn University. P.C. Stancil acknowledges support from NASA grants through University of Georgia at Athens. This research used computational resources at the National Energy Research Scientific Computing Center (NERSC) in Berkeley, CA, USA, and at the High Performance Computing Center Stuttgart (HLRS) of the University of Stuttgart, Stuttgart, Germany. The Advanced Light Source is supported by the Director, Office of Science, Office of Basic Energy Sciences, of the US Department of Energy under Contract No. DE-AC02-05CH11231.

References

1. J.F. Babb, K. Kirby, Molecule formation in dust-poor environments. in *The Molecular Astrophysics of Stars and Galaxies*, ed. by T.W. Hartquist, D.A. Williams (Clarendon, Oxford, 1998)
2. J.F. Babb, B.M. McLaughlin, Radiative charge transfer in collisions of C with He^+. J. Phys. B: At. Mol. Opt. Phys. **50**, 044003 (2017)
3. J.F. Babb, B.M. McLaughlin, Radiative association of $C(^3P)$ and H^+: Triplet States. Mon. Not. R. Astron. Soc. **468**, 2052 (2017)
4. N.R. Badnell et al., Recombination of W^{19+} ions with electrons: absolute rate coefficients from a storage-ring experiment and from theoretical calculations. Phys. Rev. A **93**, 052703 (2016)
5. C.P. Ballance, D.C. Griffin, Relativistic radiatively damped R-matrix calculation of the electron-impact excitation of W^{46+}. J. Phys. B: At. Mol. Opt. Phys. **39**, 3617 (2006)
6. G. Barinovs, M.C. van Hemert, CH^+ Radiative Association. Astrophys. J **636**, 923 (2006)
7. M. Barthel, R. Flesch, E. Rühl, B.M. McLaughlin, Photoionization of the $3s^2 3p^4$ ^3P and the $3s^2 3p^4$ $^1D,^1S$ states of sulfur: experiment and theory. Phys. Rev A **91**, 013406 (2015)
8. D.R. Bates, Rate of formation of molecules by radiative association. Mon. Not. R. Astron. Soc. **111**, 303 (1951)
9. J.C. Berengut et al., Level-resolved quantum statistical theory of electron capture into many-electron compound resonances in highly charged ions. Phys. Rev. A **92**, 062717 (2015)
10. C. Blancard et al., L-shell photoionization of Ar^+ to Ar^{3+} ions. Phys. Rev. A **85**, 043408 (2012)
11. A. Borovik Jr., B. Ebinger, D. Schury, S. Schippers, A. Müller, Electron-impact single ionization of W^{19+} ions. Phys. Rev. A **93**, 012708 (2016)
12. P.G. Burke, *R-Matrix Theory of Atomic Collisions: Application to Atomic, Molecular and Optical Processes* (Springer, New York, 2011)
13. I. Cherchneff, E. Dwek, The chemistry of population III supernova ejecta. I. Formation of molecules in the early universe. Astrophys. J. **703**, 642 (2009)
14. D.D. Clayton, Analytic approximation of carbon condensation issues in Type II Supernovae. Astrophys. J. **762**, 5 (2012)
15. A.M. Covington, A. Aguilar, I.R. Covington, G. Hinojosa, C.A. Shirley, R.A. Phaneuf, I. Álvarez, C. Cisneros, I. Dominguez-Lopez, M.M. Sant'Anna, A.S. Schlachter, C.P. Ballance, B.M. McLaughlin, Valence-shell photoionization of chlorinelike Ar^+ ions. Phys. Rev. A **84**, 013413 (2011)
16. R.D. Cowan, *The Theory of Atomic Structure and Spectra* (University of California Press, Berkeley, 1981)
17. K.G. Dyall, I.P. Grant, C.T. Johnson, E.P. Plummer, GRASP: a general-purpose relativistic atomic structure program. Comput. Phys. Commun. **55**, 425 (1989)
18. V. Fivet, M.A. Bautista, C.P. Ballance, Fine-structure photoionization cross sections of Fe II. J. Phys. B: At. Mol. Opt. Phys. **45**, 035201 (2012)
19. V.V. Flambaum, M.G. Kozlov, G.F. Gribakin, Coherent and stochastic contributions of compound resonances in atomic processes: electron recombination, photoionization, and scattering. Phys. Rev. A **91**, 052704 (2015)
20. C.J. Fontes et al., The Los Alamos suite of relativistic atomic physics codes. J. Phys. B: At. Mol. Opt. Phys. **48**, 144014 (2015)
21. S.C.O. Glover, P.C. Clark, Is atomic carbon a good tracer of molecular gas in metal-poor galaxies? Mon. Not. R. Astron. Soc. **456**, 3596 (2016)
22. I.P. Grant, *Quantum Theory of Atoms and Molecules: Theory and Computation* (Springer, New York, 2007)
23. T. Helgaker, P. Jorgesen, J. Olsen, *Molecular Electronic-Structure Theory* (Wiley, New York, 2000)
24. E.M. Hernández et al., Absolute measurements of chlorine Cl^+ cation single photoionization cross section. J. Quant. Spectrosc. Radiat. Transf. **151**, 217 (2015)

25. G. Hinojosa, A.M. Covington, G.A. Alna'Washi, M. Lu, R.A. Phaneuf, M.M. Sant'Anna, C. Cisneros, I. Álvarez, A. Aguilar, A.L.D. Kilcoyne, A.S. Schlachter, C.P. Ballance, B.M. McLaughlin, Valence-shell single photoionization of Kr^+ ions: experiment and theory. Phys. Rev. A **86**, 063402 (2012)
26. A. Jerkstrand, C. Fransson, C. Kozma, The ^{44}Ti-powered spectrum of SN 1987A. Astron. Astrophys. **762**, A45 (2011)
27. T.R. Kallman, Challenges of plasma modelling: current status and future plans. Space Sci. Rev. **157**, 177 (2010)
28. E.T. Kennedy, J.-P. Mosnier, P. Van Kampen, D. Cubaynes, S. Guilbaud, C. Blancard, B.M. McLaughlin, J.-M. Bizau, Photoionization cross sections of the aluminumlike Si^+ ion in the region of the $2p$ threshold (94–137 eV). Phys. Rev. A **90**, 063409 (2014)
29. M. Kimura et al., Rate coefficients for charge transfer of He^+ with C. Astrophys. J. **417**, 812 (1993)
30. M. Kimura et al., Electron capture and excitation in collisions of O^+ ($^4S,^2D,^2P$) ions with He atoms and He^+ ions with O atoms at energies below 10 keV. Phys. Rev. A **50**, 4854 (1994)
31. M. Kimura et al., Non radiative and radiative electron capture in collisions of He^+ ions with C atoms below 1000 eV. Phys. Rev. A **49**, 2541 (1994)
32. A.E. Kramida, Y. Ralchenko, J. Reader, NIST ASD Team, NIST Atomic Spectra Database (version 5.3), National Institute of Standards and Technology, Gaithersburg, MD, USA (2015)
33. C. Krantz, K. Spruck, N.R. Badnell, A. Becker, D. Bernhardt, M. Grieser, M. Hahn, O. Novotný, R. Repnow, D.W. Savin, A. Wolf, A. Müller, S. Schippers, Absolute rate coefficients for the recombination of open f-shell tungsten ions. J. Phys. Conf. Ser. **488**, 012051 (2014)
34. C. Krantz, N.R. Badnell, A. Müller, S. Schippers, A. Wolf, Recombination of open-f-shell tungsten ions. J. Phys. B: At. Mol. Opt. Phys. **50**, 052001 (2017)
35. S. Langhoff, E.R. Davidson, Configuration interaction calculations on the nitrogen molecule. Int. J. Quantum Chem. **8**, 61 (1974)
36. S. Lepp, A. Dalgarno, R. McCray, Molecules in the ejecta of SN 1987A. Astrophys. J. **358**, 262 (1990)
37. X.J. Liu et al., Radiative charge transfer and radiative association in He^+ + Ne collisions. Phys. Rev. A **81**, 022717 (2010)
38. D.A. Macaluso, A. Aguilar, A.L.D. Kilcoyne, E.C. Red, R.C. Bilodeau, R.A. Phaneuf, N.C. Sterling, B.M. McLaughlin, Absolute single-photoionization cross sections of Sc^{2+}: experiment and theory. Phys. Rev. A **92**, 063424 (2015)
39. B.M. McLaughlin, Inner-shell photoionization, fluorescence and auger yields, in *Spectroscopic Challenges of Photoionized Plasma*, ed. by G. Ferland, D.W. Savin. ASP *Conf*. Series, vol. 247 (Astronomical Society of the Pacific, San Francisco, 2001), p. 87
40. B.M. McLaughlin, Photoionisation of Cl^+ from the $3s^23p^4$ $^3P_{2,1,0}$ and the $3s^23p^4$ $^1D_2,^1S_0$ states in the energy range 19–28 eV. Mon. Not. R. Astron. Soc. **464**, 1990 (2017)
41. B.M. McLaughlin, C.P. Ballance, Photoionization cross section calculations for the halogen-like ions Kr^+ and Xe^+. J. Phys. B: At. Mol. Opt. Phys. **45**, 085701 (2012)
42. B.M. McLaughlin, C.P. Ballance, Photoionization cross sections for the trans-iron element Se^+ from 18 to 31 eV. J. Phys. B: At. Mol. Opt. Phys. **45**, 095202 (2012)
43. B.M. McLaughlin, C.P. Ballance, Photoionization, fluorescence and inner-shell processes, in *McGraw-Hill Yearbook of Science and Technology* (Mc Graw Hill, New York, 2013), p. 281
44. B.M. McLaughlin, C.P. Ballance, in *Petascale Computations for Large-Scale Atomic and Molecular Collisions, Sustained Simulated Performance 2014*, ed. by M.M. Resch, Y. Kovalenko, E. Fotch, W. Bez, H. Kobayashi, chap. 15 (Springer, New York, 2014)
45. B.M. McLaughlin, H.L.D. Lamb, I.C. Lane, J.F. McCann, Ultracold radiative charge transfer in hybrid Yb ion - Rb atom traps. J. Phys. B: At. Mol. Opt. Phys. **47**, 145201 (2014)
46. B.M. McLaughlin, C.P. Ballance, M.S. Pindzola, A. Müller, in *PAMOP: Petascale Atomic, Molecular and Optical Collisions: High Performance Computing in Science and Engineering'14*, ed. by W.E. Nagel, D.H. Kröner, M.M. Resch, chap. 4 (Springer, New York, 2015)

47. B.M. McLaughlin, C.P. Ballance, M.S. Pindzola, S. Schippers, A. Müller, in *PAMOP: Petascale Computations in Support Of experiments: High Performance Computing in Science and Engineering'15*, ed. by W.E. Nagel, D.H. Kröner, M.M. Resch, chap. 4 (Springer, New York, 2016)

48. B.M. McLaughlin, C.P. Ballance, S. Schippers, J. Hellhund, A.L.D. Kilcoyne, R.A. Phaneuf, A. Müller, Photoionization of tungsten ions: experiment and theory for W^{2+} and W^{3+}. J. Phys. B: At. Mol. Opt. Phys. **49**, 065201 (2016)

49. B.M. McLaughlin, C.P. Ballance, M.S. Pindzola, P. C. Stancil, S. Schippers, A. Müller, *PAMOP Project: Computations in Support of Experiments and Astrophysical Applications: High Performance Computing in Science and Engineering'16*, ed. by W.E. Nagel, D.H. Kröner, M.M. Resch, chap. 4 (Springer, New York, 2017)

50. B.M. McLaughlin, J.-M. Bizau, D. Cubaynes, S. Guilbaud, S. Douix, M.M. Al Shorman, M.O.A. El Ghazaly, I. Sahko, M.F. Gharaibeh, K-Shell photoionization of O^{4+} and O^{5+} ions: experiment and theory. Mon. Not. R. Astron. Soc. **465**, 4690 (2017)

51. A. Müller, Precision studies of deep-inner-shell photoabsorption by atomic ions. Phys. Scr. **90**, 054004 (2015)

52. A. Müller, Fusion-related ionization and recombination data for tungsten ions in low to moderately high charge states. Atoms **3**, 120 (2015)

53. A. Müller, S. Schippers, A.L.D. Kilcoyne, D. Esteves, Photoionization of tungsten ions with synchrotron radiation. Phys. Scr. **T144**, 014052 (2011)

54. A. Müller, S. Schippers, A.L.D. Kilcoyne, A. Aguilar, D. Esteves, R.A. Phaneuf, Photoionization of singly and multiply charged tungsten ions. J. Phys. Conf. Ser. **388**, 022037 (2012)

55. A. Müller, S. Schippers, D. Esteves-Macaluso, M. Habibi, A. Aguilar, A.L.D. Kilcoyne, R.A. Phaneuf, C.P. Ballance, B.M. McLaughlin, High resolution valence shell photoionization of Ag-like (Xe^{7+}) Xenon ions: experiment and theory. J. Phys. B: At. Mol. Opt. Phys. **47**, 215202 (2014)

56. A. Müller, S. Schippers, J. Hellhund, A.L.D. Kilcoyne, R.A. Phaneuf, C.P. Ballance, B.M. McLaughlin, Single and multiple photoionization of W^{q+} tungsten ions in charged states $q = 1, 2, .., 5$: experiment and theory. J. Phys. Conf. Ser. **488**, 022032 (2014)

57. A. Müller, S. Schippers, J. Hellhund, K. Holosto, A.L.D. Kilcoyne, R.A. Phaneuf, C.P. Ballance, B.M. McLaughlin, Single-photon single ionization of W^+ ions: experiment and theory. J. Phys. B: At. Mol. Opt. Phys. **48**, 2352033 (2015)

58. A. Müller, S. Schippers, J. Hellhund, A.L.D. Kilcoyne, R.A. Phaneuf, B.M. McLaughlin, Photoionization of tungsten ions: experiment and theory for W^{4+}. J. Phys. B: At. Mol. Opt. Phys. **50**, 085007 (2017)

59. R. Neu et al., Tungsten as plasma-facing material in ASDEX Upgrade. Fusion Eng. Des. **65**, 367 (2003)

60. R. Neu et al., First operation with the JET international thermonuclear experimental reactor-like wall. Phys. Plasmas **20**, 056111 (2013)

61. P.H. Norrington, I.P. Grant, Low-energy electron scattering by Fe XXIII and Fe VII using the Dirac R-matrix method. J. Phys. B: At. Mol. Opt. Phys. **20**, 4869 (1987)

62. M.S. Pindzola et al., The time-dependent close-coupling method for atomic and molecular collision processes. J. Phys. B: At. Mol. Opt. Phys. **40**, R39 (2007)

63. M.S. Pindzola, Y. Li, J.P. Colgan, Multiphoton double ionization of helium using femtosecond laser pulses. J. Phys. B: At. Mol. Opt. Phys. **49**, 215603 (2016)

64. M.S. Pindzola, Y. Li, J.P. Colgan, Multiphoton double ionization of H_2 using circularly polarized laser pulses. J. Phys. B: At. Mol. Opt. Phys. **50**, 045601 (2017)

65. T. Pütterich et al., Calculation and experimental test of the cooling factor of tungsten. Nucl. Fusion **50**, 025012 (2010)

66. R-matrix DARC and BREIT-PAULI codes (2016). http://connorb.freeshell.org

67. J. Rausch, A. Becker, K. Spruck, J. Hellhund, A. Borovik Jr., K. Huber, S. Schippers A. Müller, Electron-impact single and double ionization of W^{17+}. J. Phys. B: At. Mol. Opt. Phys. **44**, 165202 (2011)

68. R.F. Reilmann, S.T. Manson, Photoabsorption cross sections for positive atomic ions with $Z \leq$ 30. Astrophys. J. Suppl. Ser. **40**, 815 (1979)
69. M. Satta et al., Reducing Si population in the ISM by charge exchange collisions with He^+: a quantum modelling of the process. Mon. Not. R. Astron. Soc. **436**, 2722 (2017)
70. S. Schippers et al., Dielectronic recombination of xenonlike tungsten ions. Phys. Rev. A **83**, 012711 (2011)
71. K. Spruck et al., Recombination of W^{18+} ions with electrons: absolute rate coefficients from a storage-ring experiment and from theoretical calculations. Phys. Rev. A **90**, 032715 (2014)
72. P.C. Stancil, Charge transfer calculations for astrophysical modeling, in *Spectroscopic Challenges of Photoionized Plasmas*, ed. by G. Ferland, D.W. Savin. ASP Conference Series, vol. 247 (Astronomical Society of the Pacific, San Francisco, 2001), p. 3
73. P.C. Stancil, B. Zygelman, Radiative charge transfer in collisions of Li with H^+. Astrophys. J. **472**, 102 (1996)
74. P.C. Stancil et al., Charge transfer in collisions of C^+ with H and H^+ with C. Astrophys. J. **502**, 1006 (1998)
75. P.J. Storey, T. Sochi, Collision strengths and transition probabilities for Co III forbidden lines. Mon. Not. R. Astron. Soc. **459**, 2558 (2016)
76. N.B. Tyndall, C.A. Ramsbottom, C.P. Ballance, A. Hibbert, Photoionization of Co^+ and electron-impact excitation of Co^{2+} using the Dirac R - matrix method. Mon. Not. R. Astron. Soc. **462**, 3350 (2016)
77. N.B. Tyndall, C.A. Ramsbottom, C.P. Ballance, A. Hibbert, Valence and L-shell photoionization of Cl-like argon using R-matrix techniques. Mon. Not. R. Astron. Soc. **456**, 366 (2016)
78. D. Verner, G.J. Ferland, K.T. Korista, D.G. Yakovlev, Atomic data for astrophysics. II. New analytic fits for photoionization cross sections of atoms and ions. Astrophys. J. **465**, 487 (1996)
79. D.A. Williams, The chemistry of interstellar CH^+: the contribution of Bates and Spitzer (1951). Planet. Space Sci. **40**, 1683 (1992)
80. B. Yang, X.H. Wang, P. Zhang, P.C. Stancil, J.M. Bowman, N. Balakrishnan, B.M. McLaughlin, R.C. Forrey, Full-dimensional quantum dynamics in collisions of SiO with H_2. J. Phys. Chem. A (2017). Submitted for publication
81. L.B. Zhao et al., Radiative charge transfer in collisions of O with He^+. Astrophys. J. **615**, 1063 (2004)
82. B. Zygelman, A. Dalgarno, M. Kimura, N.F. Lane, Radiative and nonradiative charge transfer in He^+ + H collisions at low energy. Phys. Rev. A **40**, 2340 (1989)

Phenomenology of Strange Resonances

Constraining the Hadronic Spectrum Through QCD Thermodynamics on the Lattice

Paolo Alba, Rene Bellwied, Szabolcs Borsanyi, Zoltan Fodor, Jana Günther, Sandor D. Katz, Valentina Mantovani Sarti, Jacquelyn Noronha-Hostler, Paolo Parotto, Attila Pasztor, Israel Portillo Vazquez, and Claudia Ratti

Abstract Fluctuations of conserved charges allow to study the chemical composition of hadronic matter. A comparison between lattice simulations and the Hadron Resonance Gas (HRG) model suggested the existence of missing strange resonances. To clarify this issue we calculate the partial pressures of mesons and baryons with different strangeness quantum numbers using lattice simulations in the confined phase of QCD. In order to make this calculation feasible, we perform simulations at imaginary strangeness chemical potentials. We systematically study the effect of different hadronic spectra on thermodynamic observables in the HRG model and compare to lattice QCD results. We show that, for each hadronic sector,

P. Alba
Frankfurt Institute for Advanced Sciences, Goethe Universität Frankfurt, D-60438 Frankfurt am Main, Germany

R. Bellwied • J. Noronha-Hostler • P. Parotto • I.P. Vazquez • C. Ratti
Department of Physics, University of Houston, Houston, TX 77204, USA

S. Borsanyi • J. Günther • A. Pasztor
Department of Physics, Wuppertal University, Gaussstr. 20, 42119 Wuppertal, Germany

Z. Fodor (✉)
Department of Physics, Wuppertal University, Gaussstr. 20, 42119 Wuppertal, Germany

Institute for Theoretical Physics, Eötvös University, Pázmány P. sétány 1/A, H-1117 Budapest, Hungary

Jülich Supercomputing Centre, Forschungszentrum Jülich, 52425 Jülich, Germany
e-mail: fodor@bodri.elte.hu

S.D. Katz
Institute for Theoretical Physics, Eötvös University, Pázmány P. sétány 1/A, H-1117 Budapest, Hungary

MTA-ELTE "Lendület" Lattice Gauge Theory Research Group, Pázmány P. sétány 1/A, H-1117 Budapest, Hungary

V.M. Sarti
Department of Physics, Torino University and INFN, Sezione di Torino, via P. Giuria 1, 10125 Torino, Italy

© Springer International Publishing AG 2018
W.E. Nagel et al. (eds.), *High Performance Computing in Science and Engineering '17*, https://doi.org/10.1007/978-3-319-68394-2_4

61

the well established states are not enough in order to have agreement with the lattice results. Additional states, either listed in the Particle Data Group booklet (PDG) but not well established, or predicted by the Quark Model (QM), are necessary in order to reproduce the lattice data. For mesons, it appears that the PDG and the quark model do not list enough strange mesons, or that, in this sector, interactions beyond those included in the HRG model are needed to reproduce the lattice QCD results.

1 Introduction

The precision achieved by recent lattice simulations of QCD thermodynamics allows to extract, for the first time, quantitative predictions which provide a new insight into our understanding of strongly interacting matter. Recent examples include the precise determination of the QCD transition temperature [1–4], the QCD equation of state at zero [5–7] and small chemical potential [8–10] and fluctuations of quark flavors and/or conserved charges near the QCD transition [11–13]. The latter are particularly interesting because they can be related to experimental measurements of particle multiplicity cumulants, thus allowing to extract the freeze-out parameters of heavy-ion collisions from first principles [14–18]. Furthermore, they can be used to study the chemical composition of strongly interacting matter and identify the degrees of freedom which populate the system in the vicinity of the QCD phase transition [19–21].

The vast majority of lattice results for QCD thermodynamics can be described, in the hadronic phase, by a non-interacting gas of hadrons and resonances which includes the measured hadronic spectrum up to a certain mass cut-off. This approach is commonly known as the Hadron Resonance Gas (HRG) model [22–26]. There is basically no free parameter in such a model, the only uncertainty being the number of states, which is determined by the spectrum listed in the Particle Data Book. It has been proposed recently to use the precise lattice QCD results on specific observables, and their possible discrepancy with the HRG model predictions, to infer the existence of higher mass states [27–29], not yet measured but predicted by Quark Model (QM) calculations [30, 31] and lattice QCD simulations [32]. This leads to a better agreement between selected lattice QCD observables and the corresponding HRG curves. However, for other observables the agreement with the lattice gets worse, once the QM states are included.

Amongst experimentally measured hadronic resonances within the Particle Data Group (PDG) list, there are different confidence levels on the existence of individual resonances. The most well-established states are denoted by **** stars whereas * states indicate states with the least experimental confirmation. Furthermore, states with the fewest stars often do not have the full decay channel information known nor the branching ratios for different decay channels.

In Fig. 1 we compare, for several particle species, the states listed in the PDG2016 (including states with two, three and four stars) [33], in the PDG2016+ (including also states with one star) [33] and those predicted by the original Quark

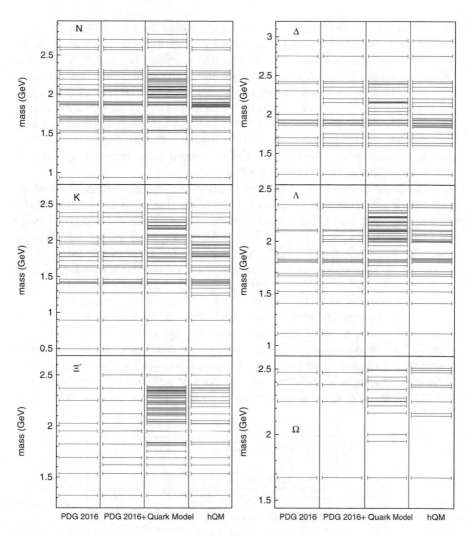

Fig. 1 Comparison of hadronic states, grouped according to the particle species, experimentally established in the PDG2016 (green), PDG2016 including one star states (red) [33] and predicted by the QM (blue) [30, 31] and the hQM (magenta) [34, 35]

Model [30, 31] and a more recent hypercentral version (hQM) [34]. The latter contains fewer states than the ones found in Refs. [30, 31], due to inclusion of an interaction term between the quarks in the bound state, and the decay modes are listed for most of the predicted states. No mass cut-off has been imposed. The total number of measured particles and anti-particles, excluding the charm and the bottom sector, increases from the 2016 to the 2016+ listing: considering particles and antiparticles and their isospin multiplicity we get 608 states with two, three and four stars and 738 states when we also include the one star states. In the QM description the overall increase is much larger: in total there are 1446 states in the non-relativistic QM [30, 31] and 1237 in the hQM [34, 35]. The QM predicts such a large number of states because they arise from all possible combinations of different quark-flavor, spin and momentum configurations. However, many of these states have not been observed in experiments so far; besides the basic QM description does not provide any information on the decay properties of such particles. As already mentioned, the hQM reduces the number of states by including an interaction term between quarks in a bound state. A more drastic reduction can be achieved by assuming a diquark structure as part of the baryonic states, although experiments and lattice QCD may disfavor such a configuration [36].

In this paper, we perform an analysis of several strangeness-related observables, by comparing the lattice QCD results to those of the HRG model based on different resonance spectra: the PDG 2016 including only the more established states (labeled with two, three and four stars), the PDG 2016 including all listed states (also the ones with one star), and the PDG 2016 with the inclusion of additional Quark Model states. This is done in order to systematically test the results for different particle species, and get differential information on the missing states, based on their strangeness content. The observables which allow the most striking conclusions are the partial pressures, namely the contribution to the total pressure of QCD from the hadrons, grouped according to their baryon number and strangeness content. The main result of this paper is a lattice determination of these partial pressures. This is a difficult task, since the partial pressures involve a cancellation of positive and negative contributions (see the next section), and they span many orders of magnitude, as can be seen in Fig. 2. From this analysis a consistent picture

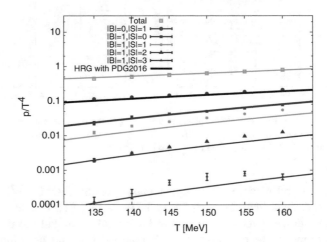

Fig. 2 Logarithmic plot illustrating the many orders of magnitude the values of the partial pressures studied in this paper cover. The total pressure is taken from Ref. [6]. Note, that the value for the $B = 0$, $|S| = 1$ sector is not a proper continuum limit, it is a continuum estimate based on the $N_t = 12$ and 16 lattices. For all other, the data are properly continuum extrapolated. In all cases, the solid lines correspond to the HRG model results based on the PDG2016 spectrum

emerges: all observables confirm the need for not yet detected or at least not yet fully established strangeness states. The full PDG2016 list provides a satisfactory description of most observables, while for others the QM states are needed in order to reproduce the lattice QCD results. Besides, it appears that the PDG and the Quark Model do not list enough strange mesons or that, in this sector, interactions beyond those included in the HRG model are needed to reproduce the lattice QCD data [37, 38].

2   HRG and the Strangeness Sectors on the Lattice

The HRG model provides an accurate description of the thermodynamic properties of hadronic matter below T_c. This is especially true for global observables such as the total pressure and other collective thermodynamic quantities. However, it was recently noticed that more differential observables which are sensitive to the flavor content of the hadrons show a discrepancy between HRG model and lattice results [29]. An example of such discrepancy is shown in Fig. 3 and will be explained below. Such observables involve the evaluation of susceptibilities of conserved charges in the system at vanishing chemical potential:

$$\chi_{lmn}^{BQS} = \left(\frac{\partial^{l+m+n} P(T, \mu_B, \mu_Q, \mu_S)/T^4}{\partial(\mu_B/T)^l \partial(\mu_Q/T)^m \partial(\mu_S/T)^n} \right)_{\mu=0}. \tag{1}$$

Cumulants of net-strangeness fluctuations and correlations with net-baryon number and net-electric charge have been evaluated on the lattice in a system of $(2 + 1)$ flavours at physical quark masses and in the continuum limit [13, 16, 39]. The same quantities can be obtained within the HRG model. In this approach, the total pressure in the thermodynamic limit for a gas of non-interacting particles in the

Fig. 3 Ratio μ_S/μ_B at leading order as a function of the temperature. The HRG results are shown for different hadronic spectra, namely by using the PDG2012 (black solid line) and the QM (dashed red line)

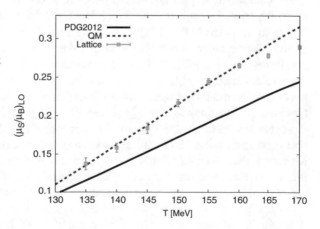

grand-canonical ensemble is given by:

$$P_{tot}(T,\mu) = \sum_k P_k(T,\mu_k) = \sum_k (-1)^{B_k+1} \frac{d_k T}{(2\pi)^3} \int d^3\mathbf{p}$$

$$\ln\left(1 + (-1)^{B_k+1} \exp\left[-\frac{(\sqrt{\mathbf{p}^2 + m_k^2} - \mu_k)}{T}\right]\right), \tag{2}$$

where the sum runs over all the hadrons and resonances included in the model. Here the single particle chemical potential is defined with respect to the global conserved charges (baryonic B, electric Q and strangeness S) as $\mu_k = B_k\mu_B + Q_k\mu_Q + S_k\mu_S$. More details on the HRG model used here can be found in Ref. [40]. In order to describe the initial conditions of the system occurring during a heavy-ion collision, we require strangeness neutrality and the proper ratio of protons to baryons given by the colliding nuclei, $n_Q = \frac{Z}{A}n_B \simeq 0.4n_B$. These conditions yield μ_S and μ_Q as functions of μ_B; their specific dependence on μ_B is affected by the amount of strange particles and charged particles included in the model. To leading order in μ_B, the ratio μ_S/μ_B reads [15, 16]:

$$\left(\frac{\mu_S}{\mu_B}\right)_{LO} = -\frac{\chi_{11}^{BS}}{\chi_2^S} - \frac{\chi_{11}^{QS}}{\chi_2^S}\frac{\mu_Q}{\mu_B}. \tag{3}$$

The inclusion of a larger number of heavy hyperons, such as Λ and Ξ, and the constraint of strangeness neutrality are reflected by a larger value of the strange chemical potential μ_S as a function of temperature and baryo-chemical potential. In Fig. 3 this ratio is shown as a function of the temperature: our new, continuum extrapolated lattice results are compared to the HRG model calculations based on the 2012 version of the PDG and on the Quark Model states (as done in Ref. [29]). One should expect agreement between HRG model and lattice calculations up to the transition temperature which has been determined independently on the lattice to be $\sim$155 MeV [1–4]. The HRG model based on the QM particle list yields a better agreement with the lattice data within error bars, while the HRG results based on the PDG2012 spectrum underestimate the data. However, for other observables such as χ_4^S/χ_2^S and χ_{11}^{us} (see the two panels of Fig. 4), the agreement between HRG model and lattice results is spoiled when including the QM states. The QM result overestimates both χ_4^S/χ_2^S and χ_{11}^{us}; χ_4^S/χ_2^S is proportional to the average strangeness squared in the system: the fact that the QM overestimates it, means that it contains either too many multi-strange states or not enough $|S| = 1$ states. Moreover, the contribution to χ_{11}^{us} is positive for baryons and negative for mesons: this observable provides the additional information that the QM list contains too many (multi-)strange baryons or not enough $|S| = 1$ mesons.

In this paper, we try to solve this ambiguity, even though we are aware that it might be difficult to resolve the contribution of high mass particles in our

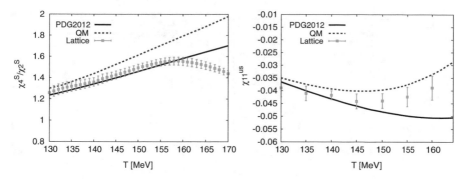

Fig. 4 Left panel: Ratio χ_4^S/χ_2^S as a function of the temperature. HRG model calculations based on the PDG2012 (black solid line) and the QM (red dashed line) spectra are shown in comparison to the lattice results from Ref. [21]. Right panel: comparison of up-strange correlator χ_{11}^{us} simulated on the lattice [13] and calculated in the HRG model using the PDG2012 (solid black line) and the QM (dashed red line) spectra

simulations. We separate the pressure of QCD as a function of the temperature into contributions coming from hadrons grouped according to their quantum numbers. This is done by assuming that, in the low temperature region we are interested in, the HRG model in the Boltzmann approximation yields a valid description of QCD thermodynamics. If this is the case, the pressure of the system can be written as [18, 20]:

$$P(\hat{\mu}_B, \hat{\mu}_S) = P_{00}^{BS} + P_{10}^{BS}\cosh(\hat{\mu}_B) + P_{01}^{BS}\cosh(\hat{\mu}_S) + P_{11}^{BS}\cosh(\hat{\mu}_B - \hat{\mu}_S)$$
$$+ P_{12}^{BS}\cosh(\hat{\mu}_B - 2\hat{\mu}_S) + P_{13}^{BS}\cosh(\hat{\mu}_B - 3\hat{\mu}_S) , \qquad (4)$$

where $\hat{\mu}_i = \mu_i/T$, and the quantum numbers can be understood as absolute values. These partial pressures are the main observables we study. Notice that we do not distinguish the particles according to their electric charge content.

Assuming this ansatz for the pressure, the partial pressures P_{ij}^{BS} can be expressed as linear combinations of the susceptibilities χ_{ij}^{BS}. An example of one such formula is:

$$P_{01}^{BS} = \chi_2^S - \chi_{22}^{BS}, \qquad (5)$$

which gives the strange meson contribution to the pressure. This means that in principle one could determine these partial pressures directly from $\mu = 0$ simulations, by evaluating linear combinations of the χ_{ij}^{BS} directly. This can be done on the lattice, by calculating fermion matrix traces, that can be evaluated with the help of random sources [13, 41].

This is not the approach we pursue here, since the noise level in the calculation would be too high, certainly for the $S = 2$ or 3 sectors, but as Fig. 5 (right) shows, probably already for $S = 1$. The higher order fluctuations are already

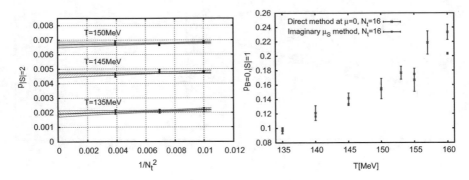

Fig. 5 Left panel: Examples of continuum limit extrapolation from our $N_t = 10, 12, 16$ lattices. Right panel: Comparison of our method to Taylor expansion from $\mu = 0$ data for P_{01}. The statistics would explain only a factor of 2 difference in the errorbars, but the improvement is much more drastic than that

quite noisy, because they involve big cancellations between positive and negative contributions [13]. In addition, when we take linear combinations to calculate the partial pressures, we introduce extra cancellations between the susceptibilities. Therefore we propose to use an imaginary strangeness chemical potential and extract the partial pressures from the $\mathrm{Im}\,\mu_S$ depdendence of low order susceptibilities. For earlier works exploiting imaginary chemical potentials, see [42–48]. A more recent work, that uses imaginary chemical potentials to estimate higher order susceptibilities is [49].

Our lattice simulations use the same 4stout staggered action as [9, 13, 50, 51]. We generate configurations at $\mu_B = \mu_S = \mu_Q = 0$ as well as $\mu_B = \mu_Q = 0$ and $\mathrm{Im}\,\mu_S > 0$, in the temperature range $135\,\mathrm{MeV} \leq T \leq 165\,\mathrm{MeV}$. All of our lattices have an aspect ratio of $LT = 4$. We run roughly 1000–2000 configurations at each simulation point, separated by 10 HMC trajectories. For the determination of the strangeness sectors we use the HRG ansatz of Eq. (4) for the pressure. With the notation $\mu_S = i\mu_I$ we get by simple differentiation:

$$\mathrm{Im}\,\chi_1^B = -P_{11}^{BS}\sin(\mu_I) - P_{12}^{BS}\sin(2\mu_I)$$
$$-P_{13}^{BS}\sin(3\mu_I),$$
$$\chi_2^B = P_{10}^{BS} + P_{11}^{BS}\cos(\mu_I) + P_{12}^{BS}\cos(2\mu_I)$$
$$+P_{13}^{BS}\cos(3\mu_I), \tag{6}$$
$$\mathrm{Im}\,\chi_1^S = (P_{01}^{BS} + P_{11}^{BS})\sin(\mu_I) + 2P_{12}^{BS}\sin(2\mu_I)$$
$$+3P_{13}^{BS}\sin(3\mu_I),$$

and similar terms for χ_2^S, χ_3^S and χ_4^S. An advantage of the imaginary chemical potential approach is that we do not have to make any extra assumptions, as

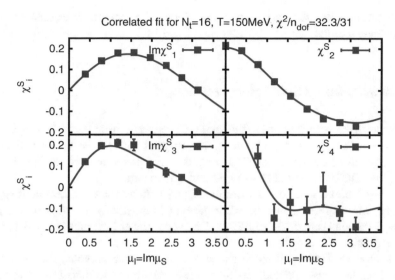

Fig. 6 An example of a correlated fit for the quantities χ_1^S, χ_2^S, χ_3^S and χ_4^S. The plot corresponds to our finest lattice near the QCD transition temperature

compared to the direct evaluation of the linear combinations, see Eq. (5), as the linear combinations there were already derived from the HRG ansatz. On the other hand, our method reduces the errors considerably, as the lower derivatives already contain the information on the higher strangeness sectors. Simulations at imaginary chemical potential are not hampered by the sign problem, so the evaluation of the lower order susceptibilities at $\text{Im}\,\mu_S > 0$ is not any harder than at $\mu = 0$.

To obtain the Fourier coefficients $(P_{01}^{BS} + P_{11}^{BS})$, P_{12}^{BS} and P_{13}^{BS} we perform a correlated fit with the previous ansatz for the observables χ_1^S, χ_2^S, χ_3^S and χ_4^S at every temperature. To obtain P_{10}^{BS} we fit χ_2^B, to obtain P_{11}^{BS} we fit χ_1^B. We note that P_{11}^{BS} could be deduced from χ_2^B as well, but with considerably higher statistical errors. To get P_{01}^{BS} we just take the difference $(P_{01}^{BS} + P_{11}^{BS}) - P_{11}^{BS}$. As an illustration that the HRG based ansatz fits our lattice data we include an example of the correlated fit for χ_1^S, χ_2^S, χ_3^S and χ_4^S in Fig. 6.

For the continuum limit we use $N_t = 10, 12$ and 16 lattices, see Fig. 5. For the $B = 0$, $S = 1$ sector, which includes a large contribution from kaons, the continuum extrapolation could not be carried out using these lattices. For this case we obtain a continuum estimate, based on the assumption that only $N_t = 10$ is not in the scaling regime, therefore using only the $N_t = 12$ and $N_t = 16$ lattices, and the same sources of systematic error as before, but now with uniform weights.

Finally, as a comparison we show in Fig. 5 (right) one of the partial pressures determined with both methods for $N_t = 16$, using the same action. The figure shows that using imaginary chemical potential improved the accuracy drastically already in the $S = 1$ sector. In the $S = 2, 3$ sectors, the direct method would be too noisy to plot, while the imaginary μ method allows for a quite accurate determination of the strangeness sectors.

3   Results and Their Interpretation

We evaluate the contributions to the total QCD pressure from the following sectors: strange mesons, non-strange baryons, and baryons with $|S| = 1, 2, 3$. For each sector, we compare the lattice QCD results to the predictions of the HRG model using the PDG2016, PDG2016+, hQM and QM spectra.

In Figs. 7 and 8 we show our results. Figure 7 shows the contribution of strange mesons, while Fig. 8 shows the contribution of non-strange baryons (upper left), $|S| = 1$ baryons (upper right), $|S| = 2$ baryons (lower left) and $|S| = 3$ baryons (lower right).

We observe that, in all cases except the non-strange baryons, the established states from the most updated version of the PDG are not sufficient to describe the lattice data. For the baryons with $|S| = 2$, a considerable improvement is achieved when the one star states from PDG2016 are included. The inclusion of the hQM states pushes the agreement with the lattice results to higher temperatures, but one has to keep in mind that the crossover nature of the QCD phase transition implies the presence of quark degrees of freedom in the system above $T \simeq 155$ MeV, which naturally yields a deviation from the HRG model curves. Notice that, in the case of

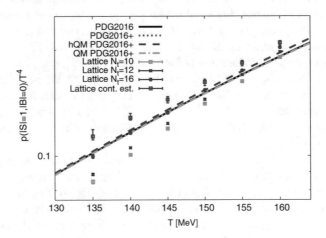

Fig. 7 Comparison between the lattice results for the partial pressures of strange mesons and the HRG model predictions. No continuum extrapolation could be carried out from our $N_t = 10, 12, 16$ data. The plot includes the lattice data at finite lattice spacings for $N_t = 10, 12, 16$ and a continuum estimate form the $N_t = 12$ and 16 lattices. The points are the lattice results, while the curves are PDG2016 (solid black), PDG2016+ (including one star states, red dotted), PDG2016+ and additional states from the hQM (blue, dashed) [34, 35], PDG2016+ and additional states from the QM (cyan, dash-dotted) [30, 31]

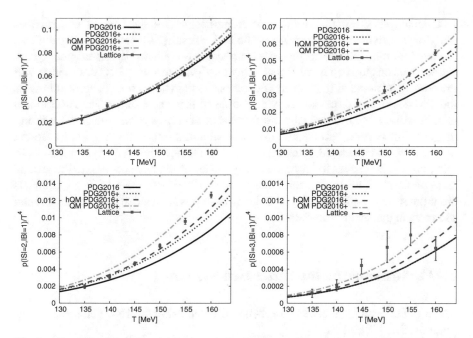

Fig. 8 Comparison between the lattice results for the partial pressures and the HRG model predictions. Upper panels: non-strange baryons (left), $|S| = 1$ baryons (right). Lower panels: $|S| = 2$ baryons (left), $|S| = 3$ baryons (right). The points are the lattice results, while the curves are PDG2016 (solid black), PDG2016+ (including one star states, red dotted), PDG2016+ and additional states from the hQM (blue, dashed) [34, 35], PDG2016+ and additional states from the QM (cyan, dash-dotted) [30, 31]

$|S| = 1$, 3 baryons, it looks like even more states than PDG2016+ with hQM are needed in order to reproduce the lattice results: the agreement improves when the resonances predicted by the QM [30, 31] are added to the spectrum. Figure 2 shows the relative contribution of the sectors to the total pressure. Notice that three orders of magnitude separate the $|S|=1$ meson contribution from the $|S| = 3$ baryon one. The method we used for this analysis, namely simulations at imaginary μ_S, was crucial in order to extract a signal for the multi-strange baryons.

We also include a continuum estimate of this quantity, based on only the $N_t = 12$ and 16 lattices, which is clearly above the HRG curves. This might mean that, for strange mesons, the interaction between particles is not well mimicked by the HRG model in the Boltzmann approximation, or that we need even more states than the ones predicted by the QM. This was already suggested in Ref. [52], based on a different analysis. In general, one should keep in mind that here we use a version of the HRG model in which particles are considered stable (no width is included). Any width effects on the partial pressures can be considered in future work. Analogously, our previous lattice QCD results did not show indications of finite volume effects for the total pressure. These effects have not been checked for the partial pressures presented here.

Our analysis shows that, for most hadronic sectors, the spectrum PDG2016, does not yield a satisfactory description of the lattice results. All sectors clearly indicate the need for more states, in some cases up to those predicted by the original Quark Model. One has to keep in mind that using the QM states in a HRG description will introduce additional difficulties in calculations used in heavy ion phenomenology, as the QM does not give us the decay properties of these new states. The HRG model is successfully used to describe the freeze-out of a heavy-ion collision, by fitting the yields of particles produced in the collision and thus extracting the freeze-out temperature and chemical potential [53–55], which are known as "thermal fits". To this purpose, one needs to know the decay modes of the resonances into the ground state particles which are reaching the detector. As of yet, the QM decay channels are unknown so predictions for their decay channels are needed first, before one can use them in thermal fits models.

4 Performance on the Hazel Hen System

Much of the presented results have been achieved using the Hazel Hen supercomputer at HLRS, Stuttgart.

The most challenging numerical effort corresponds to our finest lattice with a size of $64^3 \times 16$. These lattices had to be generated, such that an ensemble of gauge configurations is collected for each studied temperature and chemical potential. For this study 56 such ensembles have been generated. Each ensemble consists of 1000–1500 stored configurations, that are separated by 10 hybrid Monte Carlo updates (HMC). These configurations were analyzed for strangeness and baryon fluctuations with the method of random sources using several conjugate gradient inversions (analysis).

The performance of the code can be measured by the wall-clock time of one HMC update, and by the conjugate (CG) performance in gigaflop/seconds.

Our generic MPI-based code has been ported to the CRAY system in several steps of optimizations. Because of the multiple NUMA nodes that share a network interface card we use the hybrid approach, having 4 parallel processes per node, each using 6 threads. The communication goes entirely over DMAPP. The scaling of the code is mostly limited by the job placement mechanism of HAZELHEN. To code runs optimally if all nodes share a group, though this happens rarely with a larger job. Nevertheless, higher per-core performance can be achieved on HAZELHEN than on a BG/Q system.

The Cray system's AVX instructions are analogous to those of the A2 core of BG/Q, thus we use the same site fusion technique. The four dimensional lattice is communicated in three dimensions, the fourth dimension is used in the site fusion.

Parallel I/O: we use DMAPP to do a peer-to-peer communication to reorganize data from simulation layout to disk layout in chunks of 64 MB. These chunks are written by several nodes in parallel. Reads work similarly, data files are read in chunks in parallel and distributed by communication

For our $64^3 \times 16$ lattices the optimal performance is reached with 64 nodes (1536) cores. There we reach 7 Tflops/s in the multi-precision inverter.

For the full update algorithm we find the following strong scaling behaviour on HAZELHEN:

# cores	192	384	768	1536	3072
ONE update on a $64^3 \times 16$ lattice [s]	1372	658	413	173	152

The analysis of the generated configurations requires several conjugate gradient inversions. These configurations can be analyzed independently. Due to the large number of the configurations a straightforward parallelization could be used, however, due to cache limitations, the code runs more efficiently in parallel than in serial. The most optimal use of the resources we achieved by selecting a larger partitions with several sub-partitions, each sub-partition (of e.g. 64 nodes) being in charge of a distinct list of configurations.

We noticed, that the performance of the same code on the same data in the same submission depends very deterministically on which sub-partition it was running. If the nodes in one sub-partition was all from one group the performance was 25% higher than when inter-group links were also used. By selecting a larger partition and letting our own code to select the nodes that belong to one sub-partition we achieved better overall performance than by using several small jobs (each with one sub-partition only), because in the latter case the job was almost always fragmented. Our code defines the sub-partitions to minimize fragmentation.

Here we give examples of our timings for the single-source conjugate gradient solver.

256 nodes divided into 4 sub-partitions:

Layout spread over # of groups	Run time [s]	Gflops/s
1	8.2	3694.2
2	10.3	2952.4
4	10.4	2930.4
4	10.9	2799.3

When all 256 nodes were given in the same group then each sub-partition performed equally well. The best performance we achieved on 64 nodes with a local lattice of $4 \times 4 \times 16 \times 64$, at 5287.9 Gflops/s. The top CG performance we experienced was as follows (note that this performance was significantly fluctuating, probably depending on the traffic coming from nearby applications).

We note that in this 128 nodes were not in one group, in fact we almost never received enough nodes in the same group such that a scalability to 128 nodes would be practical.

# Hazel Hen cores	Gigaflops/s
3072	6483.8
1536	5287.9
768	1945.0
384	621.8

Acknowledgements We acknowledge fruitful discussions with Elena Santopinto, Igor Strakovsky, Moskov Amaryan and Mark Manley. This project was funded by the DFG grant SFB/TR55. This material is based upon work supported by the National Science Foundation under grant no. PHY-1513864 and by the U.S. Department of Energy, Office of Science, Office of Nuclear Physics, within the framework of the Beam Energy Scan Theory (BEST) Topical Collaboration. The work of R. Bellwied is supported through DOE grant DEFG02-07ER41521. An award of computer time was provided by the INCITE program. This research used resources of the Argonne Leadership Computing Facility, which is a DOE Office of Science User Facility supported under Contract DE-AC02-06CH11357. The authors gratefully acknowledge the Gauss Centre for Supercomputing (GCS) for providing computing time for a GCS Large-Scale Project on the GCS share of the supercomputer JUQUEEN [56] at Jülich Supercomputing Centre (JSC), and at HazelHen supercomputer at HLRS, Stuttgart. The authors gratefully acknowledge the use of the Maxwell Cluster and the advanced support from the Center of Advanced Computing and Data Systems at the University of Houston.

References

1. Y. Aoki, Z. Fodor, S. Katz, K. Szabo, Phys. Lett. **B643**, 46 (2006). https://doi.org/10.1016/j.physletb.2006.10.021
2. Y. Aoki, S. Borsanyi, S. Durr, Z. Fodor, S.D. Katz et al., J. High Energy Phys. **0906**, 088 (2009). https://doi.org/10.1088/1126-6708/2009/06/088
3. S. Borsanyi et al., J. High Energy Phys. **1009**, 073 (2010). https://doi.org/10.1007/JHEP09(2010)073
4. A. Bazavov, T. Bhattacharya, M. Cheng, C. DeTar, H. Ding et al., Phys. Rev. **D85**, 054503 (2012). https://doi.org/10.1103/PhysRevD.85.054503
5. S. Borsanyi, G. Endrodi, Z. Fodor, A. Jakovac, S.D. Katz et al., J. High Energy Phys. **1011**, 077 (2010). https://doi.org/10.1007/JHEP11(2010)077
6. S. Borsanyi, Z. Fodor, C. Hoelbling, S.D. Katz, S. Krieg et al., Phys. Lett. **B730**, 99 (2014). https://doi.org/10.1016/j.physletb.2014.01.007
7. A. Bazavov et al., Phys. Rev. **D90**(9), 094503 (2014). https://doi.org/10.1103/PhysRevD.90.094503
8. S. Borsanyi, G. Endrodi, Z. Fodor, S. Katz, S. Krieg et al., J. High Energy Phys. **1208**, 053 (2012). https://doi.org/10.1007/JHEP08(2012)053
9. J. Gunther, R. Bellwied, S. Borsanyi, Z. Fodor, S.D. Katz, A. Pasztor, C. Ratti, The QCD equation of state at finite density from analytical continuation. EPJ Web Conf. **137**, 07008 (2017). https://doi.org/10.1051/epjconf/201713707008
10. A. Bazavov et al., The QCD equation of state to $\mathcal{O}(\mu_B^6)$ from lattice QCD. Phys. Rev. D **95**(5), 054504 (2017). https://doi.org/10.1103/PhysRevD.95.054504
11. S. Borsanyi, Z. Fodor, S.D. Katz, S. Krieg, C. Ratti et al., J. High Energy Phys. **1201**, 138 (2012). https://doi.org/10.1007/JHEP01(2012)138
12. A. Bazavov et al., Phys. Rev. **D86**, 034509 (2012). https://doi.org/10.1103/PhysRevD.86.034509

13. R. Bellwied, S. Borsanyi, Z. Fodor, S.D. Katz, A. Pasztor, C. Ratti, K.K. Szabo, Phys. Rev. **D92**(11), 114505 (2015). https://doi.org/10.1103/PhysRevD.92.114505
14. F. Karsch, Central Eur. J. Phys. **10**, 1234 (2012). https://doi.org/10.2478/s11534-012-0074-3
15. A. Bazavov, H. Ding, P. Hegde, O. Kaczmarek, F. Karsch et al., Phys. Rev. Lett. **109**, 192302 (2012). https://doi.org/10.1103/PhysRevLett.109.192302
16. S. Borsanyi, Z. Fodor, S. Katz, S. Krieg, C. Ratti et al., Phys. Rev. Lett. **111**, 062005 (2013). https://doi.org/10.1103/PhysRevLett.111.062005
17. S. Borsanyi, Z. Fodor, S. Katz, S. Krieg, C. Ratti et al., Phys. Rev. Lett. **113**, 052301 (2014). https://doi.org/10.1103/PhysRevLett.113.052301
18. J. Noronha-Hostler, R. Bellwied, J. Gunther, P. Parotto, A. Pasztor, I.P. Vazquez, C. Ratti, Kaon fluctuations from lattice QCD (2017). arXiv:1607.02527 [hep-ph]
19. V. Koch, A. Majumder, J. Randrup, Phys. Rev. Lett. **95**, 182301 (2005). https://doi.org/10.1103/PhysRevLett.95.182301
20. A. Bazavov, H.T. Ding, P. Hegde, O. Kaczmarek, F. Karsch et al., Phys. Rev. Lett. 111, **082301**, 082301 (2013). https://doi.org/10.1103/PhysRevLett.111.082301
21. R. Bellwied, S. Borsanyi, Z. Fodor, S.D. Katz, C. Ratti, Phys. Rev. Lett. **111**, 202302 (2013). https://doi.org/10.1103/PhysRevLett.111.202302
22. R. Dashen, S.K. Ma, H.J. Bernstein, Phys. Rev. **187**, 345 (1969). https://doi.org/10.1103/PhysRev.187.345
23. R. Venugopalan, M. Prakash, Nucl. Phys. **A546**, 718 (1992). https://doi.org/10.1016/0375-9474(92)90005-5
24. F. Karsch, K. Redlich, A. Tawfik, Eur. Phys. J. **C29**, 549 (2003). https://doi.org/10.1140/epjc/s2003-01228-y
25. F. Karsch, K. Redlich, A. Tawfik, Phys. Lett. **B571**, 67 (2003). https://doi.org/10.1016/j.physletb.2003.08.001
26. A. Tawfik, Phys. Rev. **D71**, 054502 (2005). https://doi.org/10.1103/PhysRevD.71.054502
27. J. Noronha-Hostler, J. Noronha, C. Greiner, Phys. Rev. Lett. **103**, 172302 (2009). https://doi.org/10.1103/PhysRevLett.103.172302
28. A. Majumder, B. Muller, Phys. Rev. Lett. **105**, 252002 (2010). https://doi.org/10.1103/PhysRevLett.105.252002
29. A. Bazavov, H.T. Ding, P. Hegde, O. Kaczmarek, F. Karsch et al., Phys. Rev. Lett. **113**, 072001 (2014). https://doi.org/10.1103/PhysRevLett.113.072001
30. S. Capstick, N. Isgur, Phys. Rev. **D34**, 2809 (1986). https://doi.org/10.1103/PhysRevD.34.2809
31. D. Ebert, R.N. Faustov, V.O. Galkin, Phys. Rev. **D79**, 114029 (2009). https://doi.org/10.1103/PhysRevD.79.114029
32. R.G. Edwards, N. Mathur, D.G. Richards, S.J. Wallace, Phys. Rev. **D87**(5), 054506 (2013). https://doi.org/10.1103/PhysRevD.87.054506
33. C. Patrignani et al., Chin. Phys. **C40**(10), 100001 (2016). https://doi.org/10.1088/1674-1137/40/10/100001
34. M. Ferraris, M.M. Giannini, M. Pizzo, E. Santopinto, L. Tiator, Phys. Lett. **B364**, 231 (1995). https://doi.org/10.1016/0370-2693(95)01091-2
35. J. Ferretti, R. Bijker, G. Galatà, H. García-Tecocoatzi, E. Santopinto, Phys. Rev. **D94**(7), 074040 (2016). https://doi.org/10.1103/PhysRevD.94.074040
36. R.G. Edwards, J.J. Dudek, D.G. Richards, S.J. Wallace, AIP Conf. Proc. **1432**, 33 (2012). https://doi.org/10.1063/1.3701185
37. P. Alba, in *Workshop on Excited Hyperons in QCD Thermodynamics at Freeze-Out (YSTAR2016) Mini-Proceedings* (2017), pp. 148–154. http://inspirehep.net/record/1511550/files/1510585_140-146.pdf
38. D. Cabrera, L. Tolós, J. Aichelin, E. Bratkovskaya, Phys. Rev. **C90**(5), 055207 (2014). https://doi.org/10.1103/PhysRevC.90.055207
39. S. Borsanyi, Z. Fodor, S. Katz, S. Krieg, C. Ratti et al., J. Phys. Conf. Ser. **535**, 012030 (2014). https://doi.org/10.1088/1742-6596/535/1/012030
40. P. Alba, W. Alberico, R. Bellwied, M. Bluhm, V. Mantovani Sarti et al., Phys. Lett. **B738**, 305 (2014). https://doi.org/10.1016/j.physletb.2014.09.052

41. C. Allton, S. Ejiri, S. Hands, O. Kaczmarek, F. Karsch et al., Phys. Rev. **D66**, 074507 (2002). https://doi.org/10.1103/PhysRevD.66.074507
42. P. de Forcrand, O. Philipsen, Nucl. Phys. **B642**, 290 (2002). https://doi.org/10.1016/S0550-3213(02)00626-0
43. M. D'Elia, M.P. Lombardo, Phys. Rev. **D67**, 014505 (2003). https://doi.org/10.1103/PhysRevD.67.014505
44. L.K. Wu, X.Q. Luo, H.S. Chen, Phys. Rev. **D76**, 034505 (2007). https://doi.org/10.1103/PhysRevD.76.034505
45. M. D'Elia, F. Di Renzo, M.P. Lombardo, Phys. Rev. **D76**, 114509 (2007). https://doi.org/10.1103/PhysRevD.76.114509
46. S. Conradi, M. D'Elia, Phys. Rev. **D76**, 074501 (2007). https://doi.org/10.1103/PhysRevD.76.074501
47. P. de Forcrand, O. Philipsen, J. High Energy Phys. **0811**, 012 (2008). https://doi.org/10.1088/1126-6708/2008/11/012
48. M. D'Elia, F. Sanfilippo, Phys. Rev. **D80**, 014502 (2009). https://doi.org/10.1103/PhysRevD.80.014502
49. M. D'Elia, G. Gagliardi, F. Sanfilippo, Higher order quark number fluctuations via imaginary chemical potentials in $N_f = 2 + 1$ QCD. Phys. Rev. D **95**(9), 094503 (2017). https://doi.org/10.1103/PhysRevD.95.094503
50. R. Bellwied, S. Borsanyi, Z. Fodor, J. Günther, S.D. Katz, C. Ratti, K.K. Szabo, Phys. Lett. **B751**, 559 (2015). https://doi.org/10.1016/j.physletb.2015.11.011
51. S. Borsanyi et al., Nature **539**(7627), 69 (2016). https://doi.org/10.1038/nature20115
52. P.M. Lo, M. Marczenko, K. Redlich, C. Sasaki, Phys. Rev. **C92**(5), 055206 (2015). https://doi.org/10.1103/PhysRevC.92.055206
53. J. Cleymans, H. Oeschler, K. Redlich, S. Wheaton, Phys. Rev. **C73**, 034905 (2006). https://doi.org/10.1103/PhysRevC.73.034905
54. J. Manninen, F. Becattini, Phys. Rev. **C78**, 054901 (2008). https://doi.org/10.1103/PhysRevC.78.054901
55. A. Andronic, P. Braun-Munzinger, K. Redlich, J. Stachel, J. Phys. **G38**, 124081 (2011). https://doi.org/10.1088/0954-3899/38/12/124081
56. Juqueen, Ibm blue gene/q supercomputer system at the jülich supercomputing centre. Technical Report 1 A1, Jülich Supercomputing Centre, http://dx.doi.org/10.17815/jlsrf-1-18 (2015)

Real-Time Lattice Simulations of Quantum Anomalies, Topologies and Particle Production in Strongly Correlated Gauge Theories

Niklas Mueller, Oscar Garcia-Montero, Naoto Tanji, and Juergen Berges

Abstract We present results of numerical lattice simulations of anomalous and topological effects in Quantum Electrodynamics (QED) and Quantum Chromodynamics (QCD) in far from equilibrium situations. Based on the classical-statistical approximation to the Schwinger-Keldysh path integral formalism, we perform extensive numerical studies including dynamical Wilson and overlap fermions. Using advanced algorithmic techniques, we study the real-time dynamics of the axial anomaly relevant for strong field laser physics beyond the Schwinger limit and we observe novel dynamical refringence effects caused by the anomaly. Furthermore, motivated by recent interest in the physics of the Chiral Magnetic Effect in ultra-relativistic heavy ion collisions, we study the real time dynamics of fermions during and after a sphaleron transition and anomalous transport in the presence of strong magnetic fields.

1 Introduction

The main goal of this project has been to elucidate the importance of a detailed understanding of the out-of-equilibrium dynamics of topological and anomalous effects in relativistic gauge theories. Central fields of application of our theoretical studies are the physics of ultra-relativistic heavy ion collisions, matter under extreme conditions and the nature of the strong interaction.

In recent years there has been remarkable progress in the understanding of matter under extreme conditions, driven by the extremely precise and comprehensive data obtained from the Relativistic Heavy Ion Collider (RHIC) at Brookhaven National Laboratory (BNL), as well as the Large Hadron Collider (LHC) at CERN. Among many fascinating insights it was found that matter at temperatures several

N. Mueller (✉) • O. Garcia-Montero • N. Tanji • J. Berges
Institute for Theoretical Physics, Heidelberg University, Philosophenweg 16, 69120 Heidelberg, Germany
e-mail: n.mueller@thphys.uni-heidelberg.de

© Springer International Publishing AG 2018
W.E. Nagel et al. (eds.), *High Performance Computing in Science and Engineering '17*, https://doi.org/10.1007/978-3-319-68394-2_5

magnitudes larger than in the center of the sun behaves like an almost ideal fluid—the fundamental constituents of matter, quarks and gluons, thereby showing collective behavior. Simultaneously with experimental discoveries in heavy ion collisions, the understanding of the basic principles and physical laws governing the dynamics of fundamental matter has evolved very far. Nevertheless, many aspects of the theoretical descriptions of ultra-relativistic matter at particle colliders are still poorly understood. Conceptually, the description of the real-time dynamics of a heavy ion collision is one of the most elusive challenges for researchers and most phenomena crucially depend on an understanding of the early time evolution of strongly correlated non-Abelian plasmas far from thermal equilibrium.

An aspect that has risen to particular importance in recent years is the study of fundamental symmetries and their violation via quantum anomalies in the afore-mentioned experiments. While the concept of symmetry seems very abstract and not very important for the description of experimental outcome at first sight, modern physics relies on symmetry considerations in the construction of fundamental laws and equations, and symmetries have proven to be crucial for conceptual progress. Our present understanding of the universe is in fact phrased in a basic set of equations called the standard model of physics, expressed in the language of gauge theories. These equations have shown to be extremely successful in understanding nature and using symmetry as a guidance to formulate physical laws has been seen as omnipotent. In recent years however this view has drastically changed with the understanding of the existence of quantum anomalies. Their intriguing nature is the violation of fundamental symmetries not by any of the explicit basic equations that have been written down, but by the quantum property of nature itself. Related to quantum anomalies is the concept of the topology of gauge theories, which in the theory of the strong interactions manifests itself in the existence of topological transitions. Despite the fact that QCD describes nearly all of the matter that is surrounding us, the experimental access to its anomalous features have remained extremely limited.

Very recently however an intriguing perspective has been put forward: By means of the so called Chiral Magnetic Effect (CME) [1, 2] it was conjectured that quantum anomalies might manifest themselves via correlation of charged particles in heavy ion experiments. Non-central heavy ion collisions generate extremely large magnetic fields (several orders larger than anything produced in an astro-physical context), albeit restricted to the fire-ball volume of two nuclei (typically those of lead or gold atoms), which collided at nearly the speed of light. Provided there is an imbalance of chiral charge, such as is produced via anomalies, anomalous currents are generated by polarization effects. These currents comprise the CME and in fact there are a multitude of similar transport phenomena predicted in heavy ion experiments—among those the Chiral Separation Effect (CSE) and the Chiral Magnetic Wave (CMW).

The theoretical understanding of these effects however is very poor and current experimental data suggest large background contributions to the predicted signals. The main limitations in the conceptual understanding of these effects is due to the fact that these phenomena originate in the earliest moments of a heavy ion collision, where the matter and gauge fields are in a violently out-of-equilibrium stage. The understanding of the out-of-equilibrium dynamics of quantum many body systems is extremely challenging and theoretical predictions are very limited. One of the most powerful techniques to date are classical statistical real-time lattice simulations, that make use of the fact that ultra-relativistic heavy ion collisions are dominated by a quasi-classical state of highly occupied gluon fields and thus the dynamics of the system can be mapped on a statistical ensemble by use of the correspondence theorem.

The developments towards a quantitative description of out-of-equilibrium production of fermions have lead to significant advances in our understanding of the dynamics of anomalous phenomena and the CME in heavy ion collisions. Nevertheless, many challenges remain until the CME can be unambiguously identified as the proposed source of the azimuthal charged particle correlations seen by RHIC [3]. While quantitative descriptions of the initial conditions and time evolution of gluonic degrees of freedom have become remarkably accurate and allow us a nearly complete insight into various stages of a collision (semi-classical, kinetic, and hydrodynamic regimes), the phenomenology of fermions is still poorly understood. This is especially true in the very early time regime, where fermion production is dominant and where external magnetic fields are largest.

While there has been significant recent progress [4–6], important challenges remain on the way towards the quantitative descriptions of the anomalous and topological effects in heavy ion collisions: First, the time evolution of the hot matter created in a relativistic heavy ion collision is dominated by longitudinal expansion of the fire-ball. Therefore, while lattice simulations in a fixed geometry give deep insights into the far-from equilibrium behavior of Abelian and non-Abelian gauge fields and matter, quantitative descriptions of ultra-relativistic collisions require simulations with a dynamical spacetime. The conceptual groundwork for such descriptions has been given in terms of the so called Glasma model, which in turn justifies the use of classical statistical simulations to accurately describe gauge field dynamics. Real-time lattice simulations in the Glasma description have proven remarkably successful by identifying the correct weak-coupling thermalization scenario [7]. However simulating fermions has proven to be prohibitively expensive so far, while the conceptual foundations have been laid by one of the proposed project members in [8]. Our recent algorithmic advances however finally allow simulating non-equilibrium fermion dynamics using the computational resources provided by the high performance center ForHLR, as we will outline below. These simulations will prove to be a crucial ingredient in explaining experimental outcome of observables related to the CME.

A further important objective that we aim to address in future work, using our advanced simulation techniques, is the question of the life span of the magnetic field created in non-central heavy ion collisions. As magnetic fields are the crucial ingredients connecting anomalous dynamics with observable effects, their dynamics must be well understood. While it has been argued that magnetic fields are very short lived in collisions at highest energies [9, 10], it was proposed that the hot medium created in a collision might have an anomalous electric conductivity and the resulting fields might decay much slower. This possibility might be connected to the chirality to helicity transfer via the axial anomaly, which might induce helical magnetic fields, that decay slower due to their topological nature. Therefore we aim to simulate *dynamical* electromagnetic fields and determine the conductivity of the far-from-equilibrium plasma directly.

While the importance of anomalous and topological dynamics out-of-equilibrium is currently most prominent in the context of QCD, it plays an important role in many other fields. The second central aspect of the research presented here, is the study of QED in the extreme intensity limit, which might be achieved in strong-field laser experiments within the next two decades. As has been predicted by Heisenberg, Schwinger and others nearly a century ago, the vacuum of QED becomes unstable against particle-anti-particle production in the extreme limit of

$$E_c \approx \frac{m^2}{e}$$

This exciting frontier might enable researchers to perform particle physics using optical laser experiments. While these phenomena have been predicted a long time ago, they have never investigated from the perspective of recent findings: QED systems beyond the Schwinger limit represent challenging out of equilibrium systems in which the dynamics of quantum anomalies might prove crucial. In this regard strong field laser experiments might be the cleanest and most straightforward experimental set-up to understand the non-equilibrium nature of anomalous and topological effects. These experiments would not only shine light into the more involved situation in ultra-relativistic heavy ion collisions, they might also provide access to the field of baryogenesis [11–13], where it is conjectured that C- and CP-violation in combination with suitable out-of-equilibrium conditions, might be responsible for the present day abundance of matter over anti-matter in the observable universe.

2 Project Objectives Reached

With the resources provided through the Steinbruch Center in the allocation year 2016, we have been able to meet all of our objectives described in the previous project proposal [14]. Our simulation results have made an important contribution to the understanding of the non-equilibrium dynamics of anomalous and topological effects in gauge theories and our findings have made a significant impact on multiple communities, both in nuclear/high energy physics as well as in laser research. We have published the results obtained utilizing the computational resources of ForHLR I in high-impact scientific journals:

1. **N. Mueller**, S. Schlichting, S. Sharma, *Chiral magnetic effect and anomalous transport from real-time lattice simulations*, Phys. Rev. Lett. **117**, no. 14, 142301 (2016), [arXiv:1606.00342 [hep-ph]]
2. **N. Mueller**, F. Hebenstreit, **J. Berges**, *Anomaly-induced dynamical refringence in strong-field QED*, Phys. Rev. Lett. **117**, no. 6, 061601 (2016), [arXiv:1605.01413 [hep-ph]]
3. **N. Tanji, N. Mueller, J. Berges**, *Transient anomalous charge production in strong-field QCD*, Phys. Rev. D **93**, no. 7, 074507 (2016), [arXiv:1603.03331 [hep-ph]]
4. M. Mace, **N. Mueller**, S. Schlichting, S. Sharma, *Real-time lattice simulation of anomaly induced transport phenomena,—*first ever $3 + 1$ dimensional simulation of overlap fermions.
   Phys. Rev. D **95** (2017) no.3, 036023, arXiv:1612.02477 [hep-lat] |
5. M. Mace, **N. Mueller**, S. Schlichting, S. Sharma, *Chiral magnetic effect and anomalous transport from real-time lattice simulations.*
   Conference proceedings, 26th International Conference on Ultrarelativistic Nucleus-Nucleus Collisions (Quark Matter 2017), 06–11 Feb 2017. Chicago, Illinois, USA

with one further publication in preparation:

1. **N. Mueller, O. Garcia-Montero, N. Tanji, J. Berges** *Non-equilibrium photon production, in preparation.*

The great interest caused by these research achievements have made the real-time lattice simulations of gluons and quarks a prime theoretical development in nuclear science [5]. Our achievements have been presented at important conferences

1. **Talk**: Niklas Mueller, *Real-time dynamics of the Chiral Magnetic Effect*, Strong and Electroweak Matter, University of Stavanger, Jul 11–15, 2016
2. **Invited Talk**: Niklas Mueller, *Classical-statistical simulations and the Chiral Magnetic Effect*, 7th Workshop of the APS Topical Group on Hadronic Physics (GHP17), Washington D.C., Feb 1–3, 2017
3. **Talk**: Niklas Mueller *Chiral magnetic effect and anomalous transport from real-time lattice simulations*, Quark Matter 2017, Chicago, Feb 6–11, 2017, Conference proceedings: e-Print: arXiv:1704.05887 [hep-lat], Conference: C17-02-06

Below we give an overview over the scientific progress that we have achieved utilizing the computational resources of ForHLR I, both from the conceptual as well as the technical side.

2.1 Transient Anomalous Charge Production from Coherent Color Fields Out-of-Equilibrium

Our studies have made an important contribution to the understanding of the role of the axial anomaly in the far-from-equilibrium dynamics of Abelian and non-Abelian gauge theories. Conventionally anomalies are viewed from a 'global' perspective, as can be tested in a typical S-matrix scattering experiment. Global here means that experimental and theoretical questions were only concerned with the *asymptotic* outcome of an experiment, as opposed to the dynamics of an ultra-relativistic heavy ion collision. Heavy-ion experiments allow to resolve the dynamics of a quantum system *locally* in the sense of space and time resolved. This makes it possible to measure fluctuations of quantities precisely as opposed to the relatively restricted information contained in the vacuum-to-vacuum picture of previous particle collider experiments.

The most important application of this is P- and CP-violation in gauge theories, which in the context of QCD is related to the strong CP-problem. It is well known that QCD conserves P- and CP-symmetry globally, as experiments indicate that there is no explicit theta parameter in the QCD action. Nevertheless QCD is a quantum theory and therefore it can violate CP locally via fluctuations. These fluctuations can be measured in experiments at BNL and CERN, however their conceptual interpretation relies on quantitative predictions from theory. As ultra-relativistic collisions pose a far-from-equilibrium system, our techniques allow precise insight into the fundamental dynamics of these quantum systems under extreme conditions.

While CP-odd fluctuations in QCD are commonly believed to only originate from topological transitions, called sphalerons, we have shown in [15] that the role of the initial conditions of a heavy ion collision is significant for the production of axial charge. In this study, we have considered a special configuration of coherent color-electric fields $\mathbf{E}^a$ and color-magnetic fields $\mathbf{B}^a$ that carries nonzero $\mathbf{E}^a \cdot \mathbf{B}^a$ like those created in the initial moments of a heavy ion collision. Since these fields are spatially uniform, they are not related to any topological properties of the theory. Nevertheless, these coherent color fields can transiently produce significant amounts of axial charge density. By real-time gauge field simulations, we have demonstrated that part of the induced axial charges persist to be present even after the coherent fields disappear by decoherence.

The production of the axial charge density $n_5(t) = \langle \bar{\psi} \gamma^0 \gamma_5 \psi \rangle$ should satisfy the Adler-Bell-Jackiw anomaly equation

$$n_5(t) = 2m \int_0^t dt' \, \langle \overline{\psi} i \gamma_5 \psi \rangle (t') + \frac{g^2}{4\pi} \int_0^t dt' \, \mathbf{E}^a \cdot \mathbf{B}^a(t') \,, \tag{1}$$

where m is fermion mass, and g is the QCD coupling constant. However, the realization of the axial anomaly in lattice numerical simulations is nontrivial because of the fermion doubling problem [16]. We have employed the Wilson fermions to

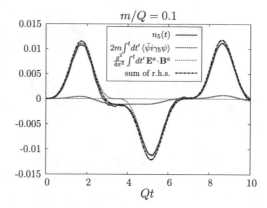

Fig. 1 Time evolution of the various terms appearing in the anomaly equation (1). The sum of the two terms on the right hand side of Eq. (1) is plotted as well. Its agreement to n_5 within the expected accuracy for the employed lattice spacing provides a crucial validity check of the employed real-time regularization scheme. The figure is taken from [15]

regulate the fermion doublers, and numerically solved the Dirac equation with a spatial Wilson term on a real-time lattice. As shown in Fig. 1, the anomaly equation (1) is indeed satisfied in our simulations, which verifies the real-time regularization scheme we employ. These results provide important bases, both technically and conceptually, for more realistic simulations of anomalous nonequilibrium processes in QCD related to heavy ion collisions.

2.2 Anomalous Transport in Ultra-Relativistic Heavy Ion Collisions

We have studied the fermionic dynamics during an isolated topological transition and the subsequent anomalous transport of axial and vector charges via the recently proposed Chiral Magnetic and Chiral Separation Effect in the presence of strong magnetic fields. These effects offer a unique experimental access to the dynamics of (local) CP-violation in QCD. Our work [4] constitutes the first ever real-time study of these anomalous effects from a first-principle field theoretical description. In our model independent study we have seen the emergence of the Chiral Magnetic and Chiral Separation Effect and their interplay in form of the Chiral Magnetic Wave dynamically. Our findings show good agreement with previous predictions based on idealized assumptions in the strong magnetic field and small fermion mass limit, but we find substantial deviations when going towards more realistic experimental situations of smaller magnetic fields and substantial quark masses. Our findings are compactly summarized in Fig. 2, which shows data for the fermionic charges (axial and vector) from a lattice simulation of a sphaleron transition in the background of strong magnetic fields. Visible are soliton-like, dissipation-less chiral density excitations that travel along the magnetic field direction, by means of the anomalous conductivities provided by the Chiral Magnetic and Chiral Separation Effects. Time runs from left to right.

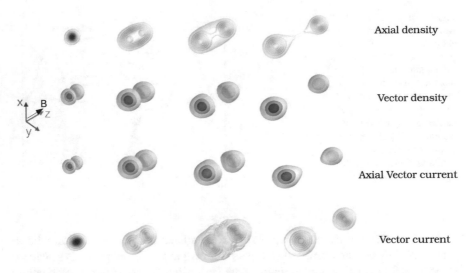

Fig. 2 Simulation results for the axial- and vector-charge generation during and after a sphaleron transition in a strong background magnetic field, as it is produced in a non-central heavy ion collision. These simulations present the first-ever real-time lattice study of topological and anomalous dynamics of fermions in $SU(2)$ and QED gauge fields and anomalous transport via the chiral magnetic wave. The figure is taken from [4]

2.2.1 Transport Phenomena from Lattice Simulations

Based on the numerical techniques described above, we will present some of our results for the dynamics of axial charge generation during a sphaleron transition in this section. Our results are compactly shown in Fig. 3 where the magnetic field and mass dependence of the CME is investigated. On the left the magnetic field dependence is shown and—as predicted by the CME—charge separation rises linearly with the applied field for small external fields, while it eventually saturates in the asymptotic limit.

As finite quark masses lead to explicit chiral violation, the anomaly equation is modified to account for finite fermion masses. Accordingly we have varied the fermion mass in our lattice simulations in a systematic way. The results are shown in the right panel of Fig. 3. Explicit axial violation leads to a significant reduction of the CME signal and differences between light and heavy fermions emerge already for modest values of the mass parameter m, measured in terms of the sphaleron scale. When converted into physical units, our findings suggest that anomalous transport is absent for strange and heavier quarks, and a two-flavour scenario is most realistic.

Furthermore our simulations have enabled us to study constitutive relations for currents and charge densities in anomalous hydrodynamics. To this end we have found that the rations between axial (vector) currents and densities are not time independent, and away from asymptotically strong magnetic fields, differ drastically from unity. Given the very short lifetime of the magnetic field in typical heavy ion

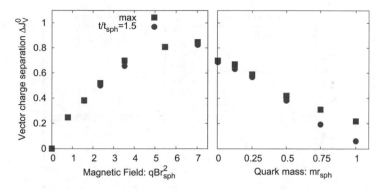

Fig. 3 Magnetic field and mass dependence of vector charge separation. The typical inverse size of the sphaleron is $r_{sph}^{-1} = 200\text{--}500$ MeV [4, 17]

collisions, a finite relaxation time of the CME currents must be included in future theoretical studies.

2.2.2 Simulating Overlap Fermions in Real Time

Presently classical-statistical real-time lattice simulations have only been performed using Wilson fermions. Unfortunately however Wilson fermions break the anomalous axial symmetry. Exact chiral symmetry on the lattice is only recovered using Overlap fermions [17], at the price of the loss of locality. In our work we have constructed a real-time evolution scheme for Overlap fermions using the Wilson kernel in its construction of the Overlap Hamiltonian:

$$\hat{H}_{ov} = \frac{1}{2}\sum_{\mathbf{x}}[\hat{\psi}_{\mathbf{x}}^{\dagger}, \gamma_0\big(-i\slashed{D}_{ov}^{s}\big)\hat{\psi}_{\mathbf{x}}] \tag{2}$$

Here $-i\slashed{D}_{ov}^{s}$ is the 3D spatial overlap Dirac operator given by

$$-i\slashed{D}_{ov}^{s} = M\left(1 + \frac{\gamma_0 H_W(M)}{\sqrt{H_W(M)^2}}\right) \tag{3}$$

and $H_W(M)$ is the original Wilson Hamiltonian kernel. The overlap Hamiltonian consists of a matrix sign function of $H_W(M)$, which contains an inverse square root of $H_W(M)^2$ and can be expressed as a Zolotarev rational function

$$\frac{1}{\sqrt{H_W(M)^2}} = \sum_{l=1}^{N_{\mathscr{O}}} \frac{b_l}{d_l + H_W(M)^2}. \tag{4}$$

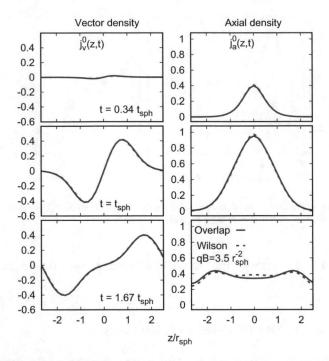

Fig. 4 Longitudinal profiles of the vector (left) and axial (right) charge densities for Wilson (NLO) fermions and overlap fermions with masses $mr_{\text{sph}} = 1.9 \cdot 10^{-2}$ in an external magnetic field $qB = 3.5r_{\text{sph}}^{-2}$ at times $t/t_{\text{sph}} = 0.34,\ 1,\ 1.67$. Figure taken from [17]

The Zolotarev expansion coefficients d_l are determined and a multi-shift conjugate gradient solver is used to calculate the inverse of $d_l + H_W(M)^2$. The lowest and the highest eigenvalues for $H_W(M)^2$ are calculated using the Kalkreuter-Simma Ritz algorithm.

Our results shown in Fig. 4 show very good agreement between Wilson fermion in the small mass limit and overlap fermions. We further observe that the lattice artifacts using overlap fermions are significantly reduced and allow us to use smaller lattices.

2.3 Anomalous Dynamical Refringence in QED Experiments Beyond the Schwinger Limits

Furthermore we have employed our simulation techniques to simulate the far-from-equilibrium dynamics of QED beyond the Schwinger limit. Future laser beam experiments will reach the non-linear regime of QED within the next two decades, thereby opening a new perspective on particle physics. Future 'laser-colliders' allow for a basically background free environment to do particle physics and to test the symmetries of nature. In this context the role of quantum anomalies have been completely overlooked, due to the 'traditional' perspective on the structure of QED: While QED so far has only been tested in S-matrix elements in the high energy regime (collider physics), laser beam experiments will allow for space and time resolved access to QED processes—much like in the case of heavy ion collisions discussed above. Furthermore, despite being topologically trivial, QED shows unexpected behaviour in the far-from-equilibrium regime. Anomalous particle production via the axial anomaly plays an important role in screening and polarization effects of the large coherent electric and magnetic field produced in laser-beams. The Abelian equivalent of the Chiral Magnetic Effect causes an anomalous rotation of the polarization of electric fields, an effect which we have predicted for the first time and we have called *dynamical refringence*. On top of this effect the systems we have investigated show non-trivial tracking behaviour, which manifest itself in self-focusing mechanisms, which bring the system into configurations which maximize the effects of the axial anomaly. Such effects might be utilized to realize non-linear optics in the regime beyond the Schwinger limit. An illustration of the anomalous rotation and tracking behaviour is seen in Fig. 5, taken from [18].

Our investigations serve as the basis for future studies and we have successfully demonstrated that the dynamics of anomalous and topological effects can be understood quantitatively from ab initio real-time simulations. After establishing the conceptual foundations, it is important to apply our techniques to the phenomenology of physical systems, such as heavy ion collisions. While our study was based on a simplified set-up, the extension towards realistic descriptions of a heavy ion collision is straightforward. To this end we aim to implement realistic initial conditions of the colliding heavy ions via the Color Glass Condensate picture [19] and we aim to simulate the subsequent space-time evolution of the *expanding* fireball directly. Expanding geometries have been investigated using classical-statistical simulations before [20], albeit neglecting dynamical fermions. Our work has shown the importance of fermions for the phenomenology of ultra-relativistic

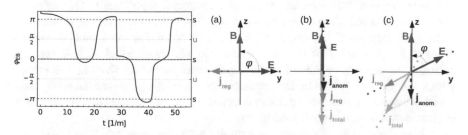

Fig. 5 Left: QED simulation results for showing the behaviour of anomaly induced dynamical refringence in laser-beam experiments beyond the Schwinger limit. A dynamical tracking behaviour is evident. Right: Illustration of the influence of the Abelian Chiral Magnetic Effect on screening effects of the electromagnetic fields. Figures taken from [18]

collisions and our studies have made their investigation possible. At this point the investigation of non-equilibrium photon production is a natural consequence. Photons originate from the produced charged fermions and the produced photon spectrum is expected to be very different than was estimated by thermal predictions.

Thus it is remarkable that very non-trivial behaviour can be found in systems that are far-from equilibrium, mainly because our intuition comes from vacuum or thermal physics. The systems under consideration here are in principle well known, as QED and QCD are established theories. Nevertheless an important aspect of their dynamics has not been considered yet and deserves further attention.

2.4 Technical Developments

Apart from the insights into the mechanisms of fundamental processes of matter under extreme conditions, our studies have significantly advanced the techniques necessary for the first-principle study of non-equilibrium gauge theories and fermion dynamics. Until now the inclusion of lattice fermions into classical-statistical simulations was prohibitively computationally expensive and has posed a serious obstacle not only in the context of QCD, but also in connection with the dynamics of the Electro-weak theory in a cosmological context [21]. In this subsection, we explain the numerical challenges behind real-time fermion simulations. Early on statistical descriptions, called 'low-cost fermion implementations' have been put forward [22], but with little success. For a wide class of situations no convergence of the statistical description towards the exact results could be found. Furthermore the fermionic backcoupling and screening effects onto the gauge field sector have been found not to be reproduced correctly at all. As the latter is expected to be an important effect, especially in Abelian theories, progress in the understanding of the real-time dynamics came to a stop.

To overcome this obstacle we have developed (tree-level) operator improvement techniques that can be applied to real-time Hamiltonian evolution [4]. By including non-local terms in the operator definition of the lattice Hamiltonian, the UV-dependence of our simulations could be reduced significantly. Our achievements have allowed us to approach the continuum limit of our lattice simulations already on very small lattices, typically $24 \times 24 \times 24$, for the typical dynamics relevant for (the bulk) of the ultra-relativistic plasma, created in a heavy ion collision. Moreover, we have introduced a technique called 'Wilson-averaging', which significantly reduce lattice artefacts when simulating the anomalous dynamics of Wilson-fermions. An extensive overview of our improved Hamiltonian time evolution techniques can be found in [4].

A key achievement of our previous studies is the first-ever real time simulation of *overlap-fermions* in $3 + 1$ dimensions, which constitute an exactly chiral fermion representation on the lattice. Chiral lattice fermions are central to the understanding of the real-time dynamics of chiral symmetry breaking and the role of fermions in chiral plasma instabilities. The latter are expected to play a crucial role in astrophysical objects such as neutron stars and supernovae [23, 24].

The technology we have developed constitutes one of the most important tools in the forefront research of the real-time dynamics of gauge theories and matter. As we have outlined and shown in our publications, the applications of our lattice techniques span various fields—from nuclear and high energy physics, to cosmology [21] and laser physics beyond the Schwinger limit [18].

2.4.1 Operator Improvements

The lattice Hamiltonian for Wilson fermions including operator improvements is given as

$$\hat{H}_W = \frac{1}{2} \sum_{\mathbf{x}} [\hat{\psi}_{\mathbf{x}}^\dagger, \gamma^0 \left(-i\mathcal{D}_W^s + m \right) \hat{\psi}_{\mathbf{x}}]. \tag{5}$$

where we introduced the following abbreviation for the gauge links

$$U_{\mathbf{x},+ni} = \prod_{k=0}^{n-1} U_{\mathbf{x}+ki,i} , \qquad U_{\mathbf{x},-ni} = \prod_{k=1}^{n} U_{\mathbf{x}-ki,i}^\dagger. \tag{6}$$

An appropriate choice of the coefficients up to C_n allows to explicitly cancel lattice artifacts $\mathcal{O}(a^{2n-1})$ in the lattice Hamiltonian. For example the first two terms are $C_1 = 4/3$ and $C_2 = -1/6$ and result an $O(a^3)$ (tree level) improvement. The next order (NNLO) is $C_1 = 3/2$, $C_2 = -3/10$, $C_3 = 1/30$ and we get an $O(a^5)$ (tree level) improvement. This improvement procedure is similar to that of Ref. [25], but not unique [26], as also improvements including Clover terms are possible. The results of tree-level improvements are shown in Fig. 6.

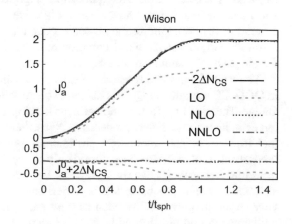

Fig. 6 Top: Comparison of the net axial charge generated during a sphaleron transition for a fixed volume of $N = 16$ using $mr_{\text{sph}} = 1.9 \cdot 10^{-2}$ Wilson fermions with different operator improvements. Bottom: Deviations relative to the topological charge are shown

2.4.2 Wilson Averaging

Lattice artifacts induce large oscillations in the lattice anomaly equation. These finite size effects however can be counteracted by the technique of Wilson averaging.

3 Outlook

Understanding the evolution of a heavy ion collision is challenging, as for most experimental probes medium effects become relevant via subsequent interactions after production. For this reason, electromagnetic probes, such as photons, whose mean free path is much larger than typical system sizes [27], are of great importance to better understand the evolution of the collision. Photons provide excellent probes for the dynamics of the space-time evolution of the fireball created, as they are produced throughout every stage of the collision. However, the production of electromagnetic probes during a heavy ion collision is not yet well-understood, as there is still significant discrepancy between theoretical predictions and experimental results. Specifically, direct photon spectra exhibit a substantial excess at low $p_\perp$, while also showing greater anisotropy than predicted by thermal production models [28]. These two effects comprise the so-called *direct photon puzzle* [29]. Our specific interest thus lies in the contribution of *direct non-thermal photons*, which are the number of electromagnetic quanta radiated from the pre-equilibrium medium. These photons might not only provide an interesting insight on the far-from equilibrium physics of the early time regime of a collision, furthermore there is strong evidence that non-thermal photons might account for the existing discrepancy between theoretical predictions and experimental outcome [30]. Until present day there exists no direct theoretical tool to investigate non-thermal photons and photon production at early times is still a relatively unexplored field. While there have been exploratory studies of the pre-equilibrium stage [31, 32], no conclusive evidence has been presented yet. Our research objective is to establish real-time lattice simulations of non-equilibrium photon production, successfully integrating gluons, fermions and photons into one common description, by making use of techniques developed by the authors of this report.

The second important research objective is the question of CP- and P-odd effects in QCD. CP-violation in QCD is related to the topological structures of non-Abelian Gauge Theories and to the anomalous violation of axial symmetry, both which have fascinated researchers for decades. Nevertheless experimental access is elusive and topological and anomalous phenomena—while being fascinating from the theoretical point of view—have played only a very limited role in phenomenological questions. Only very recently a proposal has been made of how these seemingly abstract concepts can be tested in experiment [1, 2, 33]. It was suggested that very strong magnetic fields, such as they are produced in non-central heavy ion collisions, in combination with the axial anomaly, lead to the phenomena of the

Chiral Magnetic and Chiral Separation Effects. The CME describes the mechanism that spin polarization induced by external magnetic fields generates electric currents in the presence of an axial imbalance. This in turn induces an imbalance of the measured charged particles relative to the interaction plane of a heavy ion collision on an event-by-event basis. As QCD conserves the CP-symmetry globally, the effect can only be seen, when measuring particle-particle correlations and appropriate observables have been suggested. The Chiral Separation Effect is in a sense inverse to the CME, as it couples *axial* currents to an imbalance of vector charge. The combination of both effects has been speculated to induce gap-less excitations called Chiral Magnetic Waves, which are axial-vector-density waves, which propagate dissipation-less through the hot QCD medium, because of their topological nature. Hence these objects are speculated to have unique experimental signatures.

It is in the authors research scope to explore the theoretical understanding this possible experimental signatures. In the first place, early time photon production will allow us to predict, roughly, the out-of-equilibrium evolution of the system. This will enable the relativistic hydrodynamics community to be able to secure better and more robust initial conditions. Although some analytical developments have been reached in the last year [34, 35], the complexity of the problem makes a full-fledged simulation inevitable for the grasping of the system. For this, strong computational power is required, and in this have successfully tested. It will be the next step in our research to extend our simulations to include electromagnetic radiation, and so, to achieve the first simulation of its kind. The authors will also continue the current research on the Chiral Magnetic Effect, and anomaly phenomena.

The authors would like to thank FORHLR for their support during the past year, as this research would not have possible otherwise.

References

1. D.E. Kharzeev, L.D. McLerran, H.J. Warringa, Nucl. Phys. A **803**, 227 (2008). arXiv:0711.0950 [hep-ph]
2. D.E. Kharzeev, D.T. Son, Phys. Rev. Lett. **106**, 062301 (2011). arXiv:1010.0038 [hep-ph]
3. B.I. Abelev et al., [STAR Collaboration], Phys. Rev. Lett. **103**, 251601 (2009). arXiv:0909.1739 [nucl-ex]
4. N. Mueller, S. Schlichting, S. Sharma, Phys. Rev. Lett. **117**(14), 142301 (2016). arXiv:1606.00342 [hep-ph]
5. V. Skokov, P. Sorensen, V. Koch, S. Schlichting, J. Thomas, S. Voloshin, G. Wang, H.U. Yee, arXiv:1608.00982 [nucl-th]
6. D. Kharzeev, J. Liao, S. Voloshin, G. Wang, Prog. Part. Nucl. Phys. **88**, 1 (2016). arXiv:1511.04050[hep-ph]
7. J. Berges, K. Boguslavski, S. Schlichting, R. Venugopalan, Phys. Rev. D **89**(11), 114007 (2014). arXiv:1311.3005 [hep-ph]
8. F. Gelis, N. Tanji, J. High Energy Phys. **1602**, 126 (2016). arXiv:1506.03327 [hep-ph]
9. V. Skokov, A.Y. Illarionov, V. Toneev, Int. J. Mod. Phys. A **24**, 5925 (2009). arXiv:0907.1396 [nucl-th]
10. K. Tuchin, Phys. Rev. C **88**(2), 024911 (2013). arXiv:1305.5806 [hep-ph]
11. A.D. Sakharov, Pis'ma Zh. Eksp. Teor. Fiz. **5**, 32 (1967)

12. A.G. Cohen, D.B. Kaplan, A.E. Nelson, Ann. Rev. Nucl. Part. Sci. **43**, 27 (1993)
13. V.A. Rubakov, M.E. Shaposhnikov, Usp. Fiz. Nauk **166**, 493 (1996)
14. N. Mueller, N. Tanji, O. Garcia, J. Berges, Real-time lattice simulations of quantum anomalies, topologies and particle production in strongly correlated gauge theories. ForHLR Phase I proposal, Nov 2015
15. N. Tanji, N. Mueller, J. Berges, Phys. Rev. D **93**(7), 074507 (2016). arXiv:1603.03331 [hep-ph]
16. H.B. Nielsen, M. Ninomiya, Nucl. Phys. B **185**, 20 (1981)
17. M. Mace, N. Mueller, S. Schlichting, S. Sharma, Phys. Rev. D. **95**, 036023 (2017). arXiv:1612.02477 [hep-ph]
18. N. Mueller, F. Hebenstreit, J. Berges, Phys. Rev. Lett. **117**(6), 061601 (2016). arXiv:1605.01413 [hep-ph]
19. F. Gelis, E. Iancu, J. Jalilian-Marian, R. Venugopalan, Ann. Rev. Nucl. Part. Sci. **60**, 463 (2010). arXiv:1002.0333 [hep-ph]
20. J. Berges, K. Boguslavski, S. Schlichting, R. Venugopalan, Phys. Rev. D **92**(9), 096006 (2015). arXiv:1508.03073 [hep-ph]
21. A. Tranberg, J. Smit, M. Hindmarsh, Nucl. Phys. A **785**, 102 (2007). hep-ph/0608167
22. S. Borsanyi, M. Hindmarsh, Phys. Rev. D **79**, 065010 (2009). arXiv:0809.4711 [hep-ph]
23. N. Yamamoto, Phys. Rev. D **93**(6), 065017 (2016). arXiv:1511.00933 [astro-ph.HE]
24. D. Grabowska, D.B. Kaplan, S. Reddy, Phys. Rev. D **91**(8), 085035 (2015). arXiv:1409.3602 [hep-ph]
25. T. Eguchi, N. Kawamoto, Nucl. Phys. B **237**, 609 (1984)
26. B. Sheikholeslami, R. Wohlert, Nucl. Phys. B **259**, 572 (1985)
27. P.B. Arnold, G.D. Moore, L.G. Yaffe, J. High Energy Phys. **0112**, 009 (2001). hep-ph/0111107
28. A. Adare et al., [PHENIX Collaboration], Phys. Rev. Lett. **104**, 252301 (2010). arXiv:1002.1077 [nucl-ex]
29. C. Shen, Nucl. Phys. A **956**, 184 (2016). arXiv:1601.02563[nucl-th]
30. C. Klein-Bosing, L. McLerran, Phys. Lett. B **734**, 282 (2014). arXiv:1403.1174 [nucl-th]
31. N. Tanji, Phys. Rev. D **92**(12), 125012 (2015). arXiv:1506.08442 [hep-ph]
32. L. McLerran, B. Schenke, Nucl. Phys. A **929**, 71 (2014). arXiv:1403.7462 [hep-ph]
33. K. Fukushima, D.E. Kharzeev, H.J. Warringa, Phys. Rev. D **78**, 074033 (2008). arXiv:0808.3382 [hep-ph]
34. S. Benic, K. Fukushima, O. Garcia-Montero, R. Venugopalan, J. High Energy Phys. **1701**, 115 (2017). https://doi.org/10.1007/JHEP01(2017)115, arXiv:1609.09424 [hep-ph]
35. J. Berges, K. Reygers, N. Tanji, R. Venugopalan, arXiv:1704.04032 [nucl-th]

Many-Body Effects in Fragmented, Depleted, and Condensed Bosonic Systems in Traps and Optical Cavities by MCTDHB and MCTDH-X

Ofir E. Alon, Raphael Beinke, Christoph Bruder, Lorenz S. Cederbaum, Shachar Klaiman, Axel U.J. Lode, Kaspar Sakmann, Marcus Theisen, Marios C. Tsatsos, Storm E. Weiner, and Alexej I. Streltsov

Abstract The many-body physics of trapped Bose-Einstein condensates (BECs) is very rich and demanding. During the past year of the MCTDHB project at the HLRS

O.E. Alon
Department of Mathematics, University of Haifa, Haifa 3498838, Israel

Haifa Research Center for Theoretical Physics and Astrophysics, University of Haifa, Haifa 3498838, Israel

R. Beinke • L.S. Cederbaum • S. Klaiman • M. Theisen
Theoretische Chemie, Physikalisch-Chemisches Institut, Universität Heidelberg, Im Neuenheimer Feld 229, 69120 Heidelberg, Germany

C. Bruder
Department of Physics, University of Basel, Klingelbergstrasse 82, CH-4056 Basel, Switzerland

A.U.J. Lode
Wolfgang Pauli Institute, c/o Faculty of Mathematics, University of Vienna, Oskar-Morgenstern Platz 1, 1090 Vienna, Austria

Department of Physics, University of Basel, Klingelbergstrasse 82, CH-4056 Basel, Switzerland

Vienna Center for Quantum Science and Technology, Atominstitut TU Wien, Stadionallee 2, 1020 Vienna, Austria
e-mail: axel.lode@univie.ac.at

K. Sakmann
Vienna Center for Quantum Science and Technology, Atominstitut TU Wien, Stadionallee 2, 1020 Vienna, Austria

M.C. Tsatsos
Instituto de Física de São Carlos, Universidade de São Paulo, Caixa Postal 369, São Carlos, SP 13560-970, Brazil

S.E. Weiner
Department of Physics, University of California, Berkeley, CA 94720-1462, USA

A.I. Streltsov (✉)
Theoretische Chemie, Physikalisch-Chemisches Institut, Universität Heidelberg, Im Neuenheimer Feld 229, 69120 Heidelberg, Germany

Institut für Physik, Universität Kassel, Heinrich-Plett-Str. 40, 34132 Kassel, Germany
e-mail: Alexej.Streltsov@pci.uni-heidelberg.de

© Springer International Publishing AG 2018
W.E. Nagel et al. (eds.), *High Performance Computing in Science and Engineering '17*, https://doi.org/10.1007/978-3-319-68394-2_6

we continued to shed further light on it with the help of the MultiConfigurational Time-Dependent Hartree for Bosons (MCTDHB) method and using the MCTDHB and MCTDH-X software packages. Indeed, our results on which we report below span a realm of many-body effects in fragmented, depleted, and even in fully condensed BECs. Our findings include: (1) fragmented superradiance of a BEC trapped in an optical cavity; (2) properties of phantom (fragmented) vortices in trapped BECs; (3) dynamics of a two-dimensional trapped BEC described by the Bose-Hubbard Hamiltonian with MCTDH-X; (4) overlap of exact and Gross-Pitaevskii wave-functions in trapped BECs; (5) properties of the uncertainty product of an out-of-equilibrium trapped BEC; (6) many-body excitations and de-excitations in trapped BECs and relation to variance; and (7) many-body effects in the excitation spectrum of weakly-interacting BECs in finite one-dimensional optical lattices. These are all appealing and fundamental many-body results made through the kind allocation of computer resources by the HLRS to the MCTDHB project. Finally, we put forward some future developments and research plans, as well as further many-body perspectives.

1 Introductory Notes

For more than a decade now, the multiconfigurational time-dependent Hartree for bosons (MCTDHB) method [1–9], developed, implemented, optimized, benchmarked, and applied to solve efficiently the many-boson time-dependent Schrödinger equation, has enabled us to pursue novel and deeper understanding of the many-body physics of ultra-cold trapped Bose-Einstein condensates (BECs). Many of the achievements we have reported on the past 4 years were made possible with the kind allocation of high-performance computer resources by the HLRS to the MCTDHB project [10–13]. We aim at prolonging this tradition in this year's report. In what follows, we summarize our research works made with the MCTDHB and MCTDH-X software packages [14–17] and reported in [18–24]. Our work continues and expands the crop of the previous years [7, 25–39], and enriches the knowledge on the many-body physics of trapped BECs.

2 "Social" Dynamics of Ultra-Cold Atoms Trapped in Between Two Mirrors Together with Light

Nowadays, scientists manage to routinely produce so-called BECs, a state of matter where ultra-cold atoms give up their individuality and group, like humans group in a society. Such societies of ultra-cold atoms are very sensitive to impending light. When this society is illuminated and the light it scatters is reflected back and forth between two mirrors (see upper panel of Fig. 1 for a sketch), the atoms conglomerate in the intensity minima exhibited by the scattered light, see Fig. 2.

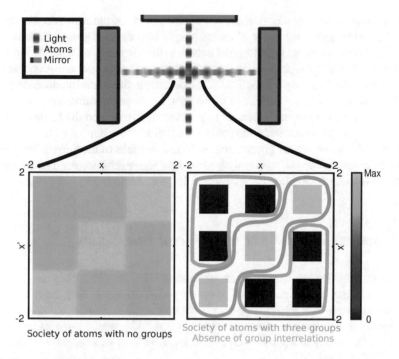

Fig. 1 Fragmented superradiance emerges as group formation in a society of atoms. Upper panel: Sketch of the experimental setup. Lower panel: quantifying group formation, i.e., fragmentation in the BEC with correlations. The absence of correlations and groups is depicted on the left while the emergence of correlations and groups is depicted on the right. See Ref. [18] for further details. The quantities shown are dimensionless. Figure material (panels) reprinted from [18]

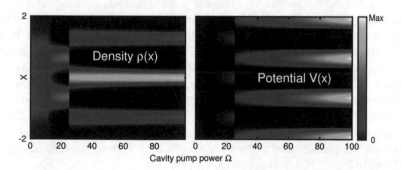

Fig. 2 Self-organization of the ground state of a BEC. The density $\rho(x)$ of the atoms and the potential $V(x) = V_{\text{1-body}}(x) + V_{\text{cavity}}(x)$ are shown as a function of the cavity pump power in the left and right panels, respectively. Once the applied pump power exceeds a critical value, the atoms self-organize because the field which is built up inside the cavity creates a periodic one-body potential (cf. left and right panels). As a result of a competition between the external and cavity potential as well as interactions, the sign of the cavity amplitude switches one time for $\Omega \approx 25$ (compare pattern in density and potential with the inset of Fig. 3a in Ref. [18]). The quantities shown are dimensionless. Figure material (panels and caption) reprinted from [18]

We show how the atoms-light interactions and the atom-atom interactions cooperate to trigger the society of atoms to split into distinct groups as soon as the intensity of the scattered light trapped between the mirrors is sufficiently large. The formation of these groups is called fragmentation and the conglomeration due to the light is referred to as superradiance. We demonstrate that both phenomena coexist and termed the new combined state of the light between the mirrors and the groups of atoms fragmented superradiance. Fragmented superradiance can be inferred from correlations in the positions of the atoms, see Fig. 1.

In a BEC, there are no groups and all atoms are interrelated. For a fragmented BEC, there are groups of atoms with almost no interrelations. As a result the off-diagonal values of the correlation function drops to zero.

3 Phantom Vortices Across Different Bose Gases

Quantized vortices are the hallmark of superfluidity. In BECs they are known to appear as nodes in the density of the observed system and they carry quantized angular momentum. We have recently shown that vortices, named *phantom vortices*, can appear in fragmented gases as well, see Ref. [12], where correlations are not constant. Our study using the MCTDH-X software [17] was recently published in *Scientific Reports* [19].

Our latest studies, with the help of the HLRS facilities, emphasized the generality and importance of phantom vortices. We examined a series of different systems with varying total particle numbers N, interaction strengths λ_0, and functional types of the interaction (finite-range versus contact inter-particle potentials.

At first, we did calculations varying the particle number, but keeping the mean-field coupling strength $\Lambda = \lambda_0(N-1)$ constant with $M = 2$ orbitals. Fragmentation always occurred, but higher particle numbers fragmented at later times. Particle numbers tested up to $N = 10^4$ do fragment during the anisotropy ramp-up or during the period of maximum anisotropy. Then we compared the dynamical behavior of a system whose interaction is modeled using a Gaussian function to the dynamical behavior of a system whose interaction is modeled by a contact potential [see upper panel of Fig. 3]. Even though quantitatively the two systems fragment at a different rate (see 'Occupation 1' in Fig. 3) the energy and angular momentum absorbed differ only marginally. The phantom vortices of the Gaussian-interaction system are shown in Fig. 3. For further details, see [19].

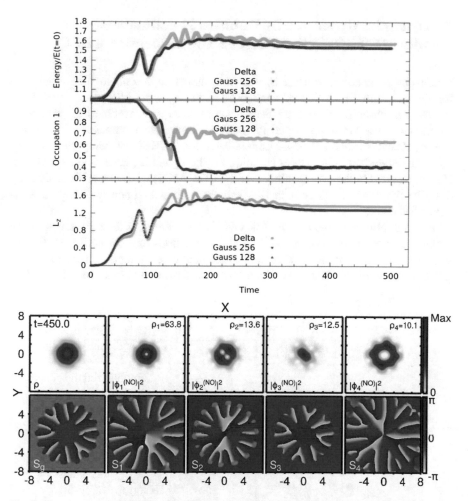

Fig. 3 Phantom vortices in rotating systems. Upper panel: The time evolution of system's observables and properties (energy, largest natural occupation, and angular momentum—upper, middle, and lower panels, respectively) are practically identical for the 128×128 and 256×256 grids. This renders our results exact with respect to the number of grid points. The energy and angular momentum of the system interacting via a delta pseudo-potential (note that the number M of time-adaptive orbitals is finite) are very close to its Gaussian counterpart while the largest occupation (at long times) is noticeably larger. Lower panel: Phantom vortices in the system interacting via a contact potential: Densities at $t = 450$ (towards the end of the rotation routine) for the system interacting via a contact potential (i.e. delta function). Qualitatively the system behaves in the same fashion as the Gaussian-interaction one; the density appears vortex-free (first panel left) while phantom vortices, of the same charges as in the case of short-range interactions, have nucleated on the first, second and fourth natural orbitals. Notably, in the case of contact interaction the first natural occupation is higher than its counterpart in the Gaussian system, resulting in a more pronounced dip in the density in the trap center. The quantities shown are dimensionless. Figure material (panels) reprinted from [19]

4 Dynamics Generated by Hubbard Hamiltonians with MCTDH-X: Splitting of a Two-Dimensional BEC

When put in a lattice potential, ultra-cold atoms represent an ubiquitous tool for the quantum simulation of condensed-matter systems [40–44]. In Ref. [20], we use an explicitly time-dependent basis to re-express Hubbard Hamiltonians containing long- or short-ranged interactions, tunneling parameters, and on-site energy offsets. In such a time-dependent basis, MCTDH-X can be applied to solve the time-dependent Schrödinger equation for the dynamics generated by Hubbard Hamiltonians. Unlike other methods in the field, MCTDH-X can self-consistently describe dynamics in lattices in two and three spatial dimensions with a well-controlled error.

As an application, we use the MCTDH-X software [17] to study the splitting dynamics of $N = 100$ bosons with on-site interactions in a 2D lattice of 50×50 sites: the particles are prepared in the ground state of a parabolic external trap, see left column of Fig. 4. Subsequently, a Gaussian barrier centered around $x = 0$ is

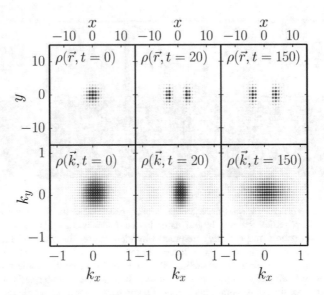

Fig. 4 Spatial and momentum densities in the splitting process with $t_{\mathrm{ramp}} = 50$. The top row depicts the spatial density $\rho(\mathbf{r}; t)$ and the bottom row depicts the momentum density $\rho(\mathbf{k}; t)$ for times $t = 0, 20$, and 150, respectively. Darker color and larger points stand for larger on-site density. The plot of $\rho(\mathbf{r}; t)$ shows that the initial Gaussian density distribution is split in two equal parts by the barrier. The left and right maxima of the density oscillate even for $t > t_{\mathrm{ramp}}$ when the barrier has fully been ramped up: the process is non-adiabatic. The momentum density $\rho(\mathbf{k}; t)$ exhibits an initial Gaussian distribution. During the splitting process, around $t = 20$, $\rho(\mathbf{k}; t)$ has three maxima which recombine to a broadened Gaussian distribution, once the splitting is complete and fragmentation has emerged [see $\rho(\mathbf{k}; t = 150)$ in the lower row and Fig. 5]. The quantities shown are dimensionless. Figure material (panel and caption) reprinted from [20]

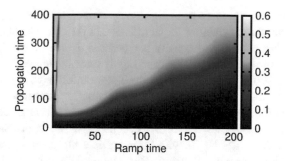

Fig. 5 Emergence of fragmentation in the splitting of a two-dimensional BEC in a lattice. The fragmentation of the system is shown as a function of the propagation time and the barrier-raising time t_{ramp}. Generally, fragmentation takes longer to set in for larger values of t_{ramp}, but there is a threshold for very fast ramps; for short barrier-raising-times $t_{ramp} \lesssim 40$, the system needs about 50 time units to become fragmented. For very short ramps $t_{ramp} \in [4, 10]$, revivals of coherence are seen where the fragmentation returns back close to zero ($F = 0$) after some propagation time, see text for discussion. The quantities shown are dimensionless. Figure material (panel and caption) reprinted from [20]

ramped up linearly within a time t_{ramp} to a fixed final height. See middle and right columns of Fig. 4 for a plot of the density for a ramp with $t_{ramp} = 50$, and Ref. [20] for further details about the Hamiltonian of the system.

We continue by analyzing the emergence of fragmentation by plotting the fraction $F(t)$ of particles outside of the lowest natural orbital of the system in the course of splitting as a function of the ramp time t_{ramp}, see Fig. 5 for a plot. Fragmentation emerges for all the investigated ramp times t_{ramp}. The delay of the emerging fragmentation for longer ramp times is proportional to t_{ramp}, but constant at propagation times of about 50 time units in the case of short ramps with $t_{ramp} \lesssim 40$. For very short ramps with $t_{ramp} \approx 5$, revivals of the initial fragmentation $F \approx 0$ are seen in the dynamics. Since short ramps are the farthest from an adiabatic change to the Hamiltonian, these revivals are counterintuitive: naively, one would expect a larger fragmentation when the modification of the Hamiltonian is more abrupt, see [20].

5 Overlap of Exact and Gross-Pitaevskii Wave-Functions in Trapped Bose-Einstein Condensates of Dilute Gases

Since the experimental discovery of BECs consisting of dilute atomic gases two decades ago [45–47], there has been vast interest in their properties [48–50]. In the respective theoretical studies, the Gross-Pitaevskii (GP) equation which is obtained by minimizing the GP energy functional [51] has played a particularly leading role. The simplicity of this mean-field equation adds much to its popularity as it can be solved rather straightforwardly and exhibits many interesting and appealing properties. Importantly, it has been rigorously proven by Lieb and Seiringer (LS theorem) [52] that for dimensions larger than one in the dilute limit the GP equation provides the exact energy and density per particle as does the full many-particle Schrödinger equation. One immediate and highly relevant consequence of this proof is that BECs are 100% condensed in the limit of an infinite particle number.

In the dilute limit, also called GP limit, the interaction parameter $\Lambda = \lambda_0(N-1)$ appearing in the GP equation, where λ_0 is the two-particle interaction strength, is kept fixed as $N \to \infty$. The LS theorem might raise the impression that the GP theory correctly describes BECs with large particle numbers at zero temperature. Nevertheless, it is well known that corrections beyond the GP theory can be relevant for experiments with typical particle numbers [53]. Does GP theory also provide an accurate wave-function of BECs in the dilute limit? This is a relevant question as, ultimately, the wave-function contains all the physical properties of the system. A first clear indication that boson correlations not included in GP theory can be relevant has been shown very recently in [28, 29]. To answer the latter question we have chosen the overlap of the GP and exact ground states as an obvious measure of the quality of the GP wave-function.

How small can the overlap S become? We present two examples, one which is analytically solvable in all dimensions and one which can only be numerically solved. An analytically solvable model of N interacting bosons in a trap exists which is very valuable in discussing the overlap $S(N)$ at large values of Λ in the dilute limit. In this model the trap is harmonic and the interaction potential too. We may call it the harmonic-interaction model (HIM). This model has been solved explicitly [54] and investigated in several scenarios [6, 55–59]. We have computed explicitly the overlap of these two functions as a function of N, ω and λ_0. For large values of Λ the overlap reads $S(N) = 2^{3D/8} \left(\frac{\Lambda}{\omega^2}\right)^{-D/8}$.

Obviously, the overlap between the GP and the exact ground state wave-functions approaches zero as Λ becomes large, and, interestingly, the approach is faster for larger dimensions. We stress that in the HIM model the energy and density (also density matrix) per particle in the dilute limit are exactly reproduced by the GP theory [54]. This holds *in any* dimension D, even in one dimension ($D = 1$) where the LS theorem is not applicable.

We would like to also study examples with short range interactions. In the absence of exactly solvable models we have to resort to a numerical solution of the full Schrödinger equation which is not an easy task for large boson numbers. We variationally solve the Schrödinger equation in imaginary time using the MCTDHB method [2, 4] to find the ground state.

To demonstrate that the MCTDHB method performs very well for HIM model, we plot in Fig. 6 the error in the energy per particle and in the overlap the mean-field and numerical MCTDHB wave-functions using $M = 2$ orbitals for different boson numbers.

For the short-range interaction examples, we investigate a 1D double-well trap potential and contact interaction $V(x - x') = \delta(x - x')$, a problem widely studied in the literature [5, 60–71]. Since the current proof of the LS theorem does not cover 1D, we also extend this example to 2D by choosing V to be a normalized Gaussian,

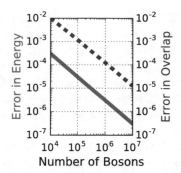

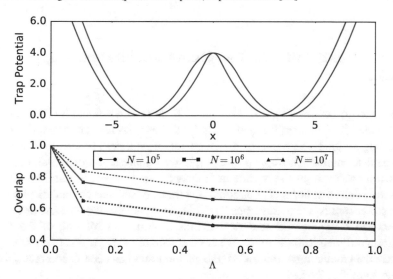

Fig. 6 Numerical error of the MCTDHB method with $M = 2$ for the exact energy per particle (solid) and overlap of the many-body and mean-field wave-functions (dashed) as a function of N. The calculations are performed for the HIM model discussed in the text. The value of $\Lambda = 100$ is used in the calculation and the exact overlap in the dilute limit is 0.704. The quantities shown are dimensionless. Figure material (panel and caption) reprinted from [21]

Fig. 7 The overlap of the GP and the numerically computed many-body wave-functions in the case of 1D and 2D double-well trap potentials. Upper panel: The two trap potentials used in the calculations (in the 2D example $\frac{1}{2}y^2$ has to be added). Lower panel: The overlap as a function of the interaction parameter for three different boson numbers N. Note that for each value of Λ the points computed for different values of N fall on top of each other. The 1D and 2D results are connected by solid and dashed lines, respectively. The quantities shown are dimensionless. Figure material (panels and caption) reprinted from [21]

see [72, 73]. Two trap potentials are studied, see upper panel of Fig. 7. The trap potential is constructed by connecting two harmonic potentials $V_{\pm}(x) = \frac{1}{2}(x \pm x_0)^2$ with a cubic spline such that the resulting barrier is of a given height V_0. For the 2D example we add the harmonic trap $\frac{1}{2}y^2$. The mass of the particles is chosen to be 1 as in the case of the HIM investigated above.

We could solve the MCTDHB with two orbitals for up to $N = 10^7$ bosons. The results for the overlap are shown in the lower panel of Fig. 7 for three particle numbers and are similar for 1D and 2D. It is clearly seen that the overlap drops as the interaction parameter grows from $\Lambda = 0$ to $\Lambda = 1$ (in all calculations $\omega = 1$). Although Λ is rather moderate, the overlap can fall below 0.5. We would like to stress that the results shown seem to saturate as N is increased: the curves for $N = 10^5$, 10^6 and 10^7 essentially fall on top of each other, i.e., the dilute limit is essentially achieved in this example.

Recapitulating, in the GP limit in which $N \to \infty$ and the interaction parameter Λ is kept fixed, the total energy as well as the density per boson are exactly reproduced by the GP theory. Nevertheless, we find that the overlap of the GP and exact many-body wave-functions is always smaller than 1, and depending on the trap and Λ, can be rather small, even vanishingly small. This in turn implies that the exact wave-function describes substantial boson correlations, by definition not present in GP theory.

6 Uncertainty Product of an Out-of-Equilibrium Trapped BEC

We wish to go beyond [28, 29], also see [13], in the investigation of the variance and uncertainty product of trapped BECs. To this end, we here study trapped bosons interacting by long-range interaction. We treat structureless bosons with harmonic inter-particle interaction trapped in a single-well anharmonic potential, and stress that we are away from the regime of fragmentation in a single trap [74–78]. The time-dependent Schrödinger equation of the trapped BEC has no analytical solution in the present study, see in this respect [21, 54], nor even the variance and uncertainty product can be computed analytically, thus a numerical solution of the out-of-equilibrium dynamics is a must. We need a suitable and proved many-body tool to make the calculations, the MCTDHB method that has been used extensively in the literature [73, 79–88].

We consider $N = 100$ and separately $N = 100,000$ trapped bosons in one spatial dimension. The one-body Hamiltonian is $-\frac{1}{2}\frac{\partial^2}{\partial x^2} + \frac{x^4}{4}$ and the inter-particle interaction is harmonic, $\lambda_0 \hat{W}(x_1 - x_2) = -\lambda_0(x_1 - x_2)^2, \lambda_0 > 0$. The system is prepared in the ground state of the trap for the interaction parameter $\Lambda = \lambda_0(N - 1) = 0.19$. At $t = 0$ the interaction parameter is suddenly quenched to $\Lambda = 0.38$ and we inquire what the out-of-equilibrium dynamics of the system would be like. Figure 8 collects the results.

Figure 8a depicts snapshots of the density per particle, $\frac{\rho(x;t)}{N}$, as a function of time. The GP and many-body results are seen to match very well. The density is seen to perform breathing dynamics [77, 89, 90]. Since the interaction is repulsive and at $t = 0$ quenched up, the density first expands at short times. In Fig. 8b the total number of

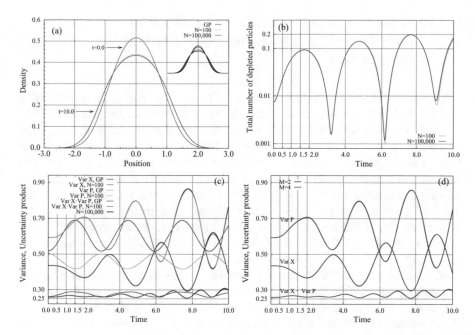

Fig. 8 Variance and uncertainty product during breathing dynamics. Shown and compared are many-body results for $N = 100$ (using $M = 4$) and $N = 100,000$ (using $M = 2$) bosons, and the mean-field results (equivalent to $M = 1$). (**a**) Snapshots of the density per particle, $\frac{\rho(x;t)}{N}$, as a function of time following a sudden increase of the interaction parameter from $\Lambda = \lambda_0(N - 1) = 0.19$ to 0.38 at $t = 0$. The inset shows the density per particle at instances, from top to bottom, $t = 0.0, 0.5, 1.0,$ and 1.5 for $N = 100$ bosons. (**b**) Total number of depleted particles outside the condensed mode as a function of time. Note the values on the y axis. (**c**) Time-dependent many-particle position variance per particle, $\frac{1}{N}\Delta_{\hat{X}}^2(t)$, momentum variance, $\frac{1}{N}\Delta_{\hat{P}}^2(t)$, and uncertainty product, $\frac{1}{N}\Delta_{\hat{X}}^2(t)\frac{1}{N}\Delta_{\hat{P}}^2(t) \equiv \Delta_{\hat{X}_{CM}}^2(t)\Delta_{\hat{P}_{CM}}^2(t)$. Note the opposite behavior of the variance at short times when computed at the many-body and mean-field level. (**d**) Convergence of the variance and uncertainty product with M for the system consisting of $N = 100$ bosons in panel (**c**). It is found that the results with $M = 2$ and $M = 4$ orbitals lie atop each other. The quantities shown are dimensionless. Figure material (panels and caption) reprinted from [22]

depleted particles outside the condensed mode are shown as a function of time. The systems are essentially fully condensed with only a fraction of a particle depleted. In Fig. 8c the time-dependent variance and uncertainty product are shown, and in Fig. 8d their convergence demonstrated. Note the opposite behavior of the variance at short times when computed at the many-body and GP level. Despite the expansion of the cloud (at short times), the time-dependent position variance increases and momentum variance decreases, which is an interesting time-dependent many-body effects.

7 Variance as a Sensitive Probe for Attribution of Many-Body Excitations and De-Excitations in Trapped BECs

In [23] the MCTDHB method has been used to study excited states of interacting BECs confined by harmonic and double-well trap potentials. Specifically, we have studied a smooth transition form a harmonic to a double-well trap by superimposing a harmonic potential and a Gaussian barrier, $V(x) = ax^2 + b\exp(-cx^2)$, see Fig. 9. Our primary goal is to access excitations. In order to reach this goal we have studied and contrasted two different approaches, a static one and a dynamic one. In the static approach the low-lying excitations have been computed by utilizing the LR-MCTDHB method [91, 92]—a linear-response (LR) theory constructed on-top of a static MCTDHB solution. In Fig. 10 we depict the static mean-field LR-GP and many-body LR-MCTDHB excitation spectra computed for $N = 10$ bosons and $\Lambda = \lambda_0(N-1) = 0.1$ as a function of the barrier height b.

Complimentary, in the dynamic approach we have addressed excitations by propagating the MCTDHB wave-function within two dynamic protocols. Namely, we propose to investigate dipole-like oscillations induced by shifting the origin of the confining potential and breathing-like excitations by quenching the frequency of the parabolic part of the trap. To contrast the static predictions and dynamic results we have computed time-evolutions and their Fourier transforms (FT) of several local and non-local observables. Namely, we study the evolution of the many-particle position operator $\langle x(t) \rangle$, its variance $\mathrm{Var}(x(t))$, and the evolution of a local density (here normalized to 1) computed at a selected position, $\rho(x = x_0, t)$, see [23] for further details. The results for the shallow double-well case are presented in Fig. 11.

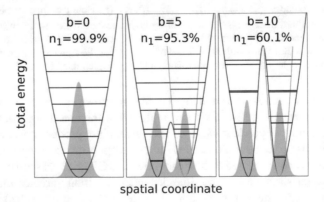

Fig. 9 Shapes of the trap potential (in red), its eigenvalues (in black), and wave-packet density (in gray). The potentials are normalized (shifted) as to vanish in the minima. For the double-wells the harmonic approximation at the potential minimum is shown (green potentials, blue eigenvalues). Occupation numbers of the first natural orbital n_1 and densities of the ground state are calculated for $N = 10$ particles with $\Lambda = \lambda_0(N-1) = 1$ at the MCTDHB($M = 2$) level. The quantities shown are dimensionless. Figure material (panel and caption) reprinted from [23]

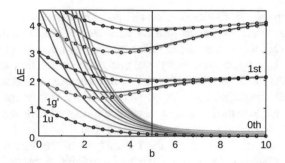

Fig. 10 Weak interaction: Excitation spectrum as a function of the barrier height b for $N = 10$ *bosons and* $\Lambda = \lambda_0(N - 1) = 0.1$. ΔE is the excitation energy with respect to the ground state. The mean-field, LR-GP≡BdG (Bogoliubov-de Gennes) results are plotted by open black circles and the many-body, LR-MCTDHB(2) predictions by colored lines. Green lines indicate *gerade* and red lines *ungerade* symmetry. States are labeled by their nodal structure and symmetry; primes are used to mark pure many-body excitations, see text for further discussion. The quantities shown are dimensionless. Figure material (panel and caption) reprinted from [23]

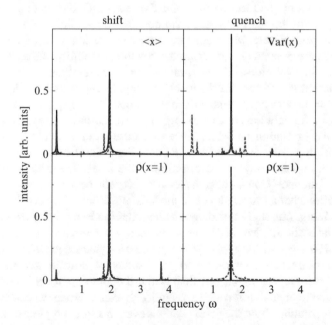

Fig. 11 Attributing excitations in a shallow double-well trap. The left panels show the frequency (FT) analysis of an oscillating BEC after a sudden shift $x_{shift} = 0.1$ of the confining harmonic potential. The quench scenario $a = 0.4 \rightarrow a = 0.5$ is depicted in the right panels. The mean-field GP (black lines) and many-body LR-MCTDHB(2) (red dashed lines) results are plotted. The intensities are normalized by the sum of all constituting peaks. The quantities shown are dimensionless. Figure material (panel and caption) reprinted from [23]

Figure 11 shows the results of the shift and quench scenarios applied to the BECs trapped in a shallow double-well case. As can be seen, from the left panels of this figure showing the studies in the shift scenario, the many-body spectra include the GP results and introduce the additional many-body excitation with pronounced intensity attributed in Table II of [23] as $1u'$. Surprisingly, this state is seen in the evolution of both the global operator $\langle x \rangle$ and the local quantity $\rho(x = 1)$. The fact that $1u'$ is also visible in the local $\rho(x = 1)$ indicates that relative and center-of-mass motions are coupled as opposed to the harmonic case.

The right panels of Fig. 11 show the FT analysis of the Var(x) and $\rho(x = 1)$ in the quench scenario. Concerning the variance, the many-body treatment in addition to the mean-field GP peaks reveals two quite intense many-body excitations attributed to $0g'$ and $1g'$ states in Table II of [23]. The most intense mean-field $1g$ peak and these two many-body excitations have comparable intensities, stressing the importance of a beyond-mean-field treatment in shallow double wells. By contrasting static and dynamic results we have managed to attribute all the lines either to excitations or to corresponding de-excitations. The later are more pronounced and, therefore, play an important role in the variance spectrum. This can be explained by the fact that the applied quench and shift manipulations pump more energy into the final system in the double-well studies in comparison with that in the harmonic case. More pumped energy means that a larger number of the excited states are populated, increasing thereby the probability of the de-excitations seen, e.g., in the evolution of the integral two-body observables.

The situation is different for the local quantity. In the quench scenario the local density $\rho(x = 1)$ does not provide information about the many-body excitations. As one can see in the right lower panel of Fig. 11, the mean-field GP excitation $1g$ is the main and only excitation available. Since we measure $\rho(x = 1)$ close to the density minimum, higher order excitations do not contribute significantly. However, taking the density at regions away from the minimum, the $2g$ excitation starts to contribute to the local dynamics. Concluding, the local density can be used for accessing many-body excitations but a proper choice of the local position is very crucial.

Summarizing, our study has demonstrated that the many-body excitations predicted by the static LR-MCTDHB approach can, alternatively, be accessed within the MCTDHB method by applying the proposed dynamic protocols, confirming thereby the correctness, validity, and consistency of both approaches and corresponding results. We found out that in the dynamic approach the variance is the most sensitive and informative quantity, because along with excitations it contains information about the de-excitations even in a linear regime of the induced dynamics.

A few words about efficiency and computational costs are appropriate. The static approach, based on the LR-MCTDHB method, boils down to the diagonalization of (very) large non-Hermitian matrices, while the dynamic approach involves long-time wave-packet propagations of several different initial conditions (scenarios) with a subsequent analysis of the many-boson wave-functions. During the last decade we have developed efficient numerical methods for the long-time propagations of the MCTDHB equations-of-motion, so, their further qualitative improvement would require considerable algorithmic efforts. In contrast, there is a large room for considerable improvements of the LR-MCTDHB approach. In the present study we have utilizes a full diagonalization of the corresponding LR matrices. This restricts the applicability of the current realization of the method to rather modest 1D systems. However, a proper parallelization of the LR algorithm and the usage of specialized block and filter diagonalizers can drastically improve the performance, which would open a road-map towards many-body excitations in bigger 1D, 2D, and 3D systems of trapped BECs. Our initial progress along this venue is described in the next section.

8 Many-Body Effects in the Excitation Spectrum of Weakly-Interacting BECs Trapped in Finite 1D Optical Lattices

Analytic results from a recent study show that the single-particle excitation energies of a trapped repulsive BEC are given exactly by the BdG predictions in the infinite-particle (or mean-field, Λ is constant) limit [93]. The energies of excited states where more than one particle is excited out of the condensate are then obtained by multiples and sums of the BdG energies. It remains unclear, however, what are the excitation energies of a BEC in a trap with only a finite number of particles, far away from the infinite-particle limit. We address this question in [24]. Besides discussing general many-body effects in the low-energy spectra, we investigate how far the BdG mean-field energies deviate from the exact many-body results, especially when the system's ground-state is almost fully condensed and a mean-field approach seems to be adequate. The results are collected in Figs. 12 and 13.

The many-body excitation spectra are obtained by constructing and diagonalizing the LR-MCTDHB(M) matrix. For $M > 1$, we perform a partial diagonalization due to the large sizes of the LR matrices. Explicitly, the LR-MCTDHB(M) matrices are of the dimensionality $N_{\text{dim}} = 2M \cdot N_{\text{DVR}} + 2N_{\text{conf}}$, where N_{DVR} is the total number of DVR grid points used and N_{conf} the number of configurations. The triple-well potentials are represented on a grid with 32 DVR points per site, yielding for $N = 10$ bosons $N_{\text{dim}} = 17,360$ for $M = 7$ and $N_{\text{dim}} = 89,244$ for $M = 9$. To this end, we had to develop a new implementation of LR-MCTDHB which is capable to construct and partly diagonalize such large matrices; Without this, it would have been impossible to obtain converged results for most of the systems in [24]. The implementation uses the *Implicitly Restarted Arnoldi Method* (IRAM) [94] and its implementation in the ARPACK numerical library [95]. The IRAM is an iterative method that employs the Arnoldi algorithm [95] to solve the LR-MCTDHB eigenvalue system with respect to a set of roots which is of special interest. In our case, we consider the lowest-in-energy positive roots, i.e., the low-energy spectrum of many-body excited states.

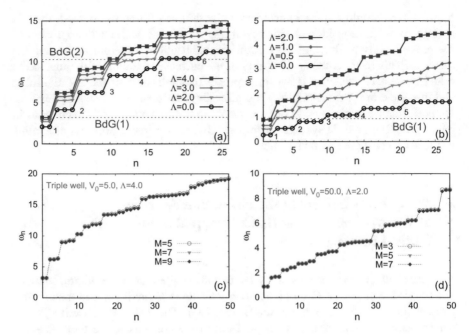

Fig. 12 Many-body effects in the excitation spectra of slightly depleted trapped BECs. (**a**) Low-energy part of the LR-MCTDHB(7) spectra of excited states for BECs consisting of $N = 10$ bosons in a shallow triple well: The one-body Hamiltonian is $\hat{h}(x) = -\frac{1}{2}\frac{d^2}{dx^2} + V_0 \cos^2(\pi x), x \in [-1.5, 1.5)$, $V_0 = 5.0$; The interaction is contact, $\lambda_0(x-x')$. Excitation energies $\omega_n = E_n - E_0$ are computed relative to the ground-state energy E_0. Results are shown for different repulsion strengths Λ. The maximal degree of ground-state depletion is $f = 1.1\%$ for $\Lambda = 4.0$. For $\Lambda = 0$, the states can be separated into distinct levels, labeled in ascending order. The two-fold degenerate states from levels 1 and 5 are single-particle excitations. They can be associated as the states with quasi-momentum $p = \pm 1$ from the first and second single-particle band. All remaining levels solely consist of many-body excitations where more than one particle is excited out of the condensate. With increasing Λ, the degeneracies between states of the same level are lifted and some levels change the order compared to the non-interacting case. The dotted red lines indicate the first two BdG lines for $\Lambda = 4.0$. (**b**) Same as in (**a**) but for the deep triple well with depth $V_0 = 50.0$. The maximal degree of depletion is $f = 8.4\%$ for $\Lambda = 2.0$, the dotted red line indicates the only BdG line for this repulsion strength in the shown energy range. Notice the different energy scales between the panels. (**c**) and (**d**) Convergence of the excitation spectra in panels (**a**) and (**b**) with the number of orbitals M. The quantities shown are dimensionless. Figure material (panels and caption) reprinted from [24]

Concluding, we discussed many-body excited states of repulsive BECs in shallow and deep triple-well potentials (in [24] finite lattices with ten sites are studied as well) and presented numerically converged results for the energy levels. By comparing the obtained many-body spectra to the corresponding BdG predictions, we demonstrated that already in the limit of weak interaction and slight ground-state depletion of the order of 1%, many-body effects occur which can not be explained by the BdG theory. This indicates the general necessity of a full many-body treatment, with LR-MCTDHB.

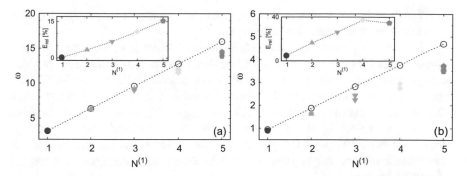

Fig. 13 Comparison of the first BdG energy BdG(1) and its multiples (open circles) with the LR-MCTDHB(7) many-body results (colored symbols) for a trapped BEC with $N = 10$ bosons. (**a**) Shallow triple well with depth $V_0 = 5.0$, $\Lambda = 4.0$, and depletion $f = 1.1\%$. Shown are the energies ω of the first five levels where only the $p = \pm 1$ states from the first single-particle band are occupied by $N^{(1)} = n^{(1)}_{+1} + n^{(1)}_{-1}$. For the many-body results, the number of points per level reflects the $(N^{(1)} + 1)$-fold degeneracy (note that some points lie atop of each other). The BdG(1) line and its multiples assign too high excitation energies ω to all levels shown. The deviation grows with $N^{(1)}$. Inset: Evolution of the relative error $E_{\mathrm{rel}} = \left| \frac{\omega_{\mathrm{BdG}}(N^{(1)}) - \omega_{\mathrm{MB}}(N^{(1)})}{\omega_{\mathrm{MB}}(N^{(1)})} \right|$ where $\omega_{\mathrm{MB}}(N^{(1)})$ denotes the LR-MCTDHB(7) energy of the state from the level $N^{(1)}$ with the largest distance to $\omega_{\mathrm{BdG}}(N^{(1)}) = N^{(1)} \cdot \mathrm{BdG}(1)$. Already for $N^{(1)} = 4$, E_{rel} exceeds 10%. (**b**) Same as in (**a**) but for the deep triple well with $V_0 = 50.0$, $\Lambda = 2.0$, and $f = 8.4\%$. For this degree of depletion, already BdG(1) itself and the corresponding many-body energy $\omega_{\mathrm{MB}}(1)$ deviate by a relative error of $E_{\mathrm{rel}} = 4.8\%$ from each other. Notice the different energy scales between the panels. The quantities shown are dimensionless. Figure material (panels and caption) reprinted from [24]

9 Brief Summary and Future Perspectives

In the sections presented above we have concisely deployed our recent findings on the many-body physics of trapped BECs obtained within the MCTDHB project supported by the HLRS. The plethora of results demonstrate that many-body effects persist in various scenarios of trapped BECs, let them be fragmented, slightly depleted, or even fully condensed. We have seen how light and ultra-cold matter conspire to give fragmented superradiance; how angular momentum in ultra-cold matter can lead to fragmented (phantom) vortices; how adiabaticity (and the lack of which) while splitting a 2D BEC is related to its fragmentation; how many-body excitations atop the mean-field excitations exist in finite BECs in double- and triple-well traps, when they are fragmented and slightly depleted; and how the variance, uncertainty product and, ultimately, the whole wave-function of a BEC are distinct at the mean-field and many-body levels, even if the BEC is fully condensed.

The key methodological message is that techniques which are able to accurately describe the many-boson time-dependent wave-function are in a great need in order to discover and faithfully quantify these many-body effects. The MCTDHB method, LR-MCTDHB, and their numerical implementations in the MCTDHB and

MCTDH-X software packages have been and will continue to serve as valuable tools in this growing research perspective.

Where are we heading to in the year to come? We list three (of many) research directions: (1) Many-body dynamics of tunneling, variance and uncertainty product, as well as excitations of 2D trapped BECs, using MCTDHB and LR-MCTDHB, would open up (further) possibilities not available in 1D; (2) Fragmentation phenomena of spin-orbit coupled BECs, see, e.g., [96], could be explored using MCTDHB for spinor bosons as implemented in MCTDH-X; and (3) Drawing ideas from multi-layer [97–101] and restricted-active-space [102] theories of bosonic systems and mixtures might prove useful in treating even larger systems of trapped interacting bosons.

Acknowledgements Financial support by the Deutsche Forschungsgemeinschaft (DFG) is gratefully acknowledged. OEA acknowledges funding by the Israel Science Foundation (Grant No. 600/15). RB acknowledges financial support by the IMPRS-QD (International Max Planck Research School for Quantum Dynamics), the Landes-graduiertenförderung Baden-Württemberg, and the Minerva Foundation. AUJL and CB acknowledge financial support by the Swiss SNF and the NCCR Quantum Science and Technology. MCT acknowledges financial support by FAPESP.

References

1. A.I. Streltsov, O.E. Alon, L.S. Cederbaum, General variational many-body theory with complete self-consistency for trapped bosonic systems. Phys. Rev. A **73**, 063626 (2006)
2. A.I. Streltsov, O.E. Alon, L.S. Cederbaum, Role of excited states in the splitting of a trapped interacting Bose-Einstein condensate by a time-dependent barrier. Phys. Rev. Lett. **99**, 030402 (2007)
3. O.E. Alon, A.I. Streltsov, L.S. Cederbaum, Unified view on multiconfigurational time propagation for systems consisting of identical particles. J. Chem. Phys. **127**, 154103 (2007)
4. O.E. Alon, A.I. Streltsov, L.S. Cederbaum, Multiconfigurational time-dependent Hartree method for bosons: many-body dynamics of bosonic systems. Phys. Rev. A **77**, 033613 (2008)
5. K. Sakmann, A.I. Streltsov, O.E. Alon, L.S. Cederbaum, Exact quantum dynamics of a bosonic Josephson junction. Phys. Rev. Lett. **103**, 220601 (2009)
6. A.U.J. Lode, K. Sakmann, O.E. Alon, L.S. Cederbaum, A.I. Streltsov, Numerically exact quantum dynamics of bosons with time-dependent interactions of harmonic type. Phys. Rev. A **86**, 063606 (2012)
7. A.U.J. Lode, The multiconfigurational time-dependent Hartree method for bosons with internal degrees of freedom: theory and composite fragmentation of multi-component Bose-Einstein condensates. Phys. Rev. A **93**, 063601 (2016)
8. H.-D. Meyer, F. Gatti, G.A. Worth (eds.), *Multidimensional Quantum Dynamics: MCTDH Theory and Applications* (Wiley-VCH, Weinheim, 2009)
9. N.P. Proukakis, S.A. Gardiner, M.J. Davis, M.H. Szymanska (eds.), *Quantum Gases: Finite Temperature and Non-equilibrium Dynamics*. Cold Atoms Series, vol. 1 (Imperial College Press, London, 2013)
10. A.U.J. Lode, K. Sakmann, R.A. Doganov, J. Grond, O.E. Alon, A.I. Streltsov, L.S. Cederbaum, Numerically-exact Schrödinger dynamics of closed and open many-boson systems with the MCTDHB package, in *High Performance Computing in Science and Engineering '13: Transactions of the High Performance Computing Center, Stuttgart (HLRS) 2013*, ed. by W.E. Nagel, D.H. Kröner, M.M. Resch (Springer, Heidelberg, 2013), pp. 81–92

11. S. Klaiman, A.U.J. Lode, K. Sakmann, O.I. Streltsova, O.E. Alon, L.S. Cederbaum, A.I. Streltsov, Quantum many-body dynamics of trapped bosons with the MCTDHB package: towards new Horizons with novel physics, in *High Performance Computing in Science and Engineering '14: Transactions of the High Performance Computing Center, Stuttgart (HLRS) 2014*, ed. by W.E. Nagel, D.H. Kröner, M.M. Resch (Springer, Heidelberg, 2015), pp. 63–86

12. O.E. Alon, V.S. Bagnato, R. Beinke, I. Brouzos, T. Calarco, T. Caneva, L.S. Cederbaum, M.A. Kasevich, S. Klaiman, A.U.J. Lode, S. Montangero, A. Negretti, R.S. Said, K. Sakmann, O.I. Streltsova, M. Theisen, M.C. Tsatsos, S.E. Weiner, T. Wells, A.I. Streltsov, MCTDHB physics and technologies: excitations and vorticity, single-shot detection, measurement of fragmentation, and optimal control in correlated ultra-cold bosonic many-body systems, in *High Performance Computing in Science and Engineering '15: Transactions of the High Performance Computing Center, Stuttgart (HLRS) 2015*, ed. by W.E. Nagel, D.H. Kröner, M.M. Resch (Springer, Heidelberg, 2016), pp. 23–50

13. O.E. Alon, R. Beinke, L.S. Cederbaum, M.J. Edmonds, E. Fasshauer, M.A. Kasevich, S. Klaiman, A.U.J. Lode, N.G. Parker, K. Sakmann, M.C. Tsatsos, A.I. Streltsov, Vorticity, variance, and the vigor of many-body phenomena in ultracold quantum systems: MCTDHB and MCTDH-X, in *High Performance Computing in Science and Engineering '16: Transactions of the High Performance Computing Center, Stuttgart (HLRS) 2016*, ed. by W.E. Nagel, D.H. Kröner, M.M. Resch (Springer, Heidelberg, 2016), pp. 79–96

14. A.I. Streltsov, K. Sakmann, A.U.J. Lode, O.E. Alon, L.S. Cederbaum, The multiconfigurational time-dependent Hartree for Bosons Package, version 2.3, Heidelberg (2013)

15. A.I. Streltsov, L.S. Cederbaum, O.E. Alon, K. Sakmann, A.U.J. Lode, J. Grond, O.I. Streltsova, S. Klaiman and R. Beinke, The multiconfigurational time-dependent Hartree for Bosons Package, version 3.x, Heidelberg/Kassel (2006-Present), http://mctdhb.org

16. A.I. Streltsov, O.I. Streltsova, The multiconfigurational time-dependent Hartree for Bosons Laboratory, version 1.5, http://MCTDHB-lab.com

17. A.U.J. Lode, M.C. Tsatsos, E. Fasshauer, MCTDH-X: The multiconfigurational time-dependent Hartree for indistinguishable particles software, http://ultracold.org; http://mctdh.bf; http://schroedinger.org (2016)

18. A.U.J. Lode, C. Bruder, Fragmented superradiance of a Bose-Einstein condensate in an optical cavity. Phys. Rev. Lett. **118**, 013603 (2017)

19. S.E. Weiner, M.C. Tsatsos, L.S. Cederbaum, A.U.J. Lode, Phantom vortices: hidden angular momentum in ultracold dilute Bose-Einstein condensates. Sci. Rep. **7**, 40122 (2017)

20. A.U.J. Lode, C. Bruder, Dynamics of Hubbard Hamiltonians with the multiconfigurational time-dependent Hartree method for indistinguishable particles. Phys. Rev. A **94**, 013616 (2016)

21. S. Klaiman, L.S. Cederbaum, Overlap of exact and Gross-Pitaevskii wave functions in Bose-Einstein condensates of dilute gases. Phys. Rev. A **94**, 063648 (2016)

22. S. Klaiman, A.I. Streltsov, O.E. Alon, Uncertainty product of an out-of-equilibrium Bose-Einstein condensate. J. Phys.: Conf. Ser. **826**, 012020 (2017)

23. M. Theisen, A.I. Streltsov, Many-body excitations and deexcitations in trapped ultracold bosonic clouds. Phys. Rev. A **94**, 053622 (2016)

24. R. Beinke, S. Klaiman, L.S. Cederbaum, A.I. Streltsov, O.E. Alon, Many-body effects in the excitation spectrum of weakly-interacting Bose-Einstein condensates in one-dimensional optical lattices. Phys. Rev. A **95**, 063602 (2017)

25. K. Sakmann, M. Kasevich, Single-shot simulations of dynamic quantum many-body systems. Nat. Phys. **12**, 451 (2016)

26. R. Beinke, S. Klaiman, L.S. Cederbaum, A.I. Streltsov, O.E. Alon, Many-body tunneling dynamics of Bose-Einstein condensates and vortex states in two spatial dimensions. Phys. Rev. A **92**, 043627 (2015)

27. M.C. Tsatsos, M.J. Edmonds, N.G. Parker, Transition from vortices to solitonic vortices in trapped atomic Bose-Einstein condensates. Phys. Rev. A **94**, 023627 (2016)

28. S. Klaiman, O.E. Alon, Variance as a sensitive probe of correlations. Phys. Rev. A **91**, 063613 (2015)

29. S. Klaiman, A.I. Streltsov, O.E. Alon, Uncertainty product of an out-of-equilibrium many-particle system. Phys. Rev. A **93**, 023605 (2016)
30. E. Fasshauer, A.U.J. Lode, Multiconfigurational time-dependent Hartree method for fermions: implementation, exactness, and few-fermion tunneling to open space. Phys. Rev. A **93**, 033635 (2016)
31. A.U.J. Lode, B. Chakrabarti, V.K.B. Kota, Many-body entropies, correlations, and emergence of statistical relaxation in interaction quench dynamics of ultracold bosons. Phys. Rev. A **92**, 033622 (2015)
32. T. Wells, A.U.J. Lode, V.S. Bagnato, M.C. Tsatsos, Vortex reconnections in anisotropic trapped three-dimensional Bose-Einstein condensates. J. Low Temp. Phys. **180**, 133 (2015)
33. M.C. Tsatsos, A.U.J. Lode, Resonances and dynamical fragmentation in a stirred Bose-Einstein condensate. J. Low Temp. Phys. **181**, 171 (2015)
34. A.U.J. Lode, S. Klaiman, O.E. Alon, A.I. Streltsov, L.S. Cederbaum, Controlling the velocities and number of emitted particles in the tunneling to open space dynamics. Phys. Rev. A **89**, 053620 (2014)
35. A.U.J. Lode, A.I. Streltsov, K. Sakmann, O.E. Alon, L.S. Cederbaum, How an interacting many-body system tunnels through a potential barrier to open space. Proc. Natl. Acad. Sci. USA **109**, 13521 (2012)
36. A.U.J. Lode, *Tunneling Dynamics in Open Ultracold Bosonic Systems* (Springer Theses, Springer, 2015). ISBN 978-3-319-07085-8
37. I. Březinová, A.U.J. Lode, A.I. Streltsov, O.E. Alon, L.S. Cederbaum, J. Burgdörfer, Wave chaos as signature for depletion of a Bose-Einstein condensate. Phys. Rev. A **86**, 013630 (2012)
38. I. Březinová, A.U.J. Lode, A.I. Streltsov, L.S. Cederbaum, O.E. Alon, L.A. Collins, B.I. Schneider, J. Burgdörfer, Elastic scattering of a Bose-Einstein condensate at a potential landscape. J. Phys. Conf. Ser. **488**, 012032 (2014)
39. O.E. Alon, A.I. Streltsov, K. Sakmann, A.U.J. Lode, J. Grond, L.S. Cederbaum, Recursive formulation of the multiconfigurational time-dependent Hartree method for fermions, bosons and mixtures thereof in terms of one-body density operators. Chem. Phys. **401**, 2 (2012)
40. D. Jaksch, C. Bruder, J.I. Cirac, C.W. Gardiner, P. Zoller, Cold bosonic atoms in optical lattices. Phys. Rev. Lett. **81**, 3108 (1998)
41. M. Greiner, O. Mandel, T. Esslinger, T.W. Hänsch, I. Bloch, Quantum phase transition from a superfluid to a Mott insulator in a gas of ultracold atoms. Nature **415**, 39 (2002)
42. R. Jördens, N. Strohmaier, K. Günter, H. Moritz, T. Esslinger, A Mott insulator of fermionic atoms in an optical lattice. Nature **455**, 204 (2008)
43. J. Struck, M. Weinberg, C. Ölschläger, P. Windpassinger, J. Simonet, K. Sengstock, R. Höppner, P. Hauke, A. Eckardt, M. Lewenstein, L. Mathey, Engineering Ising-XY spin-models in a triangular lattice using tunable artificial gauge fields. Nat. Phys. **9**, 738 (2013)
44. P. Hauke, O. Tielemann, A. Celi, C. Ölschläger, J. Simonet, J. Struck, M. Weinberg, P. Windpassinger, K. Sengstock, M. Lewenstein, A. Eckardt, Non-abelian gauge fields and topological insulators in Shaken optical lattices. Phys. Rev. Lett. **109**, 145301 (2012)
45. M.H. Anderson, J.R. Ensher, M.R. Matthews, C.E. Wieman, E.A. Cornell, Observation of Bose-Einstein condensation in a dilute atomic vapor. Science **269**, 198 (1995)
46. C.C. Bradley, C.A. Sackett, J.J. Tollett, R.G. Hulet, Evidence of Bose-Einstein condensation in an atomic gas with attractive interactions. Phys. Rev. Lett. **75**, 1687 (1995)
47. K.B. Davis, M.-O. Mewes, M.R. Andrews, N.J. van Druten, D.S. Durfee, D.M. Kurn, W. Ketterle, Bose-Einstein condensation in a gas of sodium atoms. Phys. Rev. Lett. **75**, 3969 (1995)
48. F. Dalfovo, S. Giorgini, L.P. Pitaevskii, S. Stringari, Theory of Bose-Einstein condensation in trapped gases. Rev. Mod. Phys. **71**, 463 (1999)
49. A.J. Leggett, Bose-Einstein condensation in the alkali gases: some fundamental concepts. Rev. Mod. Phys. **73**, 307 (2001)
50. I. Bloch, J. Dalibard, W. Zwerger, Many-body physics with ultracold gases. Rev. Mod. Phys. **80**, 885 (2008)

51. E.H. Lieb, R. Seiringer, J. Yngvason, Bosons in a trap: a rigorous derivation of the Gross-Pitaevskii energy functional. Phys. Rev. A **61**, 043602 (2000)
52. E.H. Lieb, R. Seiringer, Proof of Bose-Einstein condensation for dilute trapped gases. Phys. Rev. Lett. **88**, 170409 (2002)
53. N.P. Proukakis, B. Jackson, Finite-temperature models of Bose-Einstein condensation. J. Phys. B **41**, 203002 (2008)
54. L. Cohen, C. Lee, Exact reduced density matrices for a model problem. J. Math. Phys. **26**, 3105 (1985)
55. J. Yan, Harmonic interaction model and its applications in Bose-Einstein condensation. J. Stat. Phys. **113**, 623 (2003)
56. M. Gajda, Criterion for Bose-Einstein condensation in a harmonic trap in the case with attractive interactions. Phys. Rev. A **73**, 023603 (2006)
57. O.E. Alon, Many-body excitation spectra of trapped bosons with general interaction by linear response. J. Phys.: Conf. Ser. **594**, 012039 (2015)
58. C. Schilling, R. Schilling, Number-parity effect for confined fermions in one dimension. Phys. Rev. **93**, 021601(R) (2016)
59. E. Fasshauer, A.U.J. Lode, Multiconfigurational time-dependent Hartree method for fermions: implementation, exactness, and few-fermion tunneling to open space. Phys. Rev. A **93**, 033635 (2016)
60. G.J. Milburn, J. Corney, E.M. Wright, D.F. Walls, Quantum dynamics of an atomic Bose-Einstein condensate in a double-well potential. Phys. Rev. A **55**, 4318 (1997)
61. A. Smerzi, S. Fantoni, S. Giovanazzi, S.R. Shenoy, Quantum coherent atomic tunneling between two trapped Bose-Einstein condensates. Phys. Rev. Lett. **79**, 4950 (1997)
62. S. Raghavan, A. Smerzi, V.M. Kenkre, Transitions in coherent oscillations between two trapped Bose-Einstein condensates. Phys. Rev. A **60**, R1787 (1999)
63. C. Orzel, A.K. Tuchman, M.L. Fenselau, M. Yasuda, M.A. Kasevich, Squeezed states in a Bose-Einstein condensate. Science **291**, 2386 (2001)
64. A. Vardi, J.R. Anglin, Bose-Einstein condensates beyond mean field theory: quantum backreaction as decoherence. Phys. Rev. Lett. **86**, 568 (2001)
65. M. Albiez, R. Gati, J. Fölling, S. Hunsmann, M. Cristiani, M.K. Oberthaler, Direct observation of tunneling and nonlinear self-trapping in a single bosonic Josephson junction. Phys. Rev. Lett. **95**, 010402 (2005)
66. T. Schumm, S. Hofferberth, L.M. Andersson, S. Wildermuth, S. Groth, I. Bar-Joseph, J. Schmiedmayer, P. Krüger, Matter-wave interferometry in a double well on an atom chip. Nat. Phys. **1**, 57 (2005)
67. S. Levy, E. Lahoud, I. Shomroni, J. Steinhauer, The a.c. and d.c. Josephson effects in a Bose–Einstein condensate. Nature (London) **449**, 579 (2007)
68. M. Trujillo-Martinez, A. Posazhennikova, J. Kroha, Nonequilibrium Josephson oscillations in Bose-Einstein condensates without dissipation. Phys. Rev. Lett. **103**, 105302 (2009)
69. T. Zibold, E. Nicklas, C. Gross, M.K. Oberthaler, Classical bifurcation at the transition from Rabi to Josephson dynamics. Phys. Rev. Lett. **105**, 204101 (2010)
70. K. Sakmann, A.I. Streltsov, O.E. Alon, L.S. Cederbaum, Universality of fragmentation in the Schrödinger dynamics of bosonic Josephson junctions. Phys. Rev. A **89**, 023602 (2014)
71. H. Veksler, S. Fishman, Semiclassical analysis of Bose-Hubbard dynamics. New J. Phys. **17**, 053030 (2015)
72. R.A. Doganov, S. Klaiman, O.E. Alon, A.I. Streltsov, L.S. Cederbaum, Two trapped particles interacting by a finite-range two-body potential in two spatial dimensions. Phys. Rev. A **87**, 033631 (2013)
73. U.R. Fischer, A.U.J. Lode, B. Chatterjee, Condensate fragmentation as a sensitive measure of the quantum many-body behavior of bosons with long-range interactions. Phys. Rev. A **91**, 063621 (2015)
74. P. Bader, U.R. Fischer, Fragmented many-body ground states for scalar bosons in a single trap. Phys. Rev. Lett. **103**, 060402 (2009)

75. A.I. Streltsov, Quantum systems of ultracold bosons with customized interparticle interactions. Phys. Rev. A **88**, 041602(R) (2013)
76. M.-K. Kang, U.R. Fischer, Revealing single-trap condensate fragmentation by measuring density-density correlations after time of flight. Phys. Rev. Lett. **113**, 140404 (2014)
77. O.I. Streltsova, O.E. Alon, L.S. Cederbaum, A.I. Streltsov, Generic regimes of quantum many-body dynamics of trapped bosonic systems with strong repulsive interactions. Phys. Rev. A **89**, 061602(R) (2014)
78. U.R. Fischer, M.-K. Kang, "Photonic" cat states from strongly interacting matter waves. Phys. Rev. Lett. **115**, 260404 (2015)
79. J. Grond, J. Schmiedmayer, U. Hohenester, Optimizing number squeezing when splitting a mesoscopic condensate. Phys. Rev. A **79**, 021603(R) (2009)
80. J. Grond, T. Betz, U. Hohenester, N.J. Mauser, J. Schmiedmayer, T. Schumm, The Shapiro effect in atomchip-based bosonic Josephson junctions. New J. Phys. **13**, 065026 (2011)
81. M. Heimsoth, D. Hochstuhl, C.E. Creffield, L.D. Carr, F. Sols, Effective Josephson dynamics in resonantly driven Bose-Einstein condensates. New J. Phys. **15**, 103006 (2013)
82. S.I. Mistakidis, L. Cao, P. Schmelcher, Negative-quench-induced excitation dynamics for ultracold bosons in one-dimensional lattices. Phys. Rev. A **91**, 033611 (2015)
83. S. Krönke, P. Schmelcher, Many-body processes in black and gray matter-wave solitons. Phys. Rev. A **91**, 053614 (2015)
84. S. Krönke, P. Schmelcher, Two-body correlations and natural-orbital tomography in ultracold bosonic systems of definite parity. Phys. Rev. A **92**, 023631 (2015)
85. S.I. Mistakidis, T. Wulf, A. Negretti, P. Schmelcher, Resonant quantum dynamics of few ultracold bosons in periodically driven finite lattices. J. Phys. B **48**, 244004 (2015)
86. I. Brouzos, A.I. Streltsov, A. Negretti, R.S. Said, T. Caneva, S. Montangero, T. Calarco, Quantum speed limit and optimal control of many-boson dynamics. Phys. Rev. A **92**, 062110 (2015)
87. S. Dutta, S. Basu, Condensate characteristics of bosons in a tilted optical lattice. J. Phys.: Conf. Ser. **759**, 012036 (2016)
88. O.V. Marchukov, U.R. Fischer, Phase-fluctuating condensates are fragmented: an experimental benchmark for self-consistent quantum many-body calculations. arXiv:1701.06821v2 [cond-mat.quant-gas].
89. S. Bauch, K. Balzer, C. Henning, M. Bonitz, Quantum breathing mode of trapped bosons and fermions at arbitrary coupling. Phys. Rev. B **80**, 054515 (2009)
90. R. Schmitz, S. Krönke, L. Cao, P. Schmelcher, Quantum breathing dynamics of ultracold bosons in one-dimensional harmonic traps: unraveling the pathway from few- to many-body systems. Phys. Rev. A **88**, 043601 (2013)
91. J. Grond, A.I. Streltsov, A.U.J. Lode, K. Sakmann, L.S. Cederbaum, O.E. Alon, Excitation spectra of many-body systems by linear response: general theory and applications to trapped condensates. Phys. Rev. A **88**, 023606 (2013)
92. O.E. Alon, A.I. Streltsov, L.S. Cederbaum, Unified view on linear response of interacting identical and distinguishable particles from multiconfigurational time-dependent Hartree methods. J. Chem. Phys. **140**, 034108 (2014)
93. P. Grech, R. Seiringer, The excitation spectrum for weakly-interacting bosons in a trap. Commun. Math. Phys. **322**, 559 (2013)
94. Y. Saad, *Numerical Methods for Large Eigenvalue Problems* (Halstead Press, New York, 1992)
95. W.E. Arnoldi, The principle of minimized iterations in the solution of the matrix eigenvalue problem. Q. Appl. Math. **9**, 17–29 (1951). https://doi.org/10.1090/qam/42792; http://www.ams.org/journals/qam/1951-09-01/S0033-569X-1951-42792-9/
96. X.-F. Zhang, M. Kato, W. Han, S.-G. Zhang, H. Saito, Spin-orbit-coupled Bose-Einstein condensates held under a toroidal trap. Phys. Rev. A **95**, 033620 (2017)
97. H. Wang, M. Thoss, Numerically exact quantum dynamics for indistinguishable particles: the multilayer multiconfiguration time-dependent Hartree theory in second quantization representation. J. Chem. Phys. **131**, 024114 (2009)

98. S. Krönke, L. Cao, O. Vendrell, P. Schmelcher, Non-equilibrium quantum dynamics of ultra-cold atomic mixtures: the multi-layer multi-configuration time-dependent Hartree method for bosons. New J. Phys. **15**, 063018 (2013)
99. L. Cao, S. Krönke, O. Vendrell, P. Schmelcher, The multi-layer multi-configuration time-dependent Hartree method for bosons: theory, implementation, and applications. J. Chem. Phys. **139**, 134103 (2013)
100. U. Manthe, T. Weike, On the multi-layer multi-configurational time-dependent Hartree approach for bosons and fermions. J. Chem. Phys. **146**, 064117 (2017)
101. L. Cao, V. Bolsinger, S.I. Mistakidis, G.M. Koutentakis, S. Krönke, J.M. Schurer, P. Schmelcher, A unified *ab initio* approach to the correlated quantum dynamics of ultracold fermionic and bosonic mixtures. J. Chem. Phys. **147**, 044106 (2017).
102. C. Lévêque, L.B. Madsen, Time-dependent restricted-active-space self-consistent-field theory for bosonic many-body systems. New J. Phys. **19**, 043007 (2017)

Part II
Molecules, Interfaces, and Solids

Holger Fehske and Christoph van Wüllen

The following chapter reveals that science in the field of molecules, interfaces and solids has profited substantially from the computational resources provided by the High Performance Computing Center Stuttgart and the Steinbuch Centre for Computing Karlsruhe. Because of the limited space, only one-third of the generally interesting and promising contributions could be captured in the present book. We therefore selected some broadly diversified and particularly successful projects of the currently running efforts at the centres in Karlsruhe and Stuttgart in order to demonstrate the scientific progress that can be achieved by high performance computing in chemistry, physics, material science and nanotechnology.

Even a small molecular system like the Zundel cation $H_5O_2^+$ poses a computational challenge, if describing its nuclear motion needs a fully quantum mechanical description. This is so because the complexity grows exponentially with the number of coordinates. A break-through in this area is the MCTDH method developed in the group of H.-D. Meyer, where a special ansatz for the wave function has been introduced such that it can be represented in compact form. However, in order to do anything useful with the wave function one must be able to calculate the result of the Hamiltonian acting on the wave function, and this requires that one expresses the potential energy surface in the same quasi-coordinates that one uses for the wave function. The need to represent the potential this way was the bottleneck preventing MCTDH being used for larger systems, and the contribution by M. Schröder and

H. Fehske (✉)
Institut für Physik, Lehrstuhl Komplexe Quantensysteme, Ernst-Moritz-Arndt-Universität Greifswald, Felix-Hausdorff-Str. 6, 17489 Greifswald, Germany
e-mail: fehske@physik.uni-greifswald.de

C. van Wüllen (✉)
Fachbereich Chemie, Technische Universität Kaiserslautern, Erwin-Schrödinger-Str. 52, 67663 Kaiserslautern, Germany
e-mail: vanwullen@chemie.uni-kl.de

H.-D. Meyer describes how they can now transform high-dimensional potential energy surfaces into a suitable form. This is a data reduction step that requires evaluation of high-dimension integrals which are now calculated by a Monte-Carlo technique. As a by-product of importance sampling, the representation (fit) of the potential energy surface is most accurate in the low-energy regions where it matters most. Using this technique, the lowest vibrational states of the Zundel cation could be calculated.

The work of K. Remi and M. Sulpizi is targeted at the structure of water at interfaces, here the water-air interface. While water is isotropic in the bulk, a close-by interface may lead to preferred orientations, and this can be measured by Sum Frequency Generation (SFG) spectroscopy. Note that adopting a preferred orientation involves a collective rearrangement of several water molecules. The work reported here deals with how to calculate such spectra using density functional theory to describe the molecular structure and intermolecular interactions. Such calculations allow to determine which collective vibrations of supermolecular aggregates relate to different peaks in the measured spectra, and thus are of great importance to interpret them. Since the calculated spectra result from a Fourier transform of velocity-velocity correlation functions, molecular dynamics simulations of confined water (here: 128 water molecules) have to be done for long propagation times. These Born-Oppenheimer molecular dynamics calculations, where the MD trajectory is calculating using forces obtained from density functional theory, have been performed using the CP2K program package.

The presented computational treatments in the fields of solid state physics and material science were directed towards systems with high application potential. This relates in particular to the discussion of the insulator-metal phase transition of indium nanowire arrays and the simulation of ultrashort-pulse laser ablation of aluminium.

Exploiting constrained density-functional theory (DFT), the Theoretical Material Physics Group headed by W.G Schmidt at the University Paderborn performed large-scale first-principle calculations and figured out that the quite recently observed optically induced insulator-metal transition of the Si(111) substrate-stabilised In nanowire array is triggered on the microscopic scale by a non-thermal melting of a charge density wave (CDW). The CDW, being very susceptible to external perturbations, converts into the room temperature phase in the processes of laser excitation. Shortly after the excitation the nano-wires will be trapped in a metallic metastable metallic state (the so-called supercooled liquid). Using DFT within the local density approximation as included in the QUANTUM ESPRESSO package, the authors investigated the non-thermal processes leading to the destruction of the CDW. Their implementation allows for a variety of customisable parallelisation schemes (e.g., over wavenumber points and plane wave coefficients) that allows for an efficient coding on massively parallel systems such as the Cray XC40 (Hazel Hen). Having identified the driving force and the atomistic mechanism of the CDW melting, the time dynamics of the corresponding structural modification can be addressed. Here excited-state molecular dynamics (MD) in adiabatic approximation comes into play and indicates that the insulator-metal transition takes place on a time scale of a few hundred femtoseconds.

The collaborative research project on laser ablation of aluminium, conducted by scientists from the German Aerospace, the Technical University Darmstadt, University Stuttgart and the Russian Academy of Sciences, also demonstrated that ultra-short laser-matter interaction can be simulated by advanced MD methods. For picosecond pulses, the changes of the dielectric permittivity and electron thermal conductivity due to temperature, density and mean charge have to be taken into account. The dynamical recalculation of these quantities in each time step requires to adapt the IMD algorithms previously developed by the Stuttgart Institute of Functional Materials and Quantum Technologies. This can be done following the corresponding implementation in the hydrodynamic Polly-2T code, developed at the Joint Institute for High Temperatures of the Russian Academy of Sciences, which covers a wide parameter range of density and temperatures in the two-temperature model used by the authors (in such wide-range models the transition of the material from the metallic to plasma phase is assumed to take place in the vicinity of the Fermi temperature). The dynamic modelling of the permittivity and thermal conductivity with longer pulses is a prerequisite for the analysis of the subtle interplay of two counteracting effects: an enhanced absorptivity during the pulse increases the heating, whereas an increasing conductivity cooles the material which in turn decreases the absorptivity and therefore gives rise to a larger ablation threshold. To improve the performance of the present IMD treatment of the considered wide-range models the upgraded code should be parallelised in near future.

In summary, almost all projects supported during the period under review, have in common, besides a high scientific quality, the strong need for computers with high performance to achieve their results. Therefore, the supercomputing facilities at HLRS and KIT-SCC have been essential for their success.

Calculation of Global, High-Dimensional Potential Energy Surface Fits in Sum-of-Products Form Using Monte-Carlo Methods

Markus Schröder and Hans-Dieter Meyer

Abstract We have implemented a Monte-Carlo version of the well-known `potfit` algorithm. With `potfit` one can transform high-dimensional potential energy surfaces sampled on a grid into a sum-of-products form. More precisely, an fth order general tensor can be transformed into Tucker form. Using Monte-Carlo methods we avoid high-dimensional integrals that are needed to obtain optimal fits and simultaneously introduce importance sampling. The Tucker form is well suited for further use within the Heidelberg MCTDH package for solving the time-dependent as well as the time-independent Schrödinger equation of molecular systems. We demonstrate the power of the Monte-Carlo `potfit` algorithm by globally fitting the 15-dimensional potential energy surface of the Zundel cation $(H_5O_2^+)$ and subsequently calculating the lowest vibrational eigenstates of the molecule.

1 Introduction

In the field of computational, multi-dimensional quantum dynamics one seeks to accurately model the motion of microscopic particles on a quantum mechanically correct level. In the field of theoretical chemistry, it is typically the nuclei of atoms in molecules or chemical compounds which are moving in a potential energy landscape formed by chemical interactions between the different atoms. This energy landscape, also called the potential energy surface (PES), is usually obtained by applying the Born-Oppenheimer approximation to a given system, thus separating the motion of the electrons and the nuclei. The motion of the nuclei then gives raise to a number of chemical properties, such as the forming of molecules, molecular vibrations, reorganization processes, chemical reactions, etc.

M. Schröder (✉) • H.-D. Meyer
Physikalisch-Chemisches Institut, Universität Heidelberg, Im Neuenheimer Feld 229, 69120 Heidelberg, Germany
e-mail: markus.schroeder@pci.uni-heidelberg.de; hans-dieter.meyer@pci.uni-heidelberg.de

© Springer International Publishing AG 2018
W.E. Nagel et al. (eds.), *High Performance Computing in Science and Engineering '17*, https://doi.org/10.1007/978-3-319-68394-2_7

The computational modeling of such processes, however, if a big challenge. The motion of the nuclei has to be described by a multi-dimensional wavefunction that depends on the coordinates of all involved nuclei. It is hence a multi-dimensional object that needs to be stored and processed. Numerically this can be achieved by sampling the wavefunction on a product grid, i.e., by selecting a fixed number of sampling points for each physical degree of freedom (DOF) of the molecule and subsequently using these grid points as basis functions do represent the wavefunction. Numerically this can be seen as storing the wavefunction as an fth-order tensor, or, computationally, as an f-dimensional array.

For a few degrees of freedom and small grids this is easily possible. For realistic molecules, however, one needs to resort to data compression schemes to be able to store and process the wavefunction. Within the multi-configurational time-dependent Hartree method (MCTDH) [1] this is done by expressing the wave-function tensor in Tucker format: one first selects a small number of orthonormal but optimal basis functions for each DOF or for a few combined DOF (also called "modes") and subsequently expresses the wavefunction as a liner-combination of outer products of these low-dimensional basis functions. This drastically reduces the amount of memory that is needed to store the wavefunction.

The wavefunction, however, not only needs to be stored but also be processed. To solve the time-dependent or time-independent Schrödinger equation efficiently, the system Hamiltonian needs to be expressed in a sum-of-products form of terms that are defined in the same modes as the wavefunction. For the kinetic energy part this is usually the case. The potential energy, however, is usually given in terms of a complicated energy surface that is extracted from elaborate quantum chemistry calculations. For a PES one faces hence the same problems as for the wavefunction such that it needs to be transformed into a sum-of-products form. Several techniques have been employed to achieve this, such as expanding the PES into analytic functions or taking into account only low order terms. For small systems within the MCTDH package one usually employs the so-called POTFIT algorithm which transforms a given PES into Tucker form analogues to the wavefunction.

This, however, involves multiple full-dimensional integrations over the complete tensor such that this technique cannot be used for systems which are sampled on a large product grid. With Monte-Carlo POTFIT (MCPF) we could lift this restriction somewhat. Instead of using complete multi-dimensional integrals we use Monte-Carlo integrals. Of course, this comes at the cost of reduced accuracy, which we were able to counter to some extend. On the other hand, Monte-Carlo techniques allow the use of importance sampling such that instead of a globally optimal fit we are able to selectively obtain high-quality fits in regions of interest. In our case these are the low-energy regions of the potential, i.e., those regions where the wavefunction actually resides.

The successful application of the MCPF method to obtain a high-quality PES representation of the Zundel cation ($H_5O_2^+$) (15D) and the subsequent calculation of the lowest vibrational states is the main topic of this contribution. It is organized as follows. In Sect. 2 we shortly review the theory of both, MCTDH (Sect. 2.1) and the calculation if eigenstates (Sect. 2.2) and, in more detail, the Monte-Carlo

Potfit algorithm (Sect. 2.3). In Sect. 4 we present results obtained for the Zundel cation, starting with the transformation of the PES into Tucker form in Sect. 4.1 and subsequent calculation of the lowest eigenstates in Sect. 4.2. We finally summarize in Sect. 5.

2 Theory

2.1 MCTDH

As mentioned in the introduction MCTDH as well as the (Monte-Carlo) POTFIT method are both based on the decomposition of multi-dimensional functions into a sum-of-products form. The quantum mechanical state Ψ of a system is given in coordinate representation as a multi-dimensional, time-dependent function

$$\Psi = \Psi(q_1, \cdots, q_f, t) \tag{1}$$

where q_i are coordinates or internal DOF of the system (one usually separates off the three center-of-mass coordinates and three global rotational coordinates such that one ends up at $3N_{atom} - 6$ internal coordinates).

Sampled on primitive grids one can express the wavefunction in tensor format

$$\Psi_{i_1, \cdots, i_f} = \Psi(q_1^{i_1}, \cdots, q_f^{i_f}) \tag{2}$$

where $q_\kappa^{i_\kappa}$ is the i_κ-th grid point of the κth 1D-grid. If the 1D-grids are large and the number of 1D-grids is larger than 6 this easily leads to extremely large Ψ tensors. MCTDH therefore uses a data-reduction scheme, namely the Tucker format of a tensor to reduce the amount of memory needed to store the wavefunction. The wavefunction is expressed as

$$\Psi(q_1, \ldots, q_f, t) = \sum_{j_1}^{n_1} \cdots \sum_{j_f}^{n_f} A_{j_1 \cdots j_f}(t) \cdot \phi_{j_1}^{(1)}(q_1, t) \cdots \phi_{j_f}^{(f)}(q_f, t) \tag{3}$$

and analogues, when sampled on grids

$$\Psi_{i_1, \cdots, i_f}(t) = \sum_{j_1}^{n_1} \cdots \sum_{j_f}^{n_f} A_{j_1 \cdots j_f}(t) \phi_{j_1, i_1}^{(1)}(t) \cdots \phi_{j_f, i_f}^{(f)}(t). \tag{4}$$

Here, the $\phi^{(\kappa)}$ are a few, but time-dependent, basis functions, also called single-particle-functions (SPF), which are chosen optimally to represent the state vector Ψ and A are expansion coefficients, also called the core tensor of the Tucker decomposition. Note, that both quantities are time-dependent. If n_κ if much smaller then

N_κ, the size of the primitive grid of the κth coordinate, then Eq. (4) leads to a much more compact representation of the wavefunction.

To further reduce the amount of memory needed to store the wavefunction one often introduces mode combinations: a few (1–3) physical coordinates q are combined into one logical coordinate Q. In array notation this is equivalent to a reshape operation of f-dimensional array into a p-dimensional array where a number of indices are merged such that the total number of indices is reduced. Then the wavefunction can be expressed as

$$\Psi(q_1,\ldots,q_f,t) = \Psi(Q_1,\ldots,Q_p,t) = \sum_{j_1}^{n_1}\cdots\sum_{j_p}^{n_p} A_{j_1\cdots j_p}(t)\cdot\phi_{j_1}^{(1)}(Q_1,t)\cdots\phi_{j_p}^{(p)}(Q_p,t),$$

(5)

where $p < f$. This reduces the number of indices of the core tensor A such that further reduction of computational resources can be achieved.

2.2 Calculation of Eigenstates

With this ansatz one can now attempt to solve the time-dependent, or in our case, the time-independent Schrödinger equation

$$\hat{H}\Psi = E\Psi,$$

(6)

where $\hat{H} = \hat{T}+\hat{V}$ is the system Hamiltonian with kinetic part $\hat{T}$ and potential energy $\hat{V}$. Combining Eqs. (5) and (6) one arrives [2] at two separate but coupled equations which need to be fulfilled simultaneously, an eigenvalue problem for the A vector:

$$\sum_L \langle \Phi_J|\hat{H}|\Phi_L\rangle A_L = E A_J,$$

(7)

and a relaxation equation of motion for the SPP, [1]

$$\frac{\partial}{\partial\tau}\varphi_j^{(\kappa)} := -\left(1-P^{(\kappa)}\right)\sum_{k,l=1}^{n_\kappa}\left(\rho^{(\kappa)}\right)_{jk}^{-1}\left\langle\hat{H}\right\rangle_{kl}^{(\kappa)}\varphi_l^{(\kappa)} \longrightarrow 0.$$

(8)

The SPFs are to be propagated in imaginary time $\tau = -it$ until the derivatives become sufficiently small. In Eqs. (7) and (8), a number of new quantities have been introduced. Note first, that in Eq. (7) instead of lowercase indices $\{j_\kappa\}$ we have introduced the combined capital index J (and similar L) that comprises all combinations of indices $\{j_\kappa\}$ and also introduced $\Phi_J = \prod_\kappa \phi_{j_\kappa}^{(\kappa)}$. In Eq. (8) the mean-field operators

[1, 4–6] $\left(\hat{H}\right)_{k,l}^{(\kappa)}$ are introduced. These are effective Hamiltonians that act exclusively on the DOF of the κth mode and are obtained by integrating over all other DOF weighted with the wavefunction (with mode κ missing). Furthermore, in Eq. (8) the projectors $1 - P^{(\kappa)}$ are introduced which project outside the space spanned by the SPP of mode κ as well as the reduced density operators $\rho^{(\kappa)}$ of the state vector Ψ with respect to mode κ.

In particular the mean fields in Eq. (8) and the Hamiltonian matrix elements in Eq. (7) require the evaluation of multi-dimensional integrals, which is, as discussed in the introduction, extremely expensive for large primitive grids. If, however, the Hamiltonian can be written as a sum-of-products of operators that exclusively operate on the DOFs within a mode κ

$$\hat{H} = \sum_{r=1}^{s} c_r \prod_{\kappa=1}^{p} \hat{h}_r^{(\kappa)} , \tag{9}$$

then these high-dimensional integrals reduce to sums-of-products of low-dimensional integrals, for instance

$$\langle \Phi_J | \hat{H} | \Phi_L \rangle = \sum_{r=1}^{s} c_r \prod_{\kappa=1}^{p} \langle \varphi_{j_\kappa} | \hat{h}_r^{(\kappa)} | \varphi_{l_\kappa} \rangle , \tag{10}$$

in case of the matrix elements in Eq. (7). This is the key mechanism that makes MCTDH numerically efficient.

The kinetic energy part of the Hamiltonian is usually naturally in the form of Eq. (9) as it is given in terms of derivatives in coordinate space. The potential energy however is usually not. The transformation of a given PES into sum-of-products form is discussed in the following Subsection. Here we would like to emphasize that the sum in Eq. (9) can become very large, depending on the system. This means that a large number of low-dimensional integrals over the operator terms need to be evaluated to obtain the matrix elements and the mean fields. This is efficiently used for parallelization within the MCTDH software package. Here, we use both, distributed as well as shared memory parallelization using MPI and POSIX threads.

2.3 Monte-Carlo Potfit

2.3.1 Potfit

As mentioned before, one of the main challenges for realistic descriptions of molecules is the transformation of the PES into a sum-of-products form. As for the wavefunction we use the Tucker form to represent a PES sampled on a primitive grid,

$$
\begin{aligned}
V(q_1 \dots q_f) &= V(Q_1, \cdots, Q_p) \\
&\approx V^{\mathrm{PF}}(Q_1, \cdots, Q_p) \\
&= \sum_{j_1=1}^{m_1} \cdots \sum_{j_p=1}^{m_m} C_{j_1,\dots,j_p} \, v_{j_1}^{(1)}(Q_1) \cdots v_{j_p}^{(p)}(Q_p) \\
&= \sum_J C_J \Omega_J(\mathbf{Q})
\end{aligned}
\tag{11}
$$

Here we have in a first step again combined a few primitive DOF q_i into logical DOF Q_κ. The potential is then expanded using a set of optimal basis functions $v^{(\kappa)}$, also called single particle potentials (SPP), and expansion coefficients $C_{j_1,\dots,j_p}$. For numerical efficient evaluation of the matrix elements and mean fields in Eqs. (7) and (8) the mode combinations should be, the same as for the wavefunction. In a second step we have again introduced combined indices and renamed

$$
\Omega_J(\mathbf{Q}) = \prod_\kappa v_{j_\kappa}^{(\kappa)}(Q_\kappa).
\tag{12}
$$

While in Eqs. (11) and (12) are still using coordinates, let us introduce the potential V and the SPP Ω sampled on primitive grids, i.e.,

$$
\begin{aligned}
V_I &= V(Q_1^{i_1}, \cdots, Q_p^{i_p}) \\
&\approx \sum_J C_J \Omega_{I,J},
\end{aligned}
\tag{13}
$$

with

$$
\Omega_{I,J} = \prod_\kappa^p v_{i_\kappa j_\kappa}^{(\kappa)},
\tag{14}
$$

and the sampled SPP $v_{i_\kappa j_\kappa}^{(\kappa)}$ with coordinate grid index i_κ and configuration index j_κ. The coordinate dependence is hence replaced by an additional composite index I labeling grid points and $Q_\kappa^{i_\kappa}$ is the value of Q_κ at the ith grid point of the κth primitive grid.

Starting from Eq. (11) and given a potential V it is now the goal to find the optimal SPP $v^{(\kappa)}$ and the expansion coefficients C. This can be achieved by minimizing the $\mathcal{L}^2$ error between the Tucker form Eq. (11) and the original potential as

$$
\Delta^2 = \sum_I \left(V_I - \sum_J C_J \Omega_{I,J} \right)^2.
\tag{15}
$$

Variation of Eq. (19) with respect to C leads to

$$
C_J = \sum_I V_I \, \Omega_{I,J}.
\tag{16}
$$

which means that the coefficients are simply determined by overlap once the SPP are determined. The SPP, on the other hand, are given by the eigenvectors of the reduced density matrices

$$\rho^{(\kappa)}\mathbf{v}^{(\kappa)} = \mathbf{v}^{(\kappa)}\boldsymbol{\lambda}^{(\kappa)} \,, \tag{17}$$

and

$$\rho^{(\kappa)}_{i_\kappa,k_\kappa} = \sum_{I^\kappa} V_{I^\kappa,i_\kappa} V_{I^\kappa,k_\kappa} \,, \tag{18}$$

where I^κ is a composite index with the κth single index missing.

One chooses the eigenvectors with largest eigenvalues of Eq. (17) as SPP. The error of the fit can be estimated by the sum of the eigenvalues of the neglected eigenstates [7].

2.3.2 Potfit with Weights

For notational convenience we have so far assumed unweighted integrals and sums over the complete primitive grid, respectively. However, since, Eq. (11) can be seen as a lossy compression scheme if the SPP are not sufficient to fully reproduce the potential V it is useful if one could selectively increase the accuracy of the potential representation in certain areas if interest. In our case these areas are the low energy regions of the potential. Here the wavefunction usually resides and these regions are of special importance for the calculation of bound states.

We may therefore introduce a weighted error measure to be minimized as

$$\Delta^2 = \sum_I W_I \Big(V_I - \sum_J C_J \Omega_{I,J}\Big)^2 \tag{19}$$

where we introduced a positive and coordinate dependent weight function W. Re-doing the derivation of Sect. 2.3.1 one arrives at slightly modified working equations. For the coefficients one arrives (in matrix notation) at [6]

$$\mathbf{C} = \left(\boldsymbol{\Omega}^T \mathbf{W} \boldsymbol{\Omega}\right)^{-1} \boldsymbol{\Omega}^T \mathbf{W} \mathbf{V} \,, \tag{20}$$

while for the SPP one arrives at the eigenvalue equation (17), however, with a modified reduced density matrix

$$\rho^{(\kappa)}_{i_\kappa,k_\kappa} = \sum_{I_\kappa} \bar{W}^\kappa_{I^\kappa} V_{I^\kappa,i_\kappa} V_{I^\kappa,k_\kappa} \,, \tag{21}$$

Here we additionally replaced the original weight function W with a modified one $W^\kappa = \sum_{i_\kappa=1}^{N_\kappa} W_{I^\kappa,i_\kappa}$. This procedure removes the κth DOF from the weight function.

Note that, while the numerical effort to solve Eq. (21) is similar to the unweighted problem except from the multiplication with a weight function, Eq. (20) involves the inversion of a possibly large matrix, namely the matrix elements of the weight function in the SPP basis. It would therefore be advantageous to reduce the size of this matrix.

2.3.3 Mode Contraction

Within the MCTDH algorithm the precise form of the sum-of-products representation is not important. We have so far used the Tucker form to represent the PES, starting from Eq. (11). It will however be convenient to introduce a slightly modified form of this representation:

$$V_I^{\mathrm{PF}} = \sum_{J^\nu} D^{(\nu)}_{J^\nu, i_\nu} \, \Omega^\nu_{I^\nu, J^\nu} \,, \tag{22}$$

Here we have shifted the coordinate dependence of one coordinate ν into the coefficient vector. This is called contraction of one mode. It allows for the following modifications: while in Sects. 2.3.1 and 2.3.2 one needs to solve the eigenvalue equations (17) and (21) for all logical DOF, one may skip this now for the νth mode. On the other hand, Eq. (20) then becomes

$$D^{(\nu)}_{J^\nu, i_\nu} = \sum_{I^\nu} \left[\left(\mathbf{\Omega}^{\nu T} \bar{\mathbf{W}}^\nu \mathbf{\Omega}^\nu \right)^{-1} \mathbf{\Omega}^{\nu T} \bar{\mathbf{W}}^\nu \right]_{J^\nu, I^\nu} V_{I^\nu, i_\nu} \,, \tag{23}$$

where we again used a modified weight function with the νth coordinate integrated out. Note that Eq. (23) not only skips the lossy compression of mode ν in terms of SPP but also that the size of the matrix that needs to be inverted is significantly reduced as the configuration space is reduced by the logical coordinate ν. This, of course, comes at the price that Eq. (23) now needs to be solved for multiple right-hand sides, namely all primitive grid points on the contracted mode.

2.3.4 Monte-Carlo Potfit

Hawing defined the weighted potfit algorithm for complete grids one now faces the problem that the multiple integrals one has to solve to evaluate the working equations (18), (21), and (20) or (23) are numerically too expansive to accomplish because of the size of the primitive grid. For instance, the primitive grid of the Zundel cation as used below has a size of 2.2×10^{16} grid points. Storing the V_I vector alone in double precision format would therefore amount to 158 PiB. Needless to say that performing 2.2×10^{16} calculations of the PES are not possible even with modern supercomputers.

The amount of information needed to numerically represent a PES is in general much less than storage capacity of the primitive grid. The structure of the PES can in many cases be reconstructed from a small sample of grid points. The idea of Monte-Carlo Potfit therefore is to directly transform the information retrieved from a sample of grid points into a Tucker representation as in Eqs. (11) or (22) which can then be used to perform dynamical calculations using the MCTDH package.

Starting from Eqs. (21) and (23) one may interpret the weight function as a distribution of sampling points that select a subset of points from the complete primitive grid. That is, the vector W_I is either 1 if the point is selected, or 0 if it is not selected. Having selected N_c Monte-Carlo sampling points, Eq. (21) becomes

$$\rho^{(\kappa)}(i'_\kappa, i''_\kappa) = \sum_{s=1}^{N_c} V_{i'_\kappa, I^\kappa_s} V_{i''_\kappa, I^\kappa_s} , \qquad (24)$$

and Eq. (23) reads

$$D^{(\nu)}_{J^\nu, i_\nu} = \sum_{L^\nu} \left(\sum_{s=1}^{N_c} \tilde{\Omega}^\nu_{I^\nu_s, L^\nu} \tilde{\Omega}^\nu_{I^\nu_s, J^\nu} \right)^{-1} \sum_{s'=1}^{N_c} \tilde{\Omega}^\nu_{I^\nu_{s'}, L^\nu} \tilde{V}^\nu_{I^\nu_{s'}, i_\nu} , \qquad (25)$$

where the grid index I^ν_s is the subset of grid points of I^ν that has been selected as sampling points.

Note, that for both expressions the Monte-Carlo indices are only replacing the integration variables, the indices on the left hand side of Eqs. (24) and (25) are complete in the primitive basis of one mode. The matrix ρ in Eq. (24) is therefore constructed from a sum of outer products of 1D cuts through the potential, each of which traversing one sampling point. Calculating the eigenvectors of this density therefore extracts the main components of these 1D cuts. Similar, in Eq. (25), the D tensor is constructed from a set of 1D cuts, weighted with the SPP of the other modes evaluated at the sampling points and subsequently multiplied by the inverse of the Monte-Carlo overlap matrix of the SPP.

It is interesting to closer look at the error E_{MC} that is produced by the Monte-Carlo sampling compared to complete integrals. For notational simplicity we use a potfit without mode contraction i.e., with a C vector. One can shown the fit error compared to a full-grid integration is then

$$E_{MC} \propto \sum_{s'=1}^{N_c} \tilde{\Omega}_{I_{s'}, L} \tilde{\Omega}_{I_{s'}, \text{tot-}L} C_{\text{tot-}L} \qquad (26)$$

where the term tot-L means "all configurations that have been neglected in the fit". This error therefore scales with the Monte-Carlo overlap of the basis functions included in the fit with the ones not included, and the coefficient vector that represents the neglected parts of the potential due to the "lossyness" of the compression scheme.

This has two interesting consequences: first, if the sum over s in Eq. (26) were complete, the overlap would amount to zero due to the orthonormality of the SPP. That is, in the limit of complete sampling this terms goes to zero. The other interesting consequence is, that if there is no neglected part of the potential, i.e., if the SPP included in the fit are sufficient for a lossless representation of the potential ($C_{tot-L} = 0$), then Monte-Carlo Potfit also reproduces the potential exactly. Also, the less correlated the system is, that is, the fewer SPP are needed to represent the PES the more effective the algorithm will be. In particular, the numerical effort does not strongly scale with the primitive grid size as for the traditional potfit but with the size of the configuration space.

2.3.5   Restoring Molecular Symmetries

One of the drawbacks of Monte-Carlo Potfit methods is that the symmetries of the potential are usually not preserved because of the random character of the sampling points (in the traditional potfit as in Sect. 2.3.1 it is usually automatically preserved due to the complete integrals). We therefore seek a strategy to overcome this. One naive possibility would be to use all points which are equivalent to a given sampling point as well as sampling points. This would, however, not increase the amount of information extracted from the original potential but only scale the number of summation points by the number of equivalent configurations. We therefore use a different strategy.

Assume the potential remains invariant under a symmetry operation R such that

$$V(\mathbf{Q}) = V(R\mathbf{Q}) . \tag{27}$$

Having obtained a Monte-Carlo potfit this usually does not hold any longer, i.e, in general,

$$V^{\text{MCPf}}(\mathbf{Q}) \neq V^{\text{MCPF}}(R\mathbf{Q}) . \tag{28}$$

One approach to restore the property Eq. (27) for the Monte-Carlo potfit would be to average oder all symmetry operations for a given point group, that is,

$$V^{\text{MCPf, sym}}(\mathbf{Q}) = \frac{1}{N_G} \sum_{R}^{N_G} V^{\text{MCPF}}(R\mathbf{Q}) , \tag{29}$$

where N_G if the number of symmetry operations within a symmetry group.

The summation Eq. (29) can usually not be carried out straight forwardly as the SPP are different for each symmetry operation and would need to be transformed into a single (and possibly extended) set of SPP for the combined potfit on the left-hand-side. In order to be able to perform the summation anyways one can introduce symmetry adapted SPP. To obtain those, we introduce symmetrized density matrices

$$\rho_{i_\kappa,i'_\kappa}^{(\kappa,\mathrm{sym})} = \frac{1}{N_G} \sum_R \rho_{R(i_\kappa),R(i'_\kappa)}^{(\kappa)}, \tag{30}$$

Here we implicitly exploited one property that the symmetry operation must compulsorily comply to in order for Eq. (30) to be correct: the symmetry operations must not mix combined modes.

Diagonalizing the density matrices Eq. (30) then leads to eigenvectors which transform under a symmetry operation as

$$v_{R(i_\kappa),j_\kappa}^{(\kappa)} = \sum_l v_{i_\kappa,l_\kappa}^{(\kappa)} \sigma_{l_\kappa,j_\kappa}^{(\kappa)}(R), \tag{31}$$

where $\sigma^{(\kappa)}$ are (block-)diagonal matrices that contain off-diagonal elements only for degenerate eigenvalues of $\rho^{(\kappa,\mathrm{sym})}$ and have diagonal elements with values ± 1 otherwise. These matrices are simply the overlap matrices of the original with the transformed SPP. Using these relations one can then use the same sets of SPP for all symmetry transformed terms in Eq. (29) and shift the symmetry operations from the coordinate space to operations in configuration space via the $\sigma^{(\kappa)}$ matrices. Then Eq. (29) becomes

$$V_I^{\mathrm{MCPF,sym}} = \sum_{J^\nu} \Omega_{I^\nu,L^\nu}^\nu D_{L^\nu,i_\nu}^{(\nu,\mathrm{sym})}, \tag{32}$$

with

$$D_{L^\nu,i_\nu}^{(\nu,\mathrm{sym})} = \frac{1}{N_G} \sum_R \sum_{J^\nu} \sigma_{l_1,j_1}^{(1)}(R) \cdots \sigma_{l_p,j_p}^{(p)}(R) D_{J^\nu,R(i_\nu)}^{(\nu)}. \tag{33}$$

For the symmetrization of the D tensor therefore only the contracted mode needs to be symmetrized explicitly in the coordinate representation. Note also that due to the properties of the $\sigma^{(\kappa)}$ many elements in the D tensor will actually be zero. Hence, for the MCTDH calculations later on, these terms can be dropped and do not enter the calculation.

3 Monte-Carlo Potfit: Implementation Details

We have implemented the Monte-Carlo POTFIT algorithm within the MCTDH software package. Our implementation is parallelized using shared (OpenMP) and distributed (MPI) memory parallelization. The latter is used to distribute the set of Monte-Carlo sampling points included in the calculation among compute nodes.

At present, the program supports generation of sampling points with uniform and Boltzmann distribution via a Metropolis algorithm. The latter is parallelized such that multiple walkers are started in parallel and their trajectories are subsequently combined. Furthermore, arbitrary distributions can be read from external files.

In order to check convergence of the Monte-Carlo integration for the reduced density matrices, the eigenvectors (i.e., the SPP) of two densities are calculated, each one generated with one half of the sampling points. If the overlap matrix of the two sets of SPP is close to a unit matrix (except for degenerate SPP) the densities are considered converged in terms of the Monte-Carlo integral. The final SPPs are then calculated from the sum of the two densities.

The convergence of the Monte-Carlo integrals for obtaining the coefficients is checked after the generation of the Tucker form. The Monte-Carlo integration leads to optimal coefficients only on those sampling points that were used for evaluating Eq. (25) while the SPP serve as interpolating functions between those points. The convergence can be checked by comparing the $\mathcal{L}^2$-errors on those points and an independent set of points. The results should be the same within the limits of a few percent.

In most cases (but this depends on how costly the evaluation of the potential is) the calculation of the D-tensor Eq. (25) is the most demanding part due to the construction and inversion of a possibly large matrix. We have therefore implemented different solvers for Eq. (25). The first and simplest is explicitly building the SPP overlap matrix and solving Eq. (25) using the Lapack or Scalapack libraries (subroutines POSV and DPOSV, respectively) for positive definite systems. The second possibility is using the Lapack or Scalapack to solve over- or under-determined sets of equations (subroutines DGELS and PDGELS, respectively). This avoids constructing the SPP overlap matrix but uses the pseudo-inverse of the $\tilde{\Omega}^v$ matrix directly instead. This can lead to huge sets of equations to be solved such that this algorithm is probably not suitable for production jobs and has so far only been used for testing purposes. The third method is a preconditioned variant of the *BCGAdFArQ* block conjugate gradient (CG) algorithm from [8] with various optional preconditioners. Unfortunately, although CG usually shows good performance when uniform sampling is used, it performs rather poorly when other distributions of sampling points are used. In this case the SPP overlap matrix becomes often ill-conditioned resulting in a dramatical increase of the number of iterations needed for convergence. CG has, however, the advantage that the matrix of Monte-Carlo overlaps of the SPP $\sum_{s=1}^{N_c} \tilde{\Omega}_{s,L^v}^v \tilde{\Omega}_{s,J^v}^v$ never needs to be built explicitly but can be replaced by two subsequent matrix-vector multiplications of $\tilde{\Omega}^v$ and $(\tilde{\Omega}^v)^T$ with the residual D-tensor. As a last possibility we are currently implementing an explicit inversion of the SPP overlap matrix via eigenvalue or singular-value decomposition. This allows explicit regularization of the inversion which can be useful if the matrix is ill-conditioned.

Generating the sampled SPP matrices Ω^v can actually be avoided (and we do have an option in the program to do so and generate the elements on the fly as needed) but it proved useful in terms of timing to explicitly build and store them. This we do in two steps. As the algorithm requires a large number of single-number picks from random positions of the possibly large SPP arrays, in a first step (which is also done for on-the-fly generation) we fetch all needed entries in the eigenvector arrays and arrange them such that the configuration building from them can be done with little cache misses. Thereafter the configurations are built. Having generated the Ω^ν, building the SPP overlap matrix then reduces to dot products of large arrays.

4 The Zundel Cation

The Zundel cation $(H_5O_2^+)$ or protonated water dimer is a well studied molecule and plays an important role in proton diffusion. It can be described as a small water cluster which forms around a central proton. In the symmetric configuration the two water molecules are oriented with their oxygen atoms towards the central proton such that the two oxygens and the central proton for the central axis of the molecule. The Hydrogen atoms of the two waters are oriented outward such that in this configuration the Molecule has D_{2d} symmetry.

From the geometry as displayed in Fig. 1 of the molecule it can be already guessed that the molecule is very hard to model numerically. The difficulty stems from the floppiness of the molecule. The outer water molecules may undergo wagging and rocking motions or rotate against each other. The central proton moves (depending on the O-O distance) within a double well potential, attaching either to the one or the other water molecule to form hydrated hydroxonium. Also the central proton may move perpendicular to the O-O axis, thus shifting energies for the wagging and rocking motions. In addition, the internal DOF of the water molecules strongly interact with the central proton.

The floppiness of the molecule, however, makes it an ideal candidate for benchmarking the Monte-Carlo POTFIT method for obtaining a global PES fit of such a molecule.

In this contribution we follow up on a publication by Vendrell and coworkers [9] who first reported full-dimensional calculation on the protonated water dimer in a number of publications [9–12]. One of the major problems the authors were facing was an accurate representation of the PES. In this case the PES was available as a third party numerical library which was published by Huang et al. [13]. The authors in [9–12] used a so-called cluster-expansion or n-mode representation of the

Fig. 1 Structure of the Zundel cation $(H_5O_2^+)$: a central proton is bridging two water molecules

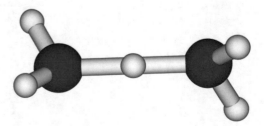

PES. The main idea here is that the total PES is expanded in n-particle interactions where one hopes to be able to truncate the expansion at low orders. The low-order terms were then transformed into a sum-of-products form using the traditional potfit algorithm as described in Sect. 2.3.1. The main disadvantage here is that this representation is not variational and its error is difficult to estimate.

4.1 PES Representation

4.1.1 Generating the PES Fit

In the following we will use a global potfit obtained with the Monte-Carlo POTFIT method. Here we will restrict ourselves to the D_{2d} case as described in [9] and refer the reader to this publication for details about the choice of the internal coordinates used here and the kinetic energy operator. In Table 1 we give an overview of the coordinates alongside with the primitive grid sizes that was used for each coordinate.

Note, that the product of the grid sizes of all primitive coordinated forms the composite grid size which amounts to 2.2×10^{16} grid points. This would therefore be the number of potential values that would have to be generated, stored and processed if the standard POTFIT method were used.

As mentioned in Sect. 2.3.1 it is often convenient to combine a few physical DOF into logical ones. In the present case we have used six logical coordinates with 2–3 physical DOF each. This is outlined in Table 2. The number of primitive grid points for each of the modes is then, of course, the product of the primitive grid sizes of the respective primitive grids. Table 2 also shows the number of basis functions or SPP that were used to generate the PES fit.

Table 1 Internal coordinates of the Zundel cation, their primitive grid size

Coordinate	Grid points	Physical meaning
R	20	O-O distance
z	19	Central proton
x	7	
y	7	
α	11	Water rotation
γ_A	19	Wagging
γ_B	19	
β_A	9	Rocking
β_B	9	
R_{1A}	9	Water A
R_{2A}	9	
θ_A	9	
R_{1B}	9	Water B
R_{2B}	9	
θ_B	9	

Table 2 Logical coordinates and their primitive grid size

Logical coordinate	Physical DOF	Primitive grid size	Number of SPP
Q_1	R, z	380	Contracted
Q_2	x, y, α	539	12
Q_3	γ_A, γ_A	361	13
Q_4	β_A, β_B	81	5
Q_5	R_{1A}, R_{2A}, θ_A	729	8
Q_6	R_{1B}, R_{2B}, θ_B	729	8

Table 3 Timings for the various sub-tasks to generate the Monte-Carlo Potfit using 19,200 CPU cores

Task	Time (min)	Remarks
Generating reduced density matrices	55	
Diagonalizing and calculating SPP	1	Not parallel, rank 0 only
Evaluating V for contracted mode	9	
Creating the SPP overlap matrix	159	
Inverting the SPP overlap matrix	2	With Scalapack
Testing the fit	3	

A critical decision for generating the fit was the selection of the sampling distribution. As mentioned above we have implemented generation of a Boltzmann distribution which generates a distribution of sampling points $\propto \exp(-V(\mathbf{Q})/k_B T)$ with k_B being the Boltzmann constant and T a given temperature. Here we faced the problem that if the temperature is chosen too low, the matrix that needs to be inverted in Eq. (25) becomes ill conditioned or even singular. If the temperature is set too high, a large fraction of random walkers will go to high energy regions where an accurate description of the PES is not required. A good compromise is a Temperature with $k_B T = 3000\,\mathrm{cm}^{-1}$ which is approximately 10 times the energy of the excited states presented below.

The length of the trajectory has been set to 5×10^7 points which has been calculated in a separate job using 64 nodes with 1536 instances of the program in total, each of which propagating a single random walker.

This calculation of the PES fit as outlined above has been done on 800 nodes on the Cray XC40 "Hazel hen" in parallel, using 1600 MPI instances, each of which with 12 OpenMP threads such that 19,200 cores were used in parallel. The calculation was done within a walltime of 227 min. Table 3 shows the approximate wall-times of the several sub-tasks to generate the Monte-Carlo Potfit. Due to the Monte-Carlo property of the algorithm it parallelizes extremely well. The distributed memory parallelization is done among the sampling points such that each instance of the program receives its private set of sampling points. Each rank then calculates its private copy of the needed quantities which are in the end reduced into rank 0. Within the tasks we use shared memory parallelization mostly also on sampling points or on configuration indices. The calculation of the 1D cuts of the potential through the sampling points required in total 1.4×10^{11} evaluations of the PES routine which is more then a factor of 10^5 less then what would be required with the standard method. This was about one quarter of the total work. About three quarters were spent for the buildup of the SPP overlap matrix which needs to be inverted. In this case this is a 49,920 by 49,920 symmetric, positive definite matrix where each entry represents a dot product of the length of the number of sampling points.

With the setup described above we were able to obtain a PES fit with a size of 174 MiB. It should be noted here that less then 1 MiB of used for the SPP such that the memory is strongly dominated by the D tensor. Note, however, that due to symmetry constraints about half of the entries of the D tensor (which are 2.2×10^7) are zero which we do not exploit for the fit generation yet. Nevertheless this is a reduction of 9 orders of magnitude compared to the potential sampled on the primitive grid.

4.1.2 Testing the PES Fit

As mentioned above the quality of the Monte-Carlo PES fit can be estimated by testing the fit on a set of sampling points which is different from the one that has been used for generating the fit. What the algorithm actually does is optimizing the fit in the $\mathcal{L}^2$ sense on the sampling points and using the SPP as interpolating functions between those points. Using an independent test set therefore can test the PES fit between the interpolation points. In Table 4 we present results of the error estimate obtained with independent Metropolis trajectories for three different temperatures. We also compare our results to the PES representation used in Ref. [9]. All tests have been performed using trajectories containing 50,000,000 points. Shown are the mean and Root-Mean-Square (RMS) errors of the exact PES less the value of the fit for this work or the Cluster expansion in case of [9].

One can clearly see that the global PES representation obtained with Monte-Carlo potfit method yields RMS errors which are one order of magnitude smaller then the ones used in [9]. The errors of the MCPF fit are smallest for a distribution of sampling points which is similar to the one used for generating the fit but are also good for a smaller temperature. For a larger temperature, the fit is also tested in areas for which it is not optimized and hence the errors are larger.

Test temperature $k_B T$ [cm^{-1}]		Errors [cm^{-1}]	
		Ref. [9]	This work
2000	Mean	−24.7	6.5
	RMS	4126.4	394.6
3000	Mean	−8.9	−1.5
	RMS	5425.9	234.5
6000	Mean	27.1	51.9
	RMS	7989.2	929.8

Table 4 Errors of the PES representation using the cluster expansion from [9] or Monte-Carlo Potfit (this work)

4.2 Calculation of Excited States

After obtaining the PES we then attempted to calculate the first few vibrationally excited states. Table 5 lists the obtained 14 lowest states within this work in comparison with those obtained in [9] alongside with state energies and assignments. The states have been calculated with MCTDH using the same mode combination scheme as was used for generating the PES fit. We used a set of SPF of $(9/25/23/13/8/8)$ for the logical coordinates while in [9] the authors used a set of $(12/25/16/12/8/8)$ which is a very similar set except for mode Q_3.

The assignments the first and second columns of Table 5 refers to the wagging modes of the water molecules (w) and the rotation of the two water molecules relative to each other (α) in the limit of separable states. In the third column we cite the state energies obtained in [9] while in the forth column we show results obtained within the current work. It can be seen that state energies are in good agreement in most cases. even though the authors of [9] used a much less accurate PES representation as in the current work. We see the main reason for this in the fact that the authors of [9] carefully tailored their PES representation to be able to obtain low lying eigenstates such that for these states we do not expect much differences in the state energies. It would be interesting to compare results for higher lying states. Here the states are not in the separable limit any more and correlation between the modes becomes much more important. We believe that the global PES obtained in this work can describe this much better.

All states have been obtained using the Block improved relaxation method [3] of the MCTDH package. The calculations were typically run on 100–200 Nodes in parallel, using both, distributed and shared memory parallelization. Here, the main numerical effort was spent in the Davidson algorithm (58%) and in generating the

Table 5 Vibrational state energies and assignments in the separable limit of the Zundel cation. In column 2 the state is assigned in a ket-notation where the kets appear in the order wagging and torsion. The energies for the ground states are given with respect to the minimum of the PES, all other energies are given with respect to the ground state

State		Energy [cm^{-1}]	
		[9]	This work
Ψ_0	$\lvert 00\rangle\lvert 0\rangle$	12398.4	12386.5
w_{1a}	$(\lvert 01\rangle + \lvert 01\rangle)\lvert 0\rangle$	102.2	109.1
w_{1b}	$(\lvert 01\rangle - \lvert 01\rangle)\lvert 0\rangle$	102.2	109.1
1α	$\lvert 00\rangle\lvert 1\rangle$	151.8	151.1
w_2	$\lvert 11\rangle\lvert 0\rangle$	231.2	238.0
$w_{1a}, 1\alpha$	$(\lvert 01\rangle + \lvert 01\rangle)\lvert 1\rangle$	255.9	254.3
$w_{1b}, 1\alpha$	$(\lvert 01\rangle - \lvert 01\rangle)\lvert 1\rangle$	255.9	254.3
2α	$\lvert 00\rangle\lvert 2\rangle$	–	279.1
$w_2, 1\alpha$	$\lvert 11\rangle\lvert 1\rangle$	379.1	372.9
w_3	$(\lvert 02\rangle - \lvert 20\rangle)\lvert 0\rangle$	386.3	378.8
$w_{1a}, 2\alpha$	$(\lvert 01\rangle + \lvert 01\rangle)\lvert 2\rangle$	–	378.9
$w_{1b}, 2\alpha$	$(\lvert 01\rangle - \lvert 01\rangle)\lvert 2\rangle$	–	378.9
3α	$\lvert 00\rangle\lvert 3\rangle$	–	404.5
w_4	$(\lvert 02\rangle + \lvert 20\rangle)\lvert 0\rangle$	447.0	437.9

mean fields and matrix elements of the Hamiltonian in the SPF basis (37%). Here parallelization is extremely useful as it is performed mainly over the operator terms which stem from the PES fit (which are approximately 2.6×10^4) not counting those that are zero due to symmetry). The walltime for one relaxation with 10 states in parallel with this setup is usually around 24 h until convergence.

5 Summary

We have implemented and demonstrated a new method for obtaining a global PES representation in terms of a sum-of-products of low-dimensional functions to be used for quantum dynamical calculations within the Heidelberg MCTDH program. The method avoids full-dimensional integrals as in the traditional potfit method but replaces them with Monte-Carlo integration techniques. This way, implementing weighted fits is naturally realized using importance sampling. The program is highly parallelized using shared and distributed memory parallelization mostly exploiting the Monte-Carlo properties of the algorithm. With this algorithm we extended the range of problems for which a global description of the PES can be obtained from 6 DOF to 15–18 DOF.

In the current case we have used Boltzmann distributed sampling points to obtain an accurate fit of the PES of the Zundel cation. The amount of memory needed to numerically represent the PES could be reduced by 9 orders of magnitude compared to sampling on a primitive grid and the number of calls to the PES routine could be reduced by 5 orders of magnitude.

We could obtain a fairly good fit of the PES in the minimal of the PES with RMS error which are a factor of 10 lower then the ones used in [9] according to Monte-Carlo testing.

With the generated PES fit we calculated the 14 lowest vibrational eigenstates using the improved relaxation algorithm of the MCTDH package and cold reproduce the state energies of [9] with slight deviations of a couple of cm^{-1}. We expect larger deviations with respect to the previous results at higher excitation energies where the much better quality of the potential representation should become visible.

References

1. M.H. Beck, A. Jäckle, G.A. Worth, H.-D. Meyer, Phys. Rep **324**, 1 (2000)
2. H.-D. Meyer, F. Le Quéré, C. Léonard, F. Gatti, Chem. Phys. **329**, 179 (2006)
3. L. Joubert-Doriol, F. Gatti, C. Iung, H.-D. Meyer, J. Chem. Phys. **129**, 224109 (2008)
4. U. Manthe, H.-D. Meyer, L.S. Cederbaum, J. Chem. Phys. **97**, 3199 (1992)
5. H.-D. Meyer, G.A. Worth, Theor. Chem. Acc. **109**, 251 (2003)
6. H.-D. Meyer, F. Gatti, G.A. Worth (eds.), *Multidimensional Quantum Dynamics: MCTDH Theory and Applications* (Wiley-VCH, Weinheim, 2009)
7. D. Peláez, H.-D. Meyer, J. Chem. Phys. **138**, 014108 (2013)

8. A.A. Dubrulle, Electron. Trans. Numer. Anal. **12**, 216 (2001)
9. O. Vendrell et al., J. Chem. Phys. **130**, 234305 (2009)
10. O. Vendrell, F. Gatti, H.-D. Meyer, Angew. Chem. Int. Ed. **46**, 6918 (2007)
11. O. Vendrell, F. Gatti, D. Lauvergnat, H.-D. Meyer, J. Chem. Phys. **127**, 184302 (2007)
12. O. Vendrell, F. Gatti, H.-D. Meyer, J. Chem. Phys. **127**, 184303 (2007)
13. X. Huang, B.J. Braams, J.M. Bowman, J. Chem. Phys. **122**, 044308 (2005)

Sum Frequency Generation Spectra from Velocity-Velocity Correlation Functions: New Developments and Applications

Khatib Rémi and Sulpizi Marialore

Abstract At the interface, the properties of water can be rather different from those observed in the bulk. In this chapter we present an overview of our computational approach to understand water structure and dynamics at the interface including atomistic and electronic structure details. In particular we show how Density Functional Theory-based molecular dynamics simulations (DFT-MD) of water interfaces can provide a microscopic interpretation of recent experimental results from surface sensitive vibrational Sum Frequency Generation spectroscopy (SFG). In our recent work we developed an expression for the calculation of the SFG spectra of water interfaces which is based on the projection of the atomic velocities on the local normal modes. Our approach permits to obtain the SFG signal from suitable velocity-velocity correlation functions, reducing the computational cost to that of the accumulation of a molecular dynamics trajectory, and therefore cutting the overhead costs associated to the explicit calculation of the dipole moment and polarizability tensor. Our method permits to interpret the peaks in the spectrum in terms of local modes, also including the bending region. The results for the water-air interface, obtained using extensive ab initio molecular dynamics simulations over 400 ns, are discussed in connection to recent phase resolved experimental data.

1 Introduction

The characterization of liquids at the interface has seen in the last years tremendous progresses thanks to the development of surface selective spectroscopic techniques, such as Sum Frequency Generation (SFG) [4, 10, 34, 43]. The necessity to provide a molecular interpretation of the experimental spectra has also fostered parallel advances in the computational community aiming to an accurate calculation and interpretation of the SFG spectra.

K. Rémi • S. Marialore (✉)
Johannes Gutenberg University Mainz, Staudinger Weg 7, 55099 Mainz, Germany
e-mail: rekhatib@uni-mainz.de; sulpizi@uni-mainz.de

© Springer International Publishing AG 2018
W.E. Nagel et al. (eds.), *High Performance Computing in Science and Engineering '17*, https://doi.org/10.1007/978-3-319-68394-2_8

141

Morita [22–24, 28] was one of the first to use atomistic molecular dynamics simulations in order to compute the SFG spectra and since his pioneer work, several groups have contributed to increase the accuracy of the calculated spectra [21, 30, 31]. The use of ab initio molecular dynamic simulations to compute SFG spectra has a more recent history, due to the much higher computational costs when compared to a force field based approach. Despite the costs, the use of electronic structure-based molecular dynamics (MD) simulations is highly desiderable since it allows an accurate description of the structure and dynamics of hydrogen bonding in highly heterogeneous environments, also including electronic polarisation.

We have recently presented a method to calculate SFG spectra from velocity-velocity correlation functions including interface specific selection rules [13, 14]. Since the velocities are a natural output in molecular dynamics (MD) simulations, they can be directly obtained without the additional cost of the direct calculation of the dipole moments and polarizabilities [38]. The advantage is that accurate Density Functional Theory (DFT)-based MD simulations can be used with no overhead cost, which is particularly useful for the description of interfaces where an accurate classical force field is not available. In the case of solid/liquid interfaces, although a very good force field may be available for water, an accurate force field for the solid is typically not available or not easily transferable from bulk materials to interfaces. The use of suitable velocity-velocity correlation functions requires the introduction of some approximations, such e.g. in the case of water, the projection of the velocities on the O-H bonds, which restricts the approach to the stretching region only. This is indeed the approximation we have used in [14] and which was also similarly adopted in a parallel work in [30]. To lift such a limitation we have recently introduced a new expression for the SFG signal, which is based on a projection of the atom velocities on the molecular normal modes [13]. The new approach permits, not only to address the stretching region, as the sum of the symmetric and antisymmetric contributions, but also to extend the SFG calculation to the bending region and to allow a discussion of the coupling between stretching and bending modes.

In this contribution we will review the new approach presented in [13] and we will also present in details an application of our method to calculate the SFG spectrum at the water-air interface with ab initio MD simulations based on DFT. Although such a system represents, probably, the most investigated one by both experimental and computational SFG approaches, a consensus is still missing on the molecular interpretation, e.g. of the hydrogen bonded region. We would like to stress that our simulations extend over a whole length of more than 400 ps, which is a quite remarkable trajectory length for a such a systems (an interface containing 128 water and 1024 electrons) and was possible thanks to the generous allocation of cpu time on Hazel Hen, the Cray XC40 at HRLS.

2 Methodology

2.1 Simulation Setup

The water/air interface was simulated by a slab of water (128 water molecules) in a $15.6404\,\text{Å} \times 15.6404\,\text{Å} \times 31.0\,\text{Å}$ cell periodically repeated in the three directions of space (Fig. 1). Eight trajectories with uncorrelated starting configurations were considered. Each trajectory consists of 10 ps equilibration followed by 40 ps of acquisition time with a time step of 0.5 fs in the NVT ensemble at 330 K. The Born-Oppenheimer molecular dynamics (MD) is carried out with the Gaussian Augmented-Plane Waves (GAPW) method [18, 19] as implemented in QUICKSTEP (a part of the CP2K package) [40]. The electronic structure is described with the BLYP functional [2, 17] including the Grimme D3 method [8], using the TZV2P basis-sets in the local space and a plane wave basis set with 280 Ry cut-off in the reciprocal space. GTH pseudopotentials [7, 9] were also employed.

We also performed a 40 ps dynamic in order to simulate a water sample. In this case the size of the box is: $15.6404\,\text{Å} \times 15.6404\,\text{Å} \times 15.6404\,\text{Å}$.

2.2 Calculating the SFG Spectra from Velocity-Velocity Correlation Functions

We aim to calculate the resonant part of the second order susceptibility tensor $\chi^{(2,\text{R})}$, which according to the classical expression [23] is:

$$\chi_{\zeta\eta\kappa}^{(2,\text{R})} = \frac{-i}{k_B T \omega_{\text{IR}}} \int_0^{+\infty} dt\, e^{i\omega_{\text{IR}}t} \langle \dot{\mathscr{A}}_{\zeta\eta}(t) \dot{\mathscr{M}}_\kappa(0) \rangle \tag{1}$$

where the indexes $\zeta\eta\kappa$ refers to the polarization of the SFG, visible and IR beams respectively, ω_{IR} is the frequency of the IR beam, $\mathscr{M}$ and $\mathscr{A}$ are the dipole moment and the polarizability of the system and $\langle \ldots \rangle$ stands for a statistical average.

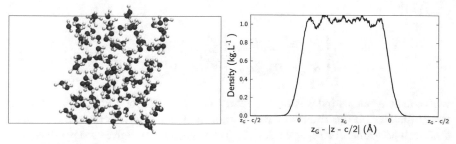

Fig. 1 (Left) A random snapshot of the water slab (128 water molecules) used in order to simulate a water-air interface. (Right) Average density (black line)

The total dipole moment and polarizability derivatives for the system can be decomposed in terms of the molecular and bond contributions:

$$
\begin{cases}
\dot{\mathscr{M}} = \sum_{i=1}^{N_{mol}} \sum_{\epsilon} \dot{\mu}_{i,\epsilon}^{l} \\
\dot{\mathscr{A}} = \sum_{i=1}^{N_{mol}} \sum_{\epsilon} \dot{\alpha}_{i,\epsilon}^{l}
\end{cases}
\tag{2}
$$

where $\dot{\mu}_{i,\epsilon}^{l}$ ($\dot{\alpha}_{i,\epsilon}^{l}$) is the dipole moment (polarizability) of the bond ϵ of the i-th molecule. Therefore, the correlation function in Eq. (1) can be rewritten as:

$$
\langle \dot{\mathscr{A}}_{\zeta\eta}(t)\dot{\mathscr{M}}_{\kappa}(0) \rangle
$$

$$
= \sum_{i=1}^{N_{mol}} \sum_{\epsilon} \left\langle \dot{\alpha}_{\zeta\eta,i,\epsilon}^{l}(t)\dot{\mu}_{\kappa,i,\epsilon}^{l}(0) \right\rangle
$$

$$
+ \sum_{i=1}^{N_{mol}} \sum_{\epsilon} \left\langle \dot{\alpha}_{\zeta\eta,i,\epsilon}^{l}(t)\dot{\mu}_{\kappa,i,-\epsilon}^{l}(0) \right\rangle
$$

$$
+ \sum_{\substack{i,j=1 \\ i\neq j}}^{N_{mol}} \sum_{\epsilon,\zeta} \left\langle \dot{\alpha}_{\zeta\eta,i,\epsilon}^{l}(t)\dot{\mu}_{\kappa,j,\zeta}^{l}(0) \right\rangle
\tag{3}
$$

The first term of the right-hand side is the bond autocorrelation, the second term accounts for the correlation between two bonds in the same water molecule, the third term is for the correlation between bonds in two different water molecules. In the following, we will refer to "intra-" and "inter-molecular" correlation to describe the sum of the first two terms and of the three terms respectively. We restrict the intermolecular correlation to the sum over the water molecules within a 4 Å distance, namely those belonging to the first solvation shell. Similarly, a cutoff was introduced in [25] for the calculation of the response function at the lipid-water interface.

The key step in our approach is a first order expansion of the dipole moment ($\mu_{i,\epsilon}$) and the polarizability ($\alpha_{i,\epsilon}$) with respect to a change in the O-H bond length ($r_{i,\epsilon}$) [14].

$$
\begin{cases}
\dot{\mu}_{i,\epsilon}^{l} = D_{m,i}D_{b,i,\epsilon}\left(\frac{\partial\mu^{b}}{\partial r}\dot{r}_{i,\epsilon}\right) \\
\dot{\alpha}_{i,\epsilon}^{l} = D_{m,i}D_{b,i,\epsilon}\left(\frac{\partial\alpha^{b}}{\partial r}\dot{r}_{i,\epsilon}\right)D_{b,i,\epsilon}^{T}D_{m,i}^{T}
\end{cases}
\tag{4}
$$

where the exponents/indexes l and b refer to the lab and bond frameworks respectively (details on the frameworks definition and the matrix connection between them can be found in a separate Sect. 2.3). The partial derivatives are parametrized (see Sect. 2.5 for details) and the velocities are obtained from the MD simulations. Therefore, the calculation of these time-derivatives do not require any additional cost with respect to that of simple trajectory accumulation. This is the main advantage over the method previously used by Sulpizi et al. [38].

2.3 Some Additional Details I: Frameworks

We use three different frameworks which are described as follows: the lab framework (x_l, y_l, z_l), the molecular framework (x_m, y_m, z_m) and the bond framework (x_b, y_b, z_b) (Fig. 2). The lab frame is defined by the simulation box and naturally by the experiment. By convention, the z_l-axis is perpendicular to the interface. The molecular frame will be used to decompose the signal into normal modes of water monomers. For the i-th molecule, the $z_{m,i}$ axis is along the bisector of the H-O-H angle, the $x_{m,i}$ axis is in the molecular plane, while the $y_{m,i}$ axis is out of the molecular plane. Finally the bond frame of the bond ϵ of the i-th molecule is such that the $z_{b,i,\epsilon}$ axis is along the bond, the $x_{b,i,\epsilon}$ is in the molecular plane and the $y_{b,i,\epsilon}$ is out of the molecular plane.

One can pass from the bond framework to the molecular framework by the direction cosine matrix:

$$
D_{b,i,\epsilon} = \begin{array}{c} \widehat{x}_{b,i,\epsilon} \ \widehat{y}_{b,i,\epsilon} \ \widehat{z}_{b,i,\epsilon} \\ \begin{bmatrix} \epsilon c & 0 & -\epsilon s \\ 0 & \epsilon & 0 \\ s & 0 & c \end{bmatrix} \begin{array}{c} \widehat{x}_{m,i} \\ \widehat{y}_{m,i} \\ \widehat{z}_{m,i} \end{array} \end{array}
\tag{5}
$$

where $c = \cos(\theta/2)$, $s = \sin(\theta/2)$ and ϵ is equals to -1 or $+1$ (see Fig. 2).

Then to go from the i-th molecular framework to the lab framework, one can use the $D_{m,i}$ direction cosine matrix which is time dependent.

2.4 Some Additional Details II: Collective Modes

We consider a water molecule with 2 O-H bonds and a H-O-H angle of $1.0\,\text{Å}$ and $105.5°$ respectively (average values from the bulk calculation). Based on this standard molecule, we have defined a collective motion (R_M) for each mode (Table 1). We assume that the mass of the O is infinite over those of the H and we have normalized the vectors in such way that $||R_M|| = 0.05\,\text{Å}$.

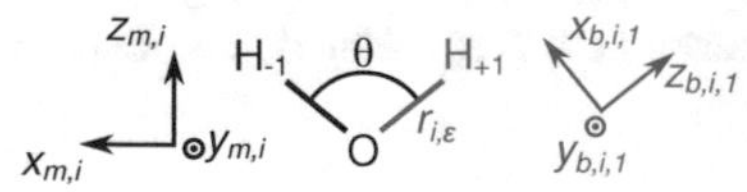

Fig. 2 Schematic representation of the molecular (left) and the bond (right) frameworks

Table 1 Collective motions (R_M in Å) associated with modes of a water molecule (B, SS, AS)

M	B			SS			AS		
	x_m	y_m	z_m	x_m	y_m	z_m	x_m	y_m	z_m
H_{-1}	0.030	0	−0.040	0.040	0	0.030	−0.040	0	−0.030
H_{+1}	−0.030	0	−0.040	−0.040	0	0.030	−0.040	0	0.030

The displacements are reported in the molecular framework

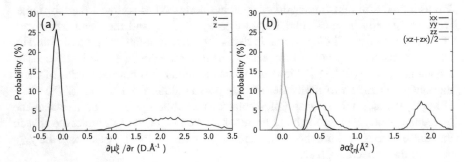

Fig. 3 Derivatives of the (**a**) dipole moment and the (**b**) polarizability according to the length of the O-H bond. The derivatives are expressed in the bond framework

2.5 Some Additional Details III: Parameters Calculations and Fresnel Factors

In order to parametrize the derivatives of the dipole moment and the polarizability, the Maximally Localized Wannier Functions [20] (MLWF) were employed according the method described in [32].

Seventeen snapshots of a bulk water sample over a 40 ps trajectory were selected leading to statistics over more than 2000 water molecules and 4000 O-H bonds. For each derivative a 2-point numerical differentiation has been performed with $\Delta r = \pm 0.05$ Å where r is either the length of the O-H bond or a collective motion (Table 1).

The probability distributions are reported in Fig. 3 and the average values are reported in Tables 2, 3 and 4. The derivative $\frac{\partial \mu_\xi^b}{\partial r}$ associated with the bond stretching is not perfectly aligned with the O-H bond but slightly outside the water molecule ($\approx 4.1°$) which is close to the observation done on ice [41] but different from the vacuum [22].

Finally, Corcelli et al. [5] pointed out that it the stretching region, the Condon approximation is not valid anymore. For this study, we use the equations that they developed to take into account the frequency dependence of the dipole moment and

Table 2 Average values for the length derivatives (dipole and polarizability) calculated thanks to the MLWF in bulk water

$\zeta =$		x_b	y_b	z_b
$\frac{\partial \mu_\zeta^b}{\partial r}$ (D Å^{-1})		-0.18	0	2.2
$\frac{\partial \alpha_{\zeta\eta}^b}{\partial r}$ (Å^2)	$\eta = x_b$	0.42		
	$\eta = y_b$	0	0.54	
	$\eta = z_b$	0.02	0	1.90

Table 3 Averages of dipole moment derivatives (D Å^{-1}) for the bending (B), the symmetric stretching (SS) and the antisymmetric stretching (AS) modes calculated thanks to the MLWF in bulk water

$\kappa =$	x_m	y_m	z_m
$\frac{\partial \mu_\kappa^m}{\partial R_B}$	0	0	-0.97
$\frac{\partial \mu_\kappa^m}{\partial R_{SS}}$	0	0	1.16
$\frac{\partial \mu_\kappa^m}{\partial R_{AS}}$	-1.83	0	0

Table 4 Averages polarizability derivatives (Å^2) for the bending (B), the symmetric stretching (SS) and the antisymmetric stretching (AS) modes calculated thanks to the MLWF in bulk water

$\zeta =$		x_m	y_m	z_m
$\frac{\partial \alpha_{\zeta\eta}^m}{\partial R_B}$	$\eta = x_m$	0.26		
	$\eta = y_m$	0	0.06	
	$\eta = z_m$	0	0	-0.18
$\frac{\partial \alpha_{\zeta\eta}^m}{\partial R_{SS}}$	$\eta = x_m$	1.34		
	$\eta = y_m$	0	0.54	
	$\eta = z_m$	0	0	0.98
$\frac{\partial \alpha_{\zeta\eta}^m}{\partial R_{AS}}$	$\eta = x_m$	0		
	$\eta = y_m$	0	0	
	$\eta = z_m$	-0.72	0	0

the polarizability:

$$\frac{\mu^{NC}}{\mu^0} = 1.377 + 53.03 \times \frac{3737.0 - \omega_{IR}}{6932.2} \tag{6}$$

$$\frac{\alpha^{NC}}{\alpha^0} = 1.271 + 5.287 \times \frac{3737.0 - \omega_{IR}}{6932.2} \tag{7}$$

where the exponent 0 stands for the property calculated with the Wannier centres, while the NC exponent stands for the property with the inclusion of the Non-Condon effects.

Since these terms are frequency dependant, they are applied as a post-treatment once in the frequency domain and in the stretching region only [27]. In order to simplify the writing, they do not explicitly appear in the equations.

In order to compare the computational results for the $\chi^{(2)}$ tensor to the experimental ones, it is necessary to take into account the Fresnel factors (L) which may drastically affect the SFG signal [15]. In our study, we will take into account exclusively the dependence in ω_{IR} of the factor associated with the IR beam ($L_{\kappa\kappa}(\omega_{IR})$) while the other two ($L_{\zeta\zeta}$ and $L_{\eta\eta}$) are considered as constant. $L_{\kappa\kappa}$ is calculated using the equation given in [44] and using the refractive index given in [3]. For the SSP polarization, we have:

$$\chi^{(2)}_{SSP} \propto L_{z_l z_l}(\omega_{IR}) \left(\chi^{(2,R)}_{\perp\perp z_l} + \chi^{(2,NR)}_{\perp\perp z_l} \right) \tag{8}$$

where $\chi^{(2,NR)}_{\perp\perp z_l}$ is the non-resonant contribution to the SFG spectrum. This term is considered to be constant and real in the stretching region [11, 12, 42].

3   Results and Discussion

In this section we review the SFG spectra calculated for the water/air interface using the velocity-velocity correlation functions [13]. In particular, in Fig. 4, we report the Im χ and Re χ for the water/air interface calculated (from a total of 8 independent trajectories of 40 ps each) with the approximation introduced in Eq. (4). In the Im χ two main features can be observed: (1) a positive peak at high frequency usually attributed to free O-H peak, (2) a negative and broad peak in the 3000–3600 cm^{-1} region. No re-crossing to positive frequencies is observed around 3100 cm^{-1}, at odd with the spectrum obtained in [38]. Such a difference is possibly due to the very limited number of configurations sampled in the 3 ps trajectory of [38]. This spectrum is overall in agreement with those obtained by other groups with different methodologies [21, 22, 30]. However, the free O-H peak does not present a shoulder on the lower frequency side [33, 37, 39]. The difference with respect to [33] is possibly due to some structural difference in the water surface as result of the different water description, which, in our case, includes the full electronic structure at DFT-BLYP D3 level.

In Fig. 4, we also compare the simulated spectra with the experimental spectra published by Nihonyanagi et al. [29]. In the comparison to the experiments, the Fresnel factors and the non-resonant background are also taken into account. Details on the Fresnel factors can be found in Sect. 2.5. The non-resonant term cannot be computed and corresponds to a frequency independent contribution which aligns the computed baseline to the experimental one. We determined that according to the following equation:

$$\chi^{(2,NR)}_{\perp\perp z_l} \approx -0.8 \max \left(\text{Im} \, \chi^{(2,R)}_{\perp\perp z_l} \right) \tag{9}$$

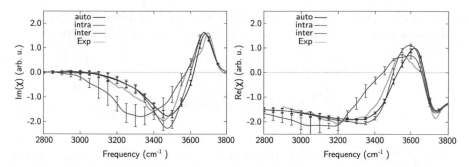

Fig. 4 SFG spectra obtained with the method based on the bond elongation including the Fresnel factors and the $\chi^{(2,NR)}_{\perp\perp z_l}$. (Left panel) Imaginary part. (Right side) Real part. The three different contributions are progressively added. The bars refer to the standard error of the mean of the resonant part only. The spectra were normalized in such way that the free O-H peak has an intensity comparable with those of Nihonyanagi et al. [29]

where $\max\left(\mathrm{Im}\,\chi^{(2,R)}_{\perp\perp z_l}\right)$ is the intensity of the free O-H peak and where $\chi_{\perp\perp z_l}$ stands for the average of the $x_l x_l z_l$ and $y_l y_l z_l$ elements. The value we use here is slightly more important than that used by Morita et al. [22] (50% of the maximal intensity). The details of all the contributions (resonant part, non-resonant part, and Fresnel factors) are discussed in the Supplementary material of [13].

The auto- (intra-) correlation is derived from the first (first two) element(s) in Eq. (3) and is represented by the black (red) line in Fig. 4. Auto- and intra-molecular correlations give comparable results in agreement with the experiment. The inter-molecular correlation introduces a red-shift of the negative band of about $150\,\mathrm{cm}^{-1}$. Such a shift had been already pointed out by Morita et al. [24] who claimed that it was necessary to consider $\chi^{(2,R)}$ as a sum of molecular hyperpolarizabilities and not as the hyperpolarizability of the whole system.

As a further step we now consider the velocities projection on the water normal modes, identified by the collective variables $\vec{R}_j$ in the molecular framework. The normal modes of a gas phase water molecule include the symmetric stretching (SS), the antisymmetric stretching (AS) and the bending (B). In the lab framework, the dipole moment and the molecular polarizability of the i-th water molecule can be written as:

$$\dot{\mu}^l_i = D_{m,i}\left(\sum_{j=Q,\,SS,\,AS} \frac{\partial \mu^m_i}{\partial R_j}\dot{R}_j\right) \tag{10}$$

$$\dot{\alpha}^l_i = D_{m,i}\left(\sum_{j=Q,\,SS,\,AS} \frac{\partial \alpha^m_i}{\partial R_j}\dot{R}_j\right) D_{m,i}{}^T \tag{11}$$

Because of the properties of the Laplace transforms, the correlation function taking into account exclusively the intra-molecular contributions can be rewritten as:

$$\chi^{(2,R)}_{\zeta\eta\kappa,\,\text{intra}}(\omega_{\text{IR}})$$

$$= \sum_i \sum_{R_1,R_2}^{N_{mol}} \frac{i\omega_{\text{IR}}}{k_B T} \int_0^{+\infty} dt\, e^{i\omega_{\text{IR}}t} \left\langle \dot{\alpha}^l_{\zeta\eta,R_1} \dot{\mu}^l_{\kappa,R_2} \right\rangle \tag{12}$$

where $\chi^{(2,R)}_{\zeta\eta\kappa,\,\text{intra}}$ is the intra-molecular contribution to the SFG signal (also noted as $\chi^{(2,\text{res})}_{ijk}$ (self) in [24]), R_1 and R_2 are one of the 3 normal modes of a water molecule in gas phase. In order to facilitate the discussion about the decomposition of the SFG signal, we will write "$R_1.R_2$" to describe the part associated with the polarizability of the mode R_1 correlated with the dipole moment of the mode R_2.

The precise description of the collective motions and the way we parametrized $\frac{\partial\alpha_i^m}{\partial R_j}$ and $\frac{\partial\mu_i^m}{\partial R_j}$ can be found in Sect. 2.5.

In Fig. 5 the $\perp\perp z_l$ spectra obtained with the two different methods (bond projection vs normal modes projection) are compared in the stretching region. The two spectra nicely overlap. The spectrum obtained with the bond projection (red dashed line) can be further analysed in terms of its normal modes contributions. The main contributions in the stretching region are associated to the SS and the AS (Fig. 5) modes while the B mode does not play a significant role (see Fig. 6). Interesting the overall spectrum broadening is the result of the "SS.AS" and the "SS.SS" contributions which are peaked at different frequencies (around 3300 and $3470\,\text{cm}^{-1}$, respectively). No direct coupling between the stretching and the bending is found to contribute to the stretching region, namely the "B.SS" and "SS.B" contributions are both zero (see Fig. 6).

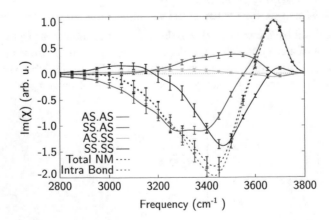

Fig. 5 Spectra associated with the $\perp\perp z_l$ polarizability. The two different methods are compared (dashed lines) and we split the spectrum obtained with the normal mode projection into its main components (plain lines)

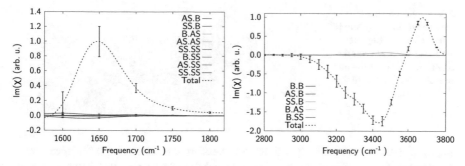

Fig. 6 SFG spectra associated with the $\perp\perp z_l$ polarization decomposed into their minor contributions in the bending (left) and in the stretching region (right)

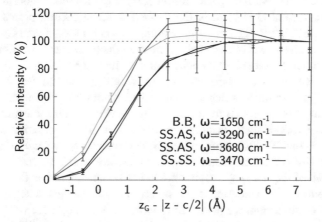

Fig. 7 Relative intensities associated with the main features of Im $\chi^{(2,R)}_{\perp\perp z_l,\,intra}$ according to the slab thickness. The normalization is done in order to reach 1 when all the water molecules are taken into account

In this sense we cannot observe here a direct effect of the Fermi resonance to the broadening of the negative peak around $3400\,cm^{-1}$ as suggested in [1, 35, 36].

A positive peak at $3500\,cm^{-1}$ and associated to the AS.AS contribution is also present. However since its intensity is relatively low it does not really give raise to a pronounced shoulder [33, 37, 39] at $3600\,cm^{-1}$ in the overall signal (black dashed line in Fig. 5).

In addition, we can also spatially localize the water molecules which contribute to the different peaks reported in Fig. 5. For this purpose we only include in the calculation of the $\chi^{(2,R)}_{\perp\perp z_l,\,intra}$ the water molecules within a certain distance (along the z_l-axis) from the surface and we observe the evolution of the peaks intensity with an increasing thickness of the included layer. In Fig. 7, we have reported the relative intensities associated with the main three contributions in the stretching region.

The two peaks of the "SS.AS" components follow a parallel evolution and are associated with the water molecules close to the water/air interface (blue lines in Fig. 7). On the contrary, the peak in the "SS.SS" component is increasing at

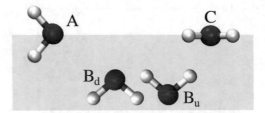

larger distances from the surface and is therefore associated with "deeper" water molecules. This is coherent with the well established idea that the water molecules at the surface have some free O-H bonds (positive peak at high frequency). Since simultaneously, the "SS.AS" peak around $3300\,cm^{-1}$ becomes more negative, we infer that the second O-H bond of the surface water molecules points toward the bulk (molecule A in Fig. 8). Since our decomposition is based on the C_{2v} symmetry of water molecules in gas phase, such orientation maximizes the contribution of the "SS.AS" term while it forbids any observation of its "SS.SS" term.

We can also suggest a molecular interpretation of the "SS.SS" contribution. The "SS.SS" contribution is negative and arises from molecules with their 2 O-H bonds pointing to the bulk (B molecule in Fig. 8). Such water molecule (1) has the orientation corresponding to the negative band of the "SS.SS" contribution in Fig. 5, (2) maximizes the "SS.SS" contribution while it doesn't contribute to the "SS.AS" element. However, these two kind of water molecules which are responsible of the main features of the $\perp\perp z_l$-spectrum should not be considered as the only kind of water molecules at the water interface. Indeed, in order to have a negative peak for the "SS.SS" contribution, we just need to have, on average, more water molecules with 2 H pointing down than with 2 H pointing up (molecule C in Fig. 8). Moreover, a water molecule laying in the plane of the interface (molecule D in Fig. 8) cannot be detected with the $\perp\perp z_l$-polarization.

In [37], the authors consider the antisymmetric mode of the "B molecules" as the origin of the shoulder observed around $3600\,cm^{-1}$. This conclusion is questionable because of the C_{2v} symmetry of water: only the $\alpha^l_{z_l\perp,\,AS}$, $\alpha^l_{\perp z_l,\,AS}$ and $\mu^l_{\perp,\,AS}$ terms are different from 0. In other words, the AS mode of the "B molecules" should be observed exclusively with the $\perp z_l\perp$ and $z_l\perp\perp$ polarizations. However, although not coming from "B molecules", the "AS.AS" term is the third most intense contribution to the total spectrum (Fig. 5), and due to its position could be the origin of a shoulder around $3600\,cm^{-1}$.

In order to complete the analysis with the normal modes approach, we now discuss the bending region. For such contribution, we work within the Condon approximation for the bending region [27]. In Fig. 9 we present the SFG spectrum with the SSP polarization of the water-vapour interface in the bending region as obtained from Eq. (12) and we compare it with the recent experimental results from [16]. We notice that the total contribution (plain black curve) corresponds entirely to the "B.B" contribution (plain red curve).

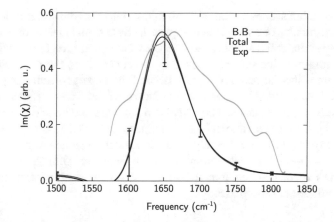

Fig. 9 SFG spectrum of the water-air interface in the bending region compared with the experimental one [16]. For the calculation of the Fresnel factors, we used the incidence angle of the IR beam used experimentally (43°)

Im $\chi^{(2,R)}_{SSP,\,intra}$ exhibits a single positive peak centred around $1650\,cm^{-1}$. The peak in the calculated spectrum is narrower than the experimental one, but its position is in agreement with the experimental one. Similarly, the agreement between the calculated and experimental real parts is very good. As we did for the stretching region, we can associate the simulated spectrum to specific orientation of the water molecules. The evolution of the "B.B" feature according to the distance from the surface, along the z_l-axis (Fig. 7) resembles very closely the behaviour associated to the "SS.SS" contribution at $3470\,cm^{-1}$. This suggests that the water molecules responsible for these two features are certainly the same. This is fully coherent with the model that we have previously proposed (Fig. 8) since the "B.B" element of molecules A and D are equal to zero while those of the molecule B is positive.

In our spectrum we only observe a positive band, as in experiments of [16], but at odd with other experimental [6] and computational [26, 27] results. We do not observe a negative band attributed in [6] to the water molecules with a free OH. Overall the weak AS-AS band at $3600\,cm^{-1}$ and the absence of a negative peak in the bending could point to a different orientation of the free OH at the interface, which in turn may depend on the water model.

4   Conclusions

In our contribution we have reviewed a method which we have recently developed to calculate the SFG spectra of water interfaces using suitable velocity-velocity correlations functions. The method is based on the projection of the atomic velocities on the single water normal modes and permits to accurately calculate the spectra using velocities correlations functions, with the advantage of no extra cost with respect to the accumulation of an ab initio molecular dynamics trajectory. Our approach permits, on one side, to decompose the SFG signal in term of contributions with a symmetric, antisymmetric, and bending character, therefore elucidating the molecular origin of the different peaks in the spectra. On the other side, it also permits to describe the bending region, beyond the bond projection approach. We have applied our method to extensive simulations of the water-air interfaces over a few hundreds of picoseconds. The calculated signal in the bending region closely resembles the recent experimental data from phase resolved SFG experiments. Our method is expected to be extremely valuable for all those heterogeneous systems where a force field parametrization is not available or difficult to obtain. Our approach is expected to enlarge the power of ab initio simulation to interpret interfacial vibrational spectroscopy.

Acknowledgements This work was supported by the DFG Research Grant SU 752/2-1 and the DFG TRR146. All the dynamics were simulated on the supercomputers of the High Performance Computing Center (HLRS) of Stuttgart (Grant 2DSFG).

References

1. S. Ashihara, N. Huse, A. Espagne, E. Nibbering, T. Elsaesser, Vibrational couplings and ultrafast relaxation of the O-H bending mode in liquid H_2O. Chem. Phys. Lett. **424**(1–3), 66–70 (2006)
2. A.D. Becke, Density-functional exchange-energy approximation with correct asymptotic behavior. Phys. Rev. A **38**, 3098–3100 (1988)
3. J.E. Bertie, M.K. Ahmed, H.H. Eysel, Infrared intensities of liquids. 5. Optical and dielectric constants, integrated intensities, and dipole moment derivatives of H_2O and D_2O at 22 °c. J. Phys. Chem. **93**(6), 2210–2218 (1989)
4. M.G. Brown, E.A. Raymond, H.C. Allen, L.F. Scatena, G.L. Richmond, The analysis of interference effects in the sum frequency spectra of water interfaces. J. Phys. Chem. A **104**(45), 10220–10226 (2000)
5. S.A. Corcelli, J.L. Skinner, Infrared and raman line shapes of dilute hod in liquid H_2O and D_2O from 10 to 90 degree. J. Phys. Chem. A **109**(28), 6154–6165 (2005)
6. C. Dutta, A.V. Benderskii, On the assignment of the vibrational spectrum of the water bend at the air/water interface. J. Phys. Chem. Lett. **8**, 801–804 (2017)
7. S. Goedecker, M. Teter, J. Hutter, Separable dual-space gaussian pseudopotentials. Phys. Rev. B **54**, 1703–1710 (1996)
8. S. Grimme, J. Antony, S. Ehrlich, H. Krieg, A consistent and accurate *ab initio* parametrization of density functional dispersion correction (DFT-D) for the 94 elements H-Pu. J. Chem. Phys. **132**(15), 154104 (2010)
9. C. Hartwigsen, S. Goedecker, J. Hutter, Relativistic separable dual-space gaussian pseudopotentials from H to Rn. Phys. Rev. B **58**, 3641–3662 (1998)
10. J. Hunt, P. Guyot-Sionnest, Y. Shen, Observation of C-H stretch vibrations of monolayers of molecules optical sum-frequency generation. Chem. Phys. Lett. **133**(3), 189–192 (1987)
11. T. Ishiyama, A. Morita, Molecular dynamics study of gas-liquid aqueous sodium halide interfaces. II. Analysis of vibrational sum frequency generation spectra. J. Chem. Phys. C **111**(2), 738–748 (2007)
12. T. Ishiyama, T. Imamura, A. Morita, Theoretical studies of structures and vibrational sum frequency generation spectra at aqueous interfaces. Chem. Rev. **114**(17), 8447–8470 (2014)

13. R. Khatib, M. Sulpizi, Sum frequency generation spectra from velocity-velocity correlation functions. J. Phys. Chem. Lett. **8**(6), 1310–1314 (2017). PMID: 28247752
14. R. Khatib, E.H.G. Backus, M. Bonn, M.-J. Perez-Haro, M.-P. Gaigeot, M. Sulpizi, Water orientation and hydrogen-bond structure at the fluorite/water interface. Sci. Rep. **6**, 24287 (2016)
15. R. Khatib, T. Hasegawa, M. Sulpizi, E.H.G. Backus, M. Bonn, Y. Nagata, Molecular dynamics simulations of SFG librational modes spectra of water at the water-air interface. J. Chem. Phys. C **120**(33), 18665–18673 (2016)
16. A. Kundu, S. Tanaka, T. Ishiyama, M. Ahmed, K.-I. Inoue, S. Nihonyanagi, H. Sawai, S. Yamaguchi, A. Morita, T. Tahara, Bend vibration of surface water investigated by heterodyne-detected sum frequency generation and theoretical study: dominant role of quadrupole. J. Chem. Phys. Lett. **7**(13), 2597–2601 (2016)
17. C. Lee, W. Yang, R.G. Parr, Development of the Colle-Salvetti correlation-energy formula into a functional of the electron density. Phys. Rev. B **37**, 785–789 (1988)
18. G. Lippert, J. Hutter, M. Parrinello, A hybrid Gaussian and plane wave density functional scheme. Mol. Phys. **92**(3), 477–488 (1997)
19. G. Lippert, J. Hutter, M. Parrinello, The Gaussian and augmented-plane-wave density functional method for *ab initio* molecular dynamics simulations. Theor. Chem. Acc. **103**(2), 124–140 (1999)
20. N. Marzari, D. Vanderbilt, Maximally localized generalized Wannier functions for composite energy bands. Phys. Rev. B **56**, 12847–12865 (1997)
21. G.R. Medders, F. Paesani, Dissecting the molecular structure of the air/water interface from quantum simulations of the sum-frequency generation spectrum. J. Am. Chem. Soc. **138**(11), 3912–3919 (2016)
22. A. Morita, J.T. Hynes, A theoretical analysis of the sum frequency generation spectrum of the water surface. Chem. Phys. **258**(2–3), 371–390 (2000)
23. A. Morita, J.T. Hynes, A theoretical analysis of the sum frequency generation spectrum of the water surface. II. Time-dependent approach. J. Phys. Chem. B **106**(3), 673–685 (2002)
24. A. Morita, T. Ishiyama, Recent progress in theoretical analysis of vibrational sum frequency generation spectroscopy. Phys. Chem. Chem. Phys. **10**, 5801–5816 (2008)
25. Y. Nagata, S. Mukamel, Vibrational sum-frequency generation spectroscopy at the water/lipid interface: molecular dynamics simulation study. J. Am. Chem. Soc. **132**(18), 6434–6442 (2010)
26. Y. Nagata, C.-S. Hsieh, T. Hasegawa, J. Voll, E.H.G. Backus, M. Bonn, Water bending mode at the water-vapor interface probed by sum-frequency generation spectroscopy: a combined molecular dynamics simulation and experimental study. J. Chem. Phys. Lett. **4**(11), 1872–1877 (2013)
27. Y. Ni, J.L. Skinner, IR and SFG vibrational spectroscopy of the water bend in the bulk liquid and at the liquid-vapor interface, respectively. J. Chem. Phys. **143**(1), 014502 (2015)
28. S. Nihonyanagi, T. Ishiyama, T.-K. Lee, S. Yamaguchi, M. Bonn, A. Morita, T. Tahara, Unified molecular view of the air/water interface based on experimental and theoretical $\chi^{(2)}$ spectra of an isotopically diluted water surface. J. Am. Chem. Soc. **133**(42), 16875–16880 (2011)
29. S. Nihonyanagi, R. Kusaka, K.-I. Inoue, A. Aniruddha, S. Yamaguchi, T. Tahara, Accurate determination of complex $\chi^{(2)}$ spectrum of the air/water interface. J. Chem. Phys. **143**(12), 124707 (2015)
30. T. Ohto, K. Usui, T. Hasegawa, M. Bonn, Y. Nagata, Toward *ab initio* molecular dynamics modeling for sum-frequency generation spectra; an efficient algorithm based on surface-specific velocity-velocity correlation function. J. Chem. Phys. **143**(12), 124702 (2015)
31. P.A. Pieniazek, C.J. Tainter, J.L. Skinner, Surface of liquid water: three-body interactions and vibrational sum-frequency spectroscopy. J. Am. Chem. Soc. **133**(27), 10360–10363 (2011)
32. M. Salanne, R. Vuilleumier, P.A. Madden, C. Simon, P. Turq, B. Guillot, Polarizabilities of individual molecules and ions in liquids from first principles. J. Phys. Condens. Matter **20**(49), 494207 (2008)

33. J. Schaefer, E.H.G. Backus, Y. Nagata, M. Bonn, Both inter- and intramolecular coupling of O-H groups determine the vibrational response of the water/air interface. J. Phys. Chem. Lett. **7**(22), 4591–4595 (2016)
34. Y.R. Shen, Surface properties probed by second-harmonic and sum-frequency generation. Nature **337**(6207), 519–525 (1989)
35. M. Sovago, R.K. Campen, G.W.H. Wurpel, M. Müller, H.J. Bakker, M. Bonn, Vibrational response of hydrogen-bonded interfacial water is dominated by intramolecular coupling. Phys. Rev. Lett. **100**, 173901 (2008)
36. M. Sovago, R.K. Campen, H.J. Bakker, M. Bonn, Hydrogen bonding strength of interfacial water determined with surface sum-frequency generation. Chem. Phys. Lett. **470**(1–3), 7–12 (2009)
37. I.V. Stiopkin, C. Weeraman, P.A. Pieniazek, F.Y. Shalhout, J.L. Skinner, A.V. Benderskii, Hydrogen bonding at the water surface revealed by isotopic dilution spectroscopy. Nature **474**(7350), 192–195 (2011)
38. M. Sulpizi, M. Salanne, M. Sprik, M.-P. Gaigeot, Vibrational sum frequency generation spectroscopy of the water liquid-vapor interface from density functional theory-based molecular dynamics simulations. J. Phys. Chem. Lett. **4**(1), 83–87 (2013)
39. C.-S. Tian, Y.R. Shen, Isotopic dilution study of the water/vapor interface by phase-sensitive sum-frequency vibrational spectroscopy. J. Am. Chem. Soc. **131**(8), 2790–2791 (2009)
40. J. VandeVondele, M. Krack, F. Mohamed, M. Parrinello, T. Chassaing, J. Hutter, Quickstep: fast and accurate density functional calculations using a mixed gaussian and plane waves approach. Comput. Phys. Commun. **167**(2), 103–128 (2005)
41. E. Whalley, D.D. Klug, Effect of hydrogen bonding on the direction of the dipole-moment derivative of the O-H bond in the water molecule. J. Chem. Phys. **84**(1), 78–80 (1986)
42. S. Yamaguchi, K. Shiratori, A. Morita, T. Tahara, Electric quadrupole contribution to the nonresonant background of sum frequency generation at air/liquid interfaces. J. Chem. Phys. **134**(18), 184705 (2011)
43. X.D. Zhu, H. Suhr, Y.R. Shen, Surface vibrational spectroscopy by infrared-visible sum frequency generation. Phys. Rev. B **35**, 3047–3050 (1987)
44. X. Zhuang, P.B. Miranda, D. Kim, Y.R. Shen, Mapping molecular orientation and conformation at interfaces by surface nonlinear optics. Phys. Rev. B **59**, 12632–12640 (1999)

Photo-Excited Surface Dynamics from Massively Parallel Constrained-DFT Calculations

A. Lücke, T. Biktagirov, A. Riefer, M. Landmann, M. Rohrmüller, C. Braun, S. Neufeld, U. Gerstmann, and W.G. Schmidt

Abstract Constrained density-functional theory (DFT) calculations show that the recently observed optically induced insulator-metal transition of the In/Si(111)(8×2)/(4×1) nanowire array (Frigge et al., Nature 544:207, 2017) corresponds to the non-thermal melting of a charge-density wave (CDW). Massively parallel numerical simulations allow for the simulation of the photo-excited nanowires and provide a detailed microscopic understanding of the CDW melting process in terms of electronic surface bands and selectively excited soft phonon modes. Excited-state molecular dynamics in adiabatic approximation shows that the insulator-metal transition can be as fast as 350 fs.

1 Introduction

Phase transitions between equilibrium states of matter as function of temperature, pressure, magnetics fields, etc. are ubiquitous. Typically, their direct atomic scale observation is not possible and they are described in terms of statistical ensembles and phenomenological models, see, e.g. [1]. For both scientific as well as techno-logical reasons, however, it is interesting to learn how phase transitions occur at the atomic scale, at what speed they evolve, and how they can be driven. In this context, the ordered array of In nanowires that self-assemble at the Si(111) surface—first described in 1965 [2]—is a particularly intriguing example. It is an extensively studied model system for phase transitions in quasi one-dimensional systems [3]: At around 120 K a charge-density wave forms and the metallic In/Si(111)(4×1) zigzag-chain structure, stable at room temperature, transforms to an insulating (8×2) In-hexagon reconstruction [4–8]. Adsorbates, depending on their species, either decrease [9] or increase the critical temperature T_C [10]. This indicates that the charge-density wave formation is very sensitive to external perturbations [11]. In fact, it has been shown that a 50 fs laser pulse with 1.55 eV photons converts the low-temperature (8×2) phase into the room-temperature (4×1) structure even at

A. Lücke • T. Biktagirov • A. Riefer • M. Landmann • M. Rohrmüller • C. Braun • S. Neufeld • U. Gerstmann • W.G. Schmidt (✉)
Lehrstuhl für Theoretische Materialphysik, Universität Paderborn, 33095 Paderborn, Germany
e-mail: W.G.Schmidt@upb.de

© Springer International Publishing AG 2018
W.E. Nagel et al. (eds.), *High Performance Computing in Science and Engineering '17*, https://doi.org/10.1007/978-3-319-68394-2_9

30 K, i.e., far below the phase transition temperature T_C [12, 13]. Subsequently to the laser excitation, the In nanowires are trapped for several hundred picoseconds in a metallic metastable state corresponding to a supercooled liquid. The CDW melting is clearly not related to surface heating effects as proven by surface temperature measurements [12]. The aim of the present study consists in the elucidation of the non-thermal processes that lead to the CDW melting.

2 Methodology

To this aim, we use DFT within the local-density approximation as implemented in the QUANTUM ESPRESSO package [14] and determine the In/Si(111)(8×2)/(4×1) ground- and excited-state atomic and electronic structure. The surface is modeled using periodic boundary conditions within a supercell that contains three bilayers of silicon, the bottom layer of which is saturated with hydrogen. Norm-conserving pseudopotentials in conjunction with a plane-wave basis set limited by a cutoff energy of 50 Ry are employed to describe the electronic structure. A 2×8×1 Monkhorst-Pack mesh [15] is used for Brillouin zone (BZ) sampling. Constrained DFT [16] is used to calculate excited-state potential energy surfaces (PESs) and for performing ab initio molecular dynamics (AIMD) within the adiabatic approximation. Thereby the occupation numbers of the electronic states were frozen such as to describe a specific excited electron configuration. The configurations explored here are restricted to charge neutral excitations, but do not necessarily conserve momentum. This allows for modeling configurations that result from electronic relaxation including scattering processes subsequent to vertical excitations.

Most excited-state electronic configurations will decay within a few femtoseconds, owing to electron-electron, electron-phonon, electron-hole and electron-defect scattering. The quasiparticle lifetimes of electrons due to electron-electron scattering can be accessed from the imaginary part of the electronic self-energy. We performed quasiparticle calculations using the G_0W_0 approach [17] within the on-mass-shell approximation [18]. Here, the one-particle Green's function G_0 and the screened Coulomb interaction W_0 were obtained from the DFT electronic structure. Owing to the large increase in the number of possible scattering events, the lifetimes are found to decrease rapidly to a few femtoseconds for energies of more than 1 eV above or below the Fermi energy, in marked contrast to states close to the Fermi edge, cf. Fig. 1. Therefore, only the states close to the Fermi edge were used for calculating the excited-state PESs.

The present Quantum Espresso constrained-DFT implementation (v6.1) is restricted to a BZ sampling with only a single **k** point. This is not sufficient for systems with strongly dispersing states such as the metallic wires studied here. Therefore, the implementation was modified to allow for an arbitrary number of **k** points. For efficiency, the implementation was written to support both the

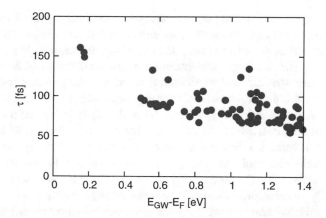

Fig. 1 Calculated lifetimes of electronic quasiparticle states in dependence on their energy relative to the Fermi level position

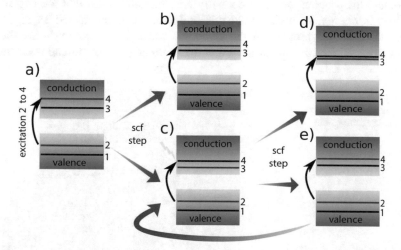

Fig. 2 Example illustrating an electronic excitation from orbital 2 to 4. The ground state is shown in (**a**), while (**b**)–(**e**) show example configurations that result from electronic relaxation upon excitation. Figures (**b**) and (**d**) represent electronically converged situations. The excitation indicated in (**d**), however, is not intended, but results from changes in the band order

parallelization with respect to **k** points (NPOOL in Quantum Espresso) as well as the parallelization with respect to the plane-wave coefficients. A newly defined input card within the standard input file can be used to define band and **k**-point specific occupation numbers.

Figure 2a shows exemplarily a system where an excitation of one electron from state 2 to state 4 is modeled. This excitation modifies the potential seen by the electrons and may lead to an energetic re-ordering of the orbitals. This may cause a situation where the system electronically relaxes into another excited state than desired. Figure 2b, c show two possible configurations that may result from electron relaxation after exciting the electron as shown in Fig. 2a. Figure 2b depicts the situation that corresponds to the intended excitation. In contrast, Fig. 2c illustrates the case where the electronic relaxation led to an interchange of bands 3 and 4. If the electronic structure calculations are based on occupation numbers simply determined from the energetic order, a wrong conductance state, i.e., here the state 4 gets occupied. The originally excited (marked red) state 4 is emptied. This re-occupation of the orbitals leads to an abrupt change in the potential which may lead to two different scenarios. Either the "wrong" occupation numbers persist, as shown in Fig. 2d, or the electronic relaxation in the self-consistent field (scf) calculations swap the orbitals again, as shown in Fig. 2e. In fact, the latter scenario may result in a numerically unstable situation due to persistent charge shuffling between different electronic orbitals. Summarizing, only the electronic structure visualized in Fig. 2b corresponds to the "correctly" excited configuration, while the other situations do either not correspond to the "correct" excitation scenario and/or are numerically unstable. This problem occurs frequently for electronic states that are degenerate or close in energy. In order avoid this problem, we implemented an algorithm where the orbital symmetry of the electronic states is monitored during the scf calculations, cf. Fig. 3. This ensures that the "correct", i.e., the intended electronic

Fig. 3 Program flow in order to guarantee a convergence to the desired excited state

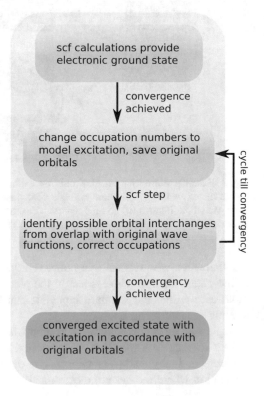

excitation is modeled in the constrained-DFT calculations and used, e.g., for the AIMD calculations.

In detail, we start by determining the electronic ground state. The ground-state orbitals are stored after they are numerically converged. At the same time, the occupation numbers are changed in order to describe the specific photo-excitation which needs to be modeled. Subsequently, the overlap with the stored ground-state wave functions is used to monitor the orbital symmetry of the wave functions updated during the scf cycles. This allows to choose the occupation numbers in such a way that they correctly describe the modeled excitation, even if changes in the energetic ordering of the states occur.

As mentioned above, our implementation allows for a variety of customizable parallelization schemes. In particular, parallelization (and data distribution) over **k** points as well as parallelization (and data distribution) over plane wave coefficients can to be used at the same time on massively parallel systems such as the Cray XC40 (Hazel Hen). This leads already to a high computational efficiency. However, the calculations can be sped up by an additional, third parallelization. Thereby we calculate the excited-state potential energy surface (PES) not point by point, but simultaneously for all reaction coordinates in parallel.

The performance of the Cray XC40 in combination with the mentioned parallelization strategies has been tested and optimized for the indium nanowire system within a (8×2) surface unit cell, containing 640 electrons. The computational techniques rely on mathematical libraries, in particular the Linear Algebra Package LAPACK or its distributed-memory implementation ScaLAPACK. The calculations discussed here have been performed with ScaLAPACK. For the (excited-state) PES calculations ten sets of reaction coordinates were calculated in parallel. The red line in Fig. 4 shows the results for taking additionally only Quantum Espresso's basis parallelization (NPOOL=1) into account which distributes the plane-wave expansion coefficients across the cores. It turns out that this parallelization is only meaningful for a number of cores of the order of about 1000.

Since Quantum Espresso supports additionally also a parallelization over **k** points (NPOOL > 1), this strategy has also been tested on Hazel Hen (cf. Fig. 4). NPOOL specifies thereby the number of **k** points that are treated in parallel. Inside a group of cores sharing the work for an individual **k** point, the other two parallelization strategies mentioned above still apply. It can be seen that the **k**-point parallelization leads to a notable saving of computation time with increasing number of cores. Especially for more than 1000 cores the **k** parallelization is mandatory to reduce the overhead caused by an increasing MPI communication. We find that this approach allows for the extension of the roughly linear scaling up to 10,000 cores for NPOOL = 8.

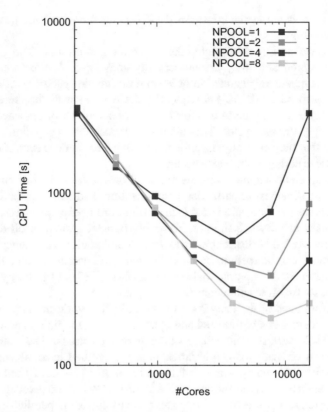

Fig. 4 CPU time on the HLRS Cray XC40 for the self-consistent calculation of the In/Si(111)(8×2) surface for various parallelization schemes. See text for details

3 Results

The photon energy used to trigger the phase transition equals 1.55 eV [12, 13]. This allows for exciting a far larger number of electronic configurations than are computationally accessible. However, not all of these configurations are long-lived enough to drive the phase transition. Their lifetimes decrease rapidly to only tens of femtoseconds with increasing distance from the Fermi energy E_F, see Fig. 1. We therefore focus on the two highest occupied (H,H−1) and the two lowest unoccupied (L,L+1) electron state of the In/Si(111)(8×2) band structure (cf. Fig. 5a) that are long-lived in the calculations and thus suitable candidates to drive the transition. These states with strong dispersion along the CDW direction correspond to strongly In-localized surface states [8].

In a second step, various excited-state PESs for charge neutral combinations between electrons in L, L+1 and holes in H, H-1 were calculated by performing constrained-DFT calculations for geometries along the minimum energy path along the (8×2) ⟶ (4×1) phase change. Thereby we sampled the surface Brillouin zone

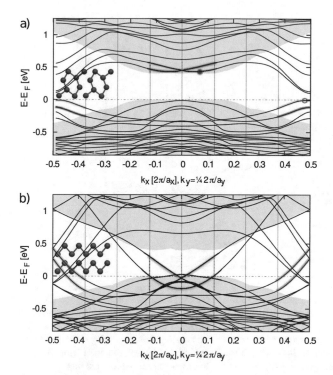

Fig. 5 (**a**) Calculated band structure of In/Si(111)(8×2) along a line within the SBZ. Blue vertical lines denote the segments used to model excitations. Exemplary a $L_5;H_8$ excitation is indicated. (**b**) Calculated band structure of In/Si(111)(4×1) along the same line as above

(SBZ) separately for electrons and holes using eight segments as shown in Fig. 5. Relevant examples of excited-state PESs are shown in Fig. 6 (lhs) in dependence on the generalized reaction coordinate. The calculated ground-state PES has its minimum at the (8×2) phase, separated from the metastable (4×1) structure by a distinct energy barrier of about 40 meV.

The (8×2) structure corresponds to the global minimum also for a number of excited-state PESs. This concerns, for example, the momentum conserving excitations $H_1;L_1$, $H_2;L_2$, $H_3;L_3$, and $H_4;L_4$. These excitations, however, decrease the energy difference between the (8×2) and (4×1) phases, and for even stronger excitations such as $H_{1,8};L_{1,8}$ and $H_{2,7};L_{2,7}$ the respective stability is even reverted, i.e., the (4×1) phase gets more stable than the (8×2) structure. Still, these excitations will not necessarily lead to a CDW melting, as there is an energy barrier separating the insulating and the metallic phase. This energy barrier disappears, however, for specific non-momentum conserving excitations such as $H_{1,8};L_{4,5}$, $H_{1,8};L+1_{4,5}$ (altogether 0.5 electron excited) or $H_{1,8}H-1_{1,8};L_{4,5}L+1_{4,5}$ (altogether 1 electron excited). Obviously, excitations that combine electrons in the lowest unoccupied surface states at the SBZ center with holes residing in the uppermost valence state at the SBZ boundary are particularly prone to cause a insulator-metal transition.

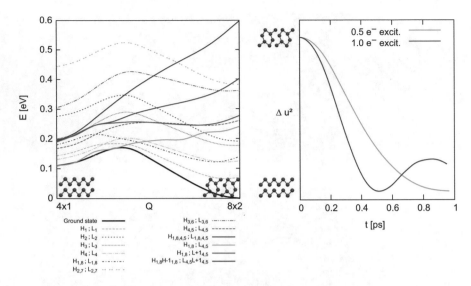

Fig. 6 (Left) Calculated ground- and excited-state potential energy surfaces vs the generalized reaction coordinate of the $(8 \times 2) \longrightarrow (4 \times 1)$ phase transition. The numbers denotes the band structure segments indicated in Fig. 5 numbered from the left, i.e. $H_1; L_1$ denotes an excited state with hole in H in segment 1 and electron in L in segment 1. (Right) Calculated time evolution of the structural deviation between the (8×2) and (4×1) phases obtained from AIMD for two different excitation strengths (corresponding to the green and red solid lines lhs)

We mention that the energetics of multiple excitations composed of several single excitations cannot simply be decomposed in the excitation of the latter. As an example we show in Fig. 6 (lhs) the $H_{1,4,5,8}; L_{1,4,5,8}$ excitation, the PES of which obviously is not given by the sum of the $H_{1,8}; L_{1,8}$ and $H_{4,5}; L_{4,5}$ PESs.

The above analysis shows that in particular emptying/occupying the In-related valence/conduction states at the SBZ boundary/center is likely to drive the phase transition. The corresponding surface bands, highlighted yellow and orange/green and blue, respectively, in Fig. 5, are known to couple strongly to the two soft In-localized surface phonons that—in combination—describe the structural deformation characteristic for the $(8 \times 2) \longrightarrow (4 \times 1)$ phase transition [4, 7, 19, 20]: The In-chain shear phonon mode at about $18/19 \, \text{cm}^{-1}$ lower the L and L+1 bands below the Fermi energy, while the the hexagon rotary mode at $27 \, \text{cm}^{-1}$ lifts the H and H-1 states above the Fermi level [7, 21]. The corresponding band-structure energy changes rationalize the steep PESs calculated here for the $H_{1,8}; L_{4,5}$ and $H_{1,8}H-1_{1,8}; L_{4,5}L+1_{4,5}$ excitations.

The coupling of these phonons, which were also detected by Raman spectroscopy [22], to the corresponding electronic states can be understood in terms of In-In bond formation: The L_4 and L_5 orbital characters are shown in Fig. 7. These states correspond to an In-In bond across parallel In-In zigzag chains. Electronic occupation of these bonds results in forces that exert an attractive interaction between the respective In atoms and excite the In-chain shear phonon mode.

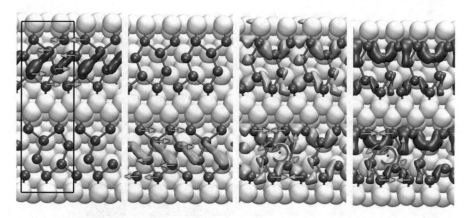

Fig. 7 Atomic structure of the In/Si(111)(8×2) surface. Dark/light balls indicate In/Si atoms. Isosurfaces indicates the orbital character of L_4, L_5, H_1, and H_8 (color coding corresponds to band structure in Fig. 5a). Arrows indicate the eigenvectors of the shear and rotary phonon modes excited upon optical excitation

Correspondingly, H_1 and H_8 are related to In hexagon bonds as well as bonds between the outermost In chain atoms, cf. Fig. 7. Emptying these bonds will prolong the corresponding In-In distances and excite the rotary phonon. The combined action of rotary and shear phonons, very effectively exited by the $H_{1,8}$;$L_{4,5}$ electronic configuration, transforms the insulating (8×2) CDW into the metallic (4×1) phase [4, 7, 19, 20]. The time scale for an optically driven phase transition can thus be expected to be set by the period of exactly these phonon modes which are known to facilitate the (8×2) ⟶ (4×1) phase change.

After having identified the driving force and the atomistic mechanism for the CDW melting, we address in a third step the time dynamics of the corresponding structural modification. For this purpose, energy conserving AIMD simulations were performed in adiabatic approximation, starting from two different electronically excited-state configurations. Specifically, we start from the $H_{1,8}$;$L_{4,5}$ (altogether 0.5 electron excited) as well as the $H_{1,8}$H-$1_{1,8}$;$L_{4,5}$L+$1_{4,5}$ configuration (altogether 1 electron excited), corresponding to the solid green and red PES shown in Fig. 6 (lhs), respectively. The calculated time-dependent structural deviation between the (8×2) and (4×1) phases is shown in Fig. 6 (rhs). Corresponding to the larger gradient of the PES excitation involving one electron compared to the half-electron excitation, the phase change in the former case occurs nearly twice as fast. In fact, in case of the one-electron excitation the CDW melting occurs with a time constant of 350 fs. This is about one quarter of the periods of the equilibrium rotational and shear phonon modes discussed above, $T_{rot} = 1.2$ ps and $T_{shear} = 1.8$ ps [7], respectively.

The AIMD shows that the energy of the excited electron configuration is at first primarily transformed to the In structural degrees of freedom, exciting the shear and rotary phonon modes. However, the velocity components soon equilibrate, first

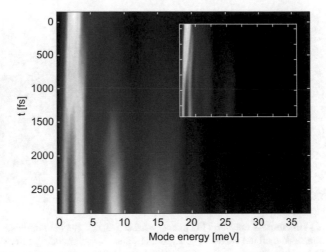

Fig. 8 Calculated distribution of the kinetic energy among the surface phonon modes of the In/Si(111) surface during the $(8\times2) \longrightarrow (4\times1)$ phase transition. Only In related surface phonon modes are considered in the inset

among all the In chain atoms and secondly among In and Si substrate atoms. About 20% of the kinetic energy is dissipated into the Si substrate after 1 ps. This can be seen in Fig. 8, where we show the distribution of the kinetic energy among the surface phonon modes of the In/Si(111) surface. The system gets trapped in the metastable (4×1) phase and cannot revert to the (8×2) ground state.

The present calculations clearly show a non-thermal melting of the CDW. A very recent macroscopic theory analysis for bulk material [23] suggests that non-thermal melting is restricted to materials having an anomalous phase diagram like ice characterized by $dT_C/dp < 0$, where T_C is the melting temperature and p is pressure. Interestingly, this condition is fulfilled for the present nanowire system: Density-functional calculations of the phase transition temperature T_C in dependence on the substrate lattice constant show indeed the respective anomaly [24]. Also the number of electrons that need to be excited for the phase transition to occur, about 5% of the In chain valence electrons, is similar to various bulk systems, see discussion in [23].

4 Conclusions

In summary, first-principles calculations were performed in order to rationalize the optically induced insulator-metal transition of the Si(111) substrate-stabilized In nanowire array. (1) It is found that the In-related surface states within the Si fundamental gap are long-lived with respect to electron-electron scattering. (2) Some photo-excited configurations involving these states, in particular electrons in empty zone-center surface states as well as holes in the uppermost In states at the

Brillouin zone boundary give rise to a potential energy surface that has its minimum at the metallic phase and allows for a barrierfree transition. (3) The energy gain realized upon relaxing these electronic states excites directly the shear and rotary phonon modes that transform between the insulating and metallic nanowire phases. (4) The phase transition may be completed within less than one picosecond. Thereby stronger electronic excitations result in shorter transition times.

Acknowledgements The Deutsche Forschungsgemeinschaft is acknowledged for financial support (FOR1700 and FOR1405). We thank the Paderborn Center for Parallel Computing (PC^2) and the Höchstleistungs-Rechenzentrum Stuttgart (HLRS) for grants of high-performance computer time. Images of molecular structures were prepared using VMD [25] and the tachyon renderer.

References

1. P.M. Chaikin, T.C. Lubensky, *Principles of Condensed Matter Physics* (Cambridge University Press, Cambridge, 2000)
2. J.J. Lander, J. Morrison, J. Appl. Phys. **36**, 1706 (1965)
3. P.C. Snijders, H.H. Weitering, Rev. Mod. Phys. **82**(1), 307 (2010)
4. C. Gonzalez, F. Flores, J. Ortega, Phys. Rev. Lett. **96**, 136101 (2006)
5. A.A. Stekolnikov, K. Seino, F. Bechstedt, S. Wippermann, W.G. Schmidt, A. Calzolari, M. Buongiorno Nardelli, Phys. Rev. Lett. **98**, 026105 (2007)
6. S. Chandola, K. Hinrichs, M. Gensch, N. Esser, S. Wippermann, W.G. Schmidt, F. Bechstedt, K. Fleischer, J.F. McGilp, Phys. Rev. Lett. **102**(22), 226805 (2009)
7. S. Wippermann, W.G. Schmidt, Phys. Rev. Lett. **105**, 126102 (2010)
8. U. Gerstmann, N.J. Vollmers, A. Lücke, M. Babilon, W.G. Schmidt, Phys. Rev. B **89**, 165431 (2014)
9. S.V. Ryjkov, T. Nagao, V.G. Lifshits, S. Hasegawa, Surf. Sci. **488**, 15 (2001)
10. H.W. Yeom, D.M. Oh, S. Wippermann, W.G. Schmidt, ACS Nano **10**, 410 (2016)
11. W.G. Schmidt, M. Babilon, C. Thierfelder, S. Sanna, S. Wippermann, Phys. Rev. B **84**, 115416 (2011)
12. S. Wall, B. Krenzer, S. Wippermann, S. Sanna, F. Klasing, A. Hanisch-Blicharski, M. Kammler, W.G. Schmidt, M.H. von Hoegen, Phys. Rev. Lett. **109**, 186101 (2012)
13. T. Frigge, B. Hafke, T. Witte, B. Krenzer, C. Streubühr, A. Samad Syed, V. Miksic Trintl, P. Avigo, I. Zhou, M. Ligges, D. von der Linde, U. Bovensiepen, M. Horn-von Hoegen, S. Wippermann, A. Lücke, S. Sanna, U. Gerstmann, W.G. Schmidt, Nature **544**, 207 (2017). https://doi.org/10.1038/nature21432
14. P. Giannozzi, S. Baroni, N. Bonini, M. Calandra, R. Car, C. Cavazzoni, D. Ceresoli, G.L. Chiarotti, M. Cococcioni, I. Dabo, A. Dal Corso, S.D. Gironcoli, S. Fabris, G. Fratesi, R. Gebauer, U. Gerstmann, C. Gougoussis, A. Kokalj, M. Lazzeri, L. Martin-Samos, N. Marzari, F. Mauri, R. Mazzarello, S. Paolini, A. Pasquarello, L. Paulatto, C. Sbraccia, S. Scandolo, G. Sclauzero, A.P. Seitsonen, A. Smogunov, P. Umari, R.M. Wentzcovitch, J. Phys. Condens. Matter **21**(39), 395502 (2009)
15. H.J. Monkhorst, J.D. Pack, Phys. Rev. B **13**(12), 5188 (1976)
16. P.H. Dederichs, S. Blügel, R. Zeller, H. Akai, Phys. Rev. Lett. **53**, 2512 (1984)
17. W.G. Aulbur, L. Jonsson, J.W. Wilkins, Quasiparticle calculations in solids, in *Solid State Physics: Advances in Research and Applications*, vol. 54 (Academic, San Diego, 2000), p. 1
18. P. Echenique, J. Pitarke, E. Chulkov, A. Rubio, Chem. Phys. **251**(1) (2000)
19. S. Riikonen, A. Ayuela, D. Sanchez-Portal, Surf. Sci. **600**, 3821 (2006)

20. C. González, J. Guo, J. Ortega, F. Flores, H.H. Weitering, Phys. Rev. Lett. **102**(11), 115501 (2009)
21. W.G. Schmidt, S. Wippermann, S. Sanna, M. Babilon, N.J. Vollmers, U. Gerstmann, Phys. Stat. Sol. (b) **249**, 343 (2012)
22. K. Fleischer, S. Chandola, N. Esser, W. Richter, J.F. McGilp, Phys. Rev. B **76**, 205406 (2007)
23. H. Hu, H. Ding, F. Liu, Sci. Rep. **5**, 8212 (2015)
24. M. Babilon, Phasenübergangsstabilität von Substrat-gestützten Indiumnanodrähten. Master's thesis, Universität Paderborn, 2013
25. W. Humphrey, A. Dalke, K. Schulten, J. Mol. Graph. **14**(1), 33 (1996)

Dynamic Material Parameters in Molecular Dynamics and Hydrodynamic Simulations on Ultrashort-Pulse Laser Ablation of Aluminum

Stefan Scharring, Marco Patrizio, Hans-Albert Eckel, Johannes Roth, and Mikhail Povarnitsyn

Abstract Molecular dynamics reveals a detailed insight into the material processes. Among various available codes, IMD features an implementation of the two-temperature model for laser-matter interaction. Reliable simulations, however, are restricted to the femtosecond regime, since a constant absorptivity is assumed. For picosecond pulses, changes of the dielectric permittivity ϵ and the electron thermal conductivity κ_e due to temperature, density and mean charge have to be considered. Therefore, IMD algorithms were modified for the dynamic recalculation of ϵ and κ_e for every timestep following the corresponding implementation in the hydrodynamic code Polly-2T. The usage of dynamic permittivity yields an enhanced absorptivity during the pulse leading to greater material heating. In contrast, increasing conductivity induces material cooling which in turn decreases absorptivity and heating resulting in a higher ablation threshold. This underlines the importance of a dynamic model for ϵ and κ_e with longer pulses which is commonly often neglected. Summarizing all simulations with respect to absorbed laser fluence, ablation depths in Polly-2T are two times higher than in IMD. This can be ascribed to the higher spallation strength in IMD stemming from the material-specific potential deviating from the equations of state used in Polly-2T.

S. Scharring • H.-A. Eckel
German Aerospace Center (DLR), Institute of Technical Physics, Stuttgart, Germany
e-mail: stefan.scharring@dlr.de; hans-albert.eckel@dlr.de

M. Patrizio
Technical University of Darmstadt, Institute for Nuclear Physics, Darmstadt, Germany
e-mail: m.patrizio@gsi.de

J. Roth (✉)
Institute of Functional Materials and Quantum Technologies (FMQ), University of Stuttgart, Stuttgart, Germany
e-mail: johannes.roth@fmq.uni-stuttgart.de

M. Povarnitsyn
Research Center for High Energy Density Physics, Joint Institute for High Temperatures (JIHT), Russian Academy of Sciences (RAS), Moscow, Russia
e-mail: povar@ihed.ras.ru

© Springer International Publishing AG 2018
W.E. Nagel et al. (eds.), *High Performance Computing in Science and Engineering '17*, https://doi.org/10.1007/978-3-319-68394-2_10

1 Introduction

Among the variety of available open-source codes for the simulation of laser ablation, Polly-2T from the Joint Institute of High Temperatures at the Russian Academy
of Sciences, Moscow, and IMD from the Institute of Functional Materials and
Quantum Technologies at the University of Stuttgart are very different simulation
tools which can partially attributed to their strongly varying numerical approaches,
namely hydrodynamics (HD) in the case of Polly-2T and molecular dynamics (MD)
for IMD. A first comparison between both programs was given in [1] revealing
similar results for a restricted parameter range as well as large discrepancies beyond
this range.

Some differences between both codes, however, stem from the unequal implementation of the physics of laser-matter interaction in both programs. Numerical
work with IMD was mainly focusing on so-called cold laser ablation with fs laser
pulses where material changes during laser irradiation are less relevant than with
long pulses. With increasing pulse length, however, the interaction of matter and
laser beam becomes more and more apparent, since optical permittivity, electron
thermal conductivity as well as thermal coupling from the electron gas to the ion
lattice depend both on density and temperature. Since large gradients of ϱ, T_e and T_i
might occur during a laser pulse, absorption of laser energy is subject to significant
changes due to the laser-induced temporal and spatial fluctuations of temperature
and density.

Models that cover a wide parameter range of parameter range of ϱ, T_e and T_i are
implemented in Polly-2T take into account for these effects. Their implementation
into IMD is the scope of this paper in order to enable a better comparison of
simulation results from both codes.

2 Theoretical Background

2.1 Two-Temperature Model

Energy deposition by an ultrafast laser pulse into a metal target can be described by
the two-temperature Model (TTM) [2] by

$$c_e (T_e) \frac{\partial T_e}{\partial t} = \nabla [\kappa_e (T_e) \nabla T_e] - \gamma_{ei} (T_e - T_i) + S (\mathbf{r}, t) \qquad (1)$$

$$c_i (T_i) \frac{\partial T_i}{\partial t} = \nabla [\kappa_i (T_i) \nabla T_i] + \gamma_{ei} (T_e - T_i) . \qquad (2)$$

where the indices e and i denote the electron and ionic subsystem, resp., c is the
specific heat capacity, T the temperature, κ represents the thermal conductivity, γ_{ei}

is the electron-phonon coupling parameter, and $S(\mathbf{r}, t)$ is the spatial and temporal distribution of the energy density. For ultrashort laser pulses, κ_i can be neglected [3] as it is done in Polly-2T and IMD.

2.2 Wide-Range Models

In the wide-range models treated in this paper it is assumed that the transition of the material from the metal phase into the plasma phase occurs in the vicinity of the Fermi temperature T_F. Under this assumption, material transport properties can be written as an interpolation between the behavior in the metal phase and the behavior in the plasma phase as sketched in the following subsections [4, 5].

2.2.1 Optical Transport Properties

The electric permittivity ϵ of the target material is described by a temperature-dependent interpolation of ϵ_{met} in the metal phase and ϵ_{pl} in the plasma phase

$$\epsilon = \epsilon_{\mathrm{pl}} + \left(\epsilon_{\mathrm{met}} - \epsilon_{\mathrm{pl}}\right) \cdot e^{-A_{\mathrm{opt}} \cdot T_e / T_F} \tag{3}$$

where A_{opt} is an empirical parameter adjusted to match experimental values. ϵ_{met} is composed of the band-to-band contribution ϵ_{bb} which is taken from tabulated data and an intraband term

$$\epsilon_{\mathrm{met}}\left(\omega_L, \varrho, T_i, T_e\right) = \epsilon_{\mathrm{bb}} + 1 - \frac{n_e}{n_{\mathrm{cr}}\left(1 + i\, \nu_{\mathrm{eff,opt}} / \omega_L\right)} \tag{4}$$

where ω_L is the frequency of the laser, ϱ is the density of the material, n_{cr} is the critical concentration of electrons and the effective collision frequency $\nu_{\mathrm{eff,opt}}$ as derived in detail in [4].

In the plasma phase, for temperatures far above T_F, ϵ_{pl} can be calculated as

$$\epsilon_{\mathrm{pl}}\left(\omega_L, \varrho, T_e\right) = 1 - \frac{n_e}{n_{\mathrm{cr}}} K\left(\frac{\nu_{\mathrm{pl}}}{\omega_L}\right) \tag{5}$$

where $K\left(\frac{\nu_{\mathrm{pl}}}{\omega_L}\right)$ is an empirical function.

2.2.2 Heat Transport Properties

Similar to Eq. (3), the thermal conductivity κ_e of the electron gas can be written as an interpolation from the metal phase to the plasma phase according to

$$\kappa_e = \kappa_{\mathrm{pl}} + \left(\kappa_{\mathrm{met}} - \kappa_{\mathrm{pl}}\right) \cdot e^{-A_{\mathrm{he}} \cdot T_e / T_F} \tag{6}$$

where A_{he} is an empirical parameter adjusted to match experimental values. In the metal phase κ_{met} can be calculated according to the Drude formalism as

$$\kappa_{met}(\varrho, T_i, T_e) = \frac{\pi^2 k_B^2 n_e}{3\, m_e \nu_{eff,he}(T_i, T_e)} T_e \tag{7}$$

where k_B is Boltzmann's constant, n_e is the electron concentration, m_e the electron mass and $\nu_{eff,he}$ the effective collision frequency as shown in [5].

For the plasma phase κ_{pl} can be calculated as

$$\kappa_{pl} = \frac{16\sqrt{2}\, k_B\, (k_B T_e)^{5/2}}{\pi^{3/2} Z\, e^4 \sqrt{m_e}\, \Lambda} \tag{8}$$

where Z is the mean charge of the ions, e is the elementary charge and Λ is the Coulomb logarithm.

Accordingly, thermal coupling from the electron gas to the ion lattice can be described in a wide-range approach by

$$\gamma_{ei}(\varrho, T_i, T_e) = \frac{3\, k_B\, m_e}{m_i} n_e \nu_{eff,ei}(T_i, T_e) \tag{9}$$

where $\nu_{eff,ei}$ represents the corresponding collision frequency.

3 Numerical Codes

3.1 Hydrodynamic Simulations

3.1.1 Target Material

The thermodynamic material properties are treated in Polly-2T using semi-empirical two-temperature multiphase equations of state (EOS) comprising the Thomas-Fermi description of thermal effects of the electron gas. The EOS provide for tabulated functions for pressure $P_e(\varrho, T_e)$, $P_i(\varrho, T_i)$ and specific energy $e_e(\varrho, T_e)$, $e_i(\varrho, T_i)$ of both subsystems which allow for the solution of the TTM equations. From the EOS, the equilibrium mean charge of ions $\langle Z \rangle$ can be derived as well.

The phase diagram of aluminum corresponding to the EOS is shown in Fig. 1.

3.1.2 Two-Temperature Model

A sound description of the specific simulation assumptions in Polly-2T is given in [4]. Laser-matter interaction is described here by the TTM in a single-fluid 1D

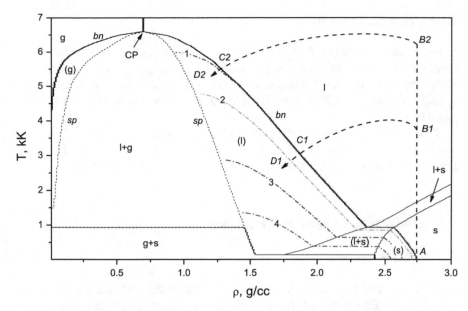

Fig. 1 Phase diagram of aluminum according to the EOS used in Polly-2T, taken from [6]. *Abbreviations:* sp: spinodal; bn: binodal; g: stable gas; l: stable liquid; s: stable solid; l + s: stable melting; l + g: liquid-gas mixture; g + s: sublimation zone; (g): metastable gas; (l): metastable liquid; (l + s): metastable melting; (s): metastable solid; CP: critical point. Dash-dot lines: isobars; 1: 0 GPa; 2: −1 GPa; 3: −3 GPa; 4: −5 GPa

Lagrangian form yielding a modification of Eqs. (1) and (2) by

$$\frac{\partial e_e}{\partial t} + P_e \frac{\partial v}{\partial m} = \frac{\partial}{\partial m}\left[\varrho \kappa_e \frac{\partial T_e}{\partial m}\right] - \gamma_{ei}\left(T_e - T_i\right)/\varrho + S\left(x, t\right)/\varrho \tag{10}$$

$$\frac{\partial e_i}{\partial t} + P_i \frac{\partial v}{\partial m} = +\gamma_{ei}\left(T_e - T_i\right)/\varrho \tag{11}$$

where e_e and e_i denote the specific energy of the electrons and ions, resp., P is the pressure, m mass, ϱ density, v velocity, and, in contrast to [4], radiation transport phenomena are neglected here. Conservation of mass and energy are granted using

$$\frac{\partial\left(1/\varrho\right)}{\partial t} - \frac{\partial v}{\partial m} = 0, \tag{12}$$

$$\frac{\partial v}{\partial t} + \frac{\partial\left(P_i + P_e\right)}{\partial m} = 0. \tag{13}$$

3.1.3 Transport Properties

In the present version of Polly-2T, which is accessible online [7] for web-based simulations as well, the wide-range models for ϵ, κ_e, and γ_{ei} are implemented. With respect to thermal transport properties, this means that the dependencies $\kappa_e(\varrho, T_i, T_e)$ and $\gamma_{ei}(\varrho, T_i, T_e)$, cf. Eqs. (6) and (9) are considered in the solver for Eqs. (10)–(13).

With respect to the permittivity ϵ, the Helmholtz equation is solved for the source term $S(x, t)$ in Eq. (10) considering the spatial and temporal fluctuations of the permittivity due to the dependencies given by $\epsilon(\omega_L, \varrho, T_i, T_e)$, cf. Eq. (3).

3.2 Molecular Dynamics Simulations

3.2.1 Target Material

Particle interactions in IMD were represented using the embedded atom model (EAM) from Ercolessi and Adams [8]. The potential takes into account the interactions between an atom and its closest neighbors while introducing a cut-off distance for longer ranges, thus creating a multi-body problem that can only be solved numerically. For any given atom inside such a system the potential can be described as

$$V_i = \frac{1}{2} \sum_{i,j,i \neq j} \phi(r_{ij}) + F\left[\sum_j \varrho_e(r_{ij})\right] \tag{14}$$

where the EAM potential ϕ contains the interaction between the cores of atoms i and j and the EAM embedding function F describes the interaction of the electron hull of atom i with all its neighbors while influenced by the local electron density $\varrho_e(r_{ij})$.

The phase diagram of aluminum corresponding to the EAM potential is shown in Fig. 2.

3.2.2 Two-Temperature Model

Whereas heat transport in the ion lattice can be described by the multi-particle interactions using the EAM potential, a supplementary system had to be added to the MD simulations to take into account energy absorption and propagation inside the electron gas. Hence, solving Eq. (1) with a Finite-Difference (FD) scheme, hybrid simulations are carried out in IMD, where electron-phonon coupling from the electronic FD subsystem into the MD core is realized by a dynamic coupling

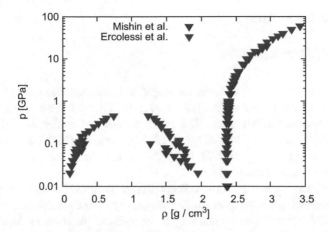

Fig. 2 Phase diagram of aluminum according to the EAM potential data from Ercolessi [8], and Mishin [9], resp.

parameter ξ following

$$m_i \frac{\partial^2 \mathbf{r}_i}{\partial t^2} = -\nabla V_i + \xi \cdot m_i \mathbf{v}_{T,i} \tag{15}$$

with

$$\xi = \frac{\frac{1}{n} \sum_{k=1}^{n} \gamma_{ei} V_N (T_{e,k} - T)}{\sum_j m_j (v_{T,j})^2} \tag{16}$$

where $v_{T,i} = v_i - v_{CMS}$ denotes the thermal velocity of the ith atom taking into account for the velocity of the center of mass (CMS) of the corresponding FD cell.

3.2.3 Transport Properties

Since applications of IMD have mainly been focused on cold ablation, i.e., laser pulses in the femtosecond regime, in the open-source version of IMD dynamic material parameters were only implemented partially, in contrast to Polly-2T. The electron heat capacity is modeled using $c_e = T_e \cdot 135 \, \text{J}/(\text{m}^3 \, \text{K}^2)$ which is rather good comparable to Polly-2T where $P_e (\varrho, T_e)$ and $e_e (\varrho, T_e)$, given by the EOS data, yield $c_e \approx T_e \cdot 100 \, \text{J}/(\text{m}^3 \, \text{K}^2)$. Electronic heat conductivity and coupling coefficient, however, are treated as constants in IMD using $\kappa_e = 235 \, \text{J}/(\text{s m K})$ and $\gamma_{ei} = 5.69 \cdot 10^{17} \, \text{J}/(\text{s m}^3 \, \text{K})$, resp., for aluminum.

In the original version of IMD, absorption of laser energy is described by the Beer-Lambert law requiring the amount of absorbed energy as input for the MD simulation.

4 Numerical Work

4.1 Code Developments

The focus of our work on IMD was the stepwise implementation of the wide-range models for ϵ $(\omega_L, \varrho, T_i, T_e)$, κ_e (ϱ, T_i, T_e) and γ_{ei} (ϱ, T_i, T_e) described in Sect. 2.2. For a better comparison of MD and HD results, wide-range models of Polly-2T were switched off if not active in IMD, i.e., the upgrades of IMD were accompanied by corresponding downgrades of Polly-2T. A detailed description of the code developments can be found in [10].

In the first step, denoted as cold reflectivity, we assumed fixed material constants and described the optical properties of aluminum by an optical absorption coefficient of $\alpha = 1.267 \cdot 10^8 \, \mathrm{cm}^{-1}$ corresponding to an optical penetration depth of $l_\alpha = \alpha^{-1} = 7.89 \, \mathrm{nm}$ and a surface reflectivity of $R = 0.94331$. Hence, in our simulations with this version of IMD, the absorbed fluence $\Phi_{\mathrm{abs}} = R \cdot \Phi$ was taken as input parameter for the absorbed laser fluence and α was used for energy allocation following Lambert-Beer. For convenience, an interface was implemented in IMD for the direct input of Φ calculating R and Φ_{abs} depending on incidence angle and polarization of the laser beam. Accordingly, the corresponding Polly-2T downgrade simulations employed a complex refractive index of $n = 1.74 + 10.72 \cdot i$ as input parameter. Moreover, in those runs electron thermal conductivity and electron-phonon coupling coefficient were set fixed to $\kappa_e = 235 \, \mathrm{W/K \, m}$ and $\gamma_{ei} = 5.69 \cdot 10^{17} \, \mathrm{W/(K \, m^3)}$, resp.

In the second step, the wide-range model for ϵ was implemented in IMD and re-activated in Polly-2T. Whereas in Polly-2T absorption of laser energy is implemented using the Helmholtz equation, α was calculated for each FD cell in IMD for every time step allowing for a dynamic allocation of absorbed energy following Lambert-Beer.

Finally, the wide-range model for κ_e has been implemented in IMD and re-activated in Polly-2T in addition to the wide-range model for ϵ. The implementation of the wide-range model for γ_{ei} in IMD, however, has not been finished successfully yet at present.

4.2 Simulation Setup

For the incident laser pulse, the temporal course of the intensity is given by

$$I(t) = I_0 \cdot \exp\left[-0.5 \cdot (t/\sigma_t)^2\right] \tag{17}$$

with

$$I_0 = \Phi / (2.507 \cdot \sigma_t \cdot \cos\vartheta) \tag{18}$$

where Φ is the incident laser fluence and ϑ is the incidence angle. In the following, we denote the pulse length τ the full width half maximum (FWHM) with $\tau\acute{R} = \acute{R}\tau_{FWHM} = 2\sqrt{2\ln 2}\cdot\sigma_t$. Simulations were started approximately at the point in time t_{in} before the laser pulse where $I(t_{in}) = 0.1$ W/cm^2. With respect to our related work on laser micropropulsion [11], $\lambda = 1064$ nm as laser wavelength and, for the sake of simplicity, an incidence angle of $\vartheta = 0°$ with linear polarized light was chosen. Pulse lengths of $\tau \approx 50$ fs, 500 ps, and 5 ps, resp., were chosen in combination with fluences of $\Phi \approx 0.19, 0.37, 0.74, 1.49$, and 2.49 J/cm^2, resp.

Since the dimension of the simulation cells in Polly-2T is in the nanometer range and therefore rather large, it is not disadvantageous with respect to the computational effort to create bulk material with a certain thickness, which was chosen as 1 mm here. The shock wave stemming from the ablation event is supposed to travel with the corresponding speed of sound and will have no impact on the rear side of the target within the simulation time t_{sim}.

In contrast to HD simulations, computational time scales linearly with the number of particles in IMD. Therefore, samples with a thickness of 650 nm have been created which is sufficient for a time span of ≈ 100 ps after the laser pulses for shock wave to travel to the rear side of the sample [1]. 20 nm have been chosen as lateral extension of the sample using periodic boundary conditions [12].

5   Results

5.1   Energy Conservation

Simulations results with the original version of IMD are shown in Fig. 3. The laser pulse energy is absorbed by the electron gas and is coupled into the ion lattice.

A comparison of the additional system energy with the absorbed laser fluence, however, shows, that in several cases the target appears to acquire more energy than delivered from the laser pulse, cf. Fig. 4. Even long after the laser pulse, the energy in the electron subsystem still increases. This unphysical behavior was already observed in [1] and can be ascribed to numerical errors in the FD computation describing the electron system in IMD since this phenomenon is especially pronounced for large laser fluences in conjunction with very short laser pulses. Code upgrades have been implemented at the FMQ yielding a significant improvement for low fluences, cf. Fig. 4. For greater fluences, however, the system energy is still divergent, albeit at a lower level, whereas for very high fluences, i.e., $\Phi_{in} = 2.97$ J/cm^2 in the most cases, simulations were aborted after computation errors.

Analysis of the system energy comprising the wide-range models for ϵ and κ_e reveals that an absorbed fluence of roughly $\Phi_{abs} \approx 0.1$ J/cm^2 seems to be the upper limit for reasonable system behavior in IMD simulations, cf. Fig. 5. Correspondingly, simulations with $\Phi_{abs} \geq 0.1$ J/cm^2 are discarded in the following.

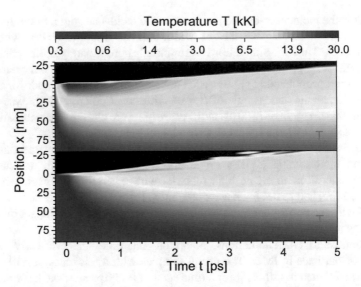

Fig. 3 Temporal and spatial evolution of the electron temperature T_e and lattice temperature T_i, results from IMD with constant ϵ, κ_e. Target material: aluminum, $\tau = 50\,\text{fs}$, $\Phi = 0.74\,\text{J/cm}^2$

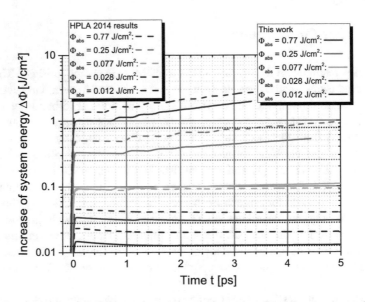

Fig. 4 Temporal course of additional system energy for various laser fluences at $\tau = 50\,\text{fs}$. The *absorbed* laser fluence Φ_{abs} which is an input parameter of IMD is compared with the actual increase of system energy due to the laser pulse. Earlier results refer to the simulations shown at the HPLA 2014 [1] whereas results from this work denote simulations with the present status of IMD prior the upgrades treated in this paper

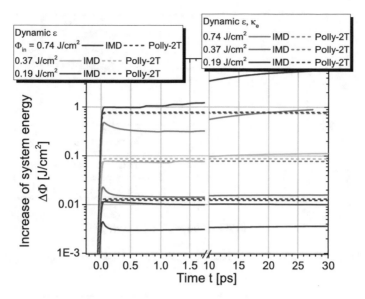

Fig. 5 Temporal course of the system energy during and after laser absorption. For higher fluences, a significantly large increase is found with IMD simulations. Results from hydrodynamic simulations with Polly-2T are given for comparison (dashed lines)

In any case, the course of the system energy shows an overshoot after the laser pulse, followed by a minimum and a slow drift upwards. This drift is rather pronounced for high fluences.

With Polly-2T which has much more coarse spatial resolution such large energy drifts have not been observed in the given laser parameter range but at fluences being one order of magnitude higher and beyond.

5.2 Permittivity

For very short laser pulses with moderate intensities, absorption of laser energy in the cold reflectivity case is comparable to the findings with implementation of the wide-range model for the permittivity ϵ, cf. Fig. 6. This might be associated with cold ablation and deduced to the thermal confinement of the system due to the relaxation time τ_e of the electron gas which prevents a significant change of the target properties during the laser pulse.

The picture turns, however, for higher fluences and/or pulse lengths which can be found most clearly in the results from HD simulations with Polly-2T, cf. Fig. 7. In this case, a deeper and faster heat dissipation in the electron gas can be detected comparing Fig. 3 with Fig. 8. This can be attributed to the temperature dependency

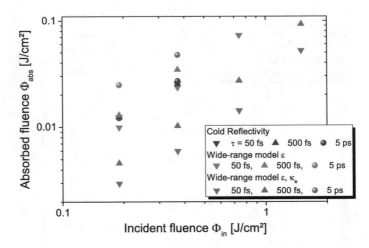

Fig. 6 Absorbed fluence vs. incident fluence, results from IMD for various upgrade steps

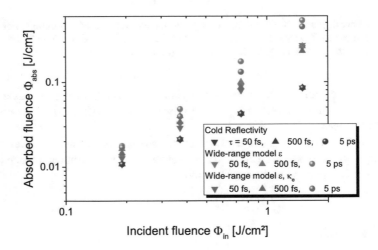

Fig. 7 Absorbed fluence vs. incident fluence, results from Polly-2T for various downgrade steps

of ϵ that is apparent in the strong changes of the refractive index which yield a higher optical penetration depth in the heat affected zone, cf. Fig. 9.

At the transition from target to vacuum, with $l_\alpha \approx 36$ nm a rather high value for the penetration depth is found. This can be ascribed to the relatively large size of the FD-cells. They comprise 4 MD-cells containing both vacuum and the surface layers of the solid target. Hence, a rather low density is calculated for the FD-cell leading to a low absorption coefficient in the respective cell following the wide-range model for the permittivity. On the whole, however, this can be regarded as a residual error with negligible impact on the simulation results which applies as well

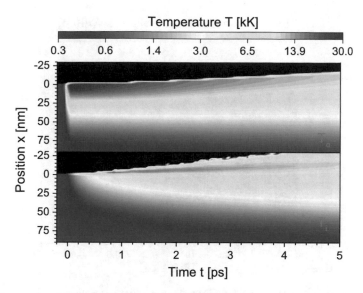

Fig. 8 Temporal and spatial evolution of the electron temperature T_e, results from IMD with $\epsilon = \epsilon(\varrho, T_e, T_i)$ according to the wide range model. Target material: aluminum, $\tau = 50$ fs, $\Phi = 0.74$ J/cm^2

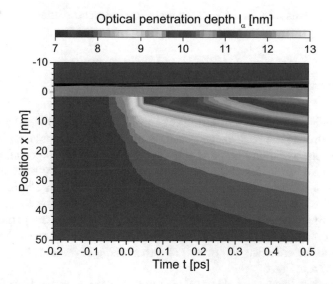

Fig. 9 Temporal and spatial evolution of the absorption length l_α, results from IMD with $\epsilon = \epsilon(\varrho, T_e, T_i)$ according to the wide range model. Target material: aluminum, $\tau = 50$ fs, $\Phi = 0.74$ J/cm^2

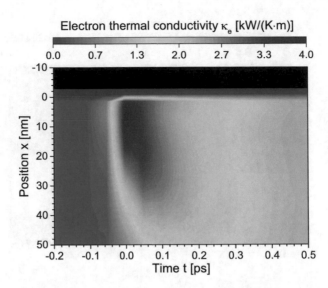

Fig. 10 Temporal and spatial evolution of the electronic heat conductivity κ_e, results from IMD with $\epsilon = \epsilon\,(\varrho, T_e, T_i)$ and $\kappa = \kappa\,(\varrho, T_e, T_i)$ according to the wide range model. Target material: aluminum, $\tau = 50\,\text{fs}$, $\Phi = 0.74\,\text{J/cm}^2$

for the calculations with the wide-range model on the electron thermal conductivity, cf. Fig. 10.

5.3   Electron Thermal Conductivity

When the electron gas is heated during laser pulse, the electron thermal conductivity raises according to the wide-range model. For the chosen example, an increase by the order of one magnitude is found at the surface layer of the target, cf. Fig. 10.

Due to the greater thermal conductivity κ_e the laser energy is dissipated into deeper layers of the material, cf. Fig. 11 in comparison with Fig. 8. Therefore, heating of the material surface as well as material expansion are less pronounced than under the assumption of constant κ_e. In turn, changes in the permittivity are rather low compared to the previous case where the wide-range model was only implemented for ϵ. Hence, the overall absorbed energy is significantly reduced as can be seen from Fig. 6.

Whereas this decrease of absorbed energy due to the dynamic behavior of κ_e is very pronounced in IMD simulations, this trend is much weaker for the results from HD simulations with Polly-2T, cf. Fig. 7. The reason for this large discrepancy is not quite clear. Since κ_e directly affects the FD scheme for the electron gas, the above-mentioned problems with the unphysical increase of energy in the electron subsystem for high fluences might have a deeper reason that is mirrored here as well.

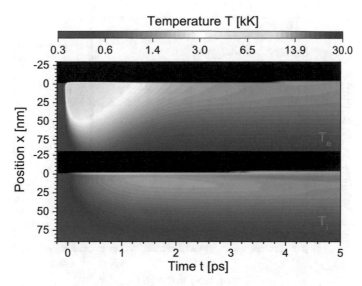

Fig. 11 Temporal and spatial evolution of the electron temperature T_e, results from IMD with $\epsilon = \epsilon\,(\varrho, T_e, T_i)$ and $\kappa = \kappa\,(\varrho, T_e, T_i)$ according to the wide range model. Target material: aluminum, $\tau = 50\,\text{fs}$, $\Phi = 0.74\,\text{J/cm}^2$

5.4 Ablation

It is evident that large deviations in absorbed energy, as shown in Figs. 6 and 7, yield a different behavior of the target material. Therefore, it is meaningful to have a look at the laser-induced processes with respect to the absorbed fluence, irrespective of the incident fluence, cf. Fig. 12. Whereas it is straightforward to determine Φ_{abs} for the Polly-2T simulations, for IMD Φ_{abs} has to be assessed from its temporal course, cf. Fig. 5. As an approximation, the point in time where the minimum of $\Delta\Phi$ occurs has been taken in order to minimize the impact of overshoot and drift. For the 5 ps pulses, however, drift is apparently present already during the laser pulse, hence, $-t_{\text{in}}$ has been chosen instead.

It can be seen that in the most cases more material is ablated in the hydrodynamic simulations than in the MD case. Correspondingly, the ablation threshold in Polly-2T is lower than in IMD, as can be seen as well from the data in Table 1. Nevertheless, all simulation data are roughly in the range of experimental results.

When the amount of ablated mass in HD and MD simulations is compared, the different models for materials have to be taken into account. Apart from issues related to the wide-range model of ϵ, κ_e and γ_{ei}, underlying EOS in Polly-2T and EAM potential in IMD, resp., imply different spall strengths for the target material which is 2 GPa for Polly-2T and 8.7 GPa for IMD, where the latter value is taken from the EAM potential of Mishin [9] being rather similar the one of Ercolessi and Adams [8] used in the present MD simulations. The higher spall strength in IMD is

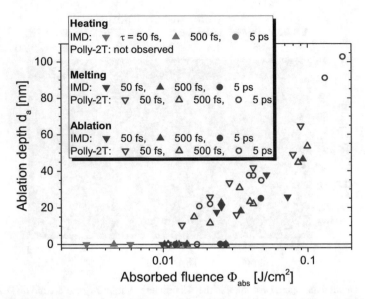

Fig. 12 Ablation depth d_a and laser-induced processes in dependence of the absorbed laser fluence Φ_{abs} for molecular dynamics simulations with IMD in comparison with results from hydrodynamic simulations with Polly-2T comprising all steps of implementation of the wide-range models of ϵ and κ_e

Table 1 Threshold Φ_0 for laser ablation of aluminum, results from hydrodynamic simulations with Polly-2T as well as IMD molecular dynamics simulations in comparison with own experimental results [16] and literature data

τ (ps)	λ (nm)	Φ_0 (J/cm^2)	Reference
0.05	1064	$0.3^c, < 0.2^\epsilon, 0.3^{\epsilon,\kappa_e}, 0.6^{\epsilon,\kappa_e,\gamma_{ei}}$	Polly-2T
0.05	1064	$0.2^c, 0.3^\epsilon, 0.7^{\epsilon,\kappa_e}$	IMD
0.1	1240	0.7	[13]
0.13	800	0.034	[14]
0.5	1064	$0.3^c, < 0.2^\epsilon, 0.3^{\epsilon,\kappa_e}, 0.4^{\epsilon,\kappa_e,\gamma_{ei}}$	Polly-2T
0.5	1064	$0.2^c, 0.2^\epsilon, 0.7^{\epsilon,\kappa_e}$	IMD
0.5	1054	≈ 0.05	[15]
5	1064	$0.3^c, < 0.2^\epsilon, 0.3^{\epsilon,\kappa_e}, 0.2^{\epsilon,\kappa_e,\gamma_{ei}}$	Polly-2T
5	1064	$0.3^c, 0.2^\epsilon, 0.3^{\epsilon,\kappa_e}$	IMD
10	1064	0.4	[16]

Indices refer to the level of implementation of the wide-range models (c: cold reflectivity)

presumably the reason for the typically lower mass removal and the higher ablation threshold in IMD compared with results from hydrodynamic simulations.

Ionization is taken into account in Polly-2T in the calculation of the cell's specific energy, but neglected in the EAM potential of IMD. This is supposed to have no significant impact on the ablated mass but rather on the velocity of the ablation jet where kinetic energy might be overestimated in IMD [1]. In the calculation of permittivity according to the wide-range model, however, ionization is implicitly taken into account for both IMD and Polly-2T being calculated from electron density via the electron temperature.

6 Conclusion and Outlook

Simulations of laser-matter interaction during an ultrashort laser pulse have been performed evaluating the usage of wide-range models for permittivity ϵ and electron thermal conductivity κ_e as strongly temperature-dependent material parameters.

Since the absorbed fluence mainly rules the laser-induced processes in the investigated cases for $\tau = 50$ and $500\,\text{fs}$, substitution of the wide-range models by the input of the effectively absorbed energy, as shown recently [1], can be a suitable workaround. For $\tau = 5\,\text{ps}$, i.e., longer pulse lengths in the range of the electron-phonon coupling time τ_i, however, the dynamic issue of laser-matter-*inter*action comes into play and this approach does not hold any more.

Future work with IMD comprising the wide-range models requires paralleliza-tion of the upgraded code. This is the main restriction up to now, whereas the standard version is already parallelized by default.

Taking into account for dynamic material properties for a wide range of temperatures and density in laser-matter-interaction, both codes a principally well-suited for the current work of our group on laser-ablative micropropulsion [11].

3D simulations of surface roughness under multiple spot ablation are needed there for the assessment of corresponding thrust noise. Such simulations are basically feasible in IMD, albeit at a much lower spatial scale, as well as with Polly-2T with a quasi-2D approach neglecting lateral side-effects [17].

However, the mentioned problems in IMD with higher fluences limit its practical use in this field of application since the anticipated working point for the future microthruster is in the range of at least 3 to 6 times of the ablation threshold [18].

For future work, implementation of laser-induced ablation comprising wide-range models presented here into a commercial finite-element method (FEM) software is planned.

References

1. S. Scharring, D.J. Foerster, H.-A. Eckel, J. Roth, M. Povarnitsyn, Open access tools for the simulation of ultrashort-pulse laser ablation, in *High Power Laser Ablation/Beamed Energy Propulsion* (2014). Available via DLR. http://elib.dlr.de/89090/
2. S.I. Anisimov, B.L. Kapeliovich, T.L. Perel'man, Electron emission from metal surfaces exposed to ultrashort laser pulses. Ann. Mat. Pura Appl. **39**, 375–377 (1974)
3. D. Baeuerle, *Laser Processing and Chemistry* (Springer, Berlin, 2000)
4. M.E. Povarnitsyn, N.E. Andreev, P.R. Levashov, K.V. Khishchenko, O.N. Rosmej, Dynamics of thin metal foils irradiated by moderate-contrast high-intensity laser beams. Phys. Plasmas **19**, 023110 (2012)

5. M.E. Povarnitsyn, N.E. Andreev, E.M. Apfelbaum, T.E. Itina, K.V. Khishchenko, O.F. Kostenko, P.R. Levashov, M.E. Veysman, A wide-range model for simulation of pump-probe experiments with metals. Appl. Surf. Sci. **258**, 9480–9483 (2012)

6. M.E. Povarnitsyn, K.V. Khishchenko, P.R. Levashov, Phase transitions in femtosecond laser ablation. Appl. Surf. Sci. **255**, 5120–5124 (2009)

7. Virtual Laser Laboratory (VLL). Joint Institute for High Temperatures (JIHT), Russian Academy of Sciences (RAS). (2012). http://vll.ihed.ras.ru/

8. F. Ercolessi, J.B. Adams, Interatomic potentials from first-principles calculations: the force-matching method. Europhys. Lett. **26**, 583–588 (1994)

9. Y. Mishin, D. Farkas, M.J. Mehl, D.A. Papaconstantopoulos, Interatomic potentials for monoatomic metals from experimental data and *ab initio* calculations. Phys. Rev. B **59**, 3393–3407 (1999)

10. M. Patrizio, Upgrade of the IMD software tool for the simulation of laser-matter interaction. Technical University of Darmstadt (2015). Available via DLR. http://elib.dlr.de/100164/

11. S. Scharring, S. Karg, R.-A. Lorbeer, N., Dahms, H.-A. Eckel, Low-noise thrust generation by laser-ablative micropropulsion, in *34th International Electric Propulsion Conference*, Kobe, Paper 2015-143 (2015). Available via DLR. http://elib.dlr.de/97576/

12. D.J. Foerster, Validation of the software package IMD for molecular dynamics simulations of laser induced ablation for micro propulsion. University of Stuttgart (2013). Available via DLR. http://elib.dlr.de/83975/

13. S.I. Anisimov, N.A. Inogamov, Y.V. Petrov, V.A. Khokhlov, V.V. Zhakhovskii, K. Nishihara, M.B. Agranat, S.I. Ashitkov, P.S. Komarov, Thresholds for front-side ablation and rear-side spallation of metal foil irradiated by femtosecond laser pulse. Appl. Phys. A **92**, 797–801 (2008)

14. C. Guo, Structural phase transition of aluminum induced by electronic excitation. Phys. Rev. Lett. **84**, 4493–4496 (2000)

15. K. Kremeyer, J. Lapeyre, S. Hamann, Compact and robust laser impulse measurement device, with ultrashort pulse laser ablation results. AIP Conf. Proc. **997**, 147–157 (2008)

16. L. Pastuschka, Optimization of material removal for laser-ablative microthrusters, Master thesis, University of Stuttgart, 2015

17. S. Scharring, R.-A. Lorbeer, S. Karg, L. Pastuschka, D.J. Foerster, H.-A. Eckel, The MICRO-LAS concept: precise thrust generation in the Micronewton range by laser ablation. IAA Book Ser. Small Satell. **6**, 27–34 (2016)

18. S. Scharring, R.-A. Lorbeer, H.-A. Eckel, Numerical simulations on laser-ablative micro-propulsion with short and ultrashort laser pulses. Trans. JSASS Aerospace Tech. Jpn. **14**, Pb 69–Pb 75 (2016)

Part III
Reactive Flows

Dietmar H. Kröner

The three contributions in the section "Reactive Flows" continue the successful work of the last year. Two projects are based on the OpenFOAM and the third one on the in-house code TASCOM3D. They increase the efficiency of the codes by implementing some parts of the Cantera-code (chemical reaction) directly into the OpenFOAM software, they apply the code to more complex applications and they improve the LES modeling, respectively.

In the first contribution by T. Zirwes, F. Zhang, J. Denev, P. Habisreuther, H. Bockhorn about "Automated code generation for maximizing performance of detailed chemistry calculations in DNS of turbulent combustion" a DNS-code for turbulent flames is considered. In the last year this group presented a code with a reaction mechanism with 18 species and 69 fundamental reactions, containing the optically active OH radical. The main goal was the investigation of the correlation between heat release rate and the luminescent species in turbulent flames. The implementation was based on the open source software OpenFOAM for the CFD part and Cantera for the chemical reaction. Now in this year they could get additional performance gains by extracting relevant classes from Cantera and implementing them directly into an OpenFOAM library, so that Cantera is not an external dependency anymore and its methods are called by the OpenFOAM solver directly. This led to further performance improvements. In this way, highly specialized code leads to a decrease of total simulation time by up to 40%, and the performance improvement increases with the complexity of the reaction mechanism. The optimization concept is applied to a realistic combustion case simulated on two high performance clusters with different network architectures,

D.H. Kröner (✉)
Mathematisches Institut, Abteilung für Angewandte Mathematik, Albert-Ludwigs-Universität Freiburg, Hermann-Herder-Str. 10, 79104 Freiburg im Breisgau, Germany
e-mail: dietmar@mathematik.uni-freiburg.de

Hazel Hen and ForHLR II, showing good parallel performance on up to 28,800 CPU cores.

The authors of the second contribution about "A resolved simulation study on the interactions between droplets and turbulent flames using OpenFOAM" are B. Wang, H. Chu, A. Kronenburg, and O.T. Stein. In the last year this group used the OpenFOAM software package for DNS. The results were compared to benchmark data obtained from a dedicated high order DNS solver, to study the effects of the lower order discretization provided by OpenFOAM. The results are provided in a concise manner and the computations performed with OpenFOAM were in good agreement with the benchmark of the more specialized code. Now in this year the group applied the code to a more complex setting and present a direct numerical simulations (DNS) of turbulent reacting flows around evaporating single fuel droplets and droplet arrays. Statistical analysis of interactions between the droplets and the turbulent flames are used to develop sub-grid scale models for mixture fraction based on approaches such as flamelet or conditional moment closure (CMC) methods. The specific challenges are posed by the effects of the evaporating spray on the composition field in inter-droplet space and by the presence of combustion. They suggest an optimal setup for fully resolved spray DNS that ensures a good balance between computational cost and solution accuracy. Furthermore, adequate scalability of OpenFOAM for the different setups is reported.

The subject of the third contribution about "Towards affordable LES of rocket combustion engines" by R. Keller, M. Grader, P. Gerlinger, and M. Aigner" is still the compressible, implicit combustion code TASCOM3D as in the year before. At that time the code was validated by nonreactive and reactive benchmark tests at high pressures. The simulation results matched experimental observations very well in a qualitative and quantitative manner. In this year the authors present new simulations with affordable LES based on different DES-models. They compare and evaluate the results on two simple test cases. The version "iDDES" shows promising results to be used in future rocket combustion engine simulations.

Automated Code Generation for Maximizing Performance of Detailed Chemistry Calculations in OpenFOAM

Thorsten Zirwes, Feichi Zhang, Jordan A. Denev, Peter Habisreuther, and Henning Bockhorn

Abstract In direct numerical simulation of turbulent combustion, the majority of the total simulation time is often spent on evaluating chemical reaction rates from detailed reaction mechanisms. In this work, an optimization method is presented for speeding up the calculation of chemical reaction rates significantly, which has been implemented into the open-source CFD code OpenFOAM. A converter tool has been developed, which translates any input file containing chemical reaction mechanisms into C++ source code. The automatically generated code allows to restructure the reaction mechanisms for efficient computation and enables more compiler optimizations. Additional performance improvements are achieved by generating densely packed data and linear access patterns that can be vectorized in order to exploit the maximum performance on HPC systems. The generated source code compiles to an OpenFOAM library, which can directly be used in simulations through OpenFOAM's runtime selection mechanism. The optimization concept has been applied to a realistic combustion case simulated on two peta-scale supercomputers, among them the fastest HPC cluster Hazel Hen (Cray XC40) in Germany. The optimized code leads to a decrease of total simulation time of up to 40% and this improvement increases with the complexity of the involved chemical reactions. Moreover, the optimized code yields good parallel performance on up to 28,800 CPU cores.

1 Introduction

Turbulent combustion remains the key technology for energy conversion. In 2035 fossil fuels will still provide 80% of the total energy for the world's economy [1]. Therefore, increasing the efficiency of combustion processes and reducing global pollutant emissions continues to be a major task. But because the interaction of

T. Zirwes • F. Zhang (✉) • J.A. Denev • P. Habisreuther • H. Bockhorn
Engler-Bunte-Institute/Combustion Technology, Karlsruhe Institute of Technology,
Engler-Bunte-Ring. 7, 76131 Karlsruhe, Germany
e-mail: thorsten.zirwes@kit.edu; feichi.zhang@kit.edu

© Springer International Publishing AG 2018
W.E. Nagel et al. (eds.), *High Performance Computing in Science
and Engineering '17*, https://doi.org/10.1007/978-3-319-68394-2_11

flames and turbulent flows is still not fully understood [2, 3], an important research goal is to gain a deeper understanding of the underlying physics and chemistry as well as their mutual interaction in combustion systems.

Numerical simulation is a well-established tool to study combustion processes [2]. With the increasing power in High Performance Computing (HPC) it became feasible to conduct direct numerical simulation (DNS) of flames in turbulent flows. In DNS, the turbulent flow and the combustion chemistry are resolved down to the smallest time and length scales, requiring enormous computational resources. Although simulation tools for turbulent combustion become more and more efficient, DNS of turbulent flames is still limited to small simulation domains. The DNS results however deliver valuable details of the complex combustion phenomena and complement experimental data, which are limited by extreme conditions in technical combustion applications like high pressures and temperatures [4].

The predominant computing time in DNS of turbulent flames is needed to evaluate chemical reaction rates if detailed chemical mechanisms are used [5, 6]. Even for the combustion of simple fuels like methane more than 50 intermediate species and over 300 reactions are typically used to give a detailed description of the chemistry [7]. For technically relevant fuels like kerosene or gasoline consisting of higher hydrocarbons, large reaction mechanisms are being developed which contain thousands of chemical species and tens of thousands of reactions [8, 9]. In addition to solving the Navier-Stokes equations, an additional conservation equation for each chemical species has to be solved, so that the total simulation time strongly depends on the number of chemical species and therefore the complexity of the combustion chemistry. The most time consuming part of the simulation is usually the computation of chemical reaction rates, which are the source terms in the additional conservation equations for the species.

Because of this, the present work introduces on an approach for speeding up the computation of chemical reaction rates. The basis for describing the combustion chemistry are "reaction mechanisms". They specify which species play a role during combustion and define all chemical reactions as well as their parameters. An example of a reaction mechanism for methane/air combustion is given in Sect. 3.2. In this work, a converter tool has been developed which reads general reaction mechanism files as input, restructures them, translates them into C++ source code and applies optimizations for faster evaluation of chemical reaction rates. The idea is that, instead of having a general program that can solve the chemical system for arbitrary reaction mechanisms, each mechanism is represented by specialized subroutines in the generated source code [10, 11].

In order to simulate combustion in turbulent flows and evaluate the performance gain achieved with the presented optimization method, the generated chemistry code is coupled with OpenFOAM [12], which is an open-source tool for computational fluid dynamics (CFD) and can be used for DNS of turbulent combustion [13].

2 Reference DNS Code

In previous works [14] a DNS solver for turbulent combustion phenomena was developed by coupling the general CFD tool OpenFOAM [12] with the thermo-chemical library Cantera [15]. OpenFOAM is an open-source toolbox written in C++. It has found widespread use in computational fluid dynamics during the last years, both in scientific and engineering applications, including DNS of flames. Cantera is a widely used, open-source chemistry library written in C++ which implements highly optimized routines for computing chemical reaction rates and also provides information about detailed molecular diffusion which plays an important role in combustion processes.

Figure 1 presents the structure of the developed coupling interface between OpenFOAM and Cantera. The OpenFOAM code has the task of solving the conservation equations for total mass, momentum, energy and the mass of each chemical species. The state parameters in terms of pressure p, temperature T and gas composition Y_k are used as input for Cantera's routines, which are called by the interface in order to calculate the reaction rates and other thermo-physical properties.

After coupling OpenFOAM with Cantera, additional performance gains were achieved by extracting relevant classes from Cantera and implementing them directly into an OpenFOAM library, so that Cantera is not an external dependency anymore and its methods are called by the OpenFOAM solver directly. This led to further performance improvements [16]. This coupling of OpenFOAM with Cantera represents the reference case for quantifying the performance gains when using the generated source code.

Profiling of the reference DNS code has been conducted in order to identify the performance bottlenecks. With the reaction mechanism by Kee et al. [17] for the combustion of methane, which contains only a small number of reactions and chemical species (see Table 1), chemistry computations take about 60% of the total simulation time with the setup described in Sect. 4.1. For the more detailed GRI 3.0 mechanism [7], almost 90% of simulation time are spent on the chemistry with the reference DNS code. These two reaction mechanisms will be used in the following

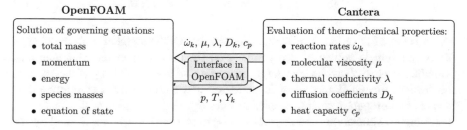

Fig. 1 Simplified chart demonstrating the coupling of OpenFOAM with Cantera in the reference case

Table 1 Reaction mechanisms for methane/air combustion used in this work

Reaction mechanism	Number of species	Number of reactions
Kee et al. [17]	17	53
GRI 3.0 [7]	58	325

sections to demonstrate the performance improvements achieved with the presented optimization method.

In order to compute the chemical reaction rates $\dot{\omega}_k$ for each species k, a number of intermediate quantities have to be calculated. First, the rates of progress $\dot{r}_r$ for each reaction r has to be computed for the forward ($'$) and reverse ($''$) reaction with

$$\dot{r}_r' = k_r' C_m \prod_k C_k^{v_{k,r}'}, \qquad \dot{r}_r'' = k_r'' C_m \prod_k C_k^{v_{k,r}''} , \qquad (1)$$

where C_m is an effective mixture concentration in case the reaction is a three-body reaction, C_k is the concentration of the species involved in the reaction and $v_{k,r}$ is the stoichiometric coefficient of the species for this reaction. The rate constant k_r' for the forward reaction is either computed via Arrhenius' law, which is explained in more detail in the next section, or a more complex falloff type formulation [17]. The rate constants k_r'' of the reverse reactions are usually obtained from equilibrium constants which are computed from thermodynamic considerations. In order to obtain the species reaction rates, all rates of progress have to be added up in the following way:

$$\dot{\omega}_k = M_k \sum_r \left(v_{k,r}'' - v_{k,r}' \right) \left(\dot{r}_r' - \dot{r}_r'' \right) . \qquad (2)$$

M_k is the molar mass of species k. These reaction rates are not used directly in the conservation equation for the species masses. Instead, they are averaged over the simulation time step Δt:

$$\bar{\dot{\omega}}_k \approx \frac{1}{\Delta t} \int_t^{t+\Delta t} \dot{\omega}_k \, dt, \qquad k = 1 \ldots N \qquad (3)$$

This system of N ordinary differential equations (ODE) together with the change of temperature is solved at every time step for each cell in the computational domain. Because N can be very large depending on the complexity of the reaction mechanism, this ODE integration is the reason why the majority of computing time is spent on computing chemical reaction rates $\dot{\omega}_k$. The advantage of this method is that a higher simulation time step Δt can be used because it is not limited by the shortest chemical reaction time scales.

Using the detailed GRI 3.0 mechanism in DNS showed that typically 20% of total simulation time is spent on evaluating the exponential function exp (measured with perf from the Linux tools). Most of the calls to exp stem from the Arrhenius law in (4). Therefore, Arrhenius' law is used in the next section to demonstrate the principles of specific optimizations that are performed during conversion from reaction mechanism input file to C++ source code.

3   Optimized Code Generation

3.1   Basic Concept of the Code Generation Approach

The basic idea in this work is to automatically generate C++ source code for each specific reaction mechanism in order to maximize performance. Two optimization steps are performed to improve the computation of chemical reaction rates:

- In the first step the structure of the chemical reaction mechanism is optimized. Redundant operations are eliminated and species and reactions are reordered and regrouped. This allows to minimize code branching and maximize reuse of cached results (see Sect. 3.2).
- The second step targets the C++ source code. The code is generated in a way that makes it easy for the compiler to optimize. For example, loops with trivial access patterns are generated to enable auto-vectorization and data is stored densely packed in memory to maximize CPU cache usage. Since all parameters from the reaction mechanisms like the number of species are compile-time constants, the compiler can make better optimization decisions with respect to inlining and loop unrolling.

In order to achieve both optimization goals, a converter tool has been developed which reads general reaction mechanism files in Cantera's ctml or xml format as input, performs the two optimization steps and automatically generates C++ source code containing all necessary routines (see Fig. 2). In contrast to Cantera's implementation which provides efficient but more general routines for computing reaction rates from arbitrary reaction mechanisms, the code generated in this way contains only routines that are specialized for one specific reaction mechanism. Instead of being scattered across different translation units, all chemistry code is visible to the compiler at once, giving it the maximum amount of information for optimizations.

Because the chemical reaction rates depend only on local mixture properties, no neighboring cell values are needed so that the generated source code does not contain any parallel communication routines. It is therefore an optimization on the node level. Compiling the generated code results in an OpenFOAM library

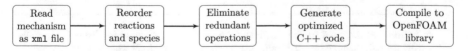

Fig. 2 Simplified overview over how the converter tool works

that can directly be used in the previously developed coupling interface described in Sect. 2 due to OpenFOAM's runtime selection mechanism. By doing this, no routines that originally came from Cantera are involved anymore in the computation of the reaction rates. But in other parts of the OpenFOAM solver, i.e. computation of thermodynamic and diffusive properties, Cantera's routines are still used. These parts are however not a performance bottleneck.

The automatically generated source code contains all necessary routines to compute the chemical reaction rates and all intermediate properties from (1) and (2). In the next section, only the part of the code regarding Arrhenius' law is shown as an example to explain the work done by the converter.

3.2 Optimized Computation of Rate Constants

As mentioned in the previous section, a considerable amount of time during the simulation is spent on evaluating the exponential function exp used in the computation of the rate constants k_r of each reaction via Arrhenius' law

$$ k_r = A_r \exp\left(b_r \log(T) - \frac{E_r}{\mathscr{R}T} \right) , \tag{4} $$

where $\mathscr{R}$ is the universal gas constant and T the temperature, which is the only variable quantity in Arrhenius' law. The parameters A_r, b_r and E_r are constants and defined for every reaction in a reaction mechanism file. Figure 3 shows a small example of the reactions defined in the GRI 3.0 reaction mechanism for methane/air combustion in standard CHEMKIN format [18], where the three Arrhenius parameters A_r, b_r and E_r are defined. In the OpenFOAM solver of the reference case, a mechanism file like this is read once at the start of each simulation.

In Cantera's original implementation, the rate constants for every reaction are computed using (4). It is however possible to compute some rate constants in a more efficient way. There are three cases, where the evaluation of the exponential function can be avoided completely. Figure 4 shows the automatically generated C++ source

```
1  Reaction                        A_r           b_r            E_r
2  ...
3  O2+CH2CHO=>OH+2HCO              2.350E+10      0.000          0.00
4  O2+CO<=>O+CO2                   2.500E+12      0.000      47800.00
5  H+O2+H2O<=>HO2+H2O            11.26E+18       -0.760          0.00
6  ...
```

Fig. 3 Example of three out of 325 reactions from the GRI 3.0 reaction mechanism in standard CHEMKIN format, defining the Arrhenius parameters A_r, b_r and E_r for each reaction

```
 1  alignas(64) static constexpr double A[354] =
 2      {3.6558396000357360e+00, 9.1726385047921717e+00, ... };
 3
 4  alignas(64) static constexpr double b[228] =
 5      {2.7000000000000002e+00, 2.0000000000000000e+00, ... };
 6
 7  alignas(64) static constexpr double E_R[228] =
 8      {3.1501544760183583e+03, 2.0128782594366505e+03, ... };
 9
10  auto invT = 1./T;
11  auto logT = std::log(T);
12  // compute the 228 rate constants for which an
13  // evaluation of the exponential function is necessary
14  for (unsigned r = 0; r != 228; ++r) {
15      auto blogT = b[r]*logT;                   // $b_r \log(T)$
16      auto E_RT = E_R[r]*invT;                  // $E_r/\mathcal{R}T$
17      auto diff = blogT - E_RT;
18      k[r] = std::exp(A[r] + diff);
19  }
20
21  // six rate constants with $E_r=0$ and $b_r$ a small integer
22  double tmp0 = invT*invT;
23  for (unsigned i = 0; i != 2; ++i)
24      k[i+228] = A[i+228]*tmp0;                 // $k_r = A_r T^{-2}$
25  double tmp1 = invT;
26  for (unsigned i = 0; i != 4; ++i)
27      k[i+230] = A[i+230]*tmp1;                 // $k_r = A_r T^{-1}$
28
29  // 100 rate constants with $E_r=0$ and $b_r=0$
30  for (unsigned i = 0; i != 100; ++i)
31      k[i+234] = A[i+234];
32
33  // 20 rate constants where $E_r$ and $b_r$ are the same as for
34  // other rate constants that have already been computed
35  for (unsigned i=0; i != 20; ++i)
36      k[i+334] = A[i+334]*k[i+209];
```

Fig. 4 Automatically generated C++ source code for the GRI 3.0 mechanism. This example shows only the part of the generated code that computes the rate constants from (4)

code for the computation of the rate constants from the GRI 3.0 mechanism, where the following cases are considered:

- If b_r and E_r for a reaction are zero, the rate constant from (4) reduces to the constant value $k_r = A_r$ and no exponential function has to be computed (lines 30–31 in Fig. 4). In total, 100 out of 354 rate constants in the GRI 3.0 reaction mechanism have $b_r = 0$ and $E_r = 0$, see e.g. the first reaction in Fig. 3.
- If different rate constants have the same values of b_r and E_r, the exponential function is evaluated once for one of them and reused for each additional

occurrence. This applies to 40 out of all 354 rate constants from the GRI 3.0 reaction mechanism (lines 35–36 in Fig. 4).

- If E_r is zero and b_r is a small integer, Arrhenius' law reduces to $k_r = A_r T^{b_r}$ (lines 22–27 in Fig. 4). Because b_r is an integer, T^{b_r} can be replaced with a small number of multiplications. This occurs for 6 rate constants in the GRI 3.0 mechanism.

After eliminating these special cases, only 228 out of 354 rate constants have to be computed using the full Arrhenius law involving the expensive exponential function in (4) (see lines 14–18 in Fig. 4).

In the generated source code, the Arrhenius parameters are stored in contiguous arrays for maximizing data cache usage. Note that A[r] in line 18 is automatically stored as $\log(A_r)$ and A[r+334] to A[r+354] are stored as $\log(A_r)/\log(A_{r,\text{duplicate}})$. They are also explicitly aligned to allow auto-vectorization by the compiler. Because the number of loop iterations is a compile-time constant in the generated source code, the compiler can make better decisions about loop unrolling. During code generation all reactions are reordered so that reactions with similar properties are grouped together in the same loops over contiguous data.

In total for the GRI 3.0 mechanism, more than a third of all exponential function evaluations are omitted in the generated code. Together with the vectorization, evaluating the exponential function with the optimized chemistry source code takes only about 5% of the total computing time instead of 20% in the reference DNS code.

Computation of the rate constants k_r is only a small subset of the code needed to compute the final chemical reaction rates. For the rest of the chemistry code indicated by (1) and (2), the basic optimization principles described so far are the same: restructure and simplify the reaction mechanism and generate code that is easy to optimize. In total, the automatically generated C++ code can speed up the chemistry computation by more than 50% (see Sect. 4.3). It should also be noted that all changes done by the converter tool to the reaction mechanism, like for the evaluation of rate constants shown in this section, are equivalent in a strict mathematical sense, so that the simulation results are almost the same as in the reference case.

3.3 Choice of Compiler

The choice of compiler has an impact on the performance of the generated chemistry code too. This is illustrated for the first loop (line 14) in Fig. 4 which contains the most expensive operations and has the highest iteration count among the depicted loops. Figure 5 shows the machine code generated by the GNU compiler (g++ 6.2.0) on the left and the Intel compiler (icpc 17.0.1) on the right on the Cray XC40 Hazel Hen cluster (see Sect. 4.2 for a description of the architecture). Both machine codes are created with the flags -std=c++14 -Ofast -march=native -mtune=native.

g++	icpc

<pre>1 // b[r]*logT
2 **vmulsd** b(%r13), %xmm7, %xmm0
3 // E_R[r]*invT
4 **vmulsd** E_R(%r13), %xmm8, %xmm1
5 // A[r]+blogT
6 **vaddsd** A(%r13), %xmm0, %xmm0
7 // A[r]+diff
8 **vsubsd** %xmm1, %xmm0, %xmm0
9 // exp(A[r]+diff)
10 **call**   exp</pre>
	<pre>// b[r]*logT
	vmulpd b(,%r15,8), %ymm8, %ymm1
	// E_R[r]*invT
	vmulpd E_R(,%r15,8), %ymm9, %ymm3
	// A[r]+blogT
	vaddpd A(,%r15,8), %ymm1, %ymm2
	// A[r]+diff
	vsubpd %ymm3, %ymm2, %ymm0
	// exp(A[r]+diff)
	call __svml_exp4</pre>

Fig. 5 Comparison of shortened machine code output on Cray XC40 Hazel Hen of the GNU g++ 6.2.0 (left) and Intel icpc 17.0.1 (right) compiler for the loop body in lines 14–18 of Fig. 4. Comments (//) show the respective C++ code

Although both machine codes look similar, the GNU compiler does not auto-vectorize the loop. This can be seen from the instructions ending in sd ("scalar double") instead of pd ("packed double"). Therefore, each instruction in the loop of the GNU compiler only operates on one double precision value whereas each iteration of the loop generated by the Intel compiler operates on four double precision values at the same time. The reason is that the Intel compiler automatically replaces the call to the exponential function std::exp with a call to svml_exp4 which is a version of the exponential function defined in Intel's short vector math library (SVML) acting on four double precision values at the same time. In our tests, the Intel compiler creates machine code that performs 15–20% faster in terms of total simulation time compared to the GNU compiler for the two investigated reaction mechanisms. An additional 2% performance gain has been achieved by using the restrict keyword wherever possible, which is a C language feature that gives the compiler additional information about memory access. The performance and scalability tests shown in Sect. 4.3 are all measured using Intel's compiler.

4   Performance Validation

4.1   Numerical Setup

In order to assess the performance benefit of using the generated and optimized chemistry code in the context of HPC simulation of turbulent combustion, the simulation of a turbulent flame has been conducted which was experimentally studied [19]. This section gives a short description of the numerical setup and the HPC clusters.

Figure 6 on the left shows the simulation setup. It consists of the burner nozzle with a diameter of 3.5 cm through which the methane/air mixture flows, and a cylindrical region with a diameter of 0.6 m and a height of 0.6 m, representing

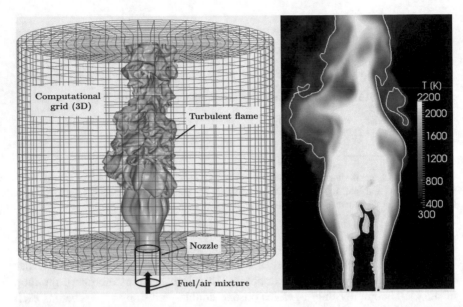

Fig. 6 Schematic drawing of the simulation setup and iso-surface of $T = 800\,\mathrm{K}$, indicating the flame surface (left), and a cut of the instantaneous temperature field (right)

the environment of the burner where the flame stabilizes. The computational grid used for the performance test is composed of 176 million cells on Hazel Hen and 76.5 million finite volumes on ForHLR II, which are refined locally where the flame burns. The governing equations are solved in OpenFOAM with the finite volume method (FVM) along with high-accuracy numerical schemes [20]. Two chemical reaction mechanisms with different complexities, as shown in Table 1, are used in order to show the general validity of the presented concept for chemistry calculations.

Figure 6 on the left illustrates the shape of the flame identified by the iso-surface of $T = 800\,\mathrm{K}$, which becomes wrinkled due to the interaction between the combustion reaction and the turbulent flow. A meridian cut though the computational domain on the right of Fig. 6 shows the temperature field. The gray line indicates the 800 K isotherm, corresponding to the contour of the flame surface depicted on the left of Fig. 6. Although not presented here, the simulation results have shown good agreement with the corresponding experimental data [19].

4.2 HPC Clusters and Software Versions

The performance gains when using the generated source code for the chemistry computations compared to the reference case are evaluated on two supercomputers.

The **Hazel Hen** cluster at the Höchstleistungsrechenzentrum Stuttgart (HLRS) is a CRAY XC40 system based on the twelve-core Intel Xeon E5-2680 v3 processor and Cray Aries network with Dragonfly network topology [21]. With its 15,424 CPUs (185,088 cores) it is currently one of the fastest supercomputers in Europe. It has 7712 dual socket nodes, each node containing a total of 24 cores and having 128 GB DDR4 memory, achieving a theoretical peak performance of 7.42 PFlops. OpenFOAM is used in version 1612+, compiled with gcc version 6.2, together with cray-mpich version 7.0.3. The OpenFOAM solver is coupled to the modified Cantera library based on Cantera 2.3.0a3. The generated chemistry source code and Cantera's implementation in the reference DNS Code are compiled with the Intel compiler version 17.0.1.

ForHLR II at the Karlsruhe Institute of Technology (KIT) has 1152 computing nodes with 64 GB RAM each [22]. Each node has 20 cores (two deca-core Intel Xeon E5-2660 v3 processors) with a total theoretical peak performance of 1 PFlops. All nodes are connected through an InfiniBand 4X EDR Interconnect. OpenFOAM is used in version 2.3.0, compiled with gcc version 4.8, together with OpenMPI version 1.10. The OpenFOAM solver is coupled to the modified Cantera library based on Cantera 2.3.0a3. The generated chemistry source code and Cantera's implementation in the reference DNS Code are compiled with the Intel compiler version 16.0.

4.3   Performance Improvement

The performance improvement by using the chemistry code generated by the converter tool instead of Cantera's implementation has first been evaluated on a single node. An interesting example of how the new chemistry code affects the performance of the simulation is shown in the table on the left of Fig. 7. Using the performance monitoring tool LIKWID, the number of cache and branch prediction misses have been recorded on ForHLR II for the detailed GRI 3.0

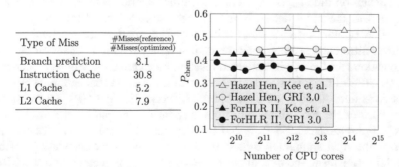

Type of Miss	$\frac{\#\text{Misses(reference)}}{\#\text{Misses(optimized)}}$
Branch prediction	8.1
Instruction Cache	30.8
L1 Cache	5.2
L2 Cache	7.9

Fig. 7 Left: Relative reduction in cache and branch prediction misses for the GRI 3.0 mechanism for a serial case. Right: Performance improvement P_{chem} for only the chemistry computations

mechanism. Branch prediction misses happen when the control flow of the program is not predictable for the CPU. Cache misses happen when either data or program instructions are not in the local CPU caches when they are needed. Because in the generated chemistry code most of the functions are inlined and all loop counters and other parameters affecting the control flow are compile time constants, the number of instruction cache misses is reduced by a factor of 30. Similarly, the number of branch prediction misses is reduced by a factor of 8. Because all data is stored and accessed linearly in memory in the optimized code, L1 cache misses are reduced by a factor of 5 and L2 cache misses by a factor of 8.

Figure 7 on the right shows the time savings of only the chemistry computations for the setup described in Sect. 4.1 on ForHLR II and Hazel Hen. On ForHLR II, the simulation was run on 32 nodes (640 cores) up to 510 nodes (10,200 cores). On Hazel Hen, it was run on 75 nodes (1800 cores) to 1200 nodes (28,800 cores). The time needed for just the chemistry t_{chem} is compared between the reference case and the generated optimized chemistry code with

$$P_{chem} = \frac{t_{chem}(\text{reference case}) - t_{chem}(\text{optimized})}{t_{chem}(\text{reference case})} . \qquad (5)$$

With the detailed GRI 3.0 reaction mechanism, the time for the chemistry computations is reduced by approximately 45% on Hazel Hen and 35–40% on ForHLR II. Using the mechanism by Kee reduces the time for the chemistry computations on ForHLR II by 42% and on Hazel Hen by 53%, thereby saving more than half of the computing time needed for the chemical reaction rates. The improvement stays nearly constant with the number of CPU cores because the optimizations performed in the chemistry routines are not affected by communication. Large portions of the generated code are not as easily vectorizable as shown for Arrhenius' law in Sect. 3.2. For example the summation in (2) cannot be vectorized so that the total speedup stays below the optimal speedup expected from perfect vectorization. Otherwise, all loops that were expected to be vectorized have been auto-vectorized by the Intel compiler.

Because the Kee mechanism is much less complex than the GRI 3.0 mechanism, total simulation time t_{tot} is on average ten times shorter with the Kee mechanism compared to the more detailed GRI 3.0 mechanism. The chemistry part of the simulation is more than 15 times faster with the smaller Kee mechanism. There is also a difference in how much of total simulation time is spent on chemistry computations for the two investigated reaction mechanisms: chemistry calculations take 60–70% of the total simulation time with the Kee mechanism in the reference DNS code and 40–50% with the optimized code. For the GRI 3.0 mechanism, about 90% of the total simulation time is spent on computing chemical reaction rates in the reference DNS code and 85% in the optimized solver. Because much more of the total simulation time is used for the chemistry calculations with the GRI 3.0 mechanism, the overall performance gain P_{tot} is greater with the GRI 3.0 mechanism compared to the Kee mechanism, although the relative improvement of only the chemistry computations P_{chem} in Fig. 7 is slightly better with the Kee mechanism.

$$P_{\text{tot}} = \frac{t_{\text{tot}}(\text{reference case}) - t_{\text{tot}}(\text{optimized})}{t_{\text{tot}}(\text{reference case})} . \tag{6}$$

Using the generated chemistry routines, the total simulation time is reduced by $P_{\text{tot}} = 40\%$ with the GRI 3.0 mechanism on Hazel Hen and 35% on ForHLR II. For example, on Hazel Hen with 28,800 CPU cores, the average time of a time step is about 8 s with the optimized chemistry but more than 13 s with the reference case implementation. The simulation with the reaction mechanism by Kee is 25–35% faster. In order to obtain statistically converged data from DNS of turbulent combustion, typically at least 10^5 simulation time steps have to be calculated. For the present case, this would require about 6.5 million CPU core hours with the GRI 3.0 mechanism on Hazel Hen with the reference DNS code. Using the optimized chemistry routines reduces the overall simulation time by almost 40%, saving over 2.5 million core hours. For future simulations, even more detailed reaction mechanisms will be employed so that higher percentages of total simulation time are used to compute the chemical reaction rates, making the performance gains from the presented optimization technique even more important.

4.4 Parallel Performance

Figure 8 shows the incremental speedup of the DNS solver using the optimized chemistry routines for strong scaling of the setup described in Sect. 4.1. The incremental speedup is defined as $S_n = t_{\text{tot}}(n)/t_{\text{tot}}(n_0)$, where n is the number of CPU cores and n_0 is 640 CPU cores on ForHLR II and 1800 CPU cores on Hazel Hen and t_{tot} is the total simulation time for a fixed number of time steps. The

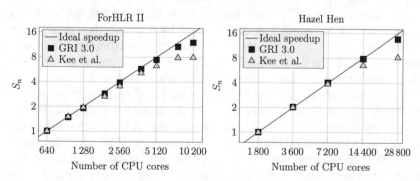

Fig. 8 Incremental speedup S_n (strong scaling): Comparison of Kee mechanism and more detailed GRI 3.0 mechanism on ForHLR II (left) with $76.5 \cdot 10^6$ cells and Hazel Hen (right) with $176 \cdot 10^6$ cells for DNS of turbulent methane/air combustion with the optimized chemistry code

simulations with the GRI 3.0 mechanism scale almost ideally. The scaling efficiency is still above 90% for 28,800 CPU cores on Hazel Hen. As mentioned before, there is no communication during the chemistry computations. Therefore, the more complex the chemistry is, the larger is the share of total simulation time without communication overhead. The simulations with the Kee mechanism scale well up to 5000 CPU cores but scaling efficiency decreases beyond that. In this case, the share of actual computation time in comparison to communication overhead decreases, leading to reduced parallel efficiency. The DNS with detailed chemistry, for example with the GRI 3.0 mechanism, is much more computationally expensive than with the Kee mechanism (see Table 1), so that a better parallel performance is achieved with the GRI 3.0 mechanism. The presented optimizations are therefore most beneficial for larger reaction mechanisms, both in terms of parallel performance and overall simulation time reduction.

It noteworthy that the scaling efficiency becomes slightly better if Cantera's chemistry implementation in the reference case is used instead of the optimized chemistry code. Because the chemistry computations take much longer with Cantera's implementation, the ratio of communication to computation becomes smaller.

5 Conclusion

This work presents an optimization technique where specialized source code is automatically generated to speed up chemistry computations in DNS of chemically reacting flows. It shows that large performance improvements are possible compared to even highly optimized libraries when source code is optimized for special cases. A converter tool has been developed which reads reaction mechanism files containing all information about the chemical reactions occurring during combustion and converts them into C++ routines. The reaction mechanism is analyzed and restructured by the converter tool to enable more efficient evaluation of the chemical reaction rates. The source code for the chemistry routines is generated in a way that is easy for the compiler to optimize and maximizes the usage of data caches and auto-vectorization. These optimizations are achieved by following the design principles of cache friendly data structures and linear data access patterns, which will become more important in the future due to increasing vector register sizes. The number of cache and branch prediction misses have been shown to be drastically reduced. The generated chemistry routines for the combustion of methane and air have been coupled with the open-source library OpenFOAM in order to perform the simulation of a realistic turbulent flame. The simulations were run on two supercomputers (Cray XC40 Hazel Hen and ForHLR II) with up to 28,800 CPU cores showing very good scalability. A decrease of up to 40% in total simulation time has been achieved compared to the reference DNS code without affecting the accuracy of the results. The method is of particular interest when applying complex chemical reaction systems instead of simplified or reduced chemistry, which is important for studying the mechanisms of flame/turbulence interaction and the generation of pollutant emissions. The performance tuning with the presented optimization technique, together with the good scalability of the widely used OpenFOAM package, will help to investigate combustion phenomena in more detail. Consequently, the time and cost required for the development of modern combustion devices may be reduced in the future.

Acknowledgements This work was supported by the German Research Council (DFG) through Research Units DFG-BO693/27 "Combustion Noise". This work was performed on the national supercomputer Cray XC40 Hazel Hen at the High Performance Computing Center Stuttgart (HLRS) under the grant with acronym 'Cnoise' and on the computational resource ForHLR II funded by the Ministry of Science, Research and the Arts Baden-Württemberg and DFG ("Deutsche Forschungsgemeinschaft").

References

1. BP Energy Outlook, British Petroleum (2016), www.bp.com/energyoutlook
2. T. Poinsot, D. Veynante, *Theoretical and Numerical Combustion* (RT Edwards, Toulouse Cedex, 2005)
3. A. Lipatnikov, *Fundamentals of Premixed Turbulent Combustion* (CRC, Boca Raton, 2012)
4. C.K. Law, *Combustion Physics* (Cambridge University Press, Cambridge, 2010)
5. J.H. Chen, Petascale direct numerical simulation of turbulent combustion—fundamental insights towards predictive models. Proc. Combust. Inst. **33**, 99–123 (2011)
6. F. Zhang, T. Zirwes, P. Habisreuther, H. Bockhorn, Numerical simulation of turbulent combustion with a multi-regional approach, in *High Performance Computing in Science and Engineering '15*, ed. by W.E. Nagel, D.B. Kröner, M.M. Resch (Springer, Berlin, Heidelberg, 2015), pp. 267–280
7. G.P. Smith, D.M. Golden, M. Frenklach, N.W. Moriarty, B. Eiteneer, M. Goldenberg, C.T. Bowman, R.K. Hanson, S. Song, W.C. Gardiner, V.V. Lissianski, Z. Qi, GRI 3.0 reaction mechanism (1999), http://www.mc.berkeley.edu/gri_mech
8. LLNL Heptane Reaction Mechanism (2012), https://combustion.llnl.gov/mechanisms/alkanes/ n-heptane-detailed-mechanism-version-3
9. T. Lu, C.K. Law, Toward accommodating realistic fuel chemistry in large-scale computations. Prog. Energy Combust. Sci. **35**(2), 192–215 (2009)
10. V. Damian, A. Sandu, M. Damian, F. Potra, G.R. Carmichael, The kinetic preprocessor KPP— a software environment for solving chemical kinetics. Comput. Chem. Eng. **26**, 1567–1579 (2002)
11. K.E. Niemeyer, N.J. Curtis, pyJac, Version 1.0.1 (2016), https://github.com/kyleniemeyer/ pyJac
12. H.G. Weller, G. Tabor, H. Jasak, C. Fureby, A tensorial approach to computational continuum mechanics using object-oriented techniques. Comput. Phys. **12**, 620–631 (1998)
13. S. Vo, A. Kronenburg, O.T. Stein, E.R. Hawkes, Direct numerical simulation of non-premixed syngas combustion using OpenFOAM, in *High Performance Computing in Science and Engineering '16*, ed. by W.E. Nagel, D.B. Kröner, M.M. Resch (Springer, Heidelberg, 2016)
14. F. Zhang, H. Bonart, T. Zirwes, P. Habisreuther, H. Bockhorn, N. Zarzalis, Direct numerical simulation of chemically reacting flows with the public domain code OpenFOAM, in *High Performance Computing in Science and Engineering '14*, ed. by W.E. Nagel, D.H. Kröner, M.M. Resch (Springer, Berlin, Heidelberg, 2015), pp. 221–236
15. D.G. Goodwin, H.K. Moffat, R.L. Speth, Cantera: an object-oriented software toolkit for chemical kinetics, thermodynamics, and transport processes. Version 2.3.0b (2016), http:// www.cantera.org

16. T. Zirwes, Weiterentwicklung und Optimierung eines auf OpenFOAM basierten DNS Lösers zur Verbesserung der Effizienz und Handhabung. Bachelor's thesis, Karlsruhe Institute of Technology, Germany, 2013
17. R.J. Kee, M.E. Coltrin, P. Glarborg, *Chemically Reacting Flow: Theory and Practice* (Wiley, Hoboken, 2005)
18. CHEMKIN 10131, Reaction Design: San Diego (2013)
19. F. Zhang, T. Zirwes, H. Nawroth, H. Bockhorn, C.O. Paschereit, Combustion generated noise: an environment related issue for future combustion systems. Energy Technol. **5**(7), 1045–1054 (2017)
20. OpenFOAM. The Open Source CFD Toolbox. User Guide (2014)
21. Cray Inc., Cray XC40 (2016), http://www.hlrs.de/systems/cray-xc40-hazel-hen
22. ForHLR II (2016), https://www.scc.kit.edu/dienste/forhlr2.php

A Resolved Simulation Study on the Interactions Between Droplets and Turbulent Flames Using OpenFOAM

B. Wang, H. Chu, A. Kronenburg, and O.T. Stein

Abstract This study presents direct numerical simulations (DNS) of turbulent reacting flows around evaporating single fuel droplets and droplet arrays. Statistical analysis of interactions between the droplets and the turbulent flames are used to develop sub-grid scale models for mixture fraction based approaches such as flamelet or conditional moment closure (CMC) methods. The specific challenges are posed by the effects of the evaporating spray on the composition field in inter-droplet space and by the presence of combustion. Here, we analyse the best possible setup for such a fully resolved DNS. The numerical constraints are given by (1) the need to resolve all small scale effects, i.e. the smallest turbulent eddies and the boundary layer thickness around the droplets, and (2) the desire to include scales covering the entire turbulence spectrum to ensure a realistic interaction between the large and small scales. The largest scales are typically limited by the size of the computational domain, and these two demands (high resolution and large domain size) can easily lead to extensive computational requirements. We suggest an optimal setup for fully resolved DNS that ensures a good balance between computational cost and solution accuracy. The optimal mesh resolution and domain size do not introduce any bias for the analysis of characteristic quantities such as mixture fraction, its PDF and conditional scalar dissipation. Further, adequate scalability of OpenFOAM for the different setups is reported.

1 Introduction

Spray combustion is usually treated in the Euler-Lagrange context where Euler equations are solved for the continuous (carrier) phase and the dispersed phase is modelled by discrete (Lagrange) particles. Most often, sub-grid scale closures

---

B. Wang • A. Kronenburg (✉) • O.T. Stein
Institut für Technische Verbrennung, Universität Stuttgart, Herdweg 51, 70174 Stuttgart, Germany
e-mail: kronenburg@itv.uni-stuttgart.de

H. Chu
Institut für Technische Verbrennung, Universität Stuttgart, Herdweg 51, 70174 Stuttgart, Germany

Present address: Institut für Technische Verbrennung, RWTH Aachen University, Templergraben 64, 52062 Aachen, Germany

© Springer International Publishing AG 2018
W.E. Nagel et al. (eds.), *High Performance Computing in Science and Engineering '17*, https://doi.org/10.1007/978-3-319-68394-2_12

are based on single droplet evaporation models [1, 2] for reacting dilute sprays [3, 4]. Appropriate modelling for an LES of combustion in relative dense sprays is less certain since droplet-turbulence-chemistry interactions are not well understood. Some common combustion models for LES, such as flamelet [5] and CMC [6], rely on the existence of a characteristic variable such as mixture fraction and the strong correlation of the reacting species on this variable. Therefore, suitable and accurate sub-grid scale closures are needed for quantities such as the mixture fraction distribution (i.e. its PDF) and its conditional scalar dissipation.

Direct numerical simulations (DNS) can be used to resolve sub-grid scale interactions between the evaporating droplets, the turbulent convective flow and combustion, and the DNS can provide data for the deduction of sub-grid scale models that can be used within the LES framework. So-called carrier-phase DNS were performed [7–9] for turbulent dilute sprays. However, in these studies only fluctuations of mixture fraction induced by turbulence are resolved but inhomogeneities of local rich mixtures in inter-droplet space primarily induced by droplet evaporation are neglected. Fully resolved DNS of reacting droplet arrays are required for a relative unambiguous statistical analysis of interactions between turbulent flames and droplets. Imaoka and Sirignano [10, 11] investigated the evaporation rate of droplet arrays with varying droplet number densities in a quiescent environment and Zoby et al. [12] extended a similar analysis to turbulent convective environments. However, the scopes of these studies do not include an analysis of the characteristic mixing quantities. Further studies by Zoby et al. [13] and Wang et al. [14] assessed scaling laws for mixture fraction, its dissipation and the PDF that were originally derived by Klimenko and Bilger [15]. Existing studies focused on non-reacting flows only. Before extending the range of parameters to evaluate the scaling laws in reacting flows, a validation of the necessary mesh resolution and domain size are needed to ensure a good balance between computational cost and solution accuracy. These issues are not new: an ideal, fully resolved DNS should include the turbulent energy spectrum from the largest to the smallest scales of turbulence. In addition, the boundary layer at the droplet surface needs to be resolved. The latter poses the most stringent requirement as evaporating droplets after secondary spray break-up are usually of similar size or even smaller than the Kolmogorov eddies leading to a necessary boundary layer resolution at least one order of magnitude higher than the turbulence spectrum would impose. These opposing demands (inclusion of large turbulence scales and resolution of the boundary layer) result in extremely high computational cost.

The present work will now provide a thorough study on the range of turbulence scales that affect mixing in inter-droplet space on one side and on the needed resolution of the boundary layer on the other side and thus provide a reference for future studies on fully resolved DNS of spray combustion. The work focuses on the prediction of characteristic quantities relevant to droplet combustion but also includes information on code performance and parallelization efficiency.

2 Governing Equations and Boundary Conditions

The assumptions of the present simulation are consistent with our earlier study [14]. Heat transfer by radiation is neglected. The gas phase mixture is considered to be ideal with unity Lewis numbers and the Schmidt and Prandtl number are specified as $Sc = Pr = 0.7$. The thermodynamic properties of the liquid phase are assumed constant. The effect of gravity and other body forces are small and can be neglected. The governing equations for the gas phase are given then by

$$\frac{\partial \rho}{\partial t} + \frac{\partial (\rho u_j)}{\partial x_j} = 0, \tag{1}$$

$$\frac{\partial (\rho u_i)}{\partial t} + \frac{\partial (\rho u_j u_i)}{\partial x_j} = -\frac{\partial p}{\partial x_i} + \frac{\partial \tau_{ij}}{\partial x_j}, \tag{2}$$

$$\frac{\partial (\rho Y_k)}{\partial t} + \frac{\partial (\rho u_j Y_k)}{\partial x_j} = \frac{\partial}{\partial x_j} \left(\frac{\mu}{Sc} \frac{\partial Y_k}{\partial x_j} \right) + \dot{\omega}_k, \tag{3}$$

$$\frac{\partial (\rho h_s)}{\partial t} + \frac{\partial (\rho u_j h_s)}{\partial x_j} = \frac{\partial}{\partial x_j} \left(\frac{\mu}{Pr} \frac{\partial h_s}{\partial x_j} \right) + \dot{\omega}_{h_s}, \tag{4}$$

$$\frac{\partial (\rho f)}{\partial t} + \frac{\partial (\rho u_j f)}{\partial x_j} = \frac{\partial}{\partial x_j} \left(\frac{\mu}{Sc} \frac{\partial f}{\partial x_j} \right), \tag{5}$$

where ρ and μ represent the density and dynamic viscosity of the gas phase, τ_{ij} denotes the viscous stress of a Newtonian fluid, Y_k, h_s and f are the mass fraction of species k, sensible enthalpy and the conserved scalar mixture fraction, $\dot{\omega}_k$ and $\dot{\omega}_{h_s}$ indicate the chemical reaction rate of species k and sensible enthalpy reaction source term.

The liquid phase is represented by a single fuel component and the mass fraction Y_{Fl} is 1. The governing equations for the liquid are

$$\frac{\partial \rho_l}{\partial t} + \frac{\partial (\rho_l u_j)}{\partial x_j} = 0, \tag{6}$$

$$\frac{\partial (\rho_l u_i)}{\partial t} + \frac{\partial (\rho_l u_j u_i)}{\partial x_j} = -\frac{\partial p}{\partial x_i} + \frac{\partial \tau_{ij}}{\partial x_j}, \tag{7}$$

$$\frac{\partial (\rho_l h_s)}{\partial t} + \frac{\partial (\rho_l u_j h_s)}{\partial x_j} = \frac{\partial}{\partial x_j} \left(\frac{\mu_l}{Pr_l} \frac{\partial h_s}{\partial x_j} \right), \tag{8}$$

where ρ_l, μ_l and Pr_l are the density, dynamic viscosity and Prandtl number of the liquid.

The mass flux of evaporation, $\dot{m}''$, is computed by a Robin type boundary condition and is conserved on both sides of the droplet surface, viz.

$$\dot{m}''Y_{Fl} = Y_{Fg,s}\dot{m}'' - \frac{\mu}{Sc}\frac{\partial Y_{Fg,s}}{\partial n}, \tag{9}$$

$$\dot{m}'' = \dot{m}''_{l,s}, \tag{10}$$

where $Y_{Fg,s}$ represents the mass fraction of fuel vapour at the droplet surface on the vapour side; $\dot{m}''_{l,s}$ denotes the mass flux on the liquid side. The surface temperature can be calculated by the balance of the energy fluxes between the liquid and gas phase

$$\frac{\mu c_p}{Sc}\frac{\partial T_{g,s}}{\partial n} = k_l\frac{\partial T_{l,s}}{\partial n} + \dot{m}''h_{fg}, \tag{11}$$

$$T_{g,s} = T_{l,s}, \tag{12}$$

where $T_{g,s}$ and $T_{l,s}$ are the surface temperature on the liquid and gas side, k_l is the conductivity of the liquid phase and h_{fg} denotes the evaporation enthalpy. The fuel vapour mass fraction at the droplet surface is determined by the liquid-gas equilibrium [16]. The boundary conditions for non-evaporating species and mixture fraction on the gas side are also given by a Robin type Boundary condition, viz.

$$\dot{m}''Y_{ig,s} - \frac{\mu}{Sc}\frac{\partial Y_{ig,s}}{\partial n} = 0, \tag{13}$$

$$\dot{m}''f_d = f_{g,s}\dot{m}'' - \frac{\mu}{Sc}\frac{\partial f}{\partial n}, \tag{14}$$

where the mixture fraction on the liquid side f_d is set to 1 because of the single component fuel. The boundary conditions at the droplet center can be expressed as

$$\frac{\partial T_l}{\partial n} = 0, \tag{15}$$

$$\frac{\partial u_l}{\partial n} = 0, \tag{16}$$

where T_l and u_l represent the temperature and velocity at the droplet center. Based on our previous study [12] and employing the well-established D^2-law [16], the evaporation time scale is estimated to be roughly one order of magnitude larger than the Kolmogorov time scales for the cases investigated here. Therefore, the different phases of one transient evaporation process can be represented by a series of quasi-steady states [17]. It is thus rational to keep the droplet size constant to approximate a quasi-steady state for the statistical analysis conducted here. To ensure the overall mass continuity, the evaporation rate at the droplet surface is replenished from the droplet center. This is realized by solving the mass and momentum equations (Eqs. (6) and (7)) in the liquid and mass conservation at the boundaries (Eqs. (10) and (16)). It is also noted that—after secondary spray breakup—the time scales for further deceleration of the droplets and for the large turbulent eddies modulating the average droplet number density is large compared to the evaporation time scale. For simplification, the positions of droplets are therefore fixed and a mean relative velocity is imposed at the inlet face. The inflow turbulent fluctuations are calculated from a modified von-Karman spectrum [18].

3   Computational Configurations and Mesh Setups

The computational domain contains a regular droplet array with three droplet layers in the flow direction. One example of the 12-droplet arrangement is visualized in Fig. 1 where d is the droplet diameter, r_c represents the inter-droplet distance, x is the axial distance from the droplet upstream and r is the radial distance from the droplet diameter on a plane perpendicular to the flow direction. Inter-droplet space 1 is defined as the space between the first and second layer while inter-droplet space 2 is located in between the second and third layer. The droplet is assumed to be composed of single-component kerosene ($C_{12}H_{23}$). The array is infinite in both cross-stream directions which is realized by periodic boundary conditions. The mean relative flow is perpendicular to the droplet array. The parameters U_∞ and U' are the inflow mean velocity and the root-mean-square (RMS) turbulent fluctuating velocity. For the cases investigated here, the inflow stream is humid air with N_2, O_2 and H_2O (76.4:23.4:0.2 on a mass basis) at a pressure of $p_\infty = 15$ bar and a temperature of $T_\infty = 1000$ K. The very high resolution needed for the boundary layers around the droplets requires computational savings elsewhere: therefore, we use a 4-step chemical mechanism with seven species [19] to approximate the combustion of kerosene. This is justified since the reduced mechanism provides accurate estimates of the laminar flame speed and of ignition delay [19]. A more detailed knowledge of the inner flame structure is beyond the scope of this paper.

Fig. 1 Schematic diagram of the computational domain containing a regular droplet array

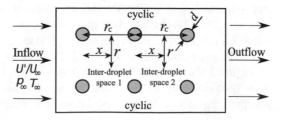

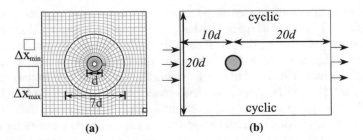

Fig. 2 Three dimensional grid and schematic diagram of one single droplet domain

The initial droplet temperature is set to $T_s = 300$ K. The thermodynamic and transport properties of the gas mixture are calculated by polynomials and semi-empirical equations [20].

Figure 2a is a schematic illustration of the mesh module for a single droplet block. A droplet array can be constructed by a succession of these modules. The mesh adjacent to the droplet on the gas side is spherical and extended by $7d$ from the droplet center. Then it is gradually transformed into a Cartesian configuration. The mesh on the liquid side is extruded from the surface mesh. The smallest cells with the size of Δx_{min} are located at the droplet surface and the largest cells with the size of Δx_{max} are located at the inlet and outlet faces of the domain. The time-step is controlled to ensure the maximum Courant number to be smaller than 0.15 and a PISO algorithm with an additional non-orthogonal correction is used to solve the pressure-velocity coupling. These options avoid potential flow instabilities originating from the mesh non-orthogonality at the conjunction between the spherical and Cartesian meshes. Since the DNS is required to resolve the smallest turbulent scales and the boundary layer at the droplet surface, the computational cost is extremely high, especially for high Reynolds number conditions. Therefore, the characteristic cell size should be determined for the single droplet block before the droplet arrays are constructed. The balance between the accuracy of solution and the computational cost can be assessed for the DNS of droplet array combustion. The mesh independence study is performed for the single droplet domain as shown in Fig. 2b. The domain is $30d$ and $20d$ (with the droplet diameter $d = 100 \, \mu$m) in the flow and cross-stream directions, respectively. The droplet is positioned at $1/3$ of the length of the domain. Three different mesh resolutions are tested for one of our setups with the highest droplet Reynolds number (inflow $Re_{d,\infty} = 17.2$ which is calculated based on the droplet diameter (d) and the inflow mean velocity ($U_{d,\infty}$)) and the smallest Kolmogorov length scale (inflow $\eta/d = 0.42$). The characteristic quantities of the three meshes are listed in Table 1. Mesh 2 is the standard mesh that was applied to the earlier study for pure evaporation [14]. Meshes 1 and 3 are coarsened and refined by a factor of 2, respectively.

Figure 3 shows instantaneous mixture fraction and temperature fields for mesh 2. A quasi-laminar wake region can be seen directly downstream of the droplet. Further away from the droplet, the wake is disturbed by small scale turbulence.

Table 1 Characteristic quantities of three different mesh resolutions

Mesh	$\Delta x_{min}/d$	$\Delta x_{max}/\eta$	# cells
1	1/10	1.5	55K
2	1/20	0.75	420K
3	1/40	0.375	3400K

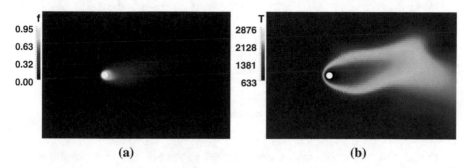

Fig. 3 Instantaneous mixture fraction (**a**) and temperature (**b**) fields in a plane containing the droplet for mesh 2

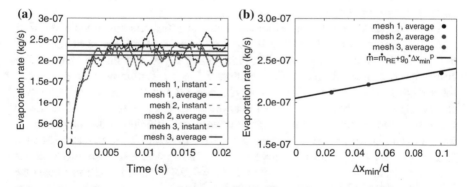

Fig. 4 Evaporation rate profiles for the three mesh resolutions. (**a**) Time evolution. (**b**) Richardson extrapolation

A thin boundary layer around the droplet surface can also be observed and the thinnest part is located near the leading edge of the droplet.

The time evolution of the evaporation rate is shown in Fig. 4a for the different cases (see dashed line). The figure includes a heat-up period until stationary conditions are reached. This period lasts approximately 5ms and is not affected by the mesh resolution. Statistics are then taken over 8 flow through times. As mentioned above, the evaporation rate is computed by a Robin type boundary condition (Eq. (9)) which is discretized by a first order approximation. The generalized Richardson extrapolation [21] can be used to assess the grid refinement study, and it is demonstrated in Fig. 4b that the time averaged evaporation rates at steady state

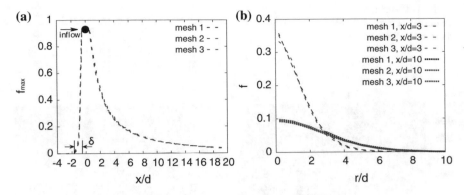

Fig. 5 The axial (**a**) and radial (**b**) variations of mixture fraction mean

for the three mesh resolutions are well captured by the estimated relationship of the Richardson extrapolation, $\dot{m} = \dot{m}_{RE} + g_0 \Delta x_{min}{}^p$. Here $\dot{m}_{RE}$ is the virtual value of the evaporation rate for Δx_{min} tending asymptotically to zero, g_0 is a constant and p is the order of the discretization for the Robin type boundary condition (Eq. (9)) and here set to 1. This indicates the boundary layer at the droplet surface is sufficiently resolved by all three mesh resolutions.

Figure 5a shows the mesh resolution does not affect the streamwise variation of the maximum value of mixture fraction (f) at each plane perpendicular to the flow direction, and the thickness of the boundary layer remains unchanged. The boundary layer at the droplet surface is around $\delta = 1d$. Within the boundary layer, the mixture fraction gradient is extremely high and the mixture fraction distribution can be approximately captured by a linear function. The comparisons of the radial variation of the mixture fraction mean in the regions dominated by the quasi-laminar wake ($x/d = 3$) and by turbulent small scale structures ($x/d = 10$) are presented in Fig. 5b. The DNS statistics agree well for the three mesh resolutions. Since the scalar dissipation rate is computed as $N_f = 2D(\nabla f)^2$, it is the most sensitive indicator for any influence by mesh resolution. Figure 6 shows the mixture fraction conditional scalar dissipation and PDF at $x/d = 3$ and $x/d = 10$. The DNS results for meshes 2 and 3 are comparable. However, the coarse mesh 1 brings notable additional fluctuations due to insufficient samples used for the statistics. In summary, considering the accuracy of our statistical analysis and the computational cost, mesh 2 presents itself as a feasible compromise for the simulation of droplet arrays subject to mixing by large turbulent scales and the then inherent need for an increased computational domain as will be demonstrated in the next section.

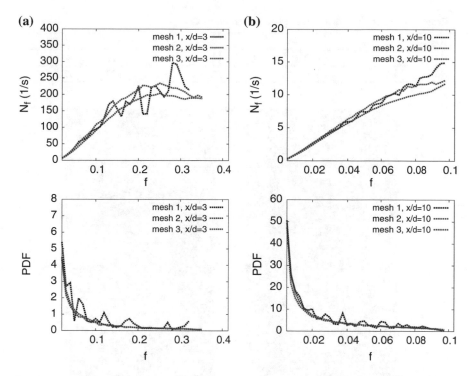

Fig. 6 The mixture fraction conditional scalar dissipation (N_f) mean and the PDF at (**a**) $x/d = 3$ and (**b**) $x/d = 10$

4 Results of Droplet Array Combustion

Two differently sized domains are investigated to analyse the effect of the integral length scales on the characteristic mixing quantities in inter-droplet space as the maximum realizable length scale depends on the domain size. The characteristic quantities are listed in Table 2 where $U'/U_{d,\infty}$ represents the inflow turbulence intensity, U' is the root-mean-square velocity fluctuation, L denotes the domain size, l_t is the inflow integral length scale that is set to $L/2$, r_c is the inter-droplet distance and n is the droplet number that is uniformly divided into three layers perpendicular to the mean flow direction. Consistent with the validation for single droplet combustion, here 4 flow through times are sufficient for the statistical analysis since the length of the domain is doubled in the streamwise direction.

Figure 7 shows the instantaneous mixture fraction and temperature fields in a plane containing the droplets. The upper row presents results from the case where the largest turbulence scale is of the order of the droplet spacing ($l_t/r_c = 1$), the lower row presents corresponding data from the case with $l_t/r_c = 4$. The quasi-laminar regions gradually expand from the first to the last layer. The area of the quasi-laminar region and the region disturbed by turbulence both seem independent

Table 2 Characteristic quantities of two investigated cases

$Re_{d,\infty}$	$U'/U_{d,\infty}$	d (μm)	r_c/d	L/r_c	l_t/r_c	n	# cells
8.6	1	100	20	2	1	12	4M
8.6	1	100	20	8	4	192	60M

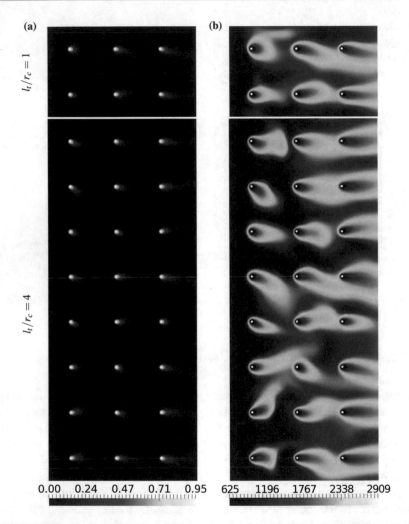

Fig. 7 Instantaneous (**a**) mixture fraction and (**b**) temperature fields in a plane containing the droplets for the two cases

of the integral length scale. Figure 7 also indicates a very weak dependence of the combustion characteristics on the integral length scale. Both simulations appear to predict the single droplet combustion regime, and the flames show a very similar spatial expansion upstream and downstream of the respective droplets.

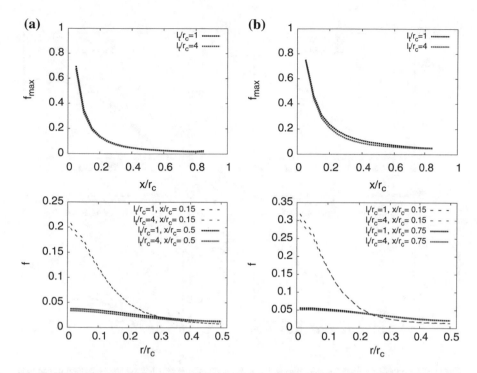

Fig. 8 The axial and radial variation of mixture fraction in (**a**) inter-droplet space 1 and (**b**) inter-droplet space 2

This rather qualitative comparison can be further corroborated by a quantitative statistical analysis of the characteristic mixing quantities. Figure 8 compares the mean mixture fraction profiles in the streamwise direction behind the droplet in inter-droplet spaces 1 and 2 for different integral length scales. In inter-droplet space 1, the axial profiles of the maximum value of mixture fraction agree quite well in the entire inter-droplet space (Fig. 8a). The radial variation at $x/r_c = 0.15$ (in the quasi-laminar wake region) and $x/r_c = 0.5$ (in the region dominated by turbulence) also present good agreement for both integral length scales. Similar profiles can be observed for inter-droplet space 2 (Fig. 8b).

The comparison of the conditional scalar dissipation for the different integral length scales is shown in Fig. 9. The conditional scalar dissipation rates agree well at $x/r_c = 0.15$ (this holds for both inter-droplet spaces) because the species diffusivities determine the mixing in the quasi-laminar region [14, 15]. The existence of additional larger integral length scales ($l_t/r_c = 4$) does not affect the scalar dissipation at $x/r_c = 0.5$ and $x/r_c = 0.75$ since the small scale dissipative structures are not varied and small scale turbulence dominates the mixing ($N_f \propto \varepsilon^{1/3}$) [14]. This is further demonstrated by Fig. 10 where the turbulent energy dissipation rate ($\varepsilon = \frac{1}{\rho}\tau_{ij}\frac{\partial u_i'}{\partial x_j}$) is compared for the streamwise positions $x/r_c = 0.5$

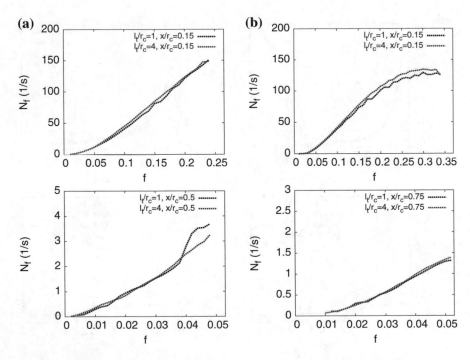

Fig. 9 The conditional scalar dissipation in (**a**) inter-droplet space 1 and (**b**) inter-droplet space 2

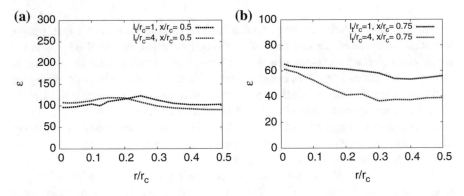

Fig. 10 The turbulent energy dissipation rate (ε) mean at $x/r_c = 0.5$ and $x/r_c = 0.75$ in (**a**) inter-droplet space 1 and (**b**) inter-droplet space 2

and $x/r_c = 0.75$. This is in contrast to the common scaling of the dissipation rate by the integral length scale, $\varepsilon \propto 1/l_t$. The reason can be attributed to the "screen effect" [22]. For an integral length scale larger than the inter-droplet distance, the flow field around a particle with fixed positions will be significantly affected by the neighbouring particles and the largest effective eddy is expected to be limited to a

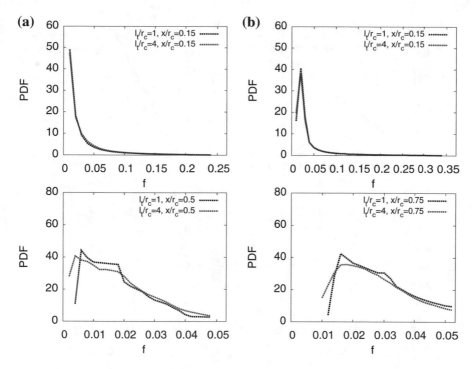

Fig. 11 The mixture fraction PDF in inter-droplet spaces (**a**) 1 and (**b**) 2

scale comparable to the inter-droplet space. The relative insensitivity of the mixing field on large scale turbulence is also shown for the mixture fraction PDF in Fig. 11. This is consistent with the results of scalar dissipation (see Fig. 9) and qualitatively demonstrates the validity of the quasi-steady state solution of the PDF transport equation for spray combustion [15]. In conclusion, the mixture fraction distribution, its conditional scalar dissipation and PDF are independent of the largest turbulent length scales that are imposed (or rather given) by the domain size. A relatively small domain size containing just 12 droplets can therefore be used as a reference for further parameter studies in [17].

5   Computational Performance

Scaling tests are performed for the 12-droplet and 192-droplet cases on the HazelHen architecture of HLRS. For the 12-droplet case, the total number of cells is 4 million and the number of computer cores is increased from 24 to 576. Figure 12a shows an almost linear speedup up to 144 cores. The efficiency decreases to around 95% for 288 cores and 55% for 576 cores. On 288 cores, around 14,000 cells are allocated on each core and a typical simulation takes approximately 48 h. For the 192-droplet case, the total number of cells comprises 60 million and the parallel

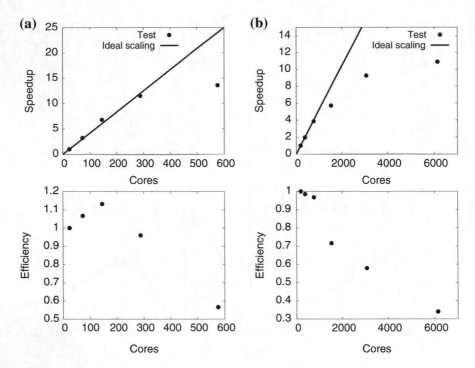

Fig. 12 The computational performance for the 12-droplet (**a**) and 192-droplet domain (**b**)

scalability is tested from 192 to 6144 cores (see Fig. 12b). It can be observed that speedup and efficiency are acceptable up to 3072 cores. The simulation costs for computations on 3072 cores with about 20,000 cells per core amount to 120 h wall clock time.

6 Conclusions

Fully resolved DNS of dynamic interactions between droplets and turbulent flames have been performed using OpenFOAM. The mesh independence study is conducted for a single droplet domain. The standard mesh with the smallest cell size at the droplet surface of $\Delta x_{min}/d = 1/20$ and the biggest cell size at the inlet and outlet faces of the domain of $\Delta x_{max}/d = 1/3$ is of sufficient resolution for the boundary layer at the droplet surface and the smallest turbulent scales. Further refinement of the mesh does not provide any significant improvement of the simulation results as demonstrated by the generalized Richardson extrapolation. The droplet array can be constructed by a succession of single droplet mesh blocks. Two domain sizes are tested to investigate the effect of the turbulent integral length scale on the mixing in inter-droplet space. It is demonstrated that an extension of the range of turbulence scales does not modify the statistics in inter-droplet space and the largest effective scale will be of the order of the droplet distance. The domain size comprising two droplets in both, the cross-stream and the spanwise directions, respectively, can therefore be used as a reference for further parameter studies.

Acknowledgements The authors acknowledge the financial support by CSC/Chinese Scholarship Council (NO 201406020093) and the computational resources provided by HLRS (University of Stuttgart).

References

1. D.B. Spalding, The combustion of liquid fuels. Symp. (Int.) Combust. **4**(1), 847–864 (1953)
2. B. Abramzon, W.A. Sirignano, Droplet vaporization model for spray combustion calculations. Int. J. Heat Mass Transf. **32**(9), 1605–1618 (1989)
3. S. Ukai, A. Kronenburg, O.T. Stein, Certain aspects of conditional moment closure for spray flame modelling, in *High Performance Computing in Science and Engineering'14* (Springer International Publishing, Cham, 2015), pp. 335–350
4. S. Ukai, A.,Kronenburg, O.T. Stein, Large eddy simulation of dilute acetone spray flames using CMC coupled with tabulated chemistry. Proc. Combust. Inst. **35**(2), 1667–1674 (2015)
5. H. Pitsch, C.M. Cha, S. Fedotov, Flamelet modelling of non-premixed turbulent combustion with local extinction and re-ignition. Combust. Theor. Model. **7**(2), 317–332 (2003)
6. S. Navarro-Martinez, A. Kronenburg, F. Di Mare, Conditional moment closure for large eddy simulations. Flow Turbul. Combust. **75**(1–4), 245–274 (2005)
7. J. Réveillon, L. Vervisch, Spray vaporization in nonpremixed turbulent combustion modeling: a single droplet model. Combust. Flame **121**(1–2), 75–90 (2000)
8. T.G. Almeida, F.A. Jaberi, Direct numerical simulations of a planar jet laden with evaporating droplets. Int. J. Heat Mass Transf. **49**(13–14), 2113–2123 (2006)
9. P. Schroll, A.P. Wandel, R.S. Cant, E. Mastorakos, Direct numerical simulations of autoignition in turbulent two-phase flows. Proc. Combust. Inst. **32**(2), 2275–2282 (2009)
10. R. Imaoka, W. Sirignano, Vaporization and combustion in three-dimensional droplet arrays. Proc. Combust. Inst. **30**(2), 1981–1989 (2005)
11. R. Imaoka, W. Sirignano, A generalized analysis for liquid-fuel vaporization and burning. Int. J. Heat Mass Transf. **48**(21–22), 4342–4353 (2005)
12. M.R.G. Zoby, A. Kronenburg, S. Navarro-Martinez, A.J. Marquis, Assessment of conventional droplet evaporation models for spray flames, in *High Performance Computing in Science and Engineering'11* (Springer, Berlin, Heidelberg, 2012), pp. 209–227
13. M.R.G. Zoby, S. Navarro-Martinez, A. Kronenburg, A.J. Marquis, Turbulent mixing in three-dimensional droplet arrays. Int. J. Heat Fluid Flow **32**(3), 499–509 (2011)
14. B. Wang, A. Kronenburg, D. Dietzel, O.T. Stein, Assessment of scaling laws for mixing fields in inter-droplet space. Proc. Combust. Inst. **36**(2), 2451–2458 (2016)
15. A.Y. Klimenko, R.W. Bilger, Conditional moment closure for turbulent combustion. Prog. Energy Combust. Sci. **25**(6), 595–687 (1999)
16. S.R. Turns, *An Introduction to Combustion* (McGraw-Hill, New York, 1999)
17. B. Wang, A. Kronenburg, G.L. Tufano, O.T. Stein, Fully resolved DNS of droplet arrays combustion in turbulent convective flows and modelling for mixing fields in inter-droplet space. Combust. Flame (2017, under review)
18. M. Billson, L.E. Eriksson, L. Davidson, Jet noise prediction using stochastic turbulence modeling, in *9th AIAA/CEAS Aeroacoustics Conference and Exhibit*, Hilton Head, SC (2003)
19. W.P. Jones, R.P. Lindstedt, Global reaction schemes for hydrocarbon combustion. Combust. Flame **73**(3), 233–249 (1988)

20. D.W. Green, R.H. Perry, *Perry's Chemical Engineers' Handbook* (McGraw-Hill, New York, 2008)
21. P.J. Roache, Perspective: a method for uniform reporting of grid refinement studies. J. Fluids Eng. **116**, 405–413 (1994)
22. L. Botto, A. Prosperetti, A fully resolved numerical simulation of turbulent flow past one or several spherical particles. Phys. Fluids **24**(1), 013303 (2012)

Towards Affordable LES of Rocket Combustion Engines

Roman Keller, Martin Grader, Peter Gerlinger, and Manfred Aigner

Abstract Due to the large disparity of length and time scales in rocket combustors corresponding LES are highly time consuming. Steps towards simulating rocket combustion engines with affordable LES with the compressible, implicit combustion code TASCOM3D are presented. DES-family (DES, DDES, iDDES) models are compared and evaluated on two simple test cases: the turbulent isotropic homogeneous turbulence and the planer channel flow. iDDES shows promising results to be used in future rocket combustion engine simulations and can make LES affordable in these kind of combustion simulations.

1 Introduction

For the precise prediction of rocket combustion phenomena, instationary simulations offer better comparison with experimental results [7, 8, 11]. Because of the large differences in the length and times scales of a rocket combustion engine, unsteady simulations require long averaging cycles. A complete LES of a combustion chamber with sufficient wall resolution for heat flux predictions is not feasible in the near future due to the large grid cell requirements in the boundary layer [14, 18]. Detached eddy simulations (DES) offer a interesting alternative for that matter. In DES, LES modeling is used in the main flow away from the walls, whereas inside the boundary layer RANS modeling is utilized. Therefore, much coarser grids can be used near the wall and the overall grid requirement can be mitigated. Unfortunately, a new problem arises: The transition from the resolved turbulence in the LES part into the modeled turbulence in the RANS area and vice versa is not straight forward [13].

In this report, two basic flows which resemble specific parts of the combustion chambers are presented, namely the decaying isotropic homogeneous turbulence and the planar channel flow. The decaying turbulence is a test case for the transport and dissipation of turbulent structures which is important throughout the entire

R. Keller (✉) • M. Grader • P. Gerlinger • M. Aigner
Institut für Verbrennungstechnik der Luft- und Raumfahrt, Pfaffenwaldring 38-40, 70569
Stuttgart, Germany
e-mail: roman.keller@dlr.de

© Springer International Publishing AG 2018
W.E. Nagel et al. (eds.), *High Performance Computing in Science and Engineering '17*, https://doi.org/10.1007/978-3-319-68394-2_13

combustion chamber. The planar channel flow addresses the LES/RANS transition problem near the wall. This report is structured in the following way: First, a brief introduction of the applied CFD code TASCOM3D and the turbulence modeling is given in Sect. 2. Then the results of the two test cases are presented in Sects. 3 and 4. In Sect. 5 a brief performance analysis regarding reaction mechanism size is presented.

2 Numerical Method

The scientific in-house code TASCOM3D (Turbulent All Speed Combustion Multi-grid Solver 3D) has been applied successfully during the last two decades to simulate reacting and non-reacting super- and subsonic flows. Reacting flows are described by solving the fully compressible Navier-Stokes, turbulence and species transport equations. Additionally, an assumed PDF (probability density function) approach is available to take turbulence-chemistry-interaction into account.

2.1 Governing Equations

The three-dimensional conservative form of the Reynolds-averaged Navier-Stokes equations is given by

$$\frac{\partial \mathbf{Q}}{\partial t} + \frac{\partial (\mathbf{F} - \mathbf{F}_v)}{\partial x} + \frac{\partial (\mathbf{G} - \mathbf{G}_v)}{\partial y} + \frac{\partial (\mathbf{H} - \mathbf{H}_v)}{\partial z} = \mathbf{S}, \tag{1}$$

where

$$\mathbf{Q} = [\rho, \rho u, \rho v, \rho w, \rho E, \rho k, \rho \omega, \rho Y_i]^T, \ i = 1, 2, \ldots, N_k - 1. \tag{2}$$

The conservative variable vector $\mathbf{Q}$ consists of the density ρ, the velocity components u, v and w, the total specific energy E, the turbulence variables k and ω and the species mass fractions Y_i. N_k is the total number of species in the gaseous phase. $\mathbf{F}$, $\mathbf{G}$, and $\mathbf{H}$ are the vectors specifying the inviscid fluxes in the x-, y- and z-direction, respectively. $\mathbf{F}_v$, $\mathbf{G}_v$, and $\mathbf{H}_v$ are their viscous counterparts. The source vector $\mathbf{S}$ includes terms from turbulence and chemistry. It is given by

$$\mathbf{S} = [0, 0, 0, 0, S_k, S_\omega, S_{Y_i}]^T, \tag{3}$$

S_K and S_ω are the source terms of the turbulence variables and S_{Y_i} the source terms of the species mass fractions due to combustion. For turbulence closure the k–ω model of Wilcox [19] is used. The equations governing the turbulence kinetic energy and specific dissipation rate are

$$\frac{\partial}{\partial t}(\rho k) + \frac{\partial}{\partial x_j}(\rho u_j k) = \rho \tau_{ij} \frac{\partial u_i}{\partial x_j} - D^k$$

$$+ \frac{\partial}{\partial x_j}\left[\left(\mu + \sigma^* \frac{\rho k}{\omega}\right)\frac{\partial k}{\partial x_j}\right], \tag{4}$$

$$\frac{\partial}{\partial t}(\rho \omega) + \frac{\partial}{\partial x_j}(\rho u_j \omega) = \alpha \frac{\omega}{k} \rho \tau_{ij} \frac{\partial u_i}{\partial x_j} - \beta \rho \omega^2$$

$$+ \sigma_d \frac{\rho}{\omega} \frac{\partial k}{\partial x_j} \frac{\partial \omega}{\partial x_j} + \frac{\partial}{\partial x_j}\left[\left(\mu + \sigma \frac{\rho k}{\omega}\right)\frac{\partial \omega}{\partial x_j}\right] \tag{5}$$

with

$$\rho \tau_{ij} = 2\mu_T \bar{S}_{ij} - \frac{2}{3}\rho k \delta_{ij}, \tag{6}$$

$$D^k = \beta_k \rho k \omega, \tag{7}$$

and

$$\bar{S}_{ij} = S_{ij} - \frac{1}{3}\frac{\partial u_k}{\partial x_k}\delta_{ij}. \tag{8}$$

2.2 DES Turbulence Modeling

For the LES modeling, the k–ω turbulence model is transferred into a classical Smagorinsky subgrid model under the assumption of equilibrium. Therefore the dissipative term of the k-transport equation is replaced by

$$D^k_{DES} = \frac{\rho k^{3/2}}{l_{DES}} \tag{9}$$

where l_{DES} is the DES length scale which is the minimum of the RANS and LES length scale

$$l_{DES} = \min\{l_{RANS}, l_{LES}\} \tag{10}$$

and blends over between RANS and LES zones. The length scales of RANS and LES are defined as

$$l_{RANS} = \frac{k^{1/2}}{\beta^* \omega}, \quad l_{LES} = C_{DES}\Delta \tag{11}$$

with C_{DES} being a model constant like the Smagorinsky constant and Δ a measure of grid spacing. Δ is defined as

$$\Delta = \max\{h_x, h_y, h_z\}, \tag{12}$$

where $h_{x,y,z}$ are the grid spacings in the coordinate directions.

2.3 DDES Turbulence Modeling

The original DES method suffers from grid induced separation which can occur when the grid is refined in a poor way [15]. For the successor named delayed detached eddy simulation (DDES) the blending function is modified to prevent too early switching from RANS to LES inside the boundary layer:

$$l_{DDES} = l_{RANS} - f_d \min(0, l_{RANS} - l_{LES}) \tag{13}$$

with

$$f_d = 1 - \tanh\left(\left[c_{dt} \frac{v_t + v}{\sqrt{U_{i,j}U_{i,j}}\kappa^2 d_w^2}\right]^3\right) \tag{14}$$

where c_{dt} is a constant, v the molecular and v_t the turbulent viscosity, $U_{i,j}$ the velocity gradient, κ the Karman constant, and d_w the distance to the wall. The blending factor f_d remains zero inside the boundary layer and delays the transition to LES.

2.4 iDDES Turbulence Modeling

While the DDES approach improves the grid induced stress depletion problem, a log-layer mismatch for wall-bound flows exists. The improved DDES (iDDES) is introduced to minimize the log-layer mismatch [13]. The measure for grid spacing Δ is changed to account for Smagorinsky constant variations

$$\Delta = \min\{\max[C_w d_w, C_w h_{\max}, h_{wn}], h_{\max}\} \tag{15}$$

with C_w being an empirical constant set to 0.15, h_{wn} is the grid step in the wall-normal direction. $h_{\max}$ is the maximal local grid spacing

$$h_{\max} = \max\{h_x, h_y, h_z\}. \tag{16}$$

Furthermore, the blending function is changed to

$$l_{iDDES} = \tilde{f}_d(1 + f_e)l_{RANS} + (1 - \tilde{f}_d)l_{LES} \tag{17}$$

where $\tilde{f}_d$ is an extension of f_d

$$\tilde{f}_d = \max\{(1 - f_d), f_B\}. \tag{18}$$

f_B is an empirical blending function using the grid spacing

$$f_B = \min\{2\exp(-9\alpha^2), 1.0\}, \quad \alpha = 0.25 - d_w/h_{\max}. \tag{19}$$

The function f_e is defined as

$$f_e = \max\{(f_{e1} - 1), 0\}f_{e2} \tag{20}$$

with f_{e1} being a variation of f_B

$$f_{e1} = \begin{cases} 2\exp(-11.09\alpha^2) & \text{if } \alpha \geq 0 \\ 2\exp(-9.0\alpha^2) & \text{if } \alpha < 0. \end{cases} \tag{21}$$

f_{e2} is a combination of two variants of f_d, f_t and f_l:

$$f_{e2} = 1.0 - \max\{f_t, f_l\} \tag{22}$$

which are defined as

$$f_t = \tanh\left(\left[c_t^2 \frac{v_t}{\sqrt{U_{i,j}U_{i,j}}\kappa^2 d_w^2}\right]^3\right), \tag{23}$$

$$f_l = \tanh\left(\left[c_l^2 \frac{v}{\sqrt{U_{i,j}U_{i,j}}\kappa^2 d_w^2}\right]^{10}\right). \tag{24}$$

2.5 Numerical Solver

The spatial discretization is performed on block structured grids based on a finite volume scheme. For the reconstruction of the cell interface values, MLP^{ld} (Multidimensional Limiting Process-low diffusion) [3, 7] with up to fifth order is used to prevent oscillations at discontinuities. MLP uses diagonal values to improve the TVD (Total Variation Diminishing) limiting behavior [20]. For LES, a sixth order central scheme with minimal dissipation is available. The central discretization scheme is stabilized by adding artificial dissipation in form of a small amount of fifth order MLP discretization. The amount is controlled by the parameter S_{min}. Thus the reconstruction of the cell side values is given by

$$q^{6.O,\lim}_{(i+1/2,j,k)} = q^{6.O}_{(i+1/2,j,k)}(1 - S_{min}) + q^{5.O,MLP}_{(i+1/2,j,k)}S_{min}. \qquad (25)$$

Using these interface values, the AUSM$^+$-up flux vector splitting [9] is employed to calculate the inviscid fluxes.

The unsteady set of Eq. (1) is solved with an implicit Lower-Upper Symmetric Gauss-Seidel (LU-SGS) [4] algorithm. Furthermore, finite-rate chemistry is treated in a fully coupled manner. The code is parallelized with Message Passing Interface (MPI). More details concerning TASCOM3D may be found in [2, 4, 16].

3   Decaying Isotropic Homogeneous Turbulence

The correct prediction of turbulent transport and decay is essential for a proper LES. The decaying isotropic homogeneous turbulence (DIHT) is a widely used test case to study grid and discretization influences in LES. The experimental setup [1] is a laminar wind channel flow which is tripped by a wire grid to become turbulent and the turbulence intensity is measure upstream of the grid. Since such an entire wind channel simulation is too costly for a LES or DNS, the numerical setup is based on the Taylor hypotheses [17]. Therefore only a small cubical volume, which is assumed to be transported in the flow field, is simulated. The initial solution is taken from a DNS [10] and the measurement positions are related to different simulation times. The energy spectrum is calculated using a three-dimensional Fourier transformation. Periodic boundary conditions are used in every direction. If not specified differently, the grid consist of 64 cells in each direction. The simulations are performed on 64 CPUs.

Figure 1 shows the computational domain and instantaneous isosurface of a vortex identification criterion $Q = 0.5(\Omega^2 - S^2)$, where Ω and S describe the vorticity and strain rate tensor, respectively. The three-dimensionality and turbulent character of the flow field can be seen nicely.

As mentioned in Sect. 2 the k–ω model is transferred into a Smagorinsky like subgrid model inside the LES domain. Figure 2 shows the energy spectra of a DES using the k–ω turbulence model as base model with $C_{DES} = 0.2$ in comparison to LES with various Smagorinsky subgrid constants C_s (0.18, 0.2, and 0.22). The energy spectra of the LES and DES with a constant value of 0.2 are very similar. This demonstrates that the assumptions used in the derivation of the DES model work very well in this test case. Furthermore, the effect of the subgrid turbulence model on the energy spectrum is observable. A higher value for the Smagorinsky

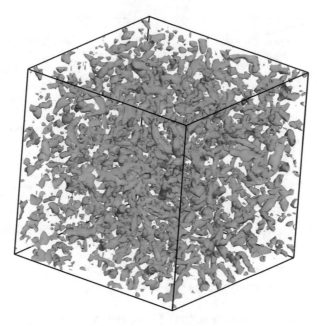

Fig. 1 Instantaneous isosurface of the Q criterion for the DIHT test case

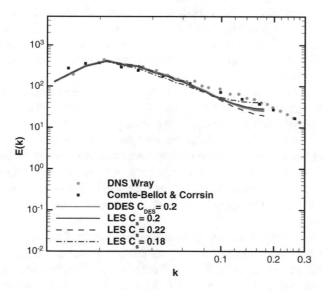

Fig. 2 Comparison of energy spectra of DES and LES solutions with different Smagorinsky constants

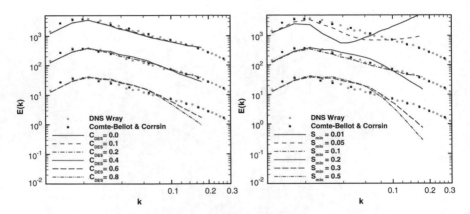

Fig. 3 Influence of C_{DES} with $S_{min} = 0.1$ (left) and S_{min} with $C_{DES} = 0.1$ (right) on the energy spectrum

constant leads to more energy in the middle of the energy spectrum and less energy in the smaller scales.

Figure 3 shows the influence of the LES subgrid parameter C_{DES} and numerical damping factor S_{min} on the energy spectrum of the DIHT. The subgrid parameter C_{DES} has a smaller impact on the energy spectrum than the numerical damping factor S_{min}. For values of $C_{DES} \leq 0.4$ excellent agreement with measurement and DNS is achieved. For S_{min} only a value of 0.1 produces a good spectrum. This shows clearly the importance of a high spatial discretization order with minimal dissipation. It is also shown that, if not enough dissipation is added, the solution degenerates quickly and becomes unstable.

This effect is also shown in the next figure. The integral kinetic energy over time for different parameters of C_{DES} and S_{min} is displayed in Fig. 4. The different curves are normalized by their respective values at $t = 10^{-5}$ s. A value of zero for S_{min} leads to an increase of the kinetic energy due to numerical oscillations which is unphysical. The two other curves show a similar dissipation behavior. This suggests that using a more dissipative scheme is not too bad for the integral kinetic energy transport and dissipation and only the distribution of the energy over the turbulent scales, thus the energy spectrum, is strongly effected by a scheme with higher dissipation.

The grid sensitivity of the energy spectrum is show in Fig. 5. For the discretization with less dissipation ($S_{min} = 0.1$) the energy spectrum's behavior is almost grid independent. For significant smaller cells in one direction (e.g. 128×64×64) the quality of the energy spectrum is derogated. For the discretization with more dissipation ($S_{min} = 0.1$) the energy spectrum, besides deviating more from the references, are strongly influenced by the grid resolution. In this case, an increase in cell number in a single direction improves the solution compared to the reference grid of 64×64×64 cells. This behavior is not fully understood. It could be related to the inevitable transformation of the non-uniform grids into a uniform grid for the Fourier transformation. This is required because the Fourier transformation only

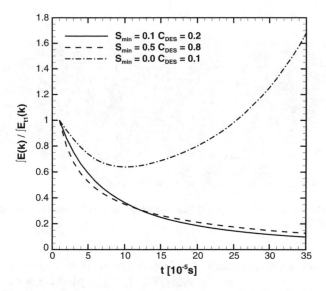

Fig. 4 Total kinetic energy temporal evolution for various numerical discretization

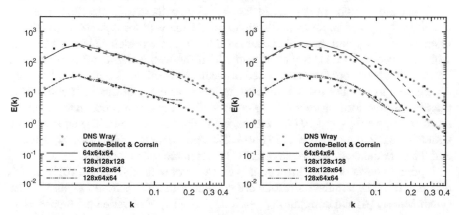

Fig. 5 Grid dependency of the numerical discretization for $S_{min} = 0.1$ (left) and $S_{min} = 0.5$ (right)

works on uniform grids. Nevertheless, these findings underscore the importance of a high-order spatial discretization with low dissipation.

4 Channel Flow

The planar channel flow is an important but also very challenging test case for hybrid LES/RANS methods. The method must ensure that the transition from modeled turbulence in the RANS domain to resolved turbulence in the LES domain is

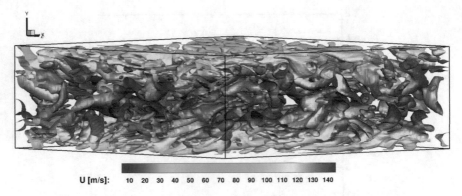

U [m/s]:   10 20 30 40 50 60 70 80 90 100 110 120 130 140

Fig. 6 Instantaneous isosurface of the Q criterion for the plane channel flow. The color indicates the streamwise velocity U

smooth. Otherwise, differences in the friction forces cause a jump in the velocity profile, the so-called log layer mismatch [13]. The numerical setup for the channel flow is the following: The channel height H is 0.03 m and the domain extents 3H into the x- and z-direction. The used grid consists of 96 cells in each direction. The grid is uniformly spaced in the x- and z-direction. At the wall the cell sizes are chosen to ensure $y^+ \leq 1$ and the grid is coarsened towards the center. This leads to cells with an aspect ratio ($\Delta x / \Delta y$) of around 220 at the wall and about 2 in the center. Periodic boundary conditions are used in x- and z-direction. The flow is artificially accelerated to account for the momentum losses at the wall due to friction. This pressure source term is closed loop controlled to achieve a predefined Reynolds number $Re_H = \frac{\rho u_b H}{\mu} = 250{,}000$ with u_b being the bulk velocity. The channel flow simulations are computed on 1728 cores. The computational domain of the channel and the instantaneous isosurface of the Q criterion are shown in Fig. 6.

Figure 7 shows the nondimensional velocity profile for DDES and iDDES with different c_{dt}. The simulation results are compared to a DNS from Lee and Moser (DNS Moser) [6] and correlation from Reichardt [12]. The varied parameter c_{dt} is present in both models (see Eq. (14)) and has a large effect on the solution. In the original model formulation of Spalart [15], using the Spalart-Allmaras turbulence model, this parameter was set to 8. In the work of Gritskevich [5], who used the k-ω SST model, c_{dt} was adjusted to 20. In the DDES presented here (left figure), even for a value of c_{dt} of 50, the velocity profiles are much different from the DNS and correlation. For the iDDES model (right figure) the velocity profiles get close but still remain different from the references. The log layer mismatch is moved towards the channel center and the jump in the profile is decreased with increasing values c_{dt}. For a value of 50 no jump in the profile is visible anymore. Due to the logarithmic scaling of y^+ it seems that there is no more LES domain. Still, large parts of the channel flow are in LES mode as can be seen in Fig. 8.

The variation of the two new parameters c_t and c_l of the iDDES model (see Eqs. (23) and (24)) are shown in Fig. 9. Both constants are used in the function f_{e2}

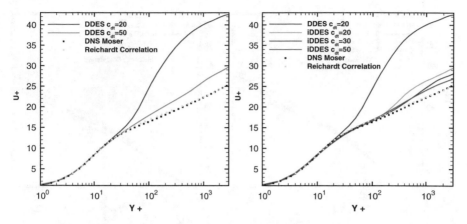

Fig. 7 Nondimensional velocity profile of DDES (left) and iDDES (right) with different c_{dt} ($c_t = 1.0$ and $c_l = 5$)

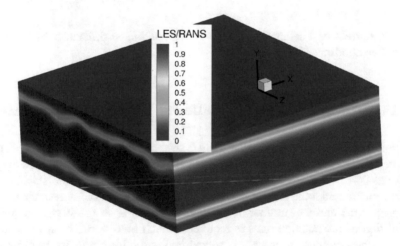

Fig. 8 LES and RANS domain in a iDDES channel flow with $c_{dt} = 50$

(see Eq. (22)). In the left figure, the variation of c_t with constant $c_l = 5$ is shown. The effect on the velocity profile of a larger deviation of c_t (1.87, blue line) is apparent around $y^+ \sim 30$, whereas for a smaller variation (0.9, red line and 1.1, green line) a deviation occurs around $y^+ \sim 600$. The larger variation deviates the slope of the velocity profile, whereas the smaller variations have an effect on the jump in the profile. Both small variations (0.9 and 1.1) decrease the jump in the velocity profile. The exact reason for this behavior is currently unclear to the authors and another parameter study is currently ongoing. In the right figure, the effect of c_l variation with a constant $c_t = 1.0$ is shown. A larger c_l increases the slope of the velocity profile and vice versa. Also, the jump in the profile is smaller for a larger value of c_l. An ideal combination of the three parameters c_{dt}, c_t, and c_l has not been found

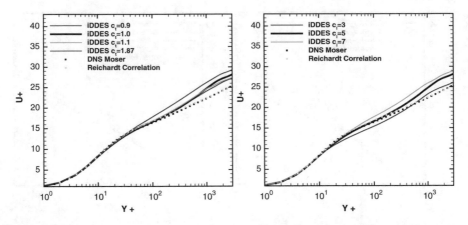

Fig. 9 Nondimensional velocity profile for parameter variation of c_t (left, $c_l = 5$) and c_l (right, $c_t = 1.0$) at $c_{dt} = 30$

yet. The current best setup $c_{dt} = 30$, $c_t = 0.9$, and $c_l = 5$ deviates by 8% in the center line nondimensional velocity.

5 Performance Analysis for Different Reaction Mechanisms

For finite rate chemistry simulations there is always a trade-off between the size and the precision of the reaction mechanism. Without the use of tabulation techniques, each reaction in the mechanism has to be calculated at every time step. From a performance standpoint, two parameters of a mechanism are of interest: the number of species (n_s) and the number of reactions (n_r). For each reaction an expensive calculation of the reaction rates is required in each time step. The species source terms, if the combustion process is solved implicitly, form a dense jacobi matrix which needs to be inversed in each time step. In this section, a brief performance comparison of two different reaction mechanisms is presented. Reaction mechanism 1 (RM1) has 21 species and 97 reactions. Reaction mechanism 2 (RM2) consist of 8 species which are involved in 21 reactions. Table 1 shows a comparison of the two mechanisms' performance on a otherwise identical setup. For the performance analysis CrayPat is used. Inside the subroutine *line* the inversion of the source term jacobi matrix is performed. The reaction coefficients are calculated in the subroutine *react*. The overall execution time, the partial execution times of the two mentioned subroutines as well as their relative time consumption are shown. The overall execution time increases by a factor of 4.67. The time spent in the subroutines increase by significant larger factors of 7.79 and 6.82. Also the percentage of time spent in the major "chemical subroutines" increases from 36% to 57%.

Table 1 Performance comparison of two different reaction mechanism

	RM1	RM2	RM1/RM2
n_s	21	8	2.625
n_r	97	21	4.62
Execution time	885.2 s	189.5 s	4.67
Time in subroutine *line*	284.5 s	36.5 s	7.79
Time in subroutine *react*	220.62 s	32.344 s	6.82
Rel. time in *line*	32.2%	19.3%	1.67
Rel. time in *react*	25.0%	17.0%	1.47

6 Conclusion

Steps towards affordable LES for rocket combustion engines have been implemented and tested in the compressible, implicit combustion code TASCOM3D. Various models of the DES-family have been implemented, tested and compared on two different simple test cases: a decaying isotropic homogeneous turbulence flow and a planar channel flow. Both cases were successfully simulated using the DES-family models. The iDDES model showed promising results for the planar channel flow and will be used further for rocket combustion engine simulations. Additionally, the performance of different reaction mechanism was compared. The size of the mechanism has a large impact on the required computational time.

Acknowledgements The presented work was performed within the framework of the SFBTR 40 funded by the Deutsche Forschungsgemeinschaft (DFG). This support is greatly appreciated. All simulations were performed on the XC40 (Hazel Hen) cluster at the High Performance Computing Center Stuttgart (HLRS) under the grant number scrcomb. The authors wish to thank for the computing time and the technical support.

References

1. G. Comte-Bellot, S. Corrsin, Simple Eulerian time correlation of full- and narrow-band velocity signals in grid-generated, 'isotropic' turbulence. J. Fluid Mech. **48**(2), 273–337 (1971)
2. P. Gerlinger, Investigation of an assumed pdf approach for finite-rate chemistry. Combust. Sci. Technol. **175**(5), 841–872 (2003)
3. P. Gerlinger, Multi-dimensional limiting for high-order schemes including turbulence and combustion. J. Comput. Phys. **231**(5), 2199–2228 (2012)
4. P. Gerlinger, H. Möbus, D. Brüggemann, An implicit multigrid method for turbulent combustion. J. Comput. Phys. **167**(2), 247–276 (2001)
5. M.S. Gritskevich, A.V. Garbaruk, J. Schütze, F.R. Menter, Development of DDES and IDDES formulations for the k-ω shear stress transport model. Flow Turbul. Combust. **88**(3), 431–449 (2012)
6. M. Lee, R.D. Moser, Direct numerical simulation of turbulent channel flow up to. J. Fluid Mech. **774**, 395–415 (2015)
7. M. Lempke, R. Keller, P. Gerlinger, Influence of spatial discretization and unsteadiness on the simulation of rocket combustors. Int. J. Numer. Methods Fluids **79**(9), 437–455 (2015)

8. C. Lian, C.L. Merkle, Contrast between steady and time-averaged unsteady combustion simulations. Comput. Fluids **44**(1), 328–338 (2011)
9. M.S. Liou, A sequel to AUSM, part II : AUSM$^+$-up for all speeds. J. Comput. Phys. **214**(1), 137–170 (2006)
10. N. Mansour, A. Wray, Decay of isotropic turbulence at low Reynolds number. Phys. Fluids **6**(2), 808–814 (1994)
11. M. Masquelet, S. Menon, Large-eddy simulation of flame-turbulence interactions in a shear coaxial injector. J. Propuls. Power **26**(5), 924–935 (2010)
12. H. Reichardt, Vollständige darstellung der turbulenten geschwindigkeitsverteilung in glatten leitungen. ZAMM-J. Appl. Math. Mech./Z. Angew. Math. Mech. **31**(7), 208–219 (1951)
13. M.L. Shur, P.R. Spalart, M.K. Strelets, A.K. Travin, A hybrid RANS-LES approach with delayed-DES and wall-modelled LES capabilities. Int. J. Heat Fluid Flow **29**(6), 1638–1649 (2008)
14. P. Spalart, W. Jou, M. Strelets, S. Allmaras et al., Comments on the feasibility of LES for wings, and on a hybrid rans/les approach, in *Advances in DNS/LES*, vol. 1 (1997), pp. 4–8
15. P. Spalart, S. Deck, M. Shur, K. Squires, M. Strelets, A. Travin, A new version of detached-eddy simulation, resistant to ambiguous grid densities. Theor. Comput. Fluid Dyn. **20**(3), 181–195 (2006)
16. P. Stoll, P. Gerlinger, D. Brüggemann, Domain decomposition for an implicit LU-SGS scheme using overlapping grids, in *13th Computational Fluid Dynamics Conference, AIAA 97-1869*, Snowmass Village, CO (1997), pp. 1–9
17. G.I. Taylor, Statistical theory of turbulence, in *Proceedings of the Royal Society of London A: Mathematical, Physical and Engineering Sciences*, vol. 151 (The Royal Society, London, 1935), pp. 421–444
18. P.K. Tucker, S. Menon, C.L. Merkle, J.C. Oefelein, V. Yang, An approach to improved credibility of CFD simulations for rocket injector design, in *43rd AIAA/ASME/SAE/ASEE Joint Propulsion Conference & Exhibit, AIAA 2007-5572*, Cincinnati, OH (2007)
19. D.C. Wilcox, Formulation of the k-w turbulence model revisited. AIAA J. **46**(11), 2823–2838 (2008)
20. S.H. Yoon, C. Kim, K.H. Kim, Multi-dimensional limiting process for three-dimensional flow physics analyses. J. Comput. Phys. **227**(12), 6001–6043 (2008)

Part IV
Computational Fluid Dynamics

Ewald Krämer

The following chapter presents a selection of research projects conducted in the field of Computational Fluid Dynamics (CFD) during the reporting period. Traditionally, CFD has a strong request for supercomputing time. At the HLRS in Stuttgart, more than 60% of the core-hours provided by the CRAY XC40 Hazel Hen in 2016 were used by researchers looking into challenging fluid dynamic problems. Two of the projects had been granted access as "Gauss Large Scale Projects" with a budget of at least 35 million core-hours for 1 year. Additionally, quite a big number of CFD simulations on a smaller scale were run on the tier-2 cluster ForHLR I and II at the Steinbuch Centre for Computing (SCC) in Karlsruhe. As in the recent years, the projects have been of excellent scientific quality, and they have in common that the results achieved for most complex flow physical problems have only been possible through the use of High Performance Computing. This year, 29 annual reports had been submitted and undergone a peer review process. Due to limited space, only ten contributions could be selected for publication in this book, which means that quite a number of highly qualified reports could not be considered. The results presented in the following articles were all achieved on the Hazel Hen at the HLRS in Stuttgart, which is currently the fastest supercomputer in Germany. The spectrum of projects covers fundamental research as well as application-oriented problems of industrial relevance. Examples for the latter are colloidal suspensions, heavy duty cooling systems, hydraulic propeller turbines, wind turbines, advanced aircraft propulsion systems, and space launchers, just to name a few. Different numerical methods, as Finite Volume, Lattice Boltzmann, or Discontinuous Galerkin methods, implemented in in-house, commercial, or open source codes, were employed.

E. Krämer (✉)
Institut für Aerodynamik und Gasdynamik, Universität Stuttgart, Pfaffenwaldring 21, 70550
Stuttgart, Germany
e-mail: kraemer@iag.uni-stuttgart.de

In the first contribution, Pandey, Chu, and Laurien of the Institute of Nuclear Technology and Energy Systems at the University of Stuttgart present results from a Direct Numerical Simulation (DNS) of a turbulent supercritical carbon dioxide flow in a cooled vertical pipe. Besides the flow direction, the strength of the buoyancy force and the heat flux were varied. The aim of the study was to investigate the cooling system heat-transfer problem for nuclear power plants or heavy-duty coolers. The open source finite volume code OpenFOAM was employed, and the main simulations were run on a grid with 80m cells using up to 1400 parallel cores. It was found that the body forces brought in by gravity affects the axial temperature profile and that the turbulence statistics is influenced by buoyancy and acceleration. The turbulence is reduced significantly at downward flow, and with the aid of an octant analysis involving three turbulent quantities, it was revealed that cold ejection events combined with hot sweeps are the main contributors to turbulent momentum flux, both being decreased in downward flow.

In the work of Dörr, Guo, Peter, and Kloker of the Institute of Aerodynamics and Gas Dynamics, University of Stuttgart, the applicability of plasma actuators to delay laminar-turbulent transition caused by traveling cross flow vortices (CFV) was explored by two techniques, the upstream flow deformation and the direct attenuation of nonlinear traveling CFVs. In the first technique, the actuators are used to excite steady narrow spaced control CFVs that induce a beneficial mean-flow distortion and weaken the primary crossflow instability. In the second technique, the actuators are positioned more downstream, where the traveling CFVs have already established. The localized unsteady forcing against the direction of the crossflow is then aimed at attenuating the amplitude of the traveling CFVs by directly tackling the three-dimensional nonlinear disturbance state. The well-established in-house DNS-code NS3D was used for this study with the DBD plasma actuators being modelled by body forces. Approximately 6000 cores were used in parallel. It was found that with both techniques, transition can be delayed, however the efficiency is significantly higher when applying upstream flow deformation. However, there are some ideas how to improve the direct attenuation approach, which is planned to be investigated in the following project phase. Besides that, the parallel efficiency of the NS3D code shall be improved further by introducing domain decomposition also in spanwise direction, since presently, the Fourier-spectral ansatz in this direction is parallelized using a shared-memory OpenMP approach, which limits the number of usable CPU cores.

In the third article, Ertl, Reutzsch, Nägel, Wittum, and Weigand of the Institute of Aerospace Thermodynamics, University of Stuttgart, address the complex physical processes associated with the breakup of a liquid jet with shear-thinning fluid and phase change. In order to resolve the broad range of spatial and time scales, very fine grids are mandatory. Results are shown for different grids with up to 1.3 billion cells using up to 32,768 parallel cores. In the second part of their contribution, the authors focus on their efforts to improve the already good performance of their employed in-house multiphase flow DNS-code FS3D, as this is considered mandatory for a further increase in Reynolds number, which goes in hand with an enlargement of the computational domain. As the code simulates incompressible flows, the momentum equation cannot be solved explicitly. Instead, the Pressure Poisson Equation has to be solved using a multigrid solver. It had turned out during scaling tests on the Hazel Hen with grids up to 8 billion cells that this task devours most of the computational costs and causes an unsatisfactory weak scaling behaviour for very large grids. Thus, a massively parallel geometric multigrid solver of the software package UG4 developed at the Goethe Center for Scientific Computing was implemented into the FS3D framework. First results show a noticeable increase in weak scaling performance, but, with respect to strong scaling, the old multigrid solver performed better. Overall, the performance of the new solver is still behind expectations. However, the authors see some potential for optimizing the implementation, and ways to further improve the results are discussed.

Loosen, Statnikov, Meinke, and Schöder from the Aerodynamic Institute of the RWTH Aachen used a zonal RANS/LES method together with an optimized dynamic mode decomposition (DMD) to gain insight into the turbulent wake characteristics of generic space launchers. The occurring wake flow modes are responsible for asymmetrical loads on the engine extension. Three different generic space launcher geometries were considered, i.e. a planar backward facing step at transonic inflow, a strut supported wind-tunnel model at supersonic speed with an air and a helium nozzle jet, respectively, and an axisymmetric free flight configuration at transonic speed. The investigated wake topologies reveal a highly unsteady behaviour of the shear layer and the separation region resulting in strongly periodic and antisymmetric wall pressure fluctuations on the nozzle surface. Reduced-order analyses based on extracted DMD modes were performed on the turbulent wake to scrutinize the underlying low frequency spatio-temporal coherent motions. Furthermore, a passive flow control device consisting of semi-circular lobes integrated at the base shoulder of the planar configuration was investigated, which reduces the reattachment length and, thus, the lever arm of the unsteady forces significantly. In addition, they partially reduce undesired low frequency pressure fluctuations on the nozzle surface. However, this reduction comes along with increased high frequency pressure fluctuations. A very good strong scaling behaviour of the code was demonstrated on Hazel Hen up to 32,256 cores.

For a long time, the Helicopter and Aeroacoustics Group at the Institute of Aerodynamics and Gas Dynamics at the University of Stuttgart have performed numerical simulations on complete rotorcraft configurations. In-depth analyses of aerodynamic phenomena, as e.g. the so-called tail shake, or of the noise emission including shielding, refraction and diffraction have been in the focus of their recent work. Such analyses require an extremely high mesh resolution up to several hundred millions of grid cells, and, thus, high computational times even on modern supercomputers. Therefore, high efforts have been put in enhancing the parallel performance of the applied flow solver, which is the FLOWer code originally developed by DLR. Letzgus, Dürrwächter, Schäferlein, Keßler, and Krämer present in their contribution recent code optimizations on that score. Priorly, a node-to-node instead of a core-to-core communication had been implemented using a shared memory, which provides a huge reduction of the number of messages. Now, a graph partitioning method was introduced to the MPI communication, which reduces the number of messages as well as the total message size again significantly leading to a further run time speed-up of 20%. Application cases shown are, firstly, a finite wing in attached and massively stalled conditions installed in a wind tunnel using URANS and DDES to investigate the influence of the wind tunnel walls, and secondly, a model Contra-Rotating Open Rotor that was simulated at various operation conditions to study the unsteady blade loadings and the noise generating mechanisms.

Wawrzinek, Lutz, and Krämer of the Institute of Aerodynamics and Gas Dynamics as well, consider in their article the effect of statically and dynamically disturbed inflow conditions on a two-element high lift airfoil installed in a wind-tunnel test section. In a first step, the two-dimensionally extruded airfoil was exposed to a static tip vortex originating from a vertically mounted finite wing ahead of the airfoil, which generates a three-dimensional inflow condition. In a second step, the static distortion was superimposed by a dynamic gust, which was created through a single pitch motion of a two-dimensional wing placed upstream of the vortex generator. Unsteady RANS simulations were performed and analysed in detail, however, not all occurrences found in the numerical results could be explained satisfactorily yet. The computational effort for such simulations is very high, on the one hand due to the small turbulence scales requiring a very fine mesh, and on the other hand due to the spectrum of frequencies, which demands small physical time steps over long recording times. Thus, a comprehensive parallel efficiency study of the used TAU-Code developed by DLR was made for different compilers and grid partitioners. Good strong scaling behaviour for a 60 m cells grid was found down to less than 10,000 cells per core.

Another working group at the Institute of Aerodynamics and Gas Dynamics is concerned with the numerical simulation of the aerodynamics, aeroelastics, and aeroacoustics of wind turbines. Fischer, Zabel, Lutz, and Krämer investigated in their project the influence of wind tunnel walls, tower, and downstream nozzle on the performance of a model wind turbine. This model wind turbine has a radius of 1.5 m and was located in a wind tunnel with a cross section of 4.2 m × 4.2 m at TU Berlin. Global loads, angle of attack distributions as well as flow fields were compared to each other to evaluate the influence of the different configurations. It was found that, due to the wind tunnel environment with a blockage ratio of more than 40%, thrust is increased by approximately 25% and power by 50% for the same inflow velocity and pitch. Compared to the influence of the walls, the joint impact of tower and nozzle on the mean values is small (approximately 1%). However, the tower leads to an increase of fluctuations. The reduction of the mean values of thrust and power, which is also an effect of the tower, is partly compensated by the nozzle, which also attenuates the fluctuation amplitudes. The conclusion of the authors is that for such a high blockage ratio, a direct transfer from the wind tunnel results to undisturbed conditions is not reasonable. When comparing the experimental to the numerical results, the latter have to include the wind tunnel walls, as well as the tower and the nozzle. The simulations were performed with the FLOWer code, typically on a 36 m cell mesh on 1296 cores considering 73 rotor revolutions.

Junginger and Riedelbauch of the Institute of Fluid Mechanics and Hydraulic Machinery, University of Stuttgart, present a numerical investigation of the draft tube flow in a low head hydraulic propeller turbine. The flow field in the draft tube is dominated by the vortex rope. A part load operating point of the turbine at equal heat and rotational speed but lower discharge factor and guide vane opening compared to the best efficiency point is considered. The commercial software package ANSYS-CFD was used for the numerical analyses. Objective of the study was to compare computational results obtained with different turbulence models at various mesh sizes up to 100 million cells. RANS simulations with the standard k-ω-SST turbulence model and hybrid RANS-LES using the SAS-SST (Scale Adaptive Simulation) and the SBES (Stress Blended Eddy Simulation), respectively, were performed to deliver steady-state as well as transient predictions. The results are compared with respect to mean velocity profiles as well as the resolved turbulent flow structures inside in the draft tube. In addition, the hydraulic losses are collated. Unfortunately, no experimental data for validation is available up to now, but these are awaited in the near future. For speed-up tests, different versions of ANSYS-CFX (16.0, 17.2, and 18.0) are compared. It is demonstrated that ANSYS-CFX v18 shows a much better parallel performance than v16 and a slightly better one than v17. At least for core numbers less than 2000, the parallel performance of the latest version for this specific application case is satisfactory.

The next contribution comes from Xie, Aouane, and Harting of the Dept. of Applied Physics, Eindhoven University of Technology, and of the Helmholtz Institute Erlangen Nürnberg for Renewable Energy. It presents simulation results for two of their latest applications of their Lattice Boltzmann method to complex fluids and interface problems. They investigated the behaviour of spherical magnetic Janus particles at a fluid-fluid interface as well as of a dense suspension of deformable capsules in a Kolmogorov flow, using a scalable algorithm that is able to resolve individual interactions on the microscopic level. For the first case, the results were obtained with their multi-component Lattice Boltzmann solver coupled to a discrete element module, and for the second case, they coupled the Lattice Boltzmann code with an Immersed Boundary Method (IBM). The authors demonstrate the strong scalability of their code for three benchmark cases on Hazel Hen. For a single-component fluid, nearly a linear speed-up was achieved up to about 100,000 cores at a domain size of $1024 \times 1200 \times 1024$ lattice nodes. For the colloidal suspension simulations, the linear scalability breaks down at an earlier stage. The coupling to IBM reduces the parallel performance further, but it is still good on 4096 cores for the case with the suspension of deformable particles.

For many years, the working group of Munz at the Institute of Aerodynamics and Gas Dynamics, University of Stuttgart, has successfully been developing a Discontinuous Galerkin (DG) based high-order simulation framework. A DG spectral element method is used in their in-house fluid dynamics code FLEXI, which has been consequently designed for scale resolving simulations in an HPC context. Beck, Bolemann, Flad, Frank, Krais, Kukuschkin, Sonntag, and Munz summarize in their contribution the latest progress in the application of this method on the Cray XC40 Hazel Hen at the HLRS. They present a large eddy simulation

(LES) of a flow around a wall mounted cylinder, an LES of a flow around an airfoil at Reynolds number 660,000 using a recently introduced kinetic energy preserving flux formulation for a better achievement of non-linear stability, and an LES of transitional flow in a low pressure turbine. Very recently, a shock-capturing mechanism employing finite volume sub-cells was implemented into the code. In the article, also results of comprehensive scaling tests are shown up to 49,152 cores, while typical production simulations have been conducted using up to 20,000 cores. According to the authors, the simulation framework FLEXI has now gained sufficient maturity to reliably and efficiently perform large-scale time-resolving simulations for industrially relevant problems on massively parallel supercomputers.

The presented selection of projects reflects the state-of-the-art in supercomputer simulation in the field of CFD. It illustrates that High Performance Computing as core enabling technology allows to tackle problem classes, which seemed unreachable only a few years ago. Furthermore, it impressively demonstrates the power of modern computer simulation methods aiming at more reliable and more efficient predictions in industrial design processes and a better understanding of sophisticated flow physical phenomena. However, it becomes more and more obvious that the engineers and natural scientists, in order to fully exploit the potential offered by modern supercomputers, must not focus solely on their physical problem, but additionally have to think about appropriate numerical algorithms as well as to get their teeth into the optimization of their codes with respect to the specific hardware architecture they are going to use. In case this exceeds the capabilities of one individual, different experts in the respective fields should work together. The very fruitful collaboration of researchers with experts from the HLRS and the SCC in many of the projects shows how this can work. Thanks are owed in this respect to the staff of both computing centres for their individual and tailored support. Without their efforts it would not have been possible and will not be possible in the future to sustain the high scientific quality we see in the projects.

Numerical Analysis of Heat Transfer During Cooling of Supercritical Fluid by Means of Direct Numerical Simulation

Sandeep Pandey, Xu Chu, and Eckart Laurien

Abstract Supercritical fluids have a wide spectrum of application, ranging from power generation to enhanced oil extraction. The sensitive nature of thermophysical properties makes the heat transfer complicated. Therefore, in this work, an investigation is made for the vertically-oriented pipe to understand the physics behind the heat transfer deterioration occurring during cooling. For that, carbon dioxide is chosen as working fluid, and direct numerical simulations with the open source finite volume code OpenFOAM have been performed with variation in the strength of body force due to buoyancy. It was found out that body force affects the axial temperature profile. Initial examination of turbulence statistics revealed that turbulence is modulated by buoyancy and deacceleration. Further investigation unveiled that long 1-dimensional structures characterized by streak elongation were present in the downward flow. In the end, Octant analysis indicates the reduction in ejection and sweep events for downward flow caused the decrease in turbulence.

1 Introduction

Supercritical fluids are widely used in different fields, their application ranges from power generation to pharmaceutical industries [1, 8]. Their use as working fluid in supercritical power plants and in transcritical refrigerators promises a better thermal performance [10, 16, 24]. In both, power generating and consuming devices, isobaric heat rejection takes place in the near-critical region. In the near-critical region, all thermophysical properties vary drastically. Figure 1 shows the variation of isobaric specific heat (c_p) and thermal conductivity (κ) for supercritical carbon dioxide (sCO$_2$). These variations bring the extra effect of buoyancy and thermal expansion

S. Pandey (✉) · E. Laurien
Institute of Nuclear Technology and Energy Systems, University of Stuttgart, Pfaffenwaldring 31, 70569 Stuttgart, Germany
e-mail: sandeep.pandey@ike.uni-stuttgart.de; eckart.laurien@ike.uni-stuttgart.de

X. Chu
Institute of Aerospace Thermodynamics, University of Stuttgart, Pfaffenwaldring 31, 70569 Stuttgart, Germany
e-mail: xu.chu@itlr.uni-stuttgart.de

© Springer International Publishing AG 2018
W.E. Nagel et al. (eds.), *High Performance Computing in Science and Engineering '17*, https://doi.org/10.1007/978-3-319-68394-2_14

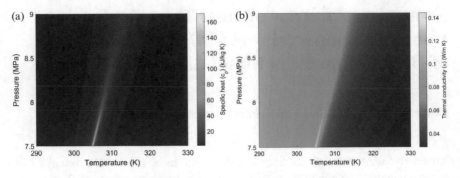

Fig. 1 Variation of thermophysical properties with pressure and temperature; left: c_p, right: κ [19]

or contraction and make the heat transfer complicated [3, 13]. Numerous attempts have been made in the past to analyze the heat transfer characteristics during the heating. These attempts include experiments [14], computational-fluid-dynamics simulation studies based on turbulence models [11, 22], and direct numerical simulation investigations [2, 4, 5, 17]. General observations from these studies were that heat transfer deterioration occurs during the upward flow and enhancement occurs during the downward flow of sCO$_2$ when the pipe is placed vertically. Also, it was reported that turbulence models are not reliable at supercritical pressure, as the peculiar behavior of heat transfer deterioration is not captured accurately with these models. While experiments only provide limited data, and understanding of deterioration and enhancement is restricted. Therefore, direct numerical simulation is an ideal candidate to analyze the heat transfer at supercritical pressure.

In the previous work under this project, heat transfer during the heating was examined for vertical [4] and horizontal [5] orientation of the pipe. In addition to this work, several other researchers have conducted DNS to study the several other phenomena. Nemati et al. [18] have studied the effect of limiting boundary conditions (i.e. isothermal and iso-heat flux) and reported that Nusselt number (**Nu**) reduced by 7% when wall temperature fluctuations are restrained. Recently, He et al. [12] investigated the effect of a streamwise body force and provided new perception for the flow relaminarization. They have categorized the body-force aided flow into three regimes: partially laminarized, fully laminarized and recovery flows based upon velocity profile and turbulent kinetic energy. None of the DNS have been conducted to investigate the heat transfer during cooling, which is equally important in transcritical refrigeration and supercritical power generation cycles.

Therefore, the work summarized in this report is the first DNS at supercritical pressure under cooling conditions at moderate inlet Reynolds number (**Re$_0$**) of 5400. Heat transfer deterioration is analyzed by varying the strength of body force. The investigation is made by examining the different turbulent statistics, streaks structure, and octant analysis. It is worth to mention here that, work under this project will provide an understanding of the heat transfer and flow characteristics of supercritical carbon dioxide. Along with it, DNS database generated here will be further used in the improvement of a semi-empirical model [20, 21] based on two-layer theory originally devised by Prandtl.

2 Computational Details

2.1 Governing Equations

Low-Mach Navier-Stokes (N-S) equations are used instead of the fully compressible N-S equations to describe the flow in this DNS. An assumption is made that during the heat transfer at supercritical pressure, the Mach number is very low so that acoustic interaction can be decoupled from the compressibility effects. This assumption was also employed in the previous DNS as well, e.g. by Bae [2], Nemati [17] and Chu [5]. In the following equations (1), (2) and (3) represent the mass, momentum and energy conservation for the current DNS, respectively:

$$\frac{\partial \rho}{\partial t} + \frac{\partial(\rho U_j)}{\partial x_j} = 0 \tag{1}$$

$$\frac{\partial \rho U_i}{\partial t} + \frac{\partial(\rho U_i U_j)}{\partial x_j} = -\frac{\partial P}{\partial x_i} + \frac{\partial}{\partial x_j}\left(\mu(\frac{\partial U_i}{\partial x_j} + \frac{\partial U_j}{\partial x_i})\right) \mp \rho g \delta_{i1} \tag{2}$$

$$\frac{\partial \rho h}{\partial t} + \frac{\partial(\rho U_j h)}{\partial x_j} = \frac{\partial}{\partial x_j}\left(\kappa \frac{\partial T}{\partial x_j}\right) \tag{3}$$

The thermophysical properties were derived from NIST REFPROP [19] and implemented by a spline function divided into different temperature ranges for a given pressure. The maximum relative error is kept below $\pm 1\%$ for the spline function. A thermophysical property (say, ϕ), enthalpy (h), inlet pressure (P_0) and temperature (T) is related as $\phi = f(h(P_0, T))$.

2.2 Numerical Method and Flow Domain

The above mentioned governing equations are discretized with the open-source finite-volume code OpenFOAM V2.4 [27]. The Pressure-Implicit-with-Splitting-of-Operators (PISO) algorithm was applied to couple the pressure and velocity term in the momentum equation. Different discretization schemes were employed for solving PDEs; spatial discretization was made with the central differencing scheme and temporal discretization was performed by a second order implicit differencing scheme. Convective terms in the energy equation were solved by the third order upwind scheme QUICK. The code is overall second-order accurate in both space

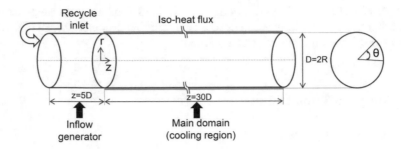

Fig. 2 Flow domain and boundary conditions

and time. To speed up the initial turbulence development, a perturbation method [23] was used.

The simulation domain consists of a tube of 2 mm diameter with a total length of 35 diameters ($35D$) as shown in Fig. 2. The tube is divided into two parts namely, the inflow generator and the main domain. The inflow generator is made up of an initial length of $5D$, and objective of this section is to provide a fully developed turbulent flow for the main domain. For this purpose, a recycling/rescaling procedure [15] is adopted with the isothermal wall. A resolution of $120\times360\times300$ was chosen for the inflow generator with a friction Reynolds number, $\mathbf{Re}_\tau = \frac{Ru_\tau}{\nu}$, of 180. Here, R is the radius of tube, ν is the kinematic viscosity and u_τ is the friction velocity calculated as: $u_\tau = \sqrt{\tau_w/\rho}$. The wall shear stress (τ_w) is based on the inlet Reynolds number $\left(\mathbf{Re}_0 = \frac{GD}{\mu_0}\right)$ of 5400 and it is calculated as: $\tau_w = \frac{1}{8}C_f\rho U^2$, with Darcy friction factor, $C_f = \frac{0.316}{\mathbf{Re}^{0.25}}$.

The main domain can be identified by a negative uniform heat flux for the remaining section of $30D$. In this domain, the no-slip boundary condition is imposed at the wall along with the convective outflow boundary condition for the velocity and other variables at the outlet. This domain is discretized with a resolution of $120\times360\times1800$. The mesh is refined in the near-wall region, which corresponds to Δy^+ between 0.1 (wall) and 0.2 (center). The dimensionless resolution in the circumferential direction $(\Delta R\theta)^+$ is 3.1 at the center. A uniform grid spacing is employed in the axial direction and it gives a dimensionless resolution of, $\Delta z^+ = 6.0$. Here, mesh resolution is given in wall units governed by $y^+ = yu_{\tau,0}/\nu_0$, based upon the inlet Reynolds number. The cylindrical pipe is discretized with a total of 80 Mio. structured hexahedral mesh. This resolution is little higher than the past DNS of Nemati [17] and Chu [5]. During the cooling, Reynolds number also decreases and it provides an opportunity to use a lesser resolution here than the heating. The mesh resolution is identical in all the simulation cases. In the post processing, the mesh coordinate transformation from Cartesian coordinate to Cylindrical coordinate is necessary. The flow statistics are obtained by averaging in time.

This numerical method has been validated previously with the air-flow experiments [6], and with other DNS [5]. Therefore, further validation is not provided in this report. In this study, attention is drawn to cooling of supercritical CO_2 while aiming to investigate the behavior of turbulence under the effects of buoyancy. Hence, different conditions are simulated through four distinct cases shown in Table 1. Here, Q^+ shows the non-dimensional cooling heat flux $\left(\frac{q_w R}{\kappa_0 T_0}\right)$. The case without gravity is referred to as the forced convection (*FC*) and the case with the large wall heat flux is referred to as the double heat flux case (*2DC*).

2.3 Inflow Turbulence

To justify the mesh resolution for the inflow generator section, a comparison with the data of KTH-FLOW project [9] is presented. They have used a high-order spectral element code based upon Galerkin approximation with 18.6×10^6 total grid points and inlet Reynolds number of 5300. A comparison is made with the root mean square (RMS) value of the velocity component. An excellent agreement can be seen in Fig. 3 and it can be concluded that inflow turbulence to the main domain is fully developed.

Table 1 Simulation conditions, identical inlet conditions at $\mathbf{Re}_0 = 5400$, $P_0 = 8\,\text{MPa}$

Case	Type	Direction	q_w (kW/m^2)	$\mathbf{Q^+}$	T_0 (K)	$U_{z,0}$ (m/s)
FC	Forced (g=0)	–	−30.87	−4.32	342.05	0.3073
UC	Mixed	Upward	−30.87	−4.32	342.05	0.3073
DC	Mixed	Downward	−30.87	−4.32	342.05	0.3073
2DC	Mixed	Downward	−61.74	−8.65	342.05	0.3073

Fig. 3 Inflow turbulence validation; symbols: KTH FLOW data [9], solid line: present DNS; top: u_z; middle: u_θ; bottom: u_r

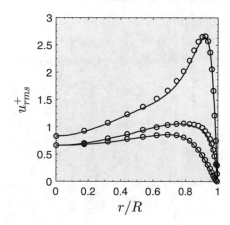

3 Results and Discussion

Direct numerical simulations have been performed with the conditions of Table 1 and various results are discussed in this section. In this section, any quantity (say ϕ) is either Reynolds averaged or Favre averaged. The Reynold averaged quantity is shown as $\overline{\phi}$ (mean part) and ϕ' (fluctuating part). On the other hand, Favre averaged quantity is decomposed in mean part as, $\widetilde{\phi}=\overline{\rho\phi}/\overline{\rho}$, and its fluctuating part is denoted by ϕ''.

3.1 Heat Transfer and Flow Characteristics

Figure 4a shows the variation in wall temperature (T_w), and it is crucially affected by the direction of fluid flow and thermal contraction. In case UC, the wall temperature drops, and it is relatively higher than all other cases. It is an indication of enhanced heat transfer. Since in case UC, buoyancy and fluid deceleration are present unlike case FC where only deceleration favors the heat transfer. Therefore, the trend of wall temperature is similar as of case UC, but with a reduced value. On the other hand, in case DC and $2DC$, T_w monotonically decreases as illustrated in Fig. 4a. This represents the heat transfer deterioration. Figure 4b shows the variation of wall shear stress. For upward flow (case UC), its value is the minimum among all and for a very short region, it shows a negative value and it indicates flow recirculation. In contrast, for forced convection, the wall shear stress remains approximately constant, and in downward flow the wall shear stress increases in the flow direction.

The external effects of buoyancy and flow deceleration significantly affect the mean velocity profile. Figure 5 shows the radial velocity profile for cases UC, FC,

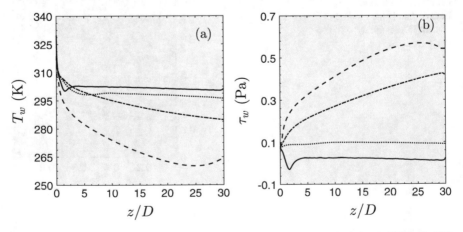

Fig. 4 Variation of (**a**) Wall temperature (T_w), (**b**) Wall shear stress (τ_w) for (solid line): UC; (dotted line): FC; (dash-dotted line): DC; (dashed line): $2DC$

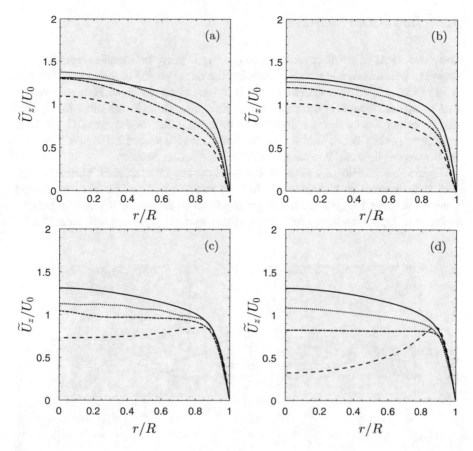

Fig. 5 Mean velocity profile (**a**) Case *UC*, (**b**) Case *FC*, (**c**) Case *DC*, (**d**) Case 2*DC*; for (solid line): $z = 0D$; (dotted line): $z = 7.5D$; (dash-dotted line): $z = 15D$; (dashed line): $z = 27.5D$

DC and 2*DC* between $r/R = 0$ (center) and $r/R = 1$ (wall). At supercritical pressure, the thermophysical-properties variation deforms the mean velocity profile significantly compared to that of constant property flow (i.e. at $z = 0$). The effects of thermal contraction can be seen in the forced convection (case *FC*) in Fig. 5b, where the fluid is decelerating. In case *DC*, the fluid also decelerates but the rate of deceleration is more than for the forced convection and it increases with the increase in the heat flux as depicted in case 2*DC*. At $z = 27.5D$, the velocity profile deforms into the 'M' shape profile for 2*DC* case as shown in Fig. 5d. The 'M' shape velocity profile indicates an increase in turbulence production and it is shown in next section.

3.2 Turbulence Statistics

Only two statistics are discussed here to show qualitative information regarding the impaired heat transfer. Figure 6 is showing the radial profile of the Reynolds shear stress ($RSS = \overline{\rho U_r'' U_z''}$) for different axial locations along the pipe. Figure 6a shows that in upward flow RSS increases in the flow direction. On the other hand, RSS decreases and it even shows a negative value at the outlet for the case DC and $2DC$ as depicted in Fig. 6c, d. This shows that turbulence is reduced in the flow direction for downward flow and it explains the heat transfer deterioration.

Figure 7 is showing the contribution of buoyancy into turbulent kinetic energy and it is termed as buoyancy production rate ($B_k = \mp g \overline{\rho' U_z'}$). For the forced convection (case FC), B_k is zero as gravity is set to zero in it. While it is present in the mixed convection and its contribution cannot be neglected. For case UC, B_k

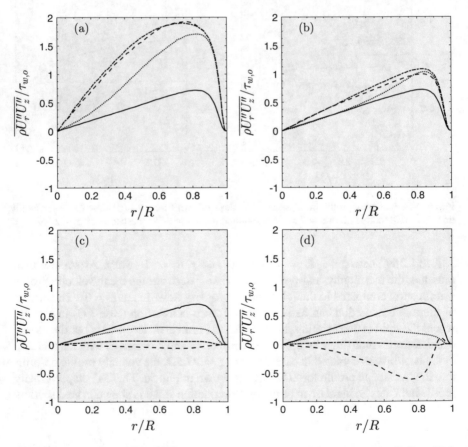

Fig. 6 Reynolds shear stress (RSS) for (**a**) Case UC, (**b**) Case FC, (**c**) Case DC, (**d**) Case $2DC$; legends are identical to Fig. 5

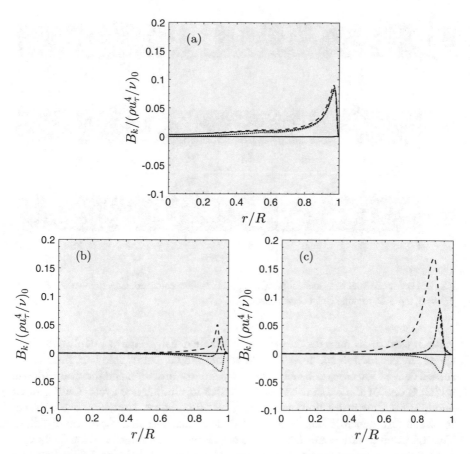

Fig. 7 Turbulent production due to buoyancy (B_k) for (**a**) Case *UC*, (**b**) Case *DC*, (**c**) Case *2DC*; legends are identical to Fig. 5

remains positive due to unstable stratification for the entire length as illustrated in Fig. 7a. On the other hand, B_k is negative in downward flow (case *DC* and *2DC*), but soon enough it becomes positive and double heat flux has higher magnitude due to larger body force. In the downward cooling, both stable and unstable stratification can occur, therefore, buoyancy can act as source or sink. This observation is contrary to the heating of sCO_2, in which downward flows are always unstably stratified and upward flows can either stably or unstably stratified [17].

3.3 Flow Visualization

To further gain the understanding of deteriorated heat transfer in the downward flow case, low-speed streaks are visualized here. For that purpose, the negative fluctuating part of the streamwise velocity ($u_z'' < 0$) is visualized and the Favre-averaged

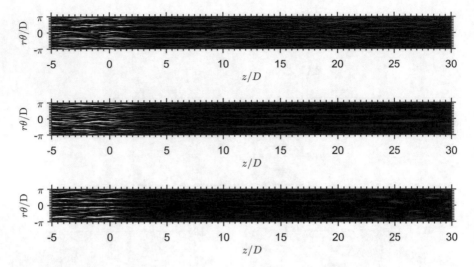

Fig. 8 Low speed streaks visualization with $u_z'' < 0$, lighter color indicates high value of $|u_z''|$ at $y^+ = 5$; top: *FC*, middle: *DC*, bottom: *2DC*

value is used to account for the effect of variable density. Figure 8 depicts the low-speed streaks for forced, downward flow with single and double heat flux. In the forced convection case, one can observe the streak formation and their breakdown, which is one of the reason for the self-sustaining turbulence cycle. On the other hand, in the downward flow with single heat flux, streak stretching can be seen, in which turbulence is impaired and 1-dimensional turbulence prevails the flow. The reduction in turbulence leads to the deteriorated heat transfer in the downward flow. If one compares the single and double heat flux for downward flow, it can be observed that streak stretching disappears near the outlet for double heat flux. It can nicely be linked with the axial temperature variation and 'M'-shape velocity profile. At $z \approx 20D$, the temperature starts decreasing again (better heat transfer) and the velocity profile also deforms into 'M' shape, and it is termed as turbulence recovery which increases the heat transfer again.

3.4 Octant Analysis

Octant analysis is a technique to conditionally average the *RSS* to investigate the eddy structure in any turbulent flow. In this technique, Reynolds shear stress (*RSS*) is decomposed based upon the sign of $\sqrt{\rho}U_r''$, $\sqrt{\rho}U_z''$ and T' [25, 26]. This decomposition gives the *RSS* value in different Octant and each Octant represents a particular eddy motion. For e.g. cold ejection event $\sqrt{\rho}U_r'' > 0$, $\sqrt{\rho}U_z'' < 0$ and $T' < 0$ occurs in octant-6 (*O6*) while hot sweep occurs when $\sqrt{\rho}U_r'' < 0$,

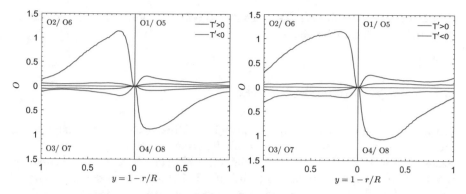

Fig. 9 Octant analysis for *RSS* for case *UC*, left for $z = 7.5D$ and right for $z = 27.5D$

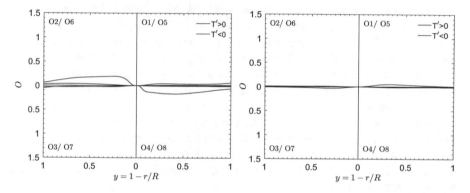

Fig. 10 Octant analysis for *RSS* for case *DC*, left for $z = 7.5D$ and right for $z = 27.5D$

$\sqrt{\rho} U_z'' > 0$ and $T' > 0$ in Octant-8 (*O8*). Figure 9 shows the Octant analysis for upward flow, where better heat transfer was observed compared with the downward case. At $z = 7.5D$, ejection by cold fluid and sweep by warm fluid have the largest contribution to the *RSS*. It implies that warm fluid is advancing towards the wall with the high streamwise velocity (sweep by the warm fluid) and at the same time, colder fluid is displacing away from the wall with a high wall-normal velocity (ejection by the cold fluid). The contribution of other quadrants also increases with the streamwise direction. Figure 10 is showing the same for the downward flow case for $z = 7.5D$ and $27.5D$. The global trend is similar to the upward flow case at $z = 7.5D$ as illustrated in Fig. 10a. As compared with the case *UC*, all the Octant show reducing magnitude in the streamwise direction and become zero at the outlet. Ejections of cold fluid dominate due to turbulent bursting and sweep by the warm fluid is equally important in any turbulent flow. This indicates that ejection moves the cold fluid away from the wall and to replace it, sweep event occurs which brings the warm fluid from the core. In the downward case, the magnitude of the ejection and sweep events decreases down significantly. This explains the heat transfer deterioration characterize by the poor heat transfer in the downward flow under cooling conditions.

3.5 Computational Performance

For the direct numerical simulations, the high-performance computer named 'Hazel Hen' was used, located at the High-Performance Computing Center (HLRS) Stuttgart. It is a Cray® XC40 system and it has 7712 computing nodes. Each node has two Intel® Haswell processors (E5-2680 v3, 12 cores) and 128 GB memory. It makes a total of 185,088 cores and a theoretical peak performance of 7.42 PFlops. The nodes are interconnected by Cray® Aries network technology, which is an innovative inter-communication technology and it uses a high-bandwidth and low-diameter network topology known as Dragonfly. In the past, scalability tests were conducted to find out the optimum number of cores to be used with the current setting of solver [7]. Figure 11 shows the scaling of the current solver and it can be seen that in the range of 400–1400 cores, a linear speed up was achieved along with the parallel efficiency in the range of 90–110%. Approximately 62,000 cells are distributed on every core with 1200 computational cores. During this study, we stored only two latest time-step data to avoid the limitation of number of files and the memory. Thus, averaging in time was done on the run and as a post-processing step, other calculations such as coordinate transformation and Favre-averaging was performed. This also saves the computational-time during reconstructing data and averaging afterward.

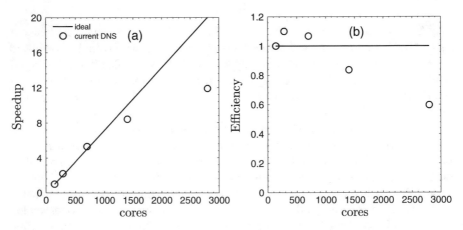

Fig. 11 Scaling characteristics of current solver: (**a**) Speedup. (**b**) Parallel efficiency [7]

4 Conclusions

In this work, an extensive analysis was performed to understand the mechanism of heat transfer during the cooling of supercritical fluids. For that, direct numerical simulation was used as a tool due to its higher accuracy. The DNS was performed at an inlet temperature of 342.05 K, inlet pressure of 8 MPa, and at a moderate Reynolds number of 5400. Four different cases were simulated to study the effects of body force brought by gravity (i.e. buoyancy), heat flux and flow direction. The results show different trends for heat transfer and flow quantities. The wall temperature remains nearly constant in upward and forced convection cases and deterioration was observed in the downward flow. The deterioration of heat transfer depends upon the strength of the buoyancy force as observed from the case with single and double heat flux. In the forced convection, fluid deceleration was noticed, and in mixed convection, buoyancy also has its significant effect on flow, and it deformed the velocity profile. Turbulence statistical analysis has shown that turbulence reduced significantly in the downward flow case as compared to forced convection. The decomposition of *RSS* was done in the octant analysis, and it revealed that cold ejections combined with hot sweep are the major contributor to turbulent momentum flux. Moreover, it has been shown that reduction in ejection and sweep events and it causes heat transfer deterioration in the downward flow.

Acknowledgements The research summarized in this report is supported by the Forschungsinstitut für Kerntechnik und Energiewandlung (KE) e.V., Stuttgart, for project DNSTHTSC. The authors are sincerely thankful to the High-Performance Computing Center (HLRS) Stuttgart and Cray team for their support.

References

1. Y. Ahn, S.J. Bae, M. Kim, S.K. Cho, S. Baik, J.I. Lee, J.E. Cha, Review of supercritical CO2 power cycle technology and current status of research and development. Nucl. Eng. Technol. **47**(6), 647–661 (2015)
2. J.H. Bae, J.Y. Yoo, H. Choi, Direct numerical simulation of turbulent supercritical flows with heat transfer. Phys. Fluids **17**, 105104 (2005)
3. M. Bazargan, D. Fraser, V. Chatoorgan, Effect of buoyancy on heat transfer in supercritical water flow in a horizontal round tube. J. Heat Transf. **127**(8), 897–902 (2005)
4. X. Chu, E. Laurien, Direct numerical simulation of heated turbulent pipe flow at supercritical pressure. J. Nucl. Eng. Radiat. Sci. **2**, 031019 (2016)
5. X. Chu, E. Laurien, Flow stratification of supercritical CO2 in a heated horizontal pipe. J. Supercrit. Fluids **116**, 172–189 (2016)
6. X. Chu, E. Laurien, D.M. McEligot, Direct numerical simulation of strongly heated air flow in a vertical pipe. Int. J. Heat Mass Transf. **101**, 1163–1176 (2016)
7. X. Chu, E. Laurien, S. Pandey, *Direct Numerical Simulation of Heated Pipe Flow with Strong Property Variation* (Springer, Cham, 2016), pp. 473–486
8. A.R.C. Duarte, J.F. Mano, R.L. Reis, Supercritical fluids in biomedical and tissue engineering applications: a review. Int. Mater. Rev. **54**(4), 214–222 (2009)

9. G.K. El Khoury, P. Schlatter, A. Noorani, P.F. Fischer, G. Brethouwer, A.V. Johansson, Direct numerical simulation of turbulent pipe flow at moderately high Reynolds numbers. Flow Turbul. Combust. **91**(3), 475–495 (2013)

10. E. Feher, The supercritical thermodynamic power cycle. Energy Convers. **8**(2), 85–90 (1968)

11. S. He, W. Kim, J. Bae, Assessment of performance of turbulence models in predicting supercritical pressure heat transfer in a vertical tube. Int. J. Heat Mass Transf. **51**(19–20), 4659–4675 (2008)

12. S. He, K. He, M. Seddighi, Laminarisation of flow at low Reynolds number due to streamwise body force. J. Fluid Mech. **809**, 31–71 (2016)

13. J.D. Jackson, Fluid flow and convective heat transfer to fluids at supercritical pressure. Nucl. Eng. Des. **264**, 24–40 (2013)

14. D.E. Kim, M.H. Kim, Experimental investigation of heat transfer in vertical upward and downward supercritical CO_2 flow in a circular tube. Int. J. Heat Fluid Flow **32**(1), 176–191 (2011)

15. T.S. Lund, X. Wu, K.D. Squires, Generation of turbulent inflow data for spatially-developing boundary layer simulations. J. Comput. Phys. **140**(2), 233–258 (1998)

16. P. Nekså, H. Rekstad, G. Zakeri, P.A. Schiefloe, Co2-heat pump water heater: characteristics, system design and experimental results. Int. J. Refrig. **21**(3), 172–179 (1998)

17. H. Nemati, A. Patel, B.J. Boersma, R. Pecnik, Mean statistics of a heated turbulent pipe flow at supercritical pressure. Int. J. Heat Mass Transf. **83**, 741–752 (2015)

18. H. Nemati, A. Patel, B.J. Boersma, R. Pecnik, The effect of thermal boundary conditions on forced convection heat transfer to fluids at supercritical pressure. J. Fluid Mech. **800**, 531–556 (2016)

19. NIST Chemistry WebBook, in *NIST Standard Reference Database Number 69*, ed. by E. Lemmon, M. McLinden, D. Friend, P. Linstrom, W. Mallard (National Institute of Standards and Technology, Gaithersburg, 2011)

20. S. Pandey, E. Laurien, Heat transfer analysis at supercritical pressure using two layer theory. J. Supercrit. Fluids **109**, 80–86 (2016)

21. S. Pandey, E. Laurien, X. Chu, A modified convective heat transfer model for heated pipe flow of supercritical carbon dioxide. Int. J. Therm. Sci. **117**, 227–238 (2017)

22. A. Pucciarelli, A. Borroni, M. Sharabi, W. Ambrosini, Results of 4-equation turbulence models in the prediction of heat transfer to supercritical pressure fluids. Nucl. Eng. Des. **281**, 5–14 (2015)

23. W. Schoppa, F. Hussain, Coherent structure dynamics in near-wall turbulence. Fluid Dyn. Res. **26**(2), 119–139 (2000)

24. T. Schulenberg, J. Starflinger, P. Marsault, D. Bittermann, C. Maráczy, E. Laurien, J.L. Nijeholt, H. Anglart, M. Andreani, M. Ruzickova, A. Toivonen, European supercritical water cooled reactor. Nucl. Eng. Des. **241**(9), 3505–3513 (2011); Seventh European Commission conference on Euratom research and training in reactor systems (Fission Safety 2009)

25. R. Volino, T. Simon, An application of octant analysis to turbulent and transitional flow data. J. Turbomach. **116**(4), 752–758 (1994)

26. J.M. Wallace, Quadrant analysis in turbulence research: history and evolution. Annu. Rev. Fluid Mech. **48**(1), 131–158 (2016)

27. H.G. Weller, G. Tabor, H. Jasak, C. Fureby, A tensorial approach to computational continuum mechanics using object-oriented techniques. Comput. Phys. **12**(6), 620–631 (1998)

Control of Traveling Crossflow Vortices Using Plasma Actuators

Philipp C. Dörr, Zhengfei Guo, Johannes M.F. Peter, and Markus J. Kloker

Abstract It has been shown recently by direct numerical simulations that plasma actuators can be used to delay laminar-turbulent transition caused by steady cross-flow vortices (CFVs) in three-dimensional boundary layers on swept aerodynamic surfaces. In the current work the applicability of such actuators to control transition caused by traveling CFVs is explored by two techniques. In the first technique, named upstream flow deformation, the actuators are used to excite steady CFV control modes. The resulting narrow spaced control CFVs induce a beneficial mean-flow distortion and weaken the primary crossflow instability, yielding delayed transition. In the second technique, the direct attenuation of nonlinear traveling CFVs, the actuators are positioned more downstream, where the traveling CFVs have already established. The localized unsteady forcing against the direction of the crossflow is then aimed at attenuating the amplitude of the traveling CFVs by directly tackling the three-dimensional nonlinear disturbance state. With both techniques transition can be delayed, however with a significantly higher efficiency for the method of upstream flow deformation.

1 Introduction

Due to the leading-edge sweep, a distinctly three-dimensional boundary-layer flow with a crossflow (CF) component perpendicular to the main flow direction develops on the wings and stabilizers of modern aircraft. The inflectional CF velocity profile invokes the so-called CF instability, yielding strong amplification of both steady and traveling CFV modes. In environments with non-negligible free-stream turbulence the traveling modes dominate. The resulting nonlinear traveling CFVs create localized high-shear layers traveling with them and trigger a secondary instability that rapidly causes transition to turbulence. For further details on this transition process see e.g. [19, 21, 22].

P.C. Dörr (✉) • Z. Guo • J.M.F. Peter • M.J. Kloker
Institut für Aero- und Gasdynamik, Universität Stuttgart, Pfaffenwaldring 21, 70550 Stuttgart, Germany
e-mail: doerr@iag.uni-stuttgart.de; guo@iag.uni-stuttgart.de; johannes.peter@iag.uni-stuttgart.de; kloker@iag.uni-stuttgart.de

© Springer International Publishing AG 2018
W.E. Nagel et al. (eds.), *High Performance Computing in Science and Engineering '17*, https://doi.org/10.1007/978-3-319-68394-2_15

Plasma actuators generate a body force that locally accelerates the surrounding fluid, making them suitable for the application in laminar flow control. Furthermore, they are fully electronic without moving parts and flush-mountable. Hence, they do not affect the surface quality when being inactive. Comprehensive overviews of current research and the actuators' working principle are given in review articles, see e.g. [3, 4, 15].

Various approaches for delaying the crossflow-induced transition have been developed, see [10, 12, 17, 18, 21]. Successful application of plasma actuators employing the techniques of base-flow stabilization [6], direct attenuation of nonlinear CFVs [7] and upstream flow deformation (UFD) [8] to control transition caused by steady CFVs has been shown recently by Dörr and Kloker. The UFD method was found to be the most efficient because the instability inherent to the flow is exploited.

In the current work we scrutinize the applicability of the UFD method and the direct attenuation of nonlinear CFVs to control transition caused by traveling CFVs. When employing the UFD method, the plasma actuators are used to excite steady CFV control modes with a spanwise wavenumber of 3/2 the wavenumber of the most unstable steady CFV mode. The resulting nonlinear steady control (UFD) CFVs induce a beneficial mean-flow distortion and the primary CF instability is weakened, yielding delayed transition. For the direct attenuation approach the plasma actuators are positioned farther downstream. The localized unsteady forcing against the direction of the CF is then aimed at decreasing the amplitude of the traveling CFVs that move over the unsteadily driven actuators by tackling the three-dimensional disturbance state. The base flow used for the current investigations resembles the recently re-designed DLR-Göttingen swept flat-plate experiment [2, 8]. It consists of a flat plate with a displacement body above to induce a chordwise varying pressure distribution on the plate, modeling the three-dimensional boundary-layer flow as it develops in the front region on the upper side of a swept wing.

The paper is structured as follows: in Sect. 2 the numerical setup for the investigations on transition control is described, including the modeling of the plasma actuators and vortical free-stream disturbances (FSDs). Section 3 presents the results using the UFD method (Sect. 3.1) and direct attenuation (Sect. 3.2). An additional code-performance study is shown in Sect. 4.

2 Numerical Setup

2.1 Basic Setup

The in-house DNS code NS3D [1] in the asymmetric base-flow setting [6, 10] is used for all simulations, solving the time-dependent, three-dimensional, compressible Navier-Stokes equations together with the continuity and energy equation in

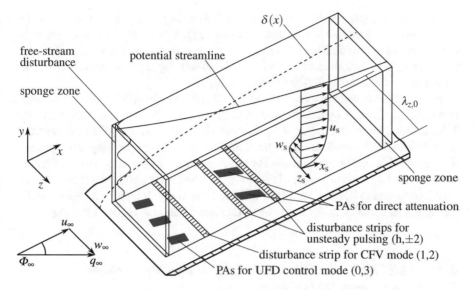

Fig. 1 Integration domain and coordinate systems

conservative formulation and with explicit 4th-order Runge-Kutta time stepping. The integration domain used for the investigations of Sect. 3 is shown Fig. 1. It consists of block-structured Cartesian grids. Explicit finite differences of 8th-order are used for the discretization in the x- and y-direction, and in the z-direction a Fourier-spectral ansatz is employed. Velocities and length scales are normalized by the chordwise reference velocity $\bar{U}_\infty$ and the reference length $\bar{L}$, respectively, the overbar denoting dimensional values. The vector $\boldsymbol{u} = [u, v, w]^T$ denotes the velocity components. Whereas the free-stream velocity $\bar{q}_{\infty,\text{exp}}$ in the experiments was set to $30\,\mathrm{m\,s^{-1}}$, yielding a Mach number $Ma_{\infty,\text{exp}} = 0.066$ based on $\bar{U}_{\infty,\text{exp}}$, the simulations are performed at $Ma_\infty = 0.20$ to allow for a reasonable time step. The ambient conditions and the Reynolds number range are kept to establish comparable flow regimes. For the DNS the following reference values result: $\bar{U}_\infty = 68.871\,\mathrm{m\,s^{-1}}$, $\bar{L} = 0.033\,\mathrm{m}$, $\bar{T}_\infty = 295.0\,\mathrm{K}$ and $\bar{\rho}_\infty = 1.181\,\mathrm{kg\,m^{-3}}$. The reference angular frequency is set to $\omega_0 = 6.0$, and the time step to $\Delta t = 8.727 \times 10^{-6}$. In the x-direction the domain covers $0.173 \leqslant x \leqslant 4.078$ with $N_x = 2604$ points, $\Delta x = 1.500 \times 10^{-3}$. In the y-direction a stretched grid with $N_y = 152$ points is used, covering $0.000 \leqslant y \leqslant 0.121$, $\Delta y_{\text{wall}} = 1.251 \times 10^{-4}$. The fundamental spanwise wavelength is $\lambda_{z,0} = 0.180$, corresponding to the fundamental spanwise wavenumber $\gamma_0 = 2\pi/\lambda_{z,0} = 35.0$. For the investigations on the UFD method, Sect. 3.1, the z-direction is discretized with $K = 21$ de-aliased Fourier modes (64 points). For the cases in Sect. 3.2 the domain width is reduced to $\lambda_{z,0}/2$ and 10 modes are used, keeping $\Delta z = 2.805 \times 10^{-3}$. A disturbance strip at the wall centered at $x = 0.8$ with synthetic blowing and suction, alternating in z, is used in Sect. 3.2 to excite the CFV mode $(1,2)$ (double-spectral notation $(h\omega_0, k\gamma_0)$ used). Farther downstream, centered at $x = 2.0$ and $x = 2.5$, respectively, two additional strips with pulse-like (background) disturbances $(h, \pm 2)$, $h = 1–50$, can be activated to initiate controlled breakdown of laminar flow downstream. For further details on the basic setup, also including information on boundary-layer parameters and stability properties of the investigated base flow, see [8, 9].

All simulations were run on the Cray XC40 (Hazel Hen) HPC system at the High Performance Computing Center Stuttgart (HLRS). The domain is decomposed into 248 sub-domains, with size $N_x \times N_y \times N_z$ of $42 \times 38 \times 32/64$ points for the cases with 10/21 Fourier-modes. Each sub-domain and its corresponding MPI process was assigned to one compute node. 24 OpenMP threads were used for the Fourier-spectral ansatz, yielding a total of 5952 CPUs. The average wall-clock time for the simulations was ≈ 12 h for the cases with 10 Fourier-modes and ≈ 25 h for the 21-modes cases, for the simulation of 14 fundamental periods/168,000 time steps.

2.2 Modeling of the Plasma Actuators and the Vortical Free-Stream Disturbance

The effect of a dielectric barrier discharge (DBD) plasma actuator is modeled by a dimensionless wall-parallel body force $f(x, y, z)$. An empirical model [16] is employed to calculate the planar force distribution, and an angle β_{PA} can be defined for rotation of the actuator about the wall-normal axis y, see Fig. 2. The dimensional force distribution $\bar{f}(x, y, z)$ corresponding to $f(x, y, z)$ can be calculated according to equation (1) given in [7]. Important actuator parameters for the current investigations are provided in Table 1. For case CNL-ACF with an unsteady body force, see Sect. 3.2, the force is modeled according to

$$f(x, y, z, t) = f(x, y, z) \left[C_s + C_u \, sin \left(\omega_{PA} t + \Theta_{PA} \right) \right], \tag{1}$$

where the constants C_s and C_u are both set to 0.5 to yield a sinusoidally varying force between zero and a positive maximum. The plasma actuator operating frequency

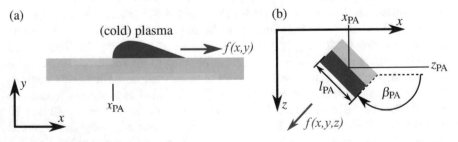

Fig. 2 (a) DBD plasma-actuator schematic and modeled body-force. (b) Rotation of the actuator about the wall-normal axis by the angle β_{PA}

Table 1 Plasma-actuator parameters for the calculations presented; $\max\{f\}$ = $\max\{[f_x^2 + f_z^2]^{1/2}\}$

Case	x_{PA}	l_{PA}	z_{PA}	β_{PA}	ω_{PA}	$\max\{f\}$	$\max\{\bar{f}\}$	c_μ
	(–)	(–)	(–)	(°)	(–)	(–)	(kN m^{-3})	(%)
UFD-ACF	0.50	0.40	0.000, 0.060, 0.120	150.0	-	0.28	1.68	0.0012
CNL-ACF	2.25	0.60	0.025, 0.115	130.2	6	0.84	5.08	0.0068

The dimensional force $\bar{f}$ corresponding to the $Ma_{\infty,\mathrm{exp}} = 0.066$ experiment is calculated according to equation (1) given in [7] ($\bar{U}_{\infty,\mathrm{exp}} = 22.657\,\mathrm{m\,s}^{-1}$, $\bar{L}_{\mathrm{exp}} = 0.1\,\mathrm{m}$, $\bar{\rho}_\infty = 1.181\,\mathrm{kg\,m}^{-3}$). The momentum coefficient c_μ is calculated according to equation (3) given in [7] (value given per actuator; for case CNL-ACF it corresponds to its maximum in time)

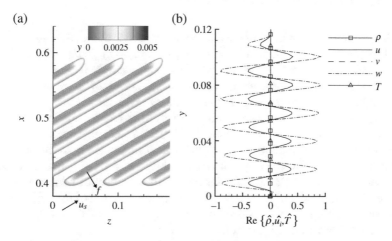

Fig. 3 (a) Plasma actuator body-force distribution for case UFD-ACF with forcing against the CF. $f_{10\%}$-isosurface shown ($f_{10\%} = \max\{[f_x^2 + f_z^2]^{1/2}\}/10 = 0.028$). The color indicates the wall-normal distance y. (b) Wall-normal shape of the free-stream mode imposed for cases REF-1 and UFD-ACF. $x_{FSD} = 0.214$, $\alpha_{r,FSD} = 157.2$, $\alpha_{i,FSD} = 2.43$, $\gamma_{FSD} = -70.0$, $\xi_{FSD} = 308.5$, $\omega_{FSD} = -6.0$

ω_{PA} is set to 6.0 and the phase Θ_{PA} to $7/4\,\pi$. Consequently, the maximum force $\max\{f\} = 0.84$ is reached at $t = 3/8\,T_0$, and the force is zero at $t = 7/8\,T_0$ ($T_0 = 2\,\pi/\omega_0$). Note that setting $\omega_{PA} = 6.0$ defines a low-frequency modulation of the body force ($\omega_{PA} = 2\,\pi\,\bar{\nu}_{PA}\,\bar{L}/\bar{U}_\infty = 6.0$ corresponds to $\bar{\nu}_{PA} = 217\,\mathrm{Hz}$ for the investigated experiment). This does not model the frequency of the AC high voltage needed to operate the actuator which can easily be set to 10 kHz or higher. According to our previous investigations [5–7, 9], the effect of high-frequency fluctuations on the flow field is negligible for the current investigations on transition control. For further details on the body-force modeling see [5–7].

For the simulations presented in Sect. 3.1 a single unsteady vortical free-stream disturbance (FSD) is forced at $x_{FSD} = 0.214$, near the inflow boundary, to excite the traveling CFV test mode $(1, 2)$. The FSD is modeled by an eigenmode from the continuous spectrum of the underlying compressible linear stability problem, see Fig. 3b. At the upper boundary a 5th-order polynomial is employed to ramp the disturbance to zero over 10 grid points. The amplitude of the FSD is set such that the maximum of $|\hat{u}_{s,FSD}| = 0.005$ and γ_{FSD} and ω_{FSD} are -70.0 and -6.0, respectively. The wall-normal wavenumber is calculated according to equation (2.14) given in [20], yielding $\xi_{FSD} = 308.5$ for the investigated FSD. The FSD wave vector in the wall-parallel plane is approximately aligned with the one of the expected, amplified traveling CFV mode.

3   Results

3.1   Upstream Flow Deformation

The downstream development of the maximum amplitude of the streamline-oriented disturbance velocity component $\tilde{u}'_s = u'_s/u_{B,s,e} = (u_s - u_{B,s})/u_{B,s,e}$ for case REF-1 with FSD but without active plasma actuators is shown in Fig. 4a. Upstream of $x \approx 0.5$ the maximum amplitude of the mode $(1, 2)$ is connected to the decaying FSD. When impinging on the boundary layer, the FSD excites the CFV mode $(1, 2)$ in the layer and exponential growth of the mode sets in at its neutral position at $x = 0.36$. At $x \approx 0.5$ the CFV mode eventually overtakes the FSD in amplitude and the disturbance development then is clearly connected to the exponential growth governed by primary instability of the CFV mode. To initiate controlled breakdown by secondary instability, the disturbance strip at $x = 2.5$ is activated. Transition to turbulence sets in directly downstream of the strip, indicated by the strong growth of the unsteady modes. For case UFD-ACF plasma actuators with forcing against the CF are activated. To excite the UFD control mode $(0, 3)$ three actuators per fundamental spanwise wavelength are used, see Fig. 3a. The resulting disturbance development is shown in Fig. 4b. The UFD control mode

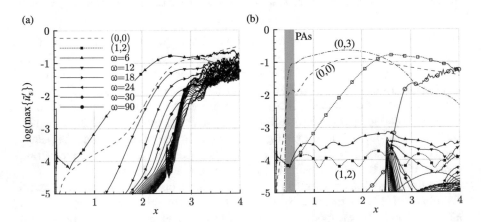

Fig. 4 Downstream development of modal $\tilde{u}'_s$, (h, k)- and $\tilde{u}'_s$, (h)-amplitudes for (**a**) case REF-1 and (**b**) case UFD-ACF from Fourier analysis in time (maximum over y or y and z, $0 \leq \omega \leq 180$, $\Delta\omega = 6$). Open symbols in (**b**) denote case REF-1

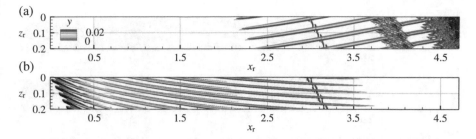

Fig. 5 Vortex visualization (snapshots at $t/T_0 = 14$, $\lambda_2 = -4$, colour indicates y) for (**a**) case REF-1 and (**b**) case UFD-ACF. Body-force distribution ($f_{10\%}$-isosurface, dark) added for case UFD-ACF. Rotated reference system with $x_0 = 0.4$, $z_0 = 0.0$, $\Phi_r = 45°$ used; $x_r = (x - x_0) \cos \Phi_r + (z - z_0) \sin \Phi_r$, $z_r = -(x - x_0) \sin \Phi_r + (z - z_0) \cos \Phi_r$; the subscript 0 labels the rotational axis. Note the compression of the x_r-axis (z_r:$x_r = 1.85$:1)

$(0, 3)$ reaches high amplitudes and induces a beneficial two-dimensional mean-flow distortion $(0, 0)$, stabilizing the flow; 3d effects are present of course, but $(0, 0)$ has the leading role. Growth of the CFV mode $(1, 2)$ is virtually fully suppressed, and transition to turbulence does not appear within the integration domain. The arising vortical structures for cases REF-1 and UFD-ACF are shown in Fig. 5a and b, respectively. The vortices dominant in case REF-1 correspond to the CFV mode $(1, 2)$. Undisturbed flow is found up to $x_r \approx 3.8$ ($x \approx 3.1$) where the first turbulent structures appear, caused by the (background) pulsing. Fully turbulent flow is found downstream of $x_r \approx 4.5$ ($x \approx 3.6$). For case UFD-ACF the vortices correspond to the UFD control mode. They die out downstream of $x_r \approx 3.7$ ($x \approx 3.0$) and no secondary structures are found. To check the effect of the UFD on the skin-friction drag, the wall-normal gradient of the spanwise mean velocity component in the direction of the oncoming flow is integrated in the x-direction, see Fig. 7b. Compared to case REF-1 the drag is significantly reduced for case UFD-ACF due to the delayed transition to turbulence.

3.2 Direct Attenuation of Nonlinear Crossflow Vortices

For the following investigations the CFV mode $(1, 2)$ is excited only using the disturbance strip at $x = 0.8$. Furthermore, both strips with background pulsing are activated. However—as expected—the arising dominant vortical structures and the disturbance-amplitude development for case REF-2 without active plasma actuators are qualitatively identical to case REF-1 with excitation of the CFV mode by the FSD impinging on the boundary layer, see Figs. 6a and 7a, respectively. For case CNL-ACF two plasma actuators per fundamental spanwise wavelength with forcing against the CF are positioned at $x_{PA} = 2.25$, in between the two disturbance strips. The body-force distribution and strength are based on previous investigations on the

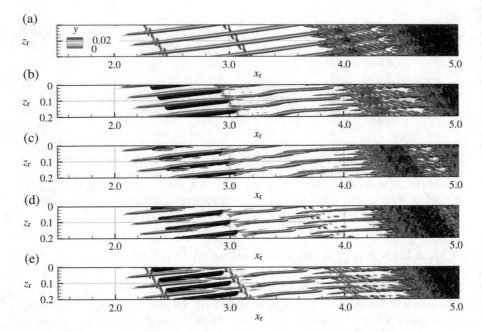

Fig. 6 Vortex visualization (snapshots, $\lambda_2 = -4$, colour indicates y) for case REF-2 at $t/T_0 = 14$ (**a**) and case CNL-ACF at $t/T_0 = 13.25$ (**b**), 13.5 (**c**), 13.75 (**d**) and 14 (**e**). Body-force distribution ($f_{10\%}$-isosurface, dark) added in (**b**)–(**e**). Rotated reference system from Fig. 5 used. Note the compression of the x_r-axis ($z_r : x_r = 1.4 : 1$)

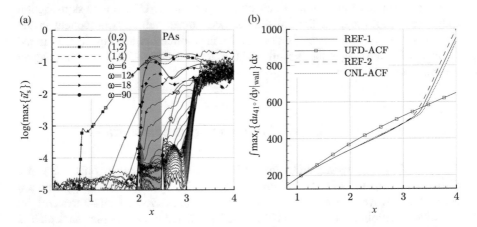

Fig. 7 (**a**) Like Fig. 4 but for case CNL-ACF. Open symbols denote case REF-2. (**b**) In the x-direction integrated evolution of the wall-normal gradient of the spanwise mean velocity component in the direction of the oncoming flow at the wall. The maximum arising over one fundamental period is used to account for a fully turbulent flow

optimum attenuation of steady CFVs using the same basic setup, see [11]. However, the force direction is adapted such that the forcing is perpendicular to the vortex axis of the dominant traveling CFVs; note that the angle between the wave vector and the potential-flow direction is somewhat lower for the traveling CFVs than for the steady ones. Compared to the optimum actuator position found for the attenuation of steady CFVs [7], the actuators are positioned slightly farther to the right, i.e. at larger z_r, relative to the traveling CFVs at $t = 3/8\,T_0$ when the force is maximum. The vortical structures arising during one fundamental period for case CNL-ACF are shown in Fig. 6b,e. The forcing clearly affects the oncoming CFVs, and for some distance downstream of the actuators their strength is decreased. Comparing the vortical structures of cases REF-2 and CNL-ACF at $t/T_0 = 14$, Fig. 6a and e, respectively, also shows that the first turbulent structures arise some distance farther downstream for case CNL-ACF. The disturbance-amplitude development for case CNL-ACF is shown in Fig. 7a. In agreement with the discussed effect on the oncoming CFVs, the amplitude of the mode $(1,2)$ is decreased in the vicinity and for some distance downstream of the actuators. But due to the temporal mean of the actuator force the steady mode $(0,2)$ reaches high amplitudes, and non-linear interaction with the mode $(1,2)$ yields the generation of further modes, e.g. the mode $(1,4)$. As simulations with a purely steady actuation indicate (not shown, see [11]), the effect of the high-amplitude steady mode counteracts the positive effect of the unsteady force component. However, compared to case REF-2 the onset of the strong growth of the unsteady modes is shifted downstream for case CNL-ACF (see the mode with $\omega = 90$), and transition to turbulence is delayed. The effect on the skin-friction drag is shown in Fig. 7b. As expected, the delayed transition yields lower drag for case CNL-ACF than for case REF-2. However, comparison to case UFD-ACF demonstrates that the UFD method is significantly more efficient. Current work checks if a modification of the time signal used to modulate the unsteady body force can improve the effect of the forcing, e.g. by employing a rectangular function instead of a sine-actuation. Furthermore, employing a second actuator row downstream is investigated. When controlling steady nonlinear CFVs, placing a second and even a third actuator row downstream of the first one yielded further, substantial delay of transition, see [7].

4 Code Performance

For the current NS3D code version, domain decomposition using MPI is only possible in the x- and y-direction where explicit 8th-order, compact 6th-order or sub-domain compact 6th-order finite differences are used for the spatial discretization (see [14] for an in-depth explanation of the sub-domain compact approach). The Fourier-spectral ansatz in the spanwise direction z is parallelized using a shared-memory OpenMP approach, limiting the number of CPU cores used for the spanwise discretization to the 24 available cores per node (then with one MPI process per node) for the current Cray XC40 at HLRS. The missing domain

decomposition in this direction limits the number of usable nodes. To investigate
the OpenMP parallelization a scaling study is performed.

The test case used for the study consists of a laminar boundary-layer flow over
a flat plate. The flow is two-dimensional but a fully three-dimensional calculation
is performed to check the code performance numbers of interest. The domain is
discretized using 6912×600 grid points in the x- and y-direction using the sub-
domain compact finite differences. To assess a possible effect of the number of grid
points used for the Fourier-spectral ansatz on the scaling, three different numbers
of resolved Fourier modes have been used, leading to 64, 128, or 256 points in the
z-direction for a total of about 265 million, 530 million and 1.06 billion grid points,
respectively. The domain is split into 384, 768, or 1536 sub-domains using MPI and
for the scaling test each MPI process uses 1 up to 24 OpenMP threads. This yields
the total number of cores used from $384 \times 1 = 384$ up to $1536 \times 24 = 36{,}864$ cores.
The nodes of the cluster are always fully used. Hence, for example, for 6 OpenMP
threads 4 MPI processes are placed on one node.

The three different grid-point numbers investigated in the z-direction showed no
significant influence on the scaling. Therefore, the following performance numbers
are averaged over all three grid sizes. Figure 8 shows the speed-up and efficiency
for the variation of the number of OpenMP threads per MPI process. Note that for
the case of 384 MPI processes (dashed line) 1 OpenMP thread means the usage
of 24 MPI processes per node and 16 nodes in total, and for 24 OpenMP threads
1 MPI process per node is run, with 384 employed nodes in total. The results for
the different domain decompositions show no significant differences. For up to 6
OpenMP threads the code shows almost perfect scaling, staying well above 90%
efficiency. For the case with 8 OpenMP threads the efficiency drops below 80%.
This is due to the ccNUMA architecture of the Cray XC40 cluster, where one node
consists of two sockets, each housing a 12-core CPU with a memory controller

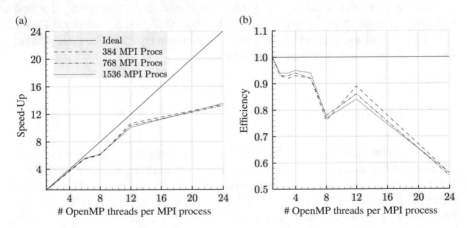

Fig. 8 (**a**) Speed-up and (**b**) efficiency for a variation of OpenMP threads (1, 2, 3, 4, 6, 8, 12 and
24) per MPI process (384, 768 and 1536)

and "socket-local" memory. While each core can access all node memory, access to memory "local" to the core's socket is much faster. For 12 OpenMP threads the placing is again optimal for the architecture and the efficiency increases again to around 85% but does not reach the level of smaller OpenMP thread counts. With 24 OpenMP threads the speed-up is slightly above 13, resulting in an efficiency of around 55% only. For this configuration the same problem as for 8 OpenMP threads shows off, as half the used cores constantly need to access memory placed on the far side of a ccNUMA node due to the global nature of the Fourier-spectral ansatz.

A detailed performance comparison of the explicit, compact, and sub-domain compact discretization schemes was presented in [13] for investigations on film cooling in a supersonic boundary-layer flow. Speed-up and efficiency were checked for a strong and a weak-scaling test with 512 MPI tasks. (We remark that the term strong scaling refers to simulations for which the total domain size remains constant, whereas the domain size per processor is decreasing with increasing number of processors. For the weak scaling the total domain size increases, while the domain size per processor is kept constant.) When employing the compact scheme a tridiagonal set of equations needs to by solved and processor idling occurs due to the use of the Thomas Algorithm. This results in significant performance loss in case a large number of MPI tasks is used. In contrast, a high parallel efficiency is obtained when using the explicit or sub-domain compact scheme. Compared to the undivided compact scheme, the specific CPU time (time per time step, grid point and CPU) was reduced by 47% when employing the sub-domain compact scheme and by 61% for the explicit scheme, see [13] for details on the numerical set-up.

5 Conclusion

Using direct numerical simulations the applicability of plasma actuators to delay laminar-turbulent transition caused by traveling CFVs has been explored. Two different techniques have been investigated: upstream flow deformation and direct attenuation of nonlinear traveling CFVs. For both techniques it is found that the transition can be delayed, but the efficiency is found to be significantly higher when employing upstream flow deformation. However, current work investigates improvements of the direct attenuation approach, with promising first tentative results.

The results of the presented scaling study have been taken as impetus to revise NS3D to allow domain decomposition also in the z-direction using the explicit or sub-domain compact finite differences.

Acknowledgements The financial support by the Deutsche Forschungsgemeinschaft, DFG, under contracts KL 890/11-1 and Collaborative Research Center SFB/TRR 40, subproject A4, are gratefully acknowledged, as well as the provision of computational resources by the High Performance Computing Center Stuttgart (HLRS) under grant GCS_Lamt (LAMTUR).

References

1. A. Babucke, J. Linn, M.J. Kloker, U. Rist, Direct numerical simulation of shear flow phenomena on parallel vector computers, in *High Performance Computing on Vector Systems 2005, Proceedings of High Performance Computing Center Stuttgart (HLRS)* (Springer, Berlin, 2006), pp. 229–247
2. H. Barth, S. Hein, R. Rosemann, Redesigned swept flat-plate experiment for crossflow-induced transition studies, in *New Results in Numerical and Experimental Fluid Dynamics XI* ed. by A. Dillmann et al. Notes on Numerical Fluid Mechanics and Multidisciplinary Design XXX, peer-reviewed contributions to the 20. STAB/DGLR-Symposium, Brunswick, Nov. 2016 (Springer, Berlin, 2017), p. 10
3. N. Benard, E. Moreau, Electrical and mechanical characteristics of surface AC dielectric barrier discharge plasma actuators applied to airflow control. Exp. Fluids **55**, 1846 (2014)
4. T.C. Corke, C.L. Enloe, S.P. Wilkinson, Dielectric barrier discharge plasma actuators for flow control. Annu. Rev. Fluid Mech. **42**, 505–29 (2010)
5. P.C. Dörr, M.J. Kloker, Numerical investigation of plasma-actuator force-term estimations from flow experiments. J. Phys. D: Appl. Phys. **48**, 395203 (2015)
6. P.C. Dörr, M.J. Kloker, Stabilisation of a three-dimensional boundary layer by base-flow manipulation using plasma actuators. J. Phys. D: Appl. Phys. **48**, 285205 (2015)
7. P.C. Dörr, M.J. Kloker, Transition control in a three-dimensional boundary layer by direct attenuation of nonlinear crossflow vortices using plasma actuators. Int. J. Heat Fluid Flow **61**, 449–465 (2016)
8. P.C. Dörr, M.J. Kloker, Crossflow transition control by upstream flow deformation using plasma actuators. J. Appl. Phys. **121**(063303), 14 (2017)
9. P.C. Dörr, M.J. Kloker, Effect of upstream flow deformation using plasma actuators on crossflow transition induced by unsteady vortical free-stream disturbances. AIAA Paper, Aviation Forum Denver, June 2017
10. T. Friederich, M.J. Kloker, Control of the secondary crossflow instability using localized suction. J. Fluid Mech. **706**, 470–495 (2012)
11. Z. Guo, Numerical simulation of travelling crossflow disturbances and their control by plasma actuators. Master's thesis, IAG, Universität Stuttgart, 2017; Supervisors: P. C. Dörr and M. J. Kloker
12. R.D. Joslin, Aircraft laminar flow control. Annu. Rev. Fluid Mech. **30**, 1–29 (1998)
13. M. Keller, M.J. Kloker, Direct numerical simulations of film cooling in a supersonic boundary-layer flow on massively-parallel supercomputers, in *Sustained Simulation Performance* (Springer, Berlin, 2013), pp. 107–128
14. M. Keller, M.J. Kloker, DNS of effusion cooling in a supersonic boundary-layer flow: influence of turbulence. AIAA-2013-2897 (2013)
15. J. Kriegseis, B. Simon, S. Grundmann, Towards in-flight applications? A review on dielectric barrier discharge-based boundary-layer control. Appl. Mech. Rev. **68**, 020802 (2016)
16. I. Maden, R. Maduta, J. Kriegseis, S. Jakirlic, C. Schwarz, S. Grundmann, C. Tropea, Experimental and computational study of the flow induced by a plasma actuator. Int. J. Heat Fluid Flow **41**, 80–89 (2013)
17. R. Messing, M.J. Kloker, Investigation of suction for laminar flow control of three-dimensional boundary layers. J. Fluid Mech. **658**, 117–147 (2010)
18. W.S. Saric, R. Carrillo, M.S. Reibert, Leading-edge roughness as a transition control mechanism. AIAA-98-0781 (1998)
19. W.S. Saric, H.L. Reed, E.B. White, Stability and transition of three-dimensional boundary layers. Annu. Rev. Fluid Mech. **35**, 413–440 (2003)
20. L.U. Schrader, L. Brandt, D.S. Henningson, Receptivity mechanisms in three-dimensional boundary-layer flows. J. Fluid Mech. **618**, 209–241 (2009)

21. P. Wassermann, M.J. Kloker, Mechanisms and passive control of crossflow-vortex-induced transition in a three-dimensional boundary layer. J. Fluid Mech. **456**, 49–84 (2002)
22. P. Wassermann, M. Kloker, Transition mechanisms induced by travelling crossflow vortices in a three-dimensional boundary layer. J. Fluid Mech. **483**, 67–89 (2003)

Towards the Implementation of a New Multigrid Solver in the DNS Code FS3D for Simulations of Shear-Thinning Jet Break-Up at Higher Reynolds Numbers

Moritz Ertl, Jonathan Reutzsch, Arne Nägel, Gabriel Wittum, and Bernhard Weigand

Abstract Liquid jet break-up appears in many technical applications, as well as in nature. It consists of complex physical processes, which happen on very small scales in space and time. This makes them hard to capture by experimental methods; and therefore a prime subject for numerical investigations. The state-of-the-art approach combines the Volume of Fluid (VOF) method with Direct Numerical Simulations (DNS) as employed in the ITLR in-house code Free Surface 3D (**FS3D**). The simulation of these jets is dependent on very fine grids, with most of the computational costs incurred by solving the Pressure Poisson Equation. In order to simulate larger computational domains, we tried to improve the performance of FS3D by the implementation of a new multigrid solver. For this we selected the solver contained in the **UG4** package developed by the Goethe Center for Scientific Computing at the University of Frankfurt. We will show simulations of the primary break-up of shear-thinning liquid jets and explain why larger computational domains are necessary. Results are preliminary. We demonstrate that the implementation of UG4 into FS3D provides a noticeable increase in weak scaling performance, while the change in strong scaling is yet detrimental. We will then discuss ways to further improve these results.

M. Ertl (✉)
Institut für Thermodynamik der Luft- und Raumfahrt, Universität Stuttgart, Pfaffenwaldring 31, 70569 Stuttgart, Germany
e-mail: moritz.ertl@itlr.uni-stuttgart.de

J. Reutzsch • B. Weigand
Institut für Thermodynamik der Luft- und Raumfahrt, Universität Stuttgart, Pfaffenwaldring 31, 70569 Stuttgart, Germany
e-mail: jonathan.reutzsch@itlr.uni-stuttgart.de; bernhard.weigand@itlr.uni-stuttgart.de

A. Nägel • G. Wittum
Goethe-Zentrum für Wissenschaftliches Rechnen, Goethe-Universität Frankfurt am Main, Kettenhofweg 139, 60325 Frankfurt am Main, Germany
e-mail: arne.naegel@gcsc.uni-frankfurt.de

© Springer International Publishing AG 2018
W.E. Nagel et al. (eds.), *High Performance Computing in Science and Engineering '17*, https://doi.org/10.1007/978-3-319-68394-2_16

269

1 Introduction

Liquid jet break-up is the process where a fluid stream is injected into a surrounding medium and thereby disintegrates into many smaller droplets. Liquid jets appear in many technical applications as well as in nature. Well known examples are fuel injection in combustion engines or gas turbines, water or foam jets for fire fighting, irrigation or pesticides in agriculture, and spray painting or ink jet printing. In some cases like spray drying, which is used to produce functional particles with well defined properties, and which is used in the production of pharmaceuticals or in food processing, the injected fluid can additionally show a Non-Newtonian behaviour [8].

Newtonian jet break-up is already under intensive investigation, especially in the context of fuel injection. A very good introduction into the fundamentals of jet break-up is given by Lefebvre [18] and some more recent developments are summarised by Lin and Reitz [20]. When it comes to the investigation of jet break-up, numerical methods are increasingly used in addition to experiments. Numerical methods are very well suited for the analysis of phenomena like jet break-up, where the time scales are fractions of seconds and the investigated structures are often in the order of micrometers, and therefore reliable experiments may be difficult to perform. In order to get rough estimates of jet behaviour, Reynolds-Averaged Navier-Stokes (RANS) methods can be used, for example to aid industrial design as shown by Beau [1]. When more accurate results are required, and even though some very interesting advances using Large Eddy Simulations (LES) have been shown by Hermann [13], the current state-of-the-art numerical approach is Direct Numerical Simulations (DNS). Klein [15] uses the Volume of Fluid (VOF) method to investigate liquid sheets at moderate Reynolds numbers. He analyses the surface deformations and compares the numerical results to experimental data. VOF DNS is also used by Sander and Weigand [34] to investigate the influence of different instability enhancing parameters on liquid sheets and round jets. Pan and Suga [22] combine the Level-Set (LS) method with DNS to investigate the break-up of laminar jets and Shinjo and Umemura [35] use one very highly resolved LS DNS to investigate the influence of vortices in the surrounding gas on the jet surface.

While many numerical studies of Newtonian liquid jet break-up exist, Non-Newtonian behaviour, such as shear-thinning, has been investigated to a lesser extend. Lakdawala [17] did 2D LS numerical simulations of shear dependent Non-Newtonian jets at low Reynolds numbers. A 3D LES with VOF and a power law model of a Non-Newtonian jet at a low Reynolds number has been investigated by Li-Ping [19] in OpenFOAM. The first author of the present report has started to do some three-dimensional numerical investigations into different aspects of the primary jet break-up of shear-thinning fluids using DNS in combination with the VOF method as implemented in the ITLR in-house code **Free Surface 3D** (**FS3D**). Striving for more accurate simulations and for investigating increasingly higher Reynolds numbers has led to a continuous increase in computational cells—from 33 millions in 2013 [42], over 450 million in 2014 [4] to 750 million during the last year [5]. The authors used the simulations to investigate the influence of different parameters, such as the Reynolds number, the turbulent intensity, the ambient pressure as well as different concentrations of the shear-thinning solution. This growth of the number of cells has gone in hand with the need for an increasing amount of processors and therefore also an increase in parallel performance. A communication imbalance (which is described in Sect. 2.2) has been identified in the utilised code FS3D. In order to improve on this, the multigrid solver of the software package **UG4** [38] has been integrated into our code. For this solver, scalability for Poisson-type equations in 3D was demonstrated for thousands of processes.

2 Fundamental Method

FS3D is a *Direct Numerical Simulation* (DNS) code for incompressible multiphase flows. It was developed at ITLR (Institute of Aerospace Thermodynamics) in Stuttgart. Due to the use of DNS, very small temporal and spatial scales are resolved, hence, no turbulence modelling is needed. FS3D is based on the *Volume of Fluid* (VOF) method and solves the incompressible Navier-Stokes equations as well as the energy equation with temperature dependent thermo-physical properties. A wide variety of phenomena have been investigated in the last 15 years. These include drop and bubble dynamic processes, for instance droplet deformation [31] or droplet impact onto a thin film [10], furthermore, droplet collisions [33], droplet wall interactions [32], bubbles [41] and also more recently rigid particle interactions [24].

2.1 Mathematical Description

The conservation equations for mass and momentum read accordingly

$$\rho_t + \nabla \cdot [\rho \mathbf{u}] = 0, \tag{1}$$

$$[\rho \mathbf{u}]_t + \nabla \cdot [\rho \mathbf{u}\mathbf{u}] = \nabla \cdot [\mathbf{S} - \mathbf{I}p] + \rho \mathbf{g} + \mathbf{f}_\gamma, \tag{2}$$

where ρ denotes the density, $\mathbf{u}$ the velocity vector, $\mathbf{S}$ the viscous stress tensor, p the pressure, $\mathbf{g}$ the gravitational acceleration and $\mathbf{f}_\gamma$ the body force which is used to model surface tension in the vicinity of the interface. The viscous stress tensor is given by

$$\mathbf{S} = \mu \left[\nabla \mathbf{u} + (\nabla \mathbf{u})^T \right]. \tag{3}$$

The shear-thinning viscosity $\mu = \mu(\dot{\gamma})$ is calculated as a function of the shear rate $\dot{\gamma}$ with the *Carreau-Yasuda model* [37]

$$\frac{\mu(\dot{\gamma}) - \mu_\infty}{\mu_0 - \mu_\infty} = [1 + (\tau \dot{\gamma})^a]^{\frac{(n-1)}{a}}. \tag{4}$$

The five constants are: the viscosity at high shear rates μ_∞ and at zero shear μ_0, as well as the three model parameters τ, a and n, which are obtained from experimental data.

2.2 Numerical Approach

In FS3D a finite volume method is used to discretise the Navier-Stokes equations. According to the *Marker and Cell method* [11] (MAC), velocities are stored on the cell faces, scalars, for instance pressure or density, at the cell centres, respectively. Additional indicator variables f_i are used to identify different phases. This method is broadly based on the VOF method [14]. The VOF variable f_i is defined as

$$f_i(\mathbf{x}, t) = \begin{cases} 0 & \text{in the continuous phase,} \\ ]0, 1[& \text{at the interface,} \\ 1 & \text{in the disperse phase.} \end{cases}$$

The index i describes different phases: f_1 for the liquid, f_2 for the vapour and f_3 for the solid phase. Due to the one-field formulation of the VOF method, local variables are defined by the local f value and the corresponding values for, e.g., the pure gaseous (index g) and liquid (index l) phases. Thus, the density and the viscosity read

$$\rho(\mathbf{x}, t) = \rho_g + [\rho_l - \rho_g] f(\mathbf{x}, t), \tag{5}$$

$$\mu(\mathbf{x}, t) = \mu_g + [\mu_l - \mu_g] f(\mathbf{x}, t). \tag{6}$$

The volume fraction f is transported across the computed domain by solving the transport equation

$$f_t + \nabla \cdot [f\mathbf{u}] = \mathfrak{S}, \tag{7}$$

where $\mathfrak{S}$ is equal to zero when no source or diffusion terms are considered. It is non zero, when phase change, for instance solidification or evaporation, is taken into account. For further details the reader is referred to [3].

To achieve a successful advection of f a sharp interface is required. For this computation FS3D makes use of the *piecewise linear interface reconstruction* (PLIC) method [29]. The reconstruction of the interface becomes necessary due to the f variable only representing the amount of the disperse phase, for instance the fluid, inside a cell; cf. Fig. 1a. In order to capture the interface a plane orthogonal to the local normal vector $\hat{\mathbf{n}}_y = \nabla f / |\nabla f|$ is placed iteratively into cells containing the interface in such a way, that the volume under the plane is in accordance with the volume fraction f. The vector is defined by the negative gradient of the volume

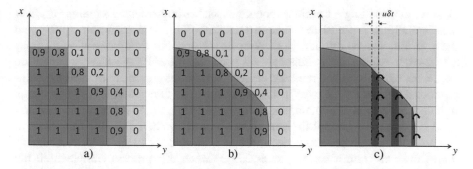

Fig. 1 (**a**) f-Field without interface information; (**b**) Interface reconstruction with the PLIC-method; (**c**) Calculation of the f-flux $u\delta t$ with δt being the timestep from a PLIC reconstructed interface

fraction from the surrounding cells. Figure 1b shows the PLIC reconstruction in 2D. Currently, there are two mass advection methods implemented in the code. On the one hand, three one-dimensional non-conservative transport equations are considered successively. Therefore, interfaces have to be reconstructed three times. With a permutation of the equation sequence [36] and a specific choice for the divergence correction second-order accuracy in space and time can be achieved. The fluid is transported parallel to the velocities perpendicular to the cell faces. This is shown in Fig. 1c. On the other hand, a new formulation [12], which makes a one-step advection possible, was implemented in FS3D. In this method a general, six-faced polyhedron is constructed for a more realistic approximation of the volume fluxes. More details regarding the new method can be found in Reitzle et al. [28]. To compute the surface tension several models are available: the conservative continuous surface stress (CSS) model by Lafaurie et al. [16], the continuum surface force model (CSF) by Brackbill et al. [2], and the balanced force approach by Popinet [23].

FS3D is fully parallelised using MPI and OpenMP; therefore, we are able to perform simulations with very high spatial resolutions, which is crucial for DNS. Heretofore, simulations with a maximum up to two billion computational cells have been conducted at the High Performance Computing Center Stuttgart (HLRS) and during the studies presented here first tests with over eight billion cells have been undertaken successfully. The code is validated and has good performance on the Cray XC40 Hornet supercomputer [25], however, the performance for very high amounts of processes, especially weak scaling, is not yet satisfying. The reasons for this, and a new approach to solve this issue is explained in the following.

Due to the infinitely fast propagation of pressure waves, which occurs because of the incompressibility of the fluids, the momentum equation cannot be solved explicitly. Therefore, the pressure term must be discretised in an implicit way. That leads to the Pressure Poisson Equation, which is defined as

$$\nabla \cdot \left[\frac{1}{\rho(f)} \nabla p \right] = \frac{\nabla \cdot \mathbf{u}}{\Delta t}. \tag{8}$$

To solve the resulting set of linear equations Rieber implemented a multigrid solver into FS3D [30]. A Red Black Gauss Seidel algorithm is used for smoothing and the algorithm can be run in a V- or W-cycle scheme. For time integration schemes a first-order explicit Euler scheme and a second order Runge Kutta scheme are implemented. Solving the pressure correction requires a great amount of the computational time of FS3D: over 70%. This is due to an unfavourable communication imbalance. The whole domain is agglomerated onto a single processor at a certain level of coarsening and all the coarse solutions are gathered on one root process. This means, that all processes need to first send, and then later receive the entire domain to and from a single process. In between this process computes all the coarser levels while all other processes are idling. Therefore, a new approach is taken into consideration, which is explained in the next subsection.

2.3 UG4 and Multigrid Solver

Besides the above described, conventional method for solving the Pressure Poisson Equation, we recently integrated a massively parallel geometric multigrid solver of the software package **UG4** [38] into our code. The software framework UG4 was developed at the Goethe Center for Scientific Computing at the Goethe University in Frankfurt, and it was designed for the solution of partial differential equations. It uses grid-based discretisation methods, such as the finite volume method, and is implemented in C++. A main focus lies on efficient and highly scalable solvers, using algebraic and geometric multigrid methods. UG4 is parallelised using MPI, thus, it is a suitable application for a fast solving of the Pressure Poisson Equation. The MPI calls are subsumed in a separate library called **pcl** (*parallel communication layer*), which yields easy structures for graph-based parallelisation. One advantage is that global IDs are redundant by storing parallel copies on each process in a well defined order in interface containers. For further details regarding pcl and the features of UG4 the reader is referred to Vogel et al. [38] or Reiter et al. [27]. Scalability and parallel performance of UG4 was tested extensively [39].

For integration into FS3D the parallel hierarchical multigrid solver of UG4 is of principal interest. A short overview of the algorithm for the parallel version can be found in [40] and is explained in the following: The discretised Pressure Poisson Equation reads $\mathbf{A}_L \mathbf{p}_L = \mathbf{b}_L$, where L denotes the index for the finest Level, $\mathbf{p}_L$ the desired solution, which is computed iteratively. With the defect $\mathbf{d}_L = \mathbf{b}_L - \mathbf{A}_L \mathbf{p}_L$ a multigrid correction $\mathbf{c}_L = \mathbf{M}_L(\mathbf{d}_L)$ is calculated. The multigrid operator $\mathbf{M}_L$ is added to the approximate solution $\mathbf{p}_L := \mathbf{p}_L + \mathbf{c}_L$. Several auxiliary coarse grid matrices $\mathbf{A}_l$, $L_B \leq l \leq L$, are used to compute the correction $\mathbf{c}_L$; L_B is the base Level. Afterwards, the multigrid cycle is defined recursively. By means of a smoothing operator the correction is partly computed with a given defect $\mathbf{d}_l$ on a certain level l. Subsequently, the latter is transferred on the next coarser level applying the algorithm to the restricted defect $\mathbf{d}_{l-1}$. Then the coarse grid correction $\mathbf{c}_{l-1}$ is prolongated to the finer level and added to the correction on

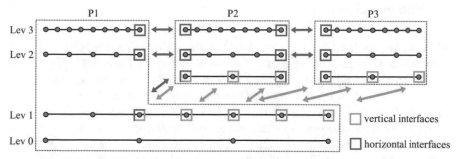

Fig. 2 Schematic illustration of a 1D parallel multigrid hierarchy distributed onto three processes (redrawn from [38]). One top process stores the coarse grid and refines to a certain level. Parallel copies are identified with horizontal (blue), respectively vertical (green) interfaces

level l. Eventually the exact correction $\mathbf{c}_L = \mathbf{A}_l^{-1}\mathbf{d}_l$ is computed on the base level L_B. The corresponding algorithm is described in detail in Reiter et al. [27]. To gain an efficient parallelisation we need an appropriate distributed multigrid hierarchy. Figure 2 (cf. [27]) shows an overview of the hierarchy and the interfaces starting from a coarse grid, which describes the domain loaded onto one process. Then the grid is refined and new levels of the multigrid hierarchy are created until the finest grid level is reached. The latter is distributed to a larger set of processes and communication structures, the vertical interfaces (green), are established. This procedure can be iterated, so that a tree structure of processes holding parts of the hierarchical grid is created. The parallelisation of the transfer between the grid levels is achieved by using the communication structures in vertical direction. The transfer operators can work completely process-locally when no vertical interface is present. On every level simple iterative schemes, also called smoothers, are applied. The horizontal interfaces (blue) are needed for the communication within these multigrid smoothers on each grid level. They are required for the computation of the level-wise correction in a consistent way. Due to the described hierarchical distribution only a smaller number of processes are needed on lower levels and most processes idle. On finer grid levels, however, the major amount of overall runtime is required. Nevertheless, a good computation vs. communication ratio is achieved by means of sophisticated parallel smoothing, prolongation and restriction operations [39].

UG4 is written in C⁺⁺. Therefore, the integration into FS3D, which is a Fortran code, is done via additional custom Fortran interfaces. The compiled UG4 library is statically linked into FS3D. A new custom module has been written in FS3D for the interaction with UG4. The solver settings and especially the layout of the parallel communication are set up in an initialisation routine, once, at the start of a simulation. During the calculation cycles of the time steps, the matrix and the right hand side of the Pressure Poisson Equation, as well as the boundary layers are set and passed to UG4. There the equation system is solved and the solution (the pressure field) is returned to FS3D.

3 Numerical Setup

All calculations have been performed on a three-dimensional regular grid. The computational domain for the jet simulations is shown schematically in Fig. 3. The nozzle is chosen with a fixed Diameter of $D = 2.5 \times 10^{-3}$ m for the injection of the jet. The domain is set up as a rectangular cuboid with a quadratic base with a width of $W = H = 16\,D$ and a length of $L = 40\,D$. The x-axis points in the direction of the jet injection, the y- and z-direction are orthogonal to it, respectively. We use three different discretisations for the domain. The amount of cells used in x- and in y- and z- direction, as well as the total number of cells are given in Table 1.

In the y-z-plane a grid refinement is applied leading to smaller cells in the center around the jet. The size of the cell edges in this region is also given in Table 1. The boundary condition at the nozzle side (gray) is a no-slip wall with an inflow boundary condition in the center. The latter is circular and has a nozzle diameter of $D = 2.5 \times 10^{-3}$ m. A block profile is chosen as the initial velocity profile at the inlet

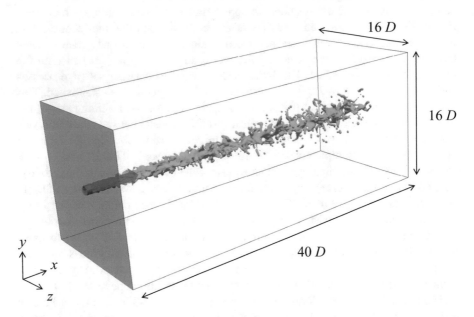

Fig. 3 Computational domain and coordinate system of jet simulation. No slip wall and inflow boundary condition on the left side. Orange arrow indicates the direction of the injection

Table 1 Discretisations of the domain

	Cells in x	Cells in y and z	$\Delta x = \Delta y = \Delta z$ in in the core region (m)	Total number of cells
Fine grid	2304	768	4.34×10^{-5}	1.3×10^{9}
Medium grid	1152	384	8.68×10^{-5}	1.7×10^{8}
Coarse grid	576	192	1.74×10^{-4}	2.1×10^{7}

Table 2 Liquid properties

	Density	Surface tension	Viscosity	Viscosity	Model parameter		
	ρ	σ	μ_0	μ_∞	n	a	τ
	(kg/m^3)	(mN/m)	(Pa s)	(Pa s)	(–)	(–)	(–)
Praestol 2500 0.3%	999.4	73.15	0.046	0.004	0.576	1.036	0.157
Air	1.19		1.8×10^{-5}				

with a velocity of $U_0 = 55.25$ m/s. The turbulent intensity is set to $Tu = 10\%$ and the turbulent length scale is $L_t = D/8 = 3.125 \times 10^{-5}$ m. The five other sides are defined as continuous (Neumann) boundary conditions. Gravitational acceleration is neglected.

As an adequate material with shear-thinning properties an aqueous solution of the polyacrylamide Praestol 2500 at 0.3% weight is chosen. The model parameters as well as the material properties of the solution and of the surrounding air at 20 °C are given in Table 2 [6]. Details regarding the viscosity model can be found in [7].

Therefore, the relevant dimensionless numbers for the simulation, the Reynolds Number and the Ohnesorge Number, can be calculated as

$$Re = \frac{\rho_l U_0 D}{\mu_0} = 3000,$$ (9)

and

$$Oh = \frac{\mu_0}{\sqrt{\rho_l \sigma D}} = 0.1.$$ (10)

The simulated jet is in the atomisation regime according to the Ohnesorge diagram [18]. The Kolmogorov length scale is calculated as $\lambda_K = 2.6 \times 10^{-5}$ m, hence, all three grids are in the order of magnitude of the smallest dissipative length scale necessary for DNS. This is essential to be able to produce physically correct results.

4 Results

We investigate the spatial development of the jet with the medium grid in time, to gain a basic understanding of the break-up process of the jet simulation. For all our analyses we use the dimensionless time

$$t^* = \frac{tU}{D}.$$ (11)

The jet from the fine grid at four different times after injection $t^* = 4.4$, $t^* = 15.5$, $t^* = 26.5$ and $t^* = 42.0$ is shown in Fig. 4. Directly after injection at $t^* = 4.4$ the jet core is still cylindrical. We can see the disturbances caused by the nozzle

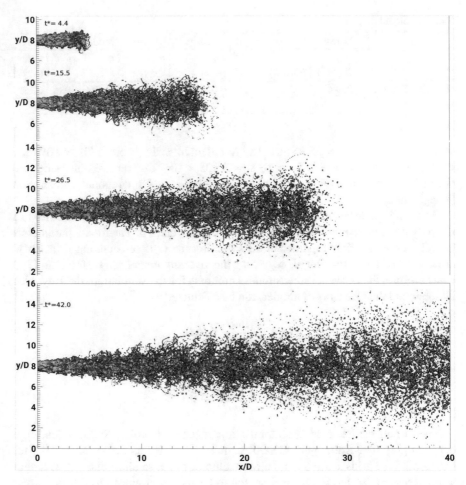

Fig. 4 The spatial development of the jet with the fine grid after injection at dimensionless times $t^* = 4.4$, $t^* = 15.5$, $t^* = 26.5$ and $t^* = 42.0$

turbulence as three-dimensional waves on the jet surface. At $t^* = 15.5$ several of the surface waves have deformed further, due to the interaction with the surrounding air, and have started to detach from the jet core, forming ligaments. Several ligaments have disintegrated further and have separated into droplets. This process continues towards $t^* = 26.5$. At this time we can observe a lot more ligaments and droplets. With increasing length the jet also shows an increasing expansion in radial direction. Particularly the separated droplets are moving away from the jet. This expansion also causes the jet core to become thinner in downstream direction—well visible for example at $x/D = 23$. At $t^* = 42.0$ the jet has reached the end of the computational domain. The disintegration has progressed even further in the regions downstream of $x/D = 21$; no jet core is visible any more. Instead we observe expanding agglomerations of ligaments. The amount of droplets has also further increased. We are observing the onset of atomisation.

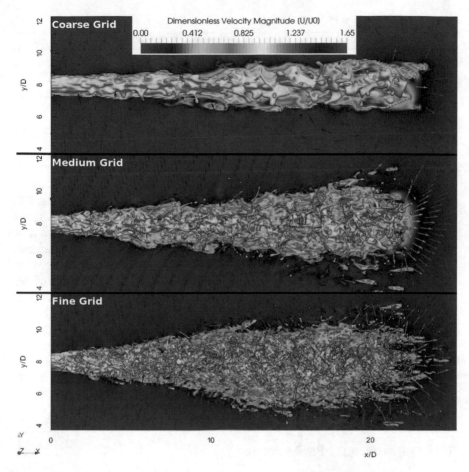

Fig. 5 Visual comparison of the three discretisations at $t^* = 22.1$. From top to bottom: coarse grid, medium grid, fine grid

In order to analyse the influence of the grid we start with a visual analysis of the simulations at $t^* = 22.1$. The impact of the grid size is shown in Fig. 5. The jet, from the simulation with the coarse grid (top), has developed a strongly disturbed surface, but only very few ligaments or droplets are visible. The jet has a cylindrical shape. When we look at the jet resulting from the medium grid, we see a smaller jet core with more ligaments around it. Towards the jet tip the core is getting displaced from the center line and exhibiting a sinusoidal shape. Additionally, towards the tip we can observe an increasing amount of detached ligaments and droplets. The jet shape is still cylindrical close to the injection, but becomes conical further downstream. The jet, which was simulated with the fine grid, looks similar to the jet from the medium grid, but breaks up into even finer structures. A continuous core is only visible up to about $x/D = 10$. Afterwards, the jet starts atomising into drops,

Table 3 Jet properties at $t^* = 22.1$

	Average jet angle ($^\circ$)	Average droplet diameter (cm)	Droplet count
Fine grid	9.37	1.98×10^{-4}	4465
Medium grid	6.91	3.35×10^{-4}	625
Coarse grid	2.46	5.79×10^{-4}	23

ligaments and larger continuous liquid parts, sometimes connected by ligaments. The jet has a conical shape.

We also used post processing, to identify all separate liquid structures, also at $t^* = 22.1$. We counted the amount of structures and calculated the mean diameter of the structures. We furthermore calculated the angle at which the jet expands. The results of these calculation are give in Table 3. The values confirm our observations from the visual analysis, showing that the size of the liquid structures, as well as the amount, are highly dependent on the resolution. This was to be expected due to the numerical methods we employ, but it is noteworthy that large differences are still visible even though all three cases are within the order of magnitude of the Kolmogorov length scale, therefore, providing a high enough resolution for DNS. It has to be noted, that while all three simulations gave different quantitative results, the simulation with the fine grid and the simulation with the medium grid exhibit a similar qualitative behaviour. Even the simulation with the medium sized grid can therefore be used to understand the basic processes of jet break-up. But it becomes obvious, that in order to obtain reliable quantitative results for jets at high Reynolds numbers fine computational grids with a high amount of cells are indispensable.

5 Computational Performance

We analysed the performance of FS3D with the newly implemented UG4 multigrid solver. We compared it to the performance of FS3D with the old multigrid solver. For the performance analysis we simulate an oscillating droplet. We use the droplet instead of the jet for two reasons: First, because it is a symmetric case which distributes the load somewhat evenly. Second and more importantly, because it can be used for the performance analysis from the beginning, as opposed to a jet simulation, which can only be sensibly analysed after the jet has passed through the computational domain once. The simulation was set up with the following parameters: An elongated droplet is initialised as an ellipsoid with the semi-principal axis $a = b = 1.357\,\text{mm}$ and $c = 0.543\,\text{mm}$ at the center of a cubic computational domain with an edge length of $x = y = z = 8\,\text{mm}$. The fluid of the droplet is an aqueous solution of Praestol modelled with the Carreau-Yasuda model as described in Sect. 3. The domain is discretised with a cubic Cartesian grid.

The results for the strong scaling are shown in Fig. 6. For the strong scaling analysis we used a fixed computational grid with 512^3 cells and a second grid with

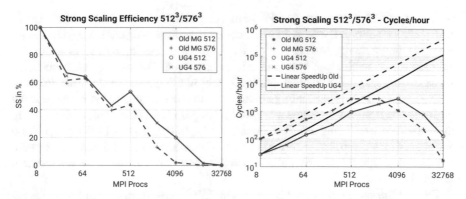

Fig. 6 Left: Strong scaling efficiency in %, for FS3D with the old multigrid solver and with UG4. Right: Strong scaling performance in cycles per hour

Table 4 Strong scaling setups

	Problem size								
	512^3					576^3			
Processors	2^3	4^3	8^3	16^3	32^3	3^3	6^3	12^3	24^3
Cells per processor	256^3	128^3	64^3	32^3	16^3	192^3	96^3	48^3	24^3
Nodes	2	4	32	256	2048	6	9	72	576
Processors per node	4	16	16	16	16	5	24	24	24

576^3 cells. We varied the number of processors from 2^3 to 32^3. The amount of processors used, as well as the resulting amount of cells which were distributed onto a processors are given in Table 4. Although FS3D has OpenMP implemented as well as MPI, and is therefore capable of hybrid operation, only the MPI parallelisation was investigated this time. For information on FS3D performance with hybrid OpenMP and MPI operation please see [26] and [9].

After running all these simulations for 1 h, we took the number of computational cycles, which were calculated during that time, and displayed the cycles per hour (cph) in Fig. 6 on the right. We then calculated the strong scaling efficiency

$$SS = \frac{t_1}{N\,t_N} \cdot 100\%, \tag{12}$$

where t_1 denotes the amount of cycles per hour, which were needed for the calculation on one processor, N is the amount of processors used in the calculation and t_N is the amount of cycles per hour, which were obtained by calculating with N processors. The strong scaling efficiency is displayed in Fig. 6 on the left.

The strong scaling efficiency of both solvers looks similar with the efficiency of UG4 slightly higher at more processors. The direct comparison of the cycles per hour shows, however, that the serial performance of the old solver was about three times better. Therefore, the performance obtained from the old solver was also much better up to 512 processors (or 64^3 cells per processor). The better efficiency of UG4 becomes only relevant at 4096 processors—while FS3D with the old multigrid solver is then actually getting slower despite the increase in processors—for UG4 an increase in cycles per hour can still be achieved. We have to conclude, that in its current state of implementation the new multigrid solver does not improve strong scaling. The strong scaling efficiency looks better, but due to the worse serial performance it doesn't achieve a higher cycles per hour performance compared to the old solver, even using more processors. At 32,768 processors UG4 does perform about ten times faster, but at this point both solvers have slowed down even further. Numerical setups that distribute less then 32^3 cells per processors are not sensible for FS3D.

One additional comment: The grid with 576^3 cells shows a somewhat worse performance, since this case was calculated with 24 processors per node (ppn), while the grid with 512^3 cells was calculated on 16 ppn, providing more bandwidth for communication. Cases with more than 128^3 cells per processor had to use a smaller amount of processors per node (and therefore more nodes) in order to provide the necessary amount of memory.

The results for the weak scaling study are shown in Fig. 7. We fixed the amount of cells per processor to 64^3 and varied the amount of processors from 2^3 to 32^3— creating problem sizes with 128^3 to 2048^3 cells. The number of processors and the corresponding problem sizes are given in Table 5. The achieved cycles per hour are

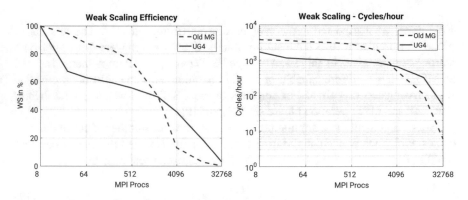

Fig. 7 Left: Weak scaling efficiency in %, for FS3D with the old multigrid solver and with UG4. Right: Weak scaling performance in cycles per hour

Table 5 Weak scaling setups

Problem size	128^3	192^3	256^3	384^3	512^3	768^3	1024^3	1536^3	2048^3
Processors	2^3	3^3	4^3	6^3	8^3	12^3	16^3	24^3	32^3
Cells per processor	64^3								
Nodes	2	3	4	9	32	72	256	576	2048
Processors per node	4	9	16	24	16	24	16	24	16

shown in Fig. 7 on the right. We also calculated the weak scaling efficiency

$$WS = \frac{t_1}{t_N} \cdot 100\%, \tag{13}$$

and displayed it in Fig. 7 on the left.

When we look at the weak scaling efficiency, we can see that for low numbers of processors both multigrid solvers scale in a similar way, but above 512 processors the efficiency of UG4 is better. When we get to a very high number of processors (32,768), the performance is bad for both multigrid solvers, with UG4 still scaling slightly better. When looking at the weak scaling performance in cycles per hour, we observe the same behaviour as seen in the strong scaling analysis—for up to 512 processors the curves run in parallel with the performance of the old multigrid solver being about three times better. From there on the performance of UG4 develops much better compared to the old solver, overtaking the old solver at 4096 processors and providing three times cph at 13,824 processors, and nine times cph at 32,768.

For jet break-up simulations we estimate, under the assumption of quadratic cells, that the necessary amount of cycles increases linearly with the domain size. The length of a cycle, or time step, for these simulations is dictated by the Courant-Friedrichs-Levy condition (CFL) [21]. Therefore, an increase in cells to obtain a finer resolution will lead to a smaller time step per cycle and thus to an increase in cycles.

We also isolated the performance of the pressure correction by timing it with calls to the MPI_Wtime function. The results are shown in Fig. 8. We can see that both strong scaling and weak scaling curves are in a good agreement with the above shown performance analysis for the entire code. This seems sensible since the pressure correction generally accounts for over 70 % of FS3D's computational time.

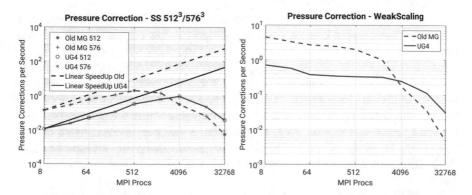

Fig. 8 Left: Strong scaling performance of the pressure correction in cycles per hour. Right: Weak scaling performance of the pressure correction in cycles per hour

6 Conclusions

We presented our multiphase flow DNS code **FS3D**, which is based on VOF and PLIC, and capable of jet break-up simulations. Moving to larger Reynolds numbers goes in hand with a demand for larger computational domains and higher resolutions. The higher resolutions become necessary due to the smaller cell size dictated by the Kolmogorov length scale and the larger domain size is due to the jet expanding more strongly in radial direction at higher Reynolds numbers. As this corresponds to an increase in size of the pressure equation sub-problem, an efficient linear solver is crucial for the overall performance.

To improve the parallel performance in such cases, we integrated the multigrid solver from UG4 into FS3D. In a preliminary performance analysis, we compared the performance of FS3D with the old multigrid solver to the performance of FS3D with replacement. However, this implementation of UG4 into FS3D was only finished shortly before submitting this report. That means no optimisation has been done on fine tuning this implementation, yet. The preliminary results are as follows:

The strong scaling analysis showed a generally better performance with the old solver. The weak scaling analysis showed the old solver to be better at low numbers of processors, while UG4 provided better performance from 4096 processors upwards. Especially around 10^4 processors we were able to obtain acceptable cph with UG4, while the old solvers performance collapsed. At 32^3 processors UG4 outperformed the old solver by a factor of 9, but the performance at this point was still deemed unacceptable for both solvers.

In these first preliminary results, we did observe an improved performance for more than 4096 processes. Since this is the region that we aim for in typical jet simulations, this is an encouraging start. However, the results show that the solver performance is still behind expectations and does not scale well. In particular, it differs from the excellent scalability (e.g., 80 % at 32^3 processors) reported for UG4 for Poisson-type problems before [27, 38, 40].

As next steps, we now seek an in-depth investigation of the observed behaviour. It must be investigated if the number of multigrid cycles for each solver call was optimal, i.e., independent of the mesh size and number of involved processes. If this is the case, the second and equally likely reason for the deficient performance is a load imbalance. As outlined in Sect. 2.3, the performance crucially depends on equally divided load among the number of involved processes at all times. At process numbers larger than 1000, it is not acceptable if just one process performs the work, while others are idling. On the other hand, it must be avoided that on coarser levels always all processes are involved, so that communication becomes the limiting factor. Both effects potentially ruin scalability, in both strong and weak sense.

In addition to these problems arising specifically in a parallel environment, the node-level performance of the UG4 solver was inferior. We have learned from the study at hand, that for low process numbers, timings were worse than for the old solver. It can be expected a tightly integrated solver provides better performance than the solver replacement being called via an external library. Yet the reason for performance differences of a factor of 5 need to be investigated more closely. At the current stage, there are lots of opportunities for optimisations. First, one can try to change the way the matrix and the operators are set up. Right now they are reinitialised once for every time step. We are planning to change this, so that they are only initialised once and then reused and only refiled with new values during the time steps. A second aspect could be moving the boundary conditions inside the main matrix. This is done in the old multigrid solver, but not yet in the UG4 implementation. This will reduce the problem size and therefore, the computational demand as well as the communication.

Acknowledgements The authors kindly acknowledge the *High Performance Computing Center Stuttgart* (HLRS) for support and supply of computational time on the Cray XC40 platform under the Grant No. FS3D/11142 and the financial support by the Deutsche Forschungsgemeinschaft (DFG) for the Collaborative Research Center SFB-TRR75.

References

1. P. Beau, M. Funk, R. Lebas, F. Demoulin, Cavitation applying quasi-multiphase model to simulate atomization in diesel engines. SAE Technical Papers 01-0220 (2005)
2. J.U. Brackbill, D.B. Kothe, C. Zemach, A continuum method for modeling surface-tension. J. Comput. Phys. **100**(2), 335–354 (1992)
3. K. Eisenschmidt, M. Ertl, H. Gomaa, C. Kieffer-Roth, C. Meister, P. Rauschenberger, M. Reitzle, K. Schlottke, B. Weigand, Direct numerical simulations for multiphase flows: an overview of the multiphase code FS3D. J. Appl. Math. Comput. **272**(2), 508–517 (2016) https://doi.org/10.1016/j.amc.2015.05.095
4. M. Ertl, B. Weigand, Investigation of the influence of atmospheric pressure on the jet breakup of a shear thinning liquid with DNS, in *ILASS 2014*, Bremen (2014)
5. M. Ertl, B. Weigand, Analysis methods for direct numerical simulations of primary breakup of shear-thinning liquid jets. Atomization Sprays **27**(4), 303–317 (2017)
6. M. Ertl, N. Roth, G. Brenn, H. Gomaa, B. Weigand, Simulations and experiments on shape oscillations of newtonian and non-Newtonian liquid droplets, in *ILASS 2013* (2013), p. 7
7. M. Ertl, G. Karch, F. Sadlo, T. Ertl, B. Weigand, Investigation and visual analysis of direct simulations of quasi-steady primary break-up of shear thinning liquids, in *Proceedings 9th International Conference on Multiphase Flow: ICMF 2016*, Firenze (2016)
8. U. Fritsching, *Process-Spray: Functional Particles Produced in Spray Processes* (Springer, Cham, 2016)
9. C. Galbiati, M. Ertl, S. Tonini, G.E. Cossali, B. Weigand, DNS investigation of the primary breakup in a conical swirled jet, in *High Performance Computing in Science and Engineering'15 Transactions of the High Performance Computing Center, Stuttgart (HLRS)* (Springer, Cham, 2016), pp. 333–347
10. H. Gomaa, I. Stotz, M. Sievers, G. Lamanna, B. Weigand, Preliminary investigation on diesel droplet impact on oil wallfilms in diesel engines, in *ILASS – Europe 2011, 24th European Conference on Liquid Atomization and Spray Systems*, Estoril, September 2011
11. F.H. Harlow, J.E. Welch, Numerical calculation of time-dependent viscous incompressible flow of fluid with free surface. Phys. Fluids **8**(12), 2182–2189 (1965)
12. J. Hernández, J. López, P. Gómez, C. Zanzi, F. Faura, A new volume of fluid method in three dimensions—part I: multidimensional advection method with face-matched flux polyhedra. Int. J. Numer. Methods Fluids **58**(8), 897–921 (2008). https://doi.org/10.1002/fld.1776

13. M. Herrmann, A dual-scale les subgrid model for turbulent liquid/gas phase interface dynamics, in *13th Triennial International Conference on Liquid Atomization and Spray Systems ICLASS 2015*, Tainan, August 23–27 (2015)
14. C.W. Hirt, B.D. Nichols, Volume of fluid (VOF) method for the dynamics of free boundaries. J. Comput. Phys. **39**(1), 201–225 (1981). https://doi.org/10.1016/0021-9991(81)90145-5
15. M. Klein, Direct numerical simulation of a spatially developing water sheet at moderate Reynolds number. Int. J. Heat Fluid Flow **26**, 722–731 (2005)
16. B. Lafaurie, C. Nardone, R. Scardovelli, S. Zaleski, G. Zanetti, Modelling merging and fragmentation in multiphase flows with SURFER. J. Comput. Phys. **113**(1), 134–147 (1994)
17. A. Lakdawala, R. Thaokar, A. Sharma, Break-up of a non-newtonian jet injected downwards in a newtonian liquid. Sadhana Indian Acad. Sci. **40**, 819–833 (2015)
18. A.H. Lefebvre, *Atomization and Sprays* (Hemisphere, New York, 1989)
19. H. Li-Ping, Z. Meng-Zheng, D. Qing, L. Ning, X. Zhen-Yan, Large eddy simulation of atomization process of non-newtonian liquid jet. Adv. Sci. Lett. **8**, 285–290 (2012)
20. S.P. Lin, R.D. Reitz, Drop and spray formation from a liquid jet. Annu. Rev. Fluid Mech. **30**, 85–105 (1998)
21. C.D. Munz, T. Westermann, *Numerische Behandlung gewöhnlicher und partieller Differenzialgleichungen* (Springer, Berlin, 2006). ISBN 978-3-540-29867-3
22. Y. Pan, H. Suga, A numerical study on the breakup process of laminar liquid jets into a gas. Phys. Fluids **18**, 052101 (2006)
23. S. Popinet, An accurate adaptive solver for surface-tension-driven interfacial flows. J. Comput. Phys. **228**(16), 5838–5866 (2009). https://doi.org/10.1016/j.jcp.2009.04.042
24. P. Rauschenberger, B. Weigand, Direct numerical simulation of rigid bodies in multiphase flow within an Eulerian framework. J. Comput. Phys. **291**, 238–253 (2015). https://doi.org/10.1016/j.jcp.2015.03.023
25. P. Rauschenberger, J. Schlottke, K. Eisenschmidt, B. Weigand, Direct numerical simulation of multiphase flow with rigid body motion in an Eulerian framework, in *ILASS - Europe 2011, 24th European Conference on Liquid Atomization and Spray Systems*, Estoril (2011)
26. P. Rauschenberger, J. Schlottke, B. Weigand, A computation technique for rigid particle flows in an Eulerian framework using the multiphase DNS code FS3D, in *High Performance Computing in Science and Engineering'11 Transactions of the High Performance Computing Center, Stuttgart (HLRS)* (2011). https://doi.org/10.1007/978-3-642-23869-7_23
27. S. Reiter, A. Vogel, I. Heppner, M. Rupp, G. Wittum, A massively parallel geometric multigrid solver on hierarchically distributed grids. Comput. Vis. Sci. **16**(4), 151–164 (2013). https://doi.org/10.1007/s00791-014-0231-x
28. M. Reitzle, C. Kieffer-Roth, H. Garcke, B. Weigand, A volume-of-fluid method for three-dimensional hexagonal solidification processes. J. Comput. Phys. **339**, 356–369 (2017). https://doi.org/10.1016/j.jcp.2017.03.001
29. W.J. Rider, D.B. Kothe, Reconstructing volume tracking. J. Comput. Phys. **141**(2), 112–152 (1998). https://doi.org/10.1006/jcph.1998.5906
30. M. Rieber, Numerische Modellierung der Dynamik freier Grenzflächen in Zweiphasenströmungen. Dissertation, Universität Stuttgart, 2004
31. M. Rieber, F. Graf, M. Hase, N. Roth, B. Weigand, Numerical simulation of moving spherical and strongly deformed droplets, in *Proceedings ILASS-Europe* (2000), pp. 1–6
32. N. Roth, J. Schlottke, J. Urban, B. Weigand, Simulations of droplet impact on cold wall without wetting, in *ILASS* (2008), pp. 1–7
33. N. Roth, H. Gomaa, B. Weigand, Droplet collisions at high weber numbers: experiments and numerical simulations, in *Proceedings of DIPSI Workshop 2010 on Droplet Impact Phenomena & Spray Investigation*, Bergamo (2010)
34. W. Sander, B. Weigand, Direct numerical simulation of primary breakup phenomena in liquid sheets, in *High-Performance Computing in Science and Engineering 2006: Transactions of the High Performance Computing Center Stuttgart (HLRS)* (Springer, Berlin, 2006), pp. 223–236
35. J. Shinjo, A. Umemura, Surface instability and primary atomization characteristics of straight liquid jet sprays. Int. J. Multiphase Flow **37**, 1294–1304 (2011)

36. G. Strang, On the construction and comparison of difference schemes. SIAM J. Numer. Anal. **5**(3), 506–517 (1968)
37. R.I. Tanner, *Engineering Rheology*. Oxford Engineering Science Series, 2nd edn. (Oxford University Press, Oxford, 2002)
38. A. Vogel, S. Reiter, M. Rupp, A. Nägel, G. Wittum, UG4: a novel flexible software system for simulating PDE based models on high performance computers. Comput. Vis. Sci. **16**(4), 165–179 (2013). https://doi.org/10.1007/s00791-014-0232-9
39. A. Vogel, A. Calotoiu, A. Strubem, S. Reiter, A. Nägel, F. Wolf, G. Wittum, 10,000 performance models per minute – scalability of the UG4 simulation framework, in *Euro-Par 2015*, ed. by J. Träff, S. Hunold, F. Versaci, vol. 9233 (2015), pp. 519–531. https://doi.org/10. 1007/978-3-662-48096-0
40. A. Vogel, A. Calotoiu, A. Nägel, S. Reiter, A. Strube, G. Wittum, F. Wolf, Automated performance modeling of the UG4 simulation framework, in *Software for Exascale Computing - SPPEXA 2013–2015*, ed. by H. Bungartz, P. Neumann, W.E. Nagel. Lecture Notes in Computational Science and Engineering, vol. 113 (Springer, Cham, 2016), pp. 467–481. https://doi.org/10.1007/978-3-319-40528-5_21
41. H. Weking, J. Schlottke, M. Boger, C.D. Munz, B. Weigand, DNS of rising bubbles using VOF and balanced force surface tension, in *High Performance Computing on Vector Systems* (Springer, Berlin, 2010)
42. C. Zhu, M. Ertl, B. Weigand, Effect of Reynolds number on the primary jet breakup of inelastic non-newtonian fluids from a duplex nozzle using direct numerical simulation (DNS), in *ILASS 2013* (2013)

Numerical Investigation of the Turbulent Wake of Generic Space Launchers

S. Loosen, V. Statnikov, M. Meinke, and W. Schröder

Abstract The turbulent wakes of generic space launchers are numerically investigated via a zonal RANS/LES method and optimized dynamic mode decomposition (DMD), to gain insight into characteristic wake flow modes being responsible for asymmetrical loads on the engine extension known as buffet loads. The considered launcher geometries range from planar space launchers up to axisymmetric free flight configurations investigated at varying free stream conditions, i.e. transonic and supersonic. The investigated wake topologies reveal a highly unsteady behavior of the shear layer and the separation region resulting in strongly periodic and antisymmetric wall pressure fluctuations on the nozzle surface. Using conventional spectral analysis and dynamic mode decomposition, several spatio-temporal coherent low frequency modes which are responsible for the detected pressure oscillations are identified. In addition, a passive flow control device consisting of semi-circular lobes integrated at the base shoulder of the planar configuration is investigated. The objective of the concept is to reduce the reattachment length and thus the lever arm of the forces as well as to stabilize the separated shear layer. The results show a significant reduction of the reattachment length by about 75%. In addition, the semi-circular lobes partially reduce undesired low frequency pressure fluctuations on the nozzle surface. However, this reduction is achieved at the expense of an increase of high frequency pressure fluctuations due to intensified small turbulent scales.

1 Introduction

The tail of a classical space launcher, e.g., ARIANE 5, TITAN 4, H-II to name a few, includes an abrupt junction between the main body and the attached rocket engine causing the boundary layer to separate on the base shoulder. According to the varying freestream (sub-, trans-, supersonic) and nozzle flow conditions (overexpanded, adapted, underexpanded), the shed turbulent shear layer subsequently reattaches on the nozzle wall or intensively interacts with the emanating jet plume.

S. Loosen (✉) • V. Statnikov • M. Meinke • W. Schröder
Institute of Aerodynamics, RWTH Aachen University, Wüllnerstraße 5a, 52062 Aachen, Germany
e-mail: s.loosen@aia.rwth-aachen.de

© Springer International Publishing AG 2018
W.E. Nagel et al. (eds.), *High Performance Computing in Science and Engineering '17*, https://doi.org/10.1007/978-3-319-68394-2_17

For an ARIANE 5-like launcher, aerodynamically most critical is the early transonic part of the flight trajectory. At this stage, the turbulent shear layer shed from the main body impinges on the nozzle just upstream of its end, which due to high dynamic pressure values leads to significant wall pressure fluctuations. The resulting unsteady aerodynamic forces known as the buffet phenomenon, can lead under unfavorable conditions to a complete loss of the vehicle. Unfortunately, unlike the rocket inner engine flow the wake flow of a real launcher cannot be analyzed in full-scale on the ground, leading to increased safety margins and consequently, a reduced launcher efficiency. Therefore, accurate numerical tools validated by high-fidelity experimental investigations are required to provide detailed insight into the wake flow phenomena, to develop methods of their controllability, and to ultimately reduce aerodynamic loads on the nozzle structure without penalizing the launcher's efficiency.

The turbulent wake flow exhibits many similarities with the separated shear flow behind a backward-facing step (BFS). The flow over a BFS has been extensively studied experimentally, e.g., by Bradshaw and Wong [2], Eaton and Johnston [6], Simpson [24], and Driver et al. [5], and numerically, e.g., by Le et al. [14], Silveira Neto et al. [23], Friedrich and Arnal [8], and Kaltenbach and Janke [12]. In all the investigations a variation of the instantaneous impingement location of the separated shear layer by about two step heights around the mean reattachment position is reported. In addition, two basic modes of characteristic frequencies were detected in nearly all of the above mentioned studies. The low frequency mode at a Strouhal number of $Sr_h = 0.012$–0.014 based on the step height and the free stream velocity reflects an overall enlarging and contraction of the separation bubble or shear-layer "flapping" as it is commonly called in the literature. The aforementioned time dependent variation of the instantaneous reattachment position can be attributed to this "flapping" motion. Besides the investigations on planar configurations a lot of research has been done on axisymmetric space launchers at transonic speed [3, 4, 15, 21]. Schrijer et al. [21] detected by proper orthogonal decomposition (POD) two dominant wake modes containing the majority of the turbulent kinetic energy. The first low frequency mode captures a oscillating growing and shrinking of the separation zone most probably being the counterpart of the shear-layer "flapping" detected in the planar BFS flows. The second higher frequency mode describes an undulating motion of the shear layer, thus, coinciding with the vortex-shedding of the BFS flow [14]. Deprés et al. [4] and Deck and Thorigny [3] performed a two-point correlation analysis in the wake of an axisymmetric configuration and detected an anti-phase oscillation with $Sr_D \approx 0.2$ in the instantaneous wall pressure signal located at opposite sides in the azimuthal direction.

For the development of efficient lightweight space launchers it is essential to decrease the low frequency dynamic loads arising due to such large scale coherent motions. One promising possibility to manipulate the flow field and to reduce these loads is passive flow control. First applications of passive flow control in conjunction with space launchers can be traced back to the second half of the last century. Modified main bodies such as boat tails were extensively investigated [22] to increase the base pressure and thereby to reduce the overall drag. However, the purpose of the majority of the control devices examined in the recent past is a reduction of the reattachment length and thus the momentum arm to decrease the dynamic bending moments. It has been demonstrated that streamwise vortices intensify the turbulent mixing in the shear layer just downstream of the BFS leading to an increasing entrainment and spreading of the shear layer and consequently to a reduced reattachment length [10, 16]. Using tabs [16], chevrons [20], and castellation-like geometries [9] the reattachment length was reduced by up to 50%. The most promising approach to decrease the reattachment length is the so-called lobed mixer initially developed to reduce the noise in jet engines [31] but more recently also applied to a BFS configuration [1]. Due to the interaction of the spanwise and streamwise vortices generated by the lobes, the flow behind the lobed mixer becomes fully three-dimensional leading to an accelerated reattachment [24].

Within the project the turbulent wake of various ARIANE 5-like configuration is investigated at transonic and supersonic freestream conditions using a zonal RANS/LES approach and Dynamic Mode Decomposition, to detect characteristic wake flow modes responsible for the buffeting phenomenon. A large number of various configurations including planar space launchers as well as axissymmetric free flight configurations are investigated. In addition, the effect of a passive flow control device, consisting of semi-circular lobes integrated at the base shoulder, on the wake of a generic planar model is investigated.

1.1 Zonal RANS/LES Flow Solver

The time-resolved numerical computations are performed using a zonal RANS-LES finite-volume method. The computational domain is split into different zones. In the zones where the flow is attached, which applies to the flow around the forebody and insight the nozzle, the RANS equations are solved. The wake flow characterized by the separation and a pronounced dynamic is determined by an LES. The Navier-Stokes equations of a three-dimensional unsteady compressible fluid are discretized second-order accurate by a mixed centered/upwind advective upstream splitting method (AUSM) scheme for the Euler terms. The non-Euler terms are approximated second-order accurate using a centered scheme. For the temporal integration an explicit 5-stage Runge-Kutta method of second-order accuracy is used. The monotone integrated LES (MILES) method determines the impact of the sub-grid scales. The solution of the RANS equations is based on the same discretization schemes. To close the time-averaged equations the one-equation turbulence model of Fares and Schröder [7] is used. For a comprehensive description of the flow solver see Statnikov et al. [26, 27].

The transition from the RANS to the LES domain is determined by the reformulated synthetic turbulence generation (RSTG) method developed by Roidl et al. [17, 18]. Using this method, a reconstruction of the time-resolved turbulent fluctuations in the LES inlet plane by the upstream RANS solution is realized. Following the synthetic eddy method (SEM) by Jarrin et al. [11], turbulence is described as a superposition of coherent structures. The structures are generated at the inlet plane by superimposed virtual eddy cores. These eddy cores are defined at random positions x_i in a virtual volume V_{virt} which encloses the inlet plane and exhibits the dimension of the turbulent length scale l_x, the boundary-layer thickness at the inlet δ_0, and the width of the computational domain L_z in the streamwise, the wall-normal, and the spanwise direction. To take the inhomogeneity of the turbulent scales in the wall-normal direction into account, the virtual eddy cores are described by different shape factors and length and time scales depending on the wall-normal distance. Having N synthetic eddies, the normalized stochastic velocity fluctuations u'_m at the LES inlet plane are determined by the sum of the contribution $u^i_m(x, t)$ of each eddy core i

$$u'_m(x, t) = \frac{1}{\sqrt{N}} \sum_{i=1}^{N} \underbrace{\epsilon_i f^i_{\sigma,m}(x - x_i)}_{u^i_m(x,t)} \tag{1}$$

with ϵ_i being a random number within the interval $[-1, 1]$, $f^i_{\sigma,m}$ being the shape function of the respective eddy, and $m = 1, 2, 3$ denoting the Cartesian coordinates in the streamwise, the wall-normal, and the spanwise direction. The final velocity components at the LES inflow plane u_m are composed of an averaged velocity component $u_{RANS,m}$ from the RANS solution and the normalized velocity fluctuations u'_m which are subjected to a Cholesky decomposition A_{mn} to assign the values of the target Reynolds-stress tensor $R_{mn} = A^T_{mn} A_{mn}$ corresponding to the turbulent eddy viscosity of the upstream RANS

$$u_m(x, t) = u_{RANS,m} + A_{mn} u'_m(x, t). \tag{2}$$

To enable a upstream information exchange, thus, a full bidirectional coupling of the two zones, the static pressure of the LES zone is imposed after a transition of three boundary-layer thicknesses at the end of the overlapping zone onto the RANS outflow boundary. The temporal window width used to computed the pressure for the RANS outflow plane is chosen such that high-frequency oscillation of the LES pressure filed are filtered out. A more detailed description of the zonal RANS/LES method specifying the shape functions and length scales is given in [17, 18].

2 Results

Within this section the results of the simulations performed within the last funding period are presented. The chapter is divided into three parts, each dealing with a different space launcher geometry and free stream conditions.

2.1 Transonic Planar Backward-Facing Step Configuration

To illustrate the flow topology along the investigated generic space launcher configuration, the instantaneous Mach number and wall pressure coefficient distributions
are presented in Fig. 1a. The incoming freestream with $Ma_\infty = 0.8$ and $Re_{D*} = 6 \cdot 10^5$ passes over the transonic nose of the model, locally accelerating to $Ma \approx 0.9$
and subsequently decelerating to $Ma_\infty \approx 0.8$ near the end of the forebody. At
the abrupt backward-facing step at the end of the forebody, the turbulent boundary
layer separates and forms a turbulent shear layer. The shed shear layer rapidly
develops downstream of the separation due to the shear layer instability, causing
the mixing layer and thus the turbulent structures to grow in size and intensity.
Further downstream, the shear layer gradually approximates the lower wall and
finally impinges on it over a relatively wide distance in the streamwise direction
between $4 \lesssim x_r/h \lesssim 9$ depending on the time instance and the spanwise position.
Moreover, as can be seen in the instantaneous skin-friction coefficient distributions
shown in Fig. 1b, despite the quasi-2D geometry of the step there are rather wedge-
shaped reattachment spots than a straight reattachment line which strongly oscillate
in the streamwise and the spanwise directions.

To provide insight into temporal periodicity aspects of the wake flow dynamics
and the resulting structural loads, the power spectral density (PSD) is computed of
the wall pressure signals at selected positions within the shear layer reattachment
region. The resulting PSD spectra are shown in Fig. 2a. At $x = 1.5D$, one
distinct peak can be clearly identified at $Sr_{D*} \approx 0.04$. The spectrum at the

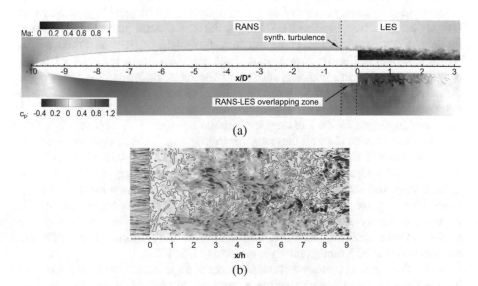

(a)

(b)

Fig. 1 (a) Flow topology of the investigated generic space launcher configuration [27]: Instantaneous Mach number (upper half) and pressure coefficient (lower half) distributions. (b) Skin-
friction coefficient distribution on the main body and the splitter plate

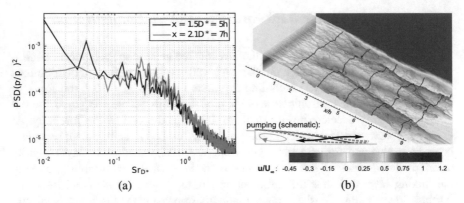

Fig. 2 (a) Power spectral density of wall pressure fluctuations p'/p_∞. (b) Reconstruction of the first DMD mode describing a longitudinal pumping motion of the recirculation region at $Sr_h \approx 0.01$ [27]

mean reattachment position $x = 2.1D^*$ reveals another dominant frequency at $Sr_{D^*} \approx 0.23$. However, besides the detected frequencies no further information about the underlying coherent fluid motion, which would simplify the intricate wake flow physics down to a few degrees of freedom, is provided. Therefore, additional post-processing is required to extract the dominant spatio-temporal modes, which is done in the following by means of dynamic mode decomposition. The dynamic mode decomposition of the three-dimensional streamwise velocity field reveals two stable DMD modes at $Sr_{D^*} = 0.037$ and $Sr_{D^*} = 0.22$ closely matching the characteristic frequency ranges of the pressure fluctuations detected by the classical spectral analysis. To visualize the DMD modes it is superimposed with the mean mode and developed in time. The resulting flow field of the first DMD mode, which is shown in Fig. 1b, reveals that at the extracted frequency $Sr_{D^*} \approx 0.037$ a coherent longitudinal pumping motion of the recirculation bubble takes place that is schematically sketched in the inlay of this figure. Thus, the mode captures a pronounced periodical low-frequency growth and subsequent collapse of the recirculation region in the streamwise direction between $4 \lesssim x_r/h \lesssim 9$, which leaves a clear trace in the wall pressure spectrum in Fig. 1a. Moreover, the obtained three-dimensional shape of the mode also explains the alternating formation of the elongated wedge-shaped coherent structures shown in Fig. 1b. Therefore, to indicate the pronounced three-dimensionality of the extracted longitudinal pumping mode given by the strong alternation in the spanwise direction the mode will be more precisely denoted as *cross-pumping*. A more detailed analysis of the performed investigations, i.e., a comparison between the presented planar BFS and an axisymmetric BFS configuration is given in [27, 28].

As outlined in the previous section, the wake of a space launcher is characterized by low frequency pressure oscillations arising due to large scale coherent motions which lead to undesired dynamic loads. For the development of future efficient lightweight space launchers it is essential to decrease these low frequency dynamic

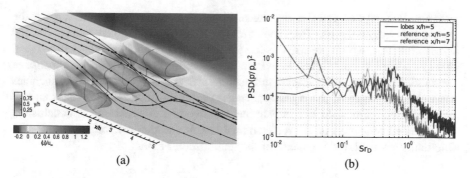

Fig. 3 (a) Flow topology of the time-averaged LES results. (b) Power spectral density of the wall pressure fluctuations p'/p_∞ at several streamwise positions for the lobe and reference configuration

loads. One promising possibility to manipulate the flow field and to reduce these loads is passive flow control. Therefore, a flow control device consisting of semi-circular lobes inserted at the trailing edge of the main body is investigated within the project. The aim of the applied loads is the generation of streamwise vortices which enhance the turbulent mixing directly behind the step and consequently, accelerates the reattachment process. The reduction of the reattachment length leads to an decreased lever arm and thus lower dynamic bending moments. Besides the reduction of the reattachment length, the lobes are to stabilize the large-scale coherent motion detected in the planar configuration described above.

To visualize the flow topology along the investigated configuration, Fig. 3a shows several characteristic streamlines and the streamwise velocity. The recirculation region is visualized by a iso-surface of the streamwise velocity $u/u_\infty = -0.01$ colored by the vertical distance from the nozzle surface. The intersection line of the velocity contour with the nozzle wall visualizes the reattachment of the flow and illustrates that averaged over time the shed shear layer impinges on the nozzle wall between $x_r/h \approx 1.3$ and $x_r/h \approx 2.3$ depending on the spanwise position. Recalling that the mean reattachment length of the planar BFS without flow control is $x/h = 7$, it is substantially reduced by more than 75% due to the lobes. In contrast to the reference case, the present configuration exhibits a pronounced three-dimensionality of the mean flow field characterized by the development of ellipsoidal reverse flow areas in the wake positioned in the spanwise direction between two adjacent lobes. A closer look at the streamlines shows that a part of the incoming boundary layer expands into the lobes, separates at the tail of the lobes, and flows straight to interact with the nozzle surface. At a streamwise position of about $x/h = 2$ this part of the flow possesses a strong spanwise velocity component such that pronounced streamwise vortices develop.

To evaluate the effect of the lobes on the pressure fluctuations, the power spectral densities (PSD) of the wall pressure signal at several streamwise positions are shown in Fig. 3b along with the previously shown spectra of the planar BFS without flow control. At the streamwise positions $x/h = 5$ and 7, where the lever arm of the pressure fluctuations and thereby the resulting dynamic moments are largest, the spectra reveal that the low frequency pressure fluctuations and hence the undesired dynamic forces, which are responsible for the buffeting phenomenon, are reduced due to the semi-circular lobes. However, this reduction is achieved at the expense of increased high frequency pressure fluctuations at $Sr_D \geq 0.4$. Since these frequencies are not in the range of the eigenfrequencies of the thrust nozzle, the loads caused by these fluctuations do not impair the stability of the structure. In conclusion, the intended reduction of the reattachment length and the low frequency pressure fluctuations is achieved by means of the semi-circular lobes.

2.2   Supersonic Strut Supported Space Launcher Configuration

In this subproject, the turbulent wake of a supported wind tunnel model of a generic ARIANE 5-like space launcher with an underexpanded air nozzle jet is investigated at $Ma_\infty = 3$ and $Re_D = 1.3 \cdot 10^6$ to gain insight into the variation of intricate wake flow phenomena of space vehicles at higher stages of the flight trajectory with increasing Mach number. Since in the configuration with an air jet the ratio between the outer flow velocity and the nozzle exit velocity is significantly smaller compared to the one of a classical space launcher, the same configuration is investigated using a helium nozzle flow. Due to the smaller molar mass of the helium the nozzle exit velocity is strongly increased without changing the Mach number. In this way, the obtained velocity ratio is nearly similar with that of a classical space launcher. Due to the use of the helium nozzle flow, a multi-species version of the zonal flow solver is used in which the non-reactive gas mixtures is described by Fick's law. For further details the reader is referred to [13].

To illustrate the flow field topology around the investigated generic rocket model, the instantaneous Mach number in the symmetry plane and the wall pressure coefficient distribution on the wall are presented in Fig. 4. The topology shows a detached bow shock around the model's fairing which deflects the supersonic

Fig. 4 Instantaneous Mach number (black and white) and wall pressure coefficient (color) distribution along the investigated generic space launcher configuration at $Ma_\infty = 3$ [29]

freestream parallel to the fairing's walls and leads to a strong increase of the wall pressure coefficient. At the junction to the cylindrical part, the flow is redirected parallel to the freestream by an expansion fan causing the pressure to decrease again. Moving further downstream to the model's tail, the turbulent supersonic boundary layer separates at the rocket model shoulder, forming a supersonic shear layer. As a result of the separation, the shear layer undergoes an expansion associated with a radial deflection towards the nozzle wall, leading to the formation of the low-pressure region and the subsonic recirculation zone. On the upper side of the configuration, a second shock and expansion wave system is formed around the double-wedge profiled strut which subsequently interacts with the shock and expansion waves emanating from the main body. A footprint of this interaction is the non-axisymmetric distribution of the pressure coefficient, shown in color in Fig. 4 on the launcher's body. To validate the numerically computed wake topology, the density gradient obtained from the LES results is compared to the experimental Schlieren pictures in Fig. 5a and b for the air jet and helium jet configuration. Note that both the experimental and the numerical results are shown for the strut-averted side and the experimental pictures are mirrored for an easier comparison. The plume's moderate afterexpansion leads to only a weak displacement effect, which results in the formation of a separation region with a triangular cross-section, having one large-scale vortex at the base and extending along the whole nozzle to its exit section. In the region close to the nozzle end the shear layer gradually realigns along the nozzle wall by weak recompression waves. The identical positions of the shock and expansion waves illustrate a good agreement between the experimental and numerical results with respect to the determined wake flow topologies. The obvious differences between the two cases is the intensity of the density gradient in the region of the nozzle exit and in the shear layer between the underexpanded jet plume and the outer flow. Due to the much larger density of the air in comparison to the helium, the density gradient is much smaller in the latter case. However, due to the supersonic outer flow the effect of the changed nozzle flow on the region further upstream is negligible. Thus, the position of the shear-layer, the expansion waves,

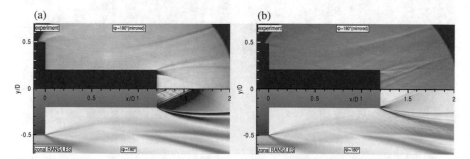

Fig. 5 Comparison of the experimentally and numerically obtained flow field topologies for the air flown nozzle (**a**) and the helium flown nozzle (**b**): Schlieren picture (top) and absolute density gradient of LES data (bottom) on the strut-averted side

and the shock, is nearly identical for the two different cases. A more extensive analysis of the supersonic strut supported space launcher configuration is given in [29]

2.3 Transonic Axisymmetric Space Launcher Configuration

After the investigation of the dynamic behavior of the wake flow of a generic planar configuration at the beginning of this chapter, the following part deals with the buffet flow of a space launchers. That is, a reduced-order analysis based on optimized DMD [19] is performed on the turbulent wake of a generic axisymmetric ARIANE 5-like configuration computed using the zonal RANS-LES method to investigate the buffet phenomenon of space launchers.

The visualize the flow topology the instantaneous spanwise vorticity distribution, time-averaged axial velocity profiles and streamlines are shown in Fig. 6a. The developed fully turbulent boundary layer separates at the axisymmetric shoulder and forms a turbulent free-shear layer. The shed shear layer rapidly evolves downstream of the separation due to the shear layer instability, causing the mixing layer and the turbulent structures to grow in size and intensity. This can be qualitatively identified in the development of the mean streamwise velocity profiles additionally shown in the lower part of Fig. 6a and in the plot of the Q-criterion iso-surface in Fig. 6b. Further downstream, the shear layer gradually approaches the nozzle and finally either impinges on its surface or passes over it, depending on the time instance and the azimuthal position, which will be discussed in the following.

The turbulent wakes of the investigated generic space launcher configurations are highly dynamic and are characterized by a manifold of spatio-temporal structures. For instance, the instantaneous streamline snapshots of the free-flight configuration

(a)   (b)

Fig. 6 Wake flow topology of the investigated free-flight configuration [30]: (**a**) instantaneous spanwise vorticity distribution (top); time-averaged axial velocity profiles and streamlines (bottom); (**b**) coherent structures in the wake visualized using Q criterion ($Q \cdot a_0^2/D^2 = 300$) and color-coded by the Mach number

taken at the same azimuthal position $\varphi = 0$ at two time steps show in Fig. 7 that sometimes the shear layer reattaches onto the nozzle close to its center more than one step height upstream of the mean reattachment position or it does not reattach onto the nozzle but passing over it. Besides, the recirculation region incorporates a manifold of large-scale spanwise vortices of different sizes such that the time-averaged topology with only two toroidal vortices, illustrated in Fig. 6a, can be hardly identified. Furthermore, despite the axisymmetric geometry and zero angle of attack, the shed shear layer approaches the nozzle not axisymmetrically resulting in a pronounced asymmetry of the induced instantaneous azimuthal pressure distribution.

To provide insight into the temporal periodicity aspects of the wall pressure fluctuations, the power spectral density (PSD) distribution of the wall pressure signals is computed at several streamwise positions. Figure 8a shows the obtained spectra for the two previously defined characteristic positions, i.e., at the nozzle center $(x/D = 0.6)$ and the mean reattachment position $x/D = 1.15$. At both

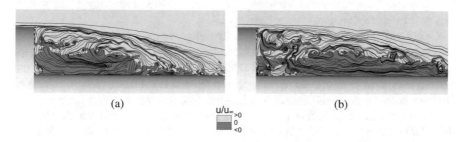

(a)   (b)

u/u_∞ >0
0
<0

Fig. 7 Two-dimensional snapshots of the projected streamlines in the near wake at the azimuthal cut $\varphi = 0°$ for two different time steps. The darker grey level in the field denotes backflow regions [30]

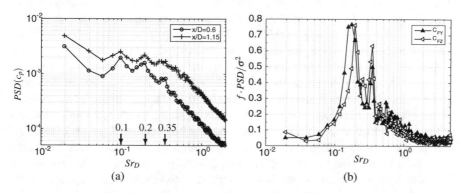

(a)   (b)

Fig. 8 Spatio-temporal spectral analysis of wall pressure fluctuations $c_{p'}$ on the nozzle [30]: (a) power spectral density (PSD); (b) premultiplied normalized power spectral density $(f \cdot PSD/\sigma^2)$ of the side load components $(\sigma_{c_{FY}} = 4.9 \cdot 10^{-3}, \sigma_{c_{FZ}} = 5.1 \cdot 10^{-3})$

streamwise positions, two distinct peaks around $Sr_D \approx 0.1$ and $Sr_D \approx 0.2$ as well as a local broad-band plateau at $Sr_D \approx 0.35 \pm 0.05$ can be identified. Furthermore, an analysis of the spatial PSD function of the transposed space-time data matrix of the previously analyzed pressure values reveals the presence of a dominant antisymmetric $m_\varphi = 1$ mode that is followed by higher wave number modes with continuously descending amplitudes. The $m_\varphi = 1$ mode causes most critical side loads on the nozzle structure, while the higher wave number modes correspond to ovalization or buckling loads, i.e., modes $m_\varphi = 2$ or $m_\varphi \geqslant 3$. To get an insight into the buffet loads arising due to the previously mentioned pressure oscillations, the instantaneous load on the nozzle is computed by the surface integral of the instantaneous wall pressure field. The buffet loads feature a pronounced temporal periodicity. Figure 8b shows the frequency-premultiplied normalized PSD function for the two Cartesian components of the side force coefficient. The energetically most dominant frequency range at $Sr_D \approx 0.2$ is clearly seen. There is another pronounced but narrower and consequently, energetically less strong range around $Sr_D \approx 0.35$. Notice that these frequencies correlate with the respective values in the wall pressure fluctuations shown in Fig. 8a, while the low frequency peaks around $Sr_D \approx 0.1$ detected previously are hardly visible.

Optimized DMD is applied to the velocity field to extract phase-averaged modes for each of the three detected characteristic frequencies. Subsequently, a reduced-order analysis based on the extracted DMD modes is performed to scrutinize the underlying spatio-temporal coherent motion responsible for the buffet phenomenon. On the whole, three dynamic modes of interest, frequencies of which closely match the characteristic peaks in the wall pressure spectra and side loads, are identified in the DMD spectrum, i.e., $Sr_D(\lambda_1) \approx 0.1$, $Sr_D(\lambda_2) \approx 0.2$ and $Sr_D(\lambda_3) \approx 0.35$. The shape of these modes contains the information about the underlying phase-averaged coherent modulation of the flow field and is analyzed in the following.

The coherent phase-averaged fluid motion is obtained by superimposing the steady mean mode with the respective dynamic modes at the given frequency. The resulting reduced-order three-dimensional streamwise velocity fields are shown in Fig. 9 for the three frequencies of interest in ascending order, i.e., $Sr_D(\lambda_1) \approx 0.1$ (a), $Sr_D(\lambda_2) \approx 0.2$ (b), $Sr_D(\lambda_3) \approx 0.35$ (c). The first (top) row of the subplots shows the time instance t_0, while the second (bottom) row illustrates the time instances $t_0 + 0.5T(\lambda_n)$, i.e., after one half of the respective period defined by $T(\lambda_n) = Sr_D(\lambda_n)^{-1}D/u_\infty$, to visualize the temporal evolution of the spatial modulation. Three types of coherent fluid motion are identified. The spatio-temporal modulation of the velocity field at $Sr_D \approx 0.1$ (Fig. 9a) indicates a longitudinal pumping motion of the recirculation region. At the second frequency of interest, $Sr_D \approx 0.2$ (Fig. 9b), a pronounced antisymmetric cross-flapping motion of the shear layer is clearly seen. At the third frequency, $Sr_D \approx 0.35$ (Fig. 9c), a quasi-swinging motion of the shear layer is extracted that occurs in phase with the previous flapping mode. In conclusion, the physical mechanism leading to the pressure oscillations and consequently the buffet loads can be traced back to the just detected motion of the recirculation region and shear layer. For a more detailed insight into the underlying mechanism of the buffet phenomenon see [25, 30]

$t = t_0$:

$t = t_0 + 0.5T(\lambda_n)$:

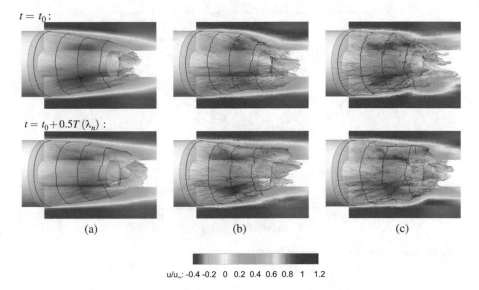

(a)　　　　　　　　　(b)　　　　　　　　　(c)

u/u$_\infty$: -0.4 -0.2　0　0.2 0.4 0.6 0.8　1　1.2

Fig. 9 Reduced-order modeled streamwise velocity field for the three characteristic frequencies [30]: (**a**) $Sr_D(\lambda_1) = 0.1$, (**b**) $Sr_D(\lambda_2) = 0.2$, (**c**) $Sr_D(\lambda_3) = 0.35$. Time instances at the beginning (top) and in the middle (bottom) of the respective period are shown

3 Computational Performance

The zonal computations as well as the subsequent DMD analyses have been performed on the CRAY XC40 (Hazel Hen) at the High-Performance Computer Center Stuttgart (HLRS,Stuttgart). The system consists of 7712 two socket nodes with 12 cores at 2.5 GHz. Each node is equipped with 128 GB of RAM, i.e., each core has 5.33 GB of memory available for the computations. In total a number of 185,088 cores with a peak performance of 7.42 Petaflops are available. The flow solver as well as the DMD postprocessor are optimized for the HLRS HPC system using hybrid parallelization based on MPI and OpenMP. Furthermore, parallel I/O procedure using HDF5 is employed.

To demonstrate the scalability of the used zonal RANS/LES flow solver, a strong scaling has been performed on the Hazel Hen. The scaling was performed on a cubic grid with 770^3 grid points. Six different core numbers were tested, i.e., 1008, 2016, 4032, 8064, 16,128 and 32,256. Each simulation runs for 1000 iteration steps and was repeated for multiple times to guarantee reproducibility. The overall speedup and the ideal speed up are given as a function of the number of cores in Fig. 10. The results prove the good scalability of the used code. For an exemplary analysis of the wake flow of an Ariane 5-like configuration [30], a zonal setup with approximately 500 Mio. grid points is used. The analyzed time-resolved data have been computed over a time interval of 409.6 t_{ref} after a transient phase of 100 t_{ref}, with $t_{ref} = D/u_\infty$ being the reference time unit needed by a particle moving with the freestream

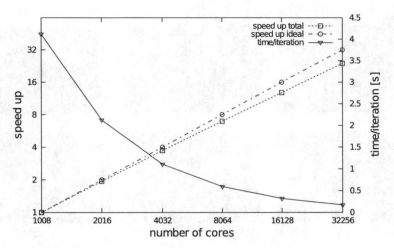

Fig. 10 Strong scaling experiment on Hazel Hen

velocity u_∞ to cover one reference length defined by the main body diameter. For this case, 14.6 Mio. core hours distributed over 9600 cores have been used in total.

4 Conclusion

Reduced-order analysis based on optimized dynamic mode decomposition has been performed on the turbulent wake of various generic space launcher geometries, including inter alia a planar backward-facing step configuration and an axisymmetric free flight launcher. The simulation has been performed using a zonal RANS/LES method. Both transonic and supersonic free stream conditions were considered. The results show a strong unsteady behavior of the recirculation region and shear layer leading to strong antisymmetric pressure loads on the nozzle extension. Using conventional statistical analysis, several characteristic frequencies were detected in the pressure signal of the different investigated configurations. To clarify the origin of the detected wake dynamics, the underlying spatio-temporal coherent modes were extracted using dynamic mode decomposition. Both, in the planar configuration and the axisymmetric free flight configuration a low frequency pumping motion of the recirculation region and a higher frequency flapping motion of the shear layer were identified which are responsible for the detected pressure fluctuations. Furthermore, a passive flow control device consisting of semi-circular lobes mounted on the base shoulder of the planar configuration was investigated. Due to this device, the reattachment length was reduced by up to 75%. In addition, the low frequency pressure oscillations which are mainly responsible for the buffet loads were slightly decreased by the lobes. However, this reduction was achieved at the expense of increased high frequency pressure fluctuations.

Acknowledgements Financial support has been provided by the German Research Foundation (Deutsche Forschungsgemeinschaft—DFG) in the framework of the Sonderforschungsbereich Transregio 40. The authors are grateful for the computing resources provided by the High Performance Computing Center Stuttgart (HLRS) and the Jülich Supercomputing Center (JSC) within a Large-Scale Project of the Gauss Center for Supercomputing (GCS).

References

1. I. Bolgar, S. Scharnowski, C.J. Kähler, Control of the reattachment length of a transonic 2d backward-facing step flow, in *International Conference on Jets, Wakes and Separated Flows* (2015)
2. P. Bradshaw, F. Wong, The reattachment and relaxation of a turbulent shear layer. J. Fluid Mech. **52**(1), 113–135 (1972)
3. S. Deck, P. Thorigny, Unsteadiness of an axisymmetric separating-reattaching flow: numerical investigation. Phys. Fluids **19**(065103), 65103 (2007)
4. D. Depres, P. Reijasse, J.P. Dussauge, Analysis of unsteadiness in afterbody transonic flows. AIAA J. **42**(12), 2541–2550 (2004)
5. D.M. Driver, H.L. Seegmiller, J.G. Marvin, Time-dependent behavior of a reattaching shear layer. AIAA J. **25**(7), 914–919 (1987)
6. J.K. Eaton, J.P. Johnston, A review of research on subsonic turbulent flow reattachment. AIAA J. **19**(9), 1093–1100 (1981)
7. E. Fares, W. Schröder, A general one-equation turbulence model for free shear and wall-bounded flows. Flow Turbul. Combust. **73**, 187–215 (2004). http://doi.org/10.1007/s10494-005-8625-y
8. R. Friedrich, M. Arnal, Analysing turbulent backward-facing step flow with the low-pass-filtered Navier-Stokes Equations. J. Wind Eng. Ind. Aerodyn. **35**, 101–128 (1990)
9. S.L. Gai, S.D. Sharma, Subsonic turbulent flow over a rearward facing segmented step. Phys. Fluids **27**, 544–546 (1984)
10. K. Isomoto, S. Honami, The effect of inlet turbulence intensity on the reattachment process over a backward-facing step. J. Fluids Eng. **111**, 87 (1989)
11. N. Jarrin, N. Benhamadouche, S. Laurence, D. Prosser, A synthetic-eddy-method for generating inflow conditions for large-eddy simulations. Int. J. Heat Fluid Flow **27**, 585–593 (2006)
12. H. Kaltenbach, G. Janke, Direct numerical simulation of flow separation behind a swept, rearward-facing step at Re_h=3000. Phys. Fluids **330**, 349–374 (2000)
13. M. Konopka, M. Meinke, W. Schröder, Large-eddy simulation of shock-cooling-film interaction at helium and hydrogen injection. Phys. Fluids **25**, 106101 (2013)
14. H. Le, P. Moin, J. Kim, Direct numerical simulation of turbulent flow over a backward-facing step. J. Fluid Mech. **330**, 349–374 (1997)
15. S. Marie, P. Druault, H. Lambare, F. Schrijer, Experimental analysis of the pressure-velocity correlations of external unsteady flow over rocket launchers. Aerosp. Sci. Technol. **30**, 83–93 (2013)
16. H. Park, W.P. Jeon, H. Choi, J.Y. Yoo, Mixing enhancement behind a backward-facing step using tabs. Phys. Fluids **19**, 105103 (2007)
17. B. Roidl, M. Meinke, W. Schröder, A reformulated synthetic turbulence generation method for a zonal RANS-LES method and its application to zero-pressure gradient boundary layers. Int. J. Heat Fluid Flow **44**, 28–40 (2013). http://doi.org/10.1016/j.ijheatfluidflow.2013.03.017
18. B. Roidl, M. Meinke, W. Schröder, Boundary layers affected by different pressure gradients investigated computationally by a zonal RANS-LES method. Int. J. Heat Fluid Flow **45**, 1–13 (2014)
19. P.J. Schmid, Dynamic mode decomposition of numerical and experimental data. J. Fluid Mech. **656**, 5–28 (2010)

20. F. Schrijer, A. Sciacchitano, F. Scarano, Experimental investigation of flow control devices for the reduction of transonic buffeting on rocket afterbodies, in *15th Symposium on Applications of Laser Techniques to Fluid Mechanics* (2010)
21. F. Schrijer, A. Sciacchitano, F. Scarano, Spatio-temporal and modal analysis of unsteady fluctuations in a high-subsonic base flow. Phys. Fluids **26**, 086101 (2014)
22. F.V. Silhan, J.M. Cubbage, Drag of conical and circular-arc boattail afterbodies at mach nnumber of 0.6 to 1.3. NACA RM L56K22 (1957)
23. A. Silveria Neto, D. Grand, O. Metais, M. Lesieur, A numerical investigation of the coherent vortices in turbulence behind a backward-facing step. J. Fluid Mech. **256**, 1–25 (1993)
24. R.L. Simpson, Turbulent boundary-layer separation. Ann. Rev. Fluid Mech. **21**, 205–2034 (1989)
25. V. Statnikov, Numerical analysis of space launcher wake flows, Ph.D. thesis, RWTH Aachen University, 2017
26. V. Statnikov, T. Sayadi, M. Meinke, P. Schmid, W. Schröder, Analysis of pressure perturbation sources on a generic space launcher after-body in supersonic flow using zonal RANS/LES and dynamic mode decomposition. Phys. Fluids **27**(016103), 1–22 (2015)
27. V. Statnikov, I. Bolgar, S. Scharnowski, M. Meinke, C. Kähler, W. Schröder, Analysis of characteristic wake flow modes on a generic transonic backward-facing step configuration. Euro. J. Mech. Fluids **59**, 124–134 (2016)
28. V. Statnikov, B. Roidl, M. Meinke, W. Schröder, Analysis of spatio-temporal wake modes of space launchers at transonic flow. AIAA Paper 2016-1116 (2016)
29. V. Statnikov, S. Stephan, K. Pausch, M. Meinke, R. Radespiel, W. Schröder, Experimental and numerical investigations of the turbulent wake flow of a generic space launcher at $M_\infty = 3$ and $M_\infty = 6$. CEAS Space J. **8**(2), 101–116 (2016)
30. V. Statnikov, M. Meinke, W. Schröder, Reduced-order analysis of buffet flow of space launchers. J. Fluid Mech. **815**, 1–25 (2017)
31. I.A. Waitz, Y.J. Qiu, T.A. Manning, A.K.S. Fung, Enhanced mixing with streamwise vorticity. Prog. Aerosp. Sci. **33**, 323–351 (1997)

Optimization and HPC-Applications of the Flow Solver FLOWer

Johannes Letzgus, Lukas Dürrwächter, Ulrich Schäferlein, Manuel Keßler, and Ewald Krämer

Abstract Recent optimizations and HPC-applications of the flow solver FLOWer are presented in this paper. A graph partitioning method is introduced to the MPI communication, which reduces the number of messages as well as the total message size, leading to a run time speed-up of 20%. A numerical investigation of a finite wing shows the influence of the wind tunnel wall only in the wing root area and agrees well with experimental data for attached flow. Both a URANS and a Delayed Detached-Eddy Simulation (DDES) of the massively stalled wing reveal difficulties in matching the experimental behaviour of flow separation. Finally, a simulation of a model Contra-Rotating Open Rotor (CROR) at various operating conditions exhibit interaction effects, blade loadings and noise emissions which agree well with expectations and results from literature.

1 Introduction

With the ongoing increase in supercomputing power, advanced numerical methods become feasible and numerical investigations in research and development of aircraft can focus on complex aerodynamic phenomena. This paper gives an overview of current work of the Helicopter and Aeroacoustics Group at the Institute of Aerodynamics and Gas Dynamics (IAG) of the University of Stuttgart. All investigations were carried out on the HLRS system Cray XC40 (Hazel Hen). At first, the latest optimization step of the numerical solver regarding high-performance computing is presented, where graph partitioning is used to reduce MPI communication. Secondly, two applications demonstrate the functionalities of the tool chain at IAG: A finite wing in a wind tunnel is investigated, showing the influence of side walls and the challenges of simulating flow separation and stalled wing with both a URANS and a DDES approach. Then the simulation of a model

J. Letzgus (✉) • L. Dürrwächter • U. Schäferlein • M. Keßler • E. Krämer
Institute of Aerodynamics and Gas Dynamics (IAG), University of Stuttgart, Pfaffenwaldring 21, 70569 Stuttgart, Germany
e-mail: letzgus@iag.uni-stuttgart.de

© Springer International Publishing AG 2018
W.E. Nagel et al. (eds.), *High Performance Computing in Science and Engineering '17*, https://doi.org/10.1007/978-3-319-68394-2_18

CROR serves as a baseline for validation processes and provides insight into the noise generating mechanisms of a CROR.

2 Numerical Method

2.1 The Flow Solver FLOWer

For the present studies, the block-structured finite volume flow solver FLOWer [1] is used. Originally developed by DLR, the code has been significantly enhanced, e.g. with a fifth-order WENO scheme, hybrid mesh capability and a new coupling library for fluid-structure interaction, and optimized for high-performance computing by IAG [2–4]. Most recently, it has been expanded by state of the art detached-eddy simulation (DES) methods like the (DDES) [5] in order to conduct hybrid RANS/LES simulations [6]. The studies presented here use an implicit dual time-stepping method of second order for time integration and second-order Jameson Schmidt Turkel (JST) [7] as spatial scheme. Convergence acceleration is achieved with a three-level multigrid method.

2.2 Optimization of Code Parallelization

The scalability of the CFD code FLOWer was recently significantly improved by the implementation of a node-to-node (n2n) communication instead of a core-to-core (c2c) communication [8]. By using a shared memory (SHM) to centrally collect the halo data of each single process on a node and a subsequent dispatch of the data as a bundled packet, a significant reduction of the number of messages could be achieved. By sending large data packages instead of many small messages, a more efficient use of the available bandwidth is achieved.

The structured code FLOWer allows the parallelization of the computational effort to several cores by dividing the computing area into blocks, which represent the smallest assignable unit. Unlike unstructured grids, the use of structured grids allows only a strongly restrictive division of the computational domain, which must meet an i-j-k index notation for each block unit. This division is to be carried out in FLOWer by the user with the aim of creating blocks with a preferably equal amount of cells per block.

The previous parallelization used a distribution of the blocks to individual computing units with the aim to achieve optimal load balance by equalizing the workload. For this purpose, several blocks may also be placed on one core, but at least one. By providing more fine-grained block units and thus degrees of freedom for the parallelization, an ideal load balance can be better approximated. However, too extensive decomposition becomes partially inefficient by, for example, loop initialization overhead. In the previous approach, the communication paths and expense of the halo cells have been taken into account as additional workload, but not optimized. In order to remedy this weakness, a graph partitioning has been introduced, which aims at an optimization of the communication paths for the reduction of the MPI data stream.

To carry out the graph partitioning, the widely used METIS library [9] was integrated into the code. A graph is set up whose vertices correspond to blocks, and edges to the communication paths to their neighbor blocks. The vertices and edges are weighted with the block cells and the number of halo cells.

For ordinary simulations, block numbers are in total by a factor of 3–4 greater than the number of computing units/cores used. Thus, the degree of freedom available for the graph partitioning is severely restricted. Since the exchange of the halo cells by the node-to-node communication is the critical size for the run time and only to be considered over a node boundary, the graph partitioning of the computational domain is carried out over the nodes used. Subsequently, the blocks assigned to a node are distributed over the node's cores, whereby only the best possible load balance is taken into account. However, due to the low number of degrees of freedom, the use of the multilevel algorithm in METIS results in a too weak consideration of the load balance over the nodes compared to the communication. To counteract this, a subsequent minor redistribution of individual blocks is performed after the graph partitioning to reduce load imbalance. The resulting, slightly increased communication paths are accepted. With this approach an optimization between minimal communication and ideal load balance is achieved.

Figure 1 shows the communication paths between individual nodes for an application-oriented computation with 50 nodes. The previous parallelization and the new implementation using the described graph partitioning are illustrated. Due to the stronger weighting of the communication in graph partitioning, there are

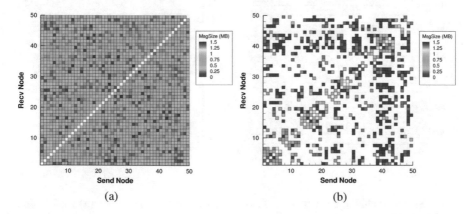

Fig. 1 Comparison of MPI communication pattern using different parallelization schemes in FLOWer. (**a**) Load balance parallelization (**b**) Graph partitioning

mainly partitions consisting of spatially adjacent blocks. The chronological assign-ment of these partitions to the computing nodes results in a concentration of the communication paths near the diagonal (see Fig. 1b). Communication with nodes more remote results from the subsequent redistribution of the graph partitioning to reduce the load imbalance. However, these messages are much smaller in their data size. Compared to the old communication pattern, the graph partitioning results in a reduction of the number of communication messages and consolidates communication towards larger messages. In this test case, the overall size of the MPI data stream is reduced by factor 18.3 and the message amount by factor 3.5, with a maximal load imbalance of 2.3%.

The advances in the optimization of the FLOWer code with regard to its scalability are shown in Fig. 2a, b. A strong and weak scaling depict the benefit of the implementations compared to the previous versions. *c2c* and *n2n* denotes the core-to-core and node-to-node communication scheme and *graph* the use of the new parallelization method. In addition to the clear advantage of node-to-node compare to the core-to-core communication, the new parallelization method also offers further improvements in the run time as well as scalability of the code. For weak scaling (see Fig. 2a) the new parallelization scheme causes a nearly constant offset of the wall time towards the ideal line by a run time speed-up of 20% on average. The scaling shows a similar behavior as the previous parallelization, however, with a slightly reduced gradient at more than 50,000 cores indicating a more efficient MPI communication. In the case of strong scaling, a comparable improvement could be achieved, which allows a further acceleration of the simulations by a smaller workload per core. Overall, very satisfactory results are recorded with a beneficial use of the new graph parallelization method.

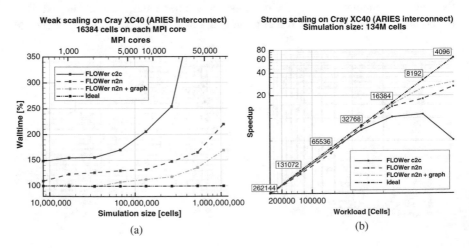

Fig. 2 Scaling studies of the flow solver FLOWer on Cray XC40 (Hazel Hen). (**a**) Weak scaling (**b**) Strong scaling

3   Static Stall on a Finite Wing

In the field of aerodynamics, flow separation plays an important role since aero-dynamic loads and general flow characteristics significantly change, depending on whether the flow is attached or separated. This is crucial especially in aviation, where flow separation mainly occurs if the angle of attack of a wing or a rotor blade is increased beyond a critical maximum angle. Then, the occurring adverse pressure gradient is too strong, slowing down the near-wall flow in the boundary layer (BL) to zero velocity and eventually even reversing its direction. Now, the flow detaches from the aerodynamic surface and highly unsteady, three-dimensional and chaotic vortical structures evolve. This process is called *stall*. It leads to a more or less sudden reduction in lift and increase in drag and thus limits the flight envelope of an aircraft. First results of a numerical investigation of a stalled finite wing in a wind tunnel environment are here presented and compared to experimental data.

3.1   Experimental Setup and Flow Conditions

The experiments [10] were conducted in the Cross Wind Simulation Facility (SWG) at DLR Göttingen. A finite wing in the shape of a rotor blade tip model was mounted in a test section of the closed-return wind tunnel with a cross section of 2.4×1.6 m. The wing has a chord length c of 0.27 m and a span b of 1.62 m. The wing tip is parabolic and a DSA-9A airfoil is used. A positive linear twist of $5.5°$ was applied to shift maximum lift and stall onset from the wing root towards the wing tip. Among others, unsteady pressure transducers of type Kulite XCQ-093 were used to obtain surface pressure data in three spanwise sections, at $z/b = 0.49$ (denoted by s1), 0.68 (s2) and 0.86 (s3), with z being the local coordinate in spanwise direction. The model was pitched around quarter-chord to investigate static and dynamic stall. The free stream velocity was 55 m/s, leading to a Reynolds number of about 900,000 and a Mach number of 0.16.

3.2   Numerical Setup

The numerical setup includes the finite wing which is positioned in a rectangular wind tunnel (see Fig. 3). The rigid wing is placed in the middle of the 16 m long wind tunnel, which uses a far field boundary condition as in- and outlet. Two wing grids were investigated, a typical URANS grid, denoted by *baseline*, and a DES-optimized grid, denoted by *fine*. The wing grids are of C-H type and embedded into a Cartesian background grid using the overset technique, which can be seen in Fig. 4. An overview of the wing grid spacings can be found in Table 1. The cell and block count of the two investigated setups are listed in Table 2. The physical time

Fig. 3 Positioning of the
finite wing in the wind tunnel
and spanwise sections of exp.
data

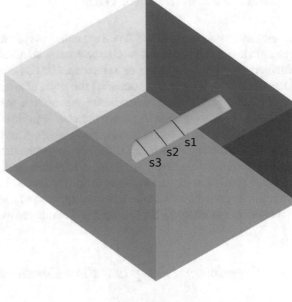

Fig. 4 Slices through
Cartesian background
grid (blue) with CH-type
wing grid (red)

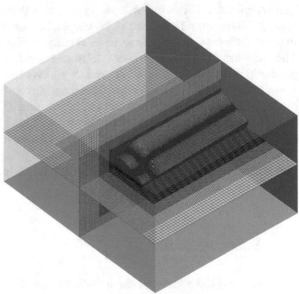

step size is $\Delta t = 4.75 \times 10^{-6}$ s, roughly representing 1/1000 of one convective time
unit, which is the time, it takes the flow to convect from leading to trailing edge. For
turbulence modelling, Menter SST, SA and SSG-ω RSM are used, while the flow is
fully turbulent.

Table 1 Spacings of the two wing grids (x is streamwise, z is spanwise direction)

Wing grid type	Baseline	Fine
$\Delta x/c$ at leading edge	0.066%	0.05%
$\Delta x/c$ at trailing edge	0.066%	0.076%
Max. $\Delta x/c$ suction side	2%	0.3%
$\Delta z/c$ (root to tip)	2%	1.8–0.25%
Growth rate in BL	1.13	1.1
y+	≤ 1	≤ 1
Based on/ reference	6. DPW [11]	Richez et al. [12]

Table 2 Cell and block count of the two investigated grids

Grid setup	Baseline	Fine
Cells wing	21×10^6	140×10^6
Cells wind tunnel	11×10^6	11×10^6
Cells in total	32×10^6	151×10^6
Number of blocks	1902	6785
Cores used	1632	6000

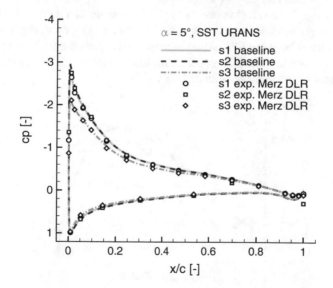

Fig. 5 Comparison of pressure distribution in three spanwise sections for attached flow, exp. data: [10]

3.3 Moderate Angle of Attack $\alpha = 5°$: Attached Flow

To gain confidence in the numerical setup, a case with a moderate angle of attack $\alpha = 5°$ is investigated at first. Here, the flow is fully attached, since trailing edge stall starts at about $\alpha = 11°$. Figure 5 shows the pressure distribution in the three spanwise sections s1, s2 and s3. For all three sections there is very good overall agreement between experiment and simulation. All computed pressure distributions shown in this study are obtained from simulations with inviscid wind tunnel walls and no modelled gap between wing root and side wall. Having the good agreement of the pressure distributions in mind, this seems reasonable, although there is a 0.4 mm wide gap in the experiments to enable the pitching motion of the wing. To further investigate this issue, the baseline setup is modified in two ways: At first, the gap is modelled but all wind tunnel walls remain inviscid, and secondly, the gap is modelled and the boundary layer of the side wall the wing is attached to is resolved. The boundary layer thickness at the wing's position was found to be around 0.1 m [13], which is reproduced in the numerical setup. The spanwise distribution of local lift coefficients found in Fig. 6 reveals that resolving the boundary layer of the side wall has a much stronger impact on the inboard flow than the gap. If only the gap is modelled, a rather weak vortex evolves, as shown on the left-hand side of Fig. 7. Comparable to the impact of a tip vortex, the effective angle of attack and consequently lift is reduced in the neighbouring region. With the resolved boundary layer, a horseshoe vortex forms around the leading edge and the wing tip-like vortex stays closer to the wing surface and grows in size. On the right-hand side of Fig. 7 the effect becomes apparent: The suction peak just downstream the leading edge becomes weaker, thus lift is reduced. However, Fig. 6 also shows that influences of both modifications decrease as the distance to the root increases. At the first section s1, the differences in lift are already small. Further outboard at section s2 and s3, no more differences are visible.

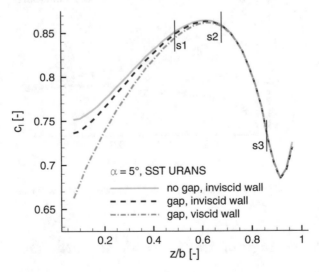

Fig. 6 Spanwise distribution of local lift coefficients over the wing shows the influence of wall and gap treatment

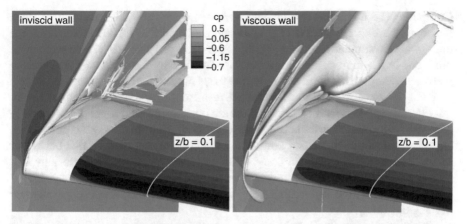

Fig. 7 Visualization of the flow in the wing root region by means of λ_2-isosurfaces and surface pressure. Left: Gap and inviscid side wall. Right: Gap and viscous side wall with resolved BL

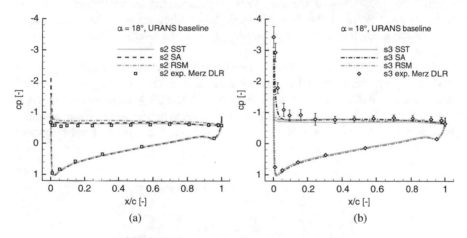

Fig. 8 Mean pressure distributions of URANS baseline setup and experiment in the sections s2 and s3 for a high angle of attack case with separated flow. Error bars represent standard deviation, exp. data: [10]. (**a**) s2: z/b = 0.68. (**b**) s3: z/b = 0.86

3.4 High Angle of Attack $\alpha = 18°$: Separated Flow

To investigate static stall with massively separated flow, the angle of attack is now increased to $\alpha = 18°$. The mean experimental pressure distribution in section s2 (see Fig. 8a) shows no more suction peak at the leading edge. The surface pressure is basically constant over the entire suction side, indicating flow separation directly at the leading edge. This is captured fairly well in the URANS simulation by the SST turbulence model. SA and RSM yield the separation point slightly too far downstream. Since separated flow is very unsteady, the exact pressure value of the plateau on the suction side highly depends on the sample size of time averaging.

Comparing the numerical results with each other, the sample size slightly differs. However, the experimental data is time-averaged over about 1000 convective time units, whereas simulation results are only time-averaged over 10–20 convective time units due to the high computational effort. This is why the offsets of the pressure plateaus seem plausible and are not in the focus of this study.

Both the experimental and numerical results in the innermost section s1 are qualitatively the same as in section s2 and are not shown here for brevity. Regarding the mean pressure distribution in the outermost section s3 Fig. 8b reveals an experimental suction peak and flow separation not until $x/c \approx 0.1$. This behaviour is not captured at all with SST and RSM turbulence modeling, where flow still separates close to the leading edge. The SA solution shows a somewhat bigger suction peak and flow separation at $x/c \approx 0.05$, but also matches the experiment in this region poorly.

The influence of only the grid resolution on the mean pressure distribution can be seen in Fig. 9. In section s2, small differences are only visible at the pressure plateau, again, most likely due to the time averaging issue. In the outermost section s3, the separation point of the fine grid is slightly shifted towards the leading edge, contradicting the experimental result. In general, the solution seems sufficiently grid-converged.

The DDES results shown here are computed on the fine grid using SA as the underlying turbulence model in the URANS regime. The resulting mean pressure distribution of the DDES approach is compared to the URANS results in Fig. 10. In both sections, DDES shows a slightly more upstream flow separation and a smaller suction peak. At this point, it cannot be ruled out that grid-induced separation

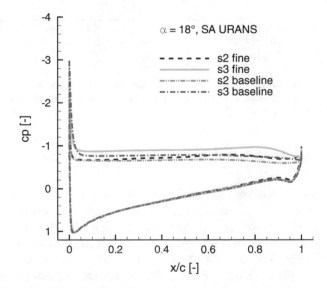

Fig. 9 Mean pressure distributions showing the influence of the grid resolution

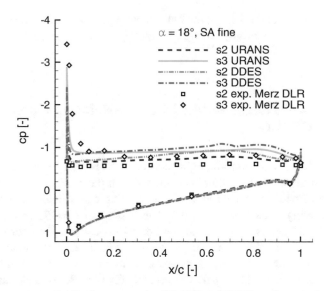

Fig. 10 Mean pressure distributions comparing URANS and DDES results, exp. data: [10]

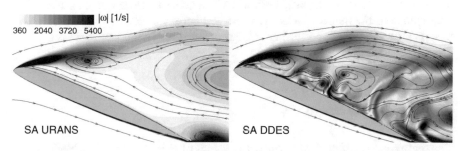

Fig. 11 Comparison of instantaneous vorticity and in-plane streamlines between URANS (left) and DDES (right) computation in section s2 (z/b = 0.68) for $\alpha = 18°$

[14]—a major issue of hybrid RANS-LES approaches in general—plays a role here. Further downstream, in both sections, DDES yields lower surface pressure in the pressure plateau region. So regarding the mean pressure distribution of this high angle of attack case, the DDES approach cannot improve the numerical results. Looking at the instantaneous flow field, however, the DDES solution is believed to be superior and closer to reality: As Fig. 11 shows, the URANS computation exhibits large-scale vortical structures evolving from the free shear layer in a periodic pattern and the in-plane streamlines indicate rather two-dimensional flow. The vorticity contours of the DDES flow field show several smaller vortical structures distributed over the entire upper side of the wing. Including the streamlines, the flow appears highly chaotic and much more three-dimensional.

In summary, all simulations seem to be unable to capture precisely the spanwise change in flow separation location seen in the experiment. While there is—to some extent—good agreement in the inner section s2, the computed mean pressure distributions in the outermost section s3 clearly differ from each other and the experiment.

4   CFD Simulation of a Model Contra-Rotating Open Rotor (CROR)

CRORs are a fixed wing aircraft engine concept with two large contra-rotating fans operating at transsonic Mach numbers developed in the 1980s. The concept is now taken on again with the recent research concerning eco-efficient engines to meet future fuel saving aims (see e.g. [15]). The benefit of CRORs is their low fuel consumption and hence reduced pollutant emission compared to conventional jet engines, which is enabled by the high by-pass ratio. One of the main drawbacks is the high noise emission caused by the contributions of single rotor noise and interaction noise. As a huge part of research on CRORs is undertaken by the industry, recent experimental data is not publicly available. A test rig featuring a model scale CROR is currently built at IAG in order to close this gap. First CFD simulations of the model CROR are presented here, which are going to serve as a baseline for validation and further enhancement of the numerical process chain in this application.

4.1   Numerical Setup

The setup consists of the two rotors with a radius of $R = 0.19$ m, seven blades in the front and six blades in the back rotor, each mounted on the cylindrical hub which is completed by a spinner at the front and extends far downstream, i.e. until the end of the computational domain in the CFD setup (see also Fig. 12). The model CROR is not optimized for efficiency and low noise levels but to provide useful data for CFD validation. Compared to an optimized CROR the blade count is reduced to keep frequencies low, the rotors are close to each other to produce high interaction noise levels and no aft rotor clipping is applied for the same reason.

For the simulations the geometry is represented by a block-structured computational mesh set up using the Chimera technique. Each blade is surrounded by its own mesh created with IAG's in-house blade meshing script Automesh and some modifications at the tip and the root, where the blade geometry is blended into the cylindrical hub. The boundary layer is resolved by 24 cells with the lowest cell height leading to a y^+ value of approximately 0.75. For the hub, an inviscid wall boundary condition is used as the flow phenomena associated with friction in this area are expected to have minor influence on performance and acoustics of the CROR due to the low blade velocity at the hub. The whole setup is immersed in

Fig. 12 CROR setup with acoustic hull surface and λ_2 vortex visualisation at 20 m/s axial inflow, 8400 rpm

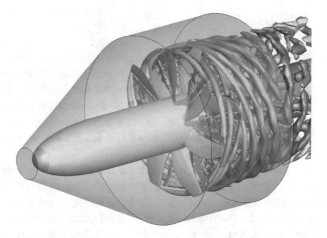

Table 3 Grid size

Front rotor	7×2.3 million
Rear rotor	6×2.3 million
Hub	2.2 million
Background	19.8 million

a Cartesian background grid automatically generated with an in-house tool. In the vicinity of the rotors the resolution is 2 mm ($\approx 0.01\,R$) and hanging grid nodes are used for coarsening towards the boundaries of the computational domain, which are placed more than $50\,R$ away from the CROR in each direction. In total, the grid size sums up to about 52 million cells, an overview can be found in Table 3. Turbulence is accounted for by the Wilcox k-ω turbulence model.

Four operating conditions are compared, with the CROR running at 6000 rpm and 8400 rpm, leading to a blade tip mach number of 0.35 and 0.49, respectively. A state with no farfield inflow and another one with an axial inflow velocity of 20 m/s is considered for each rotational speed, the former representing the beginning of the start phase and the latter resembling takeoff conditions. Each case is run for 10–15 revolutions at 1° of rotation per time step to obtain a converged flow field. Another three revolutions are added at 0.5° per time step for aerodynamic and acoustic evaluation. The computation uses 1920 cores with a maximum of 30,000 cells per core, enabling computation of six rotor revolutions in 24 h at 1° time step.

Acoustic evaluation is carried out using the in-house Ffowcs Williams-Hawkings solver ACCO [16]. During the last two revolutions of the evaluation phase the flow variables are interpolated onto an acoustic hull surface surrounding the two rotors as well as the spinner (see Fig. 12). The data from two revolutions are merged into one using a linear blending function to smooth out small periodicity imperfections that would otherwise have a large impact on the acoustic signal. The result is then processed by ACCO calculating the acoustic pressure in the time domain at prescribed observer positions in the far field. A resolution of 2 mm is chosen on the acoustic surface according to the resolution of the background mesh.

4.2 Results

Figure 12 gives a qualitative impression of the flow field at 20 m/s, 8400 rpm with pressure contours and a λ_2 iso-surface for visualisation of the tip vortices. The tip vortices originating from the front rotor are convected downstream where they hit the aft rotor in the blade tip region. Downstream of the rear rotor the flow field is dominated by the aft tip vortices, however a regular disturbance pattern caused by the wake and tip vortices of the front rotor is visible. Overall good vortex conservation is achieved.

Some time averaged results on the CROR's performance are summarized in Table 4a–d. As expected, the thrust and required power are considerably higher for the cases without inflow than for the 20 m/s cases due to the greater angles of attack on the blades. While in the 0 m/s case the front rotor contributes significantly more thrust, contributions are almost equal at 20 m/s and the ratio of power requirement even inverts. This behaviour is probably caused by a strong acceleration of the flow between the front and rear rotor in the 0 m/s case, induced by the high thrust of the front rotor and leading to comparatively low angles of attack and reduced blade loading on the rear rotor.

Figure 13a shows the range of thrust distributions on a blade of the front rotor occurring during one revolution for the two 6000 rpm cases, Fig. 13b shows the same for the rear rotor. The local thrust increases from the root of the blade towards the tip due to the increasing circumferential velocity. An extra peak near the tip is caused by the blade's own tip vortex generating an area of very low pressure on the blade suction side. Unsteadiness of the blade loading is characterized by the width of the coloured areas covering the thrust coefficient range. Unsteady thrust distribution is evident for all cases. On the front rotor blades it is caused by interaction with the downstream blades' potential field, on the rear blades wakes and tip vortices from

Table 4 CROR performance data

	Front	Rear	Total
(a) 0 m/s, 6000 rpm			
Thrust (N)	89.8	74.2	163.9
Power (kW)	2.64	2.47	5.10
(b) 0 m/s, 8400 rpm			
Thrust (N)	177.9	148.4	326.2
Power (kW)	7.31	6.89	14.20
(c) 20 m/s, 6000 rpm			
Thrust (N)	43.7	42.6	86.2
Power (kW)	1.54	1.64	3.18
Efficiency (%)	56.5	52.0	54.2
(d) 20 m/s, 8400 rpm			
Thrust (N)	117.4	106.0	223.3
Power (kW)	5.28	5.34	10.62
Efficiency (%)	44.4	39.7	42.1

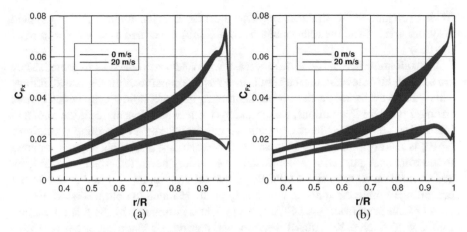

Fig. 13 Thrust coefficient ranges of a front (**a**) and a rear (**b**) rotor blade over one revolution

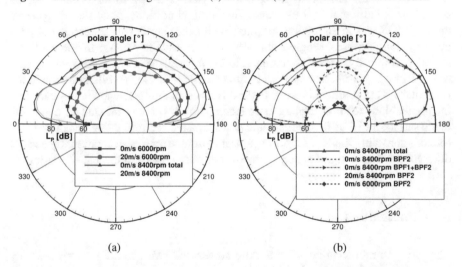

Fig. 14 Polar noise directivity at $r = 10$ m. (**a**) Overall sound pressure level. (**b**) Selected tones

the front rotor add further unsteady loads. The effect of the tip vortex impinging on the rear blade can be observed between $r/R = 0.55$ and $r/R = 0.9$ for the 0 m/s case in Fig. 13b. Two regions of increased unsteadiness are visible to either side of the vortex center which hits the rear blade around $r/R = 0.75$. In the 20 m/s case they are not as pronounced since the tip vortex is weaker and located further outward because the stream tube through the rotors is not contracted as much as in the 0 m/s case.

Figure 14a, b show the polar noise characteristics of the model CROR obtained with ACCO at a distance of 10 m from the center of the CROR. The sound pressure level L_P is shown as a function of the polar angle, which is the angle between the connecting line from the CROR's center to the observer and the rotational axis

(0° is upstream, 90° is in the rotor plane and 180° is downstream). The sound field is symmetric around the rotational axis so it is fully described by one polar from 0° to 180°.

Overall noise levels (Fig. 14a) increase with rotational speed as expected. They are also slightly elevated for the zero inflow cases compared to the cases with inflow. This can be attributed to the higher loading in the zero inflow case which causes stronger pressure fluctuations, wakes and tip vortices and thus more single rotor and interaction noise. The directivity patterns are similar for all operating conditions: Noise is mainly radiated in upstream and downstream directions slightly inclined to the rotational axis, with the downstream emission louder than the upstream lobe. This behaviour is typical for CRORs in takeoff conditions [17] and is due to the prevailing of interaction noise, which is typically radiated in oblique directions.

As can be seen from Fig. 14b, the sound field is dominated by the first interaction frequency (which is the sum of the front and rear rotor's blade passing frequencies BPF1 and BPF2). The 8400 rpm zero inflow case is shown as an example but the behaviour is similar for all examined operating conditions. Note that the present model CROR was intentionally designed to produce significant interaction noise for investigation purposes. Single rotor noise (aft rotor blade passing frequency BPF2 shown exemplarily) is mainly radiated in the rotor plane. For the zero inflow case minor BPF2 lobes can be observed around the rotational axis while radiation in the rotor plane is approximately equal to the 20 m/s case. This can again be explained by the higher blade loading causing more loading noise which is radiated normal to the blade surface and thus in upstream and downstream directions while thickness noise remains constant. Single rotor noise gains importance for higher rotational speed but does not contribute significantly to the overall sound pressure level at 6000 rpm.

5 Conclusions

The MPI communication of the flow solver FLOWer could be improved by introducing graph partitioning. For the present test case, this reduces the number of messages to be sent by factor 3.5 and the overall size of the MPI data stream by factor 18.3. This new parallelization method gives a run time speed-up of 20% compared to the already established node-to-node communication method.

The numerical investigation of a finite wing showed a strong influence of the boundary layer of the wind tunnel side wall on lift, but only in the near-wall region at the wing root. For attached flow, the computed pressure distributions match those of the experiment very well. In case of massive flow separation, all simulations fail to capture the spanwise change in flow separation location. Moreover, neither a finer grid resolution nor a DDES approach improves the numerical results.

An isolated model CROR was simulated at various operating conditions. Unsteady blade loadings were investigated as well as noise emission, both show significant interaction effects between the two rotors. The observations agree well with expectations and results from the literature and the simulations are therefore believed to form a reliable basis for further investigations on integrated CRORs.

Acknowledgements The provided supercomputing time and technical support of the High Performance Computing Center Stuttgart (HLRS) of the University of Stuttgart within the *HELISIM* project is gratefully acknowledged. Parts of the research presented in this study were supported by Deutsche Forschungsgemeinschaft (DFG) within the projects *Untersuchung der dreidimensionalen dynamischen Strömungsablösung an Rotorblättern* and *Numerical Investigation of the Noise Emission of Integrated Counter-Rotating Open Rotors*.

References

1. J. Raddatz, J.K. Fassbender, Block structured Navier-Stokes solver flower, in *MEGAFLOW-Numerical Flow Simulation for Aircraft Design* (Springer, Berlin, 2005), pp. 27–44
2. U. Kowarsch, T. Hofmann, M. Keßler, E. Krämer, Adding hybrid mesh capability to a CFD-solver for helicopter flows, in *High Performance Computing in Science and Engineering' 16* (Springer, Berlin, 2016), pp. 461–471
3. P.P. Kranzinger, U. Kowarsch, M. Schuff, M. Keßler, E. Krämer, Advances in parallelization and high-fidelity simulation of helicopter phenomena, in *High Performance Computing in Science and Engineering' 15* (Springer, Berlin, 2016), pp. 479–494
4. C. Stanger, B. Kutz, U. Kowarsch, E.R. Busch, M. Keßler, E. Krämer, Enhancement and applications of a structural URANS solver, in *High Performance Computing in Science and Engineering' 14* (Springer, Berlin, 2015), pp. 433–446
5. P.R. Spalart, S. Deck, M.L. Shur, K.D. Squires, M.Kh Strelets, A. Travin, A new version of detached-eddy simulation, resistant to ambiguous grid densities. Theor. Comput. Fluid Dyn. **20**(3), 181–195 (2006)
6. P. Weihing, J. Letzgus, G. Bangga, T. Lutz, E. Krämer, Hybrid RANS/LES capabilities of the flow solver flower - application to flow around wind turbines, in *6th Symposium on Hybrid RANS-LES Methods* (2016)
7. A. Jameson, W. Schmidt, E. Turkel, Numerical solution of the Euler equations by finite volume methods using Runge-Kutta time-stepping schemes. AIAA Paper 1259 (1981)
8. P.P. Kranzinger, U. Kowarsch, M. Schuff, M. Keßler, E. Krämer, Advances in parallelization and high-fidelity simulation of helicopter phenomena, in *High Performance Computing in Science and Engineering' 15* (Springer, Berlin, 2016), pp. 479–494
9. G. Karypis, V. Kumar, METIS–unstructured graph partitioning and sparse matrix ordering system, version 2.0 (1995)
10. C.B. Merz, C.C. Wolf, K. Richter, K. Kaufmann, A. Mielke, M. Raffel, Spanwise differences in static and dynamic stall on a pitching rotor blade tip model. J. Am. Helicopter Soc. **62**, 1–11 (2017)
11. American Institute of Aeronautics and Astronautics (AIAA), *6th Drag Prediction Workshop. Gridding Guidelines* (2015). https://aiaa-dpw.larc.nasa.gov/Workshop6/DPW6_gridding_Guidelines_2015-08-28.pdf. Accessed 12 April 2017
12. F. Richez, A. Le Pape, M. Costes, Zonal detached-eddy simulation of separated flow around a finite-span wing. AIAA J. **53**(11), 3157–3166 (2015)
13. C.B. Merz, Der dreidimensionale dynamische Strömungsabriss an einer schwingenden Rotorblattspitze. PhD thesis, Gottfried Wilhelm Leibniz Universität Hannover, 2016
14. F.R. Menter, M. Kuntz, Adaptation of eddy-viscosity turbulence models to unsteady separated flow behind vehicles, in *The Aerodynamics of Heavy Vehicles: Trucks, Buses, and Trains* (Springer, Berlin, 2004), pp. 339–352

15. D.E. Van Zante, Progress in open rotor research: a U.S. perspective, in *Proceedings of ASME Turbo Expo 2015* (2015)
16. M. Keßler, S. Wagner, Source-time dominant aeroacoustics. Comput. Fluids **33**, 791–800 (2004)
17. R.P. Woodward, E.P. Gordon, Noise of a model counterrotation propeller with reduced aft rotor diameter at simulated takeoff/approach conditions, in *AIAA 26th Aerospace Sciences Meeting* (1988)

Numerical Simulations of Artificial Disturbance Influence on a High Lift Airfoil

Katharina Wawrzinek, Thorsten Lutz, and Ewald Krämer

Abstract Numerical simulation results of the DLR-F15 high lift airfoil with statically and dynamically disturbed inflow conditions are presented and analysed. The two-dimensionally extruded high-lift airfoil, consisting of main element and flap, is exposed to the influence of a static tip vortex in order to generate a three-dimensional inflow condition. In a second step the static distortion is superimposed by an additional, artificial and dynamic gust. The set-up is a wind tunnel test section, where a pitching airfoil, which serves as a gust generator, and a finite wing, which serves as a vortex generator, are located upstream of the high lift airfoil. Unsteady RANS simulations with the inflow Mach number $M_\infty = 0.15$ and Reynolds number $Re = 2.0$ million were performed. The high lift airfoil angle of attack is $\alpha = 0°$. The results give an insight into the different emerging effects and the reaction of the high lift airfoil.

1 Introduction

Aircraft wings with their increasing number of elements are becoming more and more complex as each element influences and interacts with the inflow and the aerodynamic characteristics of the other elements. Especially during take-off and landing the high lift airfoils with their slats and flaps are exposed to inflow disturbances. These distortions may originate from buildings near the airport, from the wake vortices of other aircraft and from the atmosphere itself. Atmospheric turbulence and gusts cause load fluctuations at the wings and influence the aerodynamic characteristics of the wing. Especially close to stall these distortions can lead to critical situations.

According to "Aviation Certification Specifications" CS-25 the analysis of aircraft designs must also take into account the effects of gust and turbulence loads. Flight tests are expensive and take place in a late stage during the development process. Hence it is necessary to perform simulations in which the influence

K. Wawrzinek (✉) • T. Lutz • E. Krämer
Institute of Aerodynamics and Gas Dynamics (IAG), University of Stuttgart, Pfaffenwaldring 21, 70569 Stuttgart, Germany
e-mail: wawrzinek@iag.uni-stuttgart.de; lutz@iag.uni-stuttgart.de; kraemer@iag.uni-stuttgart.de

© Springer International Publishing AG 2018　　　　　　　　　　　　　　　　　　323
W.E. Nagel et al. (eds.), *High Performance Computing in Science
and Engineering '17*, https://doi.org/10.1007/978-3-319-68394-2_19

of turbulence is included. For instance, [1] and [3] cover this topic. However, investigations on the influence of inflow distortions on high lift airfoils are not widespread.

Even without atmospheric turbulence, it is a challenge to predict the maximum lift and stall characteristics of an airfoil. A multi element airfoil provides several additional demanding areas. Aerodynamic phenomena, which occur for example in the cove region or in the mixing of free shear and boundary layers, are difficult to predict for traditional RANS methods. It is challenging to simulate such phenomena in combination with inflow turbulence.

In the presented test case a high lift airfoil interacts with a dynamic and static inflow distortion. A two-dimensional artificial atmospheric gust is generated by a pitching airfoil. For three-dimensional effects a static vortex, lateral to the gust and generated by a finite wing, interacts with the multi element airfoil and the gust. This complex set-up creates a three-dimensional aerodynamic basis of the high lift airfoil and a three-dimensional dynamic inflow situation. For this test case experimental data is provided by the TU Braunschweig [5]. The paper presents the simulation results of the set-up described above. There are results with and without dynamically varying inflow.

2   Set-Up

2.1   Geometry

Figure 1 shows the set-up. The wind tunnel test section where three airfoils are placed is a cuboid (length: $c_l = 7.55$ m; height = width: $c_t = 1.3$ m). The DLR-F15 airfoil was used in a high lift configuration FS#1. The multi element airfoil consists

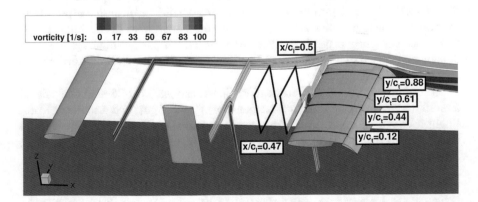

Fig. 1 Simulation set-up; left: gust generator; middle: vortex generator; right: DLR-F15; vorticity plot shows the pathway and interaction of the wakes and the wing tip vortex; turbulence model: SST

of a main element and a flap. In the configuration FS#1 the flap angle is $\alpha_F = 40°$. In the retracted airfoil the chord length is $c = 0.6$ m. A detailed description of the multi element airfoil can be found in [2]. A NACA 0021 airfoil with a chord length of $c_g = 0.3$ m is placed 2.35 m upstream to the high lift airfoil in the test section. This gust generator performs a single pitch down motion with an angle of attack increment of $\Delta\alpha = 10°$ in $t = 0.03$ s. The starting angle of attack is $\alpha_0 = 0°$ and the center of rotation is at half height of the test section. As vortex generator, a finite wing is placed between the gust generator and the multi element airfoil. The chord length of this NACA 4412 airfoil is $c_v = 0.3$ m. The leading edge is placed 0.8 m downstream of the gust generator's trailing edge at the bottom of the test section. The span is $b = 0.6$ m and the angle of attack $\alpha_v = 2.5°$. In the experiments of TU Braunschweig an inflow velocity of $u_\infty = 50$ m/s was used, however, a better match between numerical and experimental results is found for a simulation inflow velocity of $u_\infty = 50.5$ m/s.

2.2 CFD Grid

For the purpose of reducing the computational costs the test section was simulated without boundary layers along the walls. However, in order to keep the set-up similar to the experiments of the TU Braunschweig, the widening of the wind tunnel walls was neglected in the simulation to compensate the boundary layer displacement effect. In [6] a test case without gust and vortex generator was investigated and it has been shown that the simplifications meet the requirements for a simple case. The span b and the corresponding height of the tip end of the finite wing was adapted in a way, that the distances to each other element in the test section match the experiment. The leading edge of the gust generator is placed 1.6 m behind the numerical inflow plane.

The computational grid was generated using the commercial software Gridgen by Pointwise, Inc.. The mesh including the test section, the gust generator and multi element airfoil is a two-dimensional grid in x-z-plane, which is extruded in y-direction with a step size of $\Delta s = 0.005$ m to cover the test section width. The gust transport region downstream of the gust generator is resolved with $\Delta s = 0.005$ m in x- and z-direction. The vortex generator is added using the overlapping grid technique. This way a cartesian and equidistant mesh can be retained in the gust transport region and the streamwise vortex region of the vortex generator. Hence, numerical losses caused by unstructured meshes can be reduced.

The surfaces of the gust generator and the DLR-F15 airfoil are resolved by 260 cells in spanwise direction. The main element of the high lift airfoil is represented by 408 points and the flap by 232 points in chordwise direction. The high number of points of the main element is required for the higher resolution in the cove region with unstable flow characteristics. In order to provide good results during the pitching motion of the gust generator, the surface of the NACA 0021 is resolved by 400 points in chordwise direction. The vortex generator is resolved by 80 cells in spanwise direction, because of the less sensitive area which is not subject to the gust interaction. 264 points represent the surface of the vortex generator in chordwise direction. The dimensionless wall distance of the first cell of $y^+ < 1$ was satisfied. Overall, the grid consists of about 58,700,000 cells.

2.3 Numerical Set-Up

The unsteady RANS simulations were performed using the unstructured finite volume solver TAU developed by DLR [8]. The mesh deformation technique enables the motion of the gust generator. Euler walls represent the walls of the wind tunnel. There is no benefit in simulating the gust using for instance Detached-Eddy Simulation, as investigations in [7] showed. It can not be ruled out, that better results can be obtained for the vortex generator but this would require an intense mesh refinement in a large region related to an extensive increase in computational cost. For time stepping the implicit Backward Euler scheme is used. A physical time step size of $\Delta t = 0.0003$ s divides the motion of the gust generator in 1000 time steps. The pitch motion of the experiment is represented accurately and the convective time cycle of the high lift element is represented by about 390 time steps. Each time step consists of at least 100 inner iterations. A second order central approach discretises the convective fluxes. The turbulence model Menter SST was used. Due to the low Mach number the Rossow Swanson preconditioning was activated. The laminar-turbulent transition matches the prescribed experimental data. The convergence behaviour was estimated by monitoring the density residual and the force coefficients, see Fig. 2.

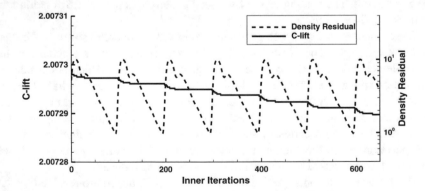

Fig. 2 Example of lift coefficient and density residual of the simulations

3   Results

3.1   Inflow Conditions Without Dynamic Disturbances

The test case without dynamic inflow turbulence shows five different superimposed phenomena, which are influencing the three inflow velocity components of the high lift airfoil in a steady state. Those phenomena are the wakes of the gust generator and the vortex generator, the circulations of the gust generator and the vortex generator and the wing tip vortex of the vortex generator. The particular effects are varying in their magnitude and affect different regions of the high lift airfoil.

Influence of the Gust Generator   The wake of the gust generator causes a reduction of the u-velocity in a small area, see Fig. 3 (left). Without the pitching motion of the gust generator, the wake is located above the high lift airfoil and does not interact with the DLR-F15 wing. The circulation effects of the gust generator are negligible because of the starting position with an angle of attack close to $\alpha = 0°$ and the symmetry of the NACA 0021 airfoil.

Influence of the Vortex Generator   The vortex generator wake reduces the u-velocity in a small vertical region, see Fig. 3 (left). This area is not exactly in the middle of the test section. Further downstream it is slightly shifted to lower y-coordinates because of the cambered NACA 4412 airfoil and its angle of attack of $\alpha = 2.5°$. The circulation of the vortex generator increases the u-velocity to higher y-coordinates and reduces the u-velocity at lower y-coordinates, compared at the given inflow velocity, see Fig. 3 (left). The vortex generator end creates a streamwise wing tip vortex with anti-clockwise rotation (viewing direction: x-direction). This vortex influences the v- and w-velocity, depending on the y- and z-coordinates, see Fig. 3 (middle and right). These additional velocities affect the wake of the gust generator. As can be seen in Fig. 3 (left), on one side the gust generator wake is shifted upwards and on the other side the wake is shifted downwards.

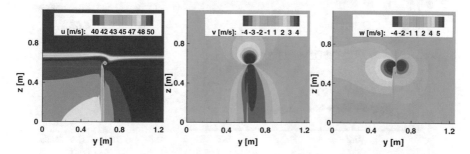

Fig. 3 Inflow conditions of the steady disturbed simulations of the DLR-F15. The figures depict the velocity components at the position $x/c_l = 0.47$ m. Left: u-velocity; middle: v-velocity; right: w-velocity. Viewing direction: -x-direction

3.2   Pressure Distribution with Static Inflow Disturbances

Figure 4 shows three chordwise surface pressure coefficient distributions at various spanwise locations. The differences among those results are rather small. They are limited to the leading edges of the main element and the flap. The main element's suction peak is increased for lower y-coordinates and reduced for higher ones, respectively. This results from the wing tip vortex of the vortex generator. The rotation direction induces a v-velocity component, which increases or reduces, depending of the side of the test section, the actual inflow velocity angle experienced by the high lift airfoil. Figure 5 (left) allows a closer look at the spanwise effects at different × locations. Regarding the main element's leading edge (dark blue line), the pressure coefficient levels at both sides of the test section are relatively constant. The drop in the middle is limited to an area of 0.5–0.7 m. Within the drop a small wiggle is visible. This results from the vortex generator wake, which is interacting with the high lift airfoil at this position reducing the u-velocity. The behaviour of the flap's leading edge (dashed red line) is not comparable to the main element and can not easily be divided into left and right, as can be seen in Fig. 5 (left). The suction peak of the flap's leading edge is significantly influenced by the separation region at the flap and the inflow from the cove region, see Fig. 5 (right). The pressure coefficient distribution varies strongly in spanwise direction and shows an irregular pattern. This behaviour can also be found in the cove and on the separation line on the flap. On average the flap's side $y/c_t < 0.5$ provides slightly lower pressure

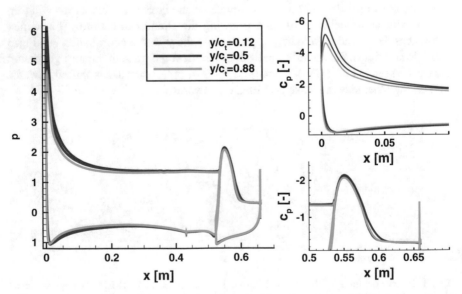

Fig. 4 Surface pressure coefficient distribution of the DLR-F15. Left: entire chordwise pressure coefficient distribution for three different spanwise locations; top right: close up leading edge main element; bottom right: close up leading edge flap

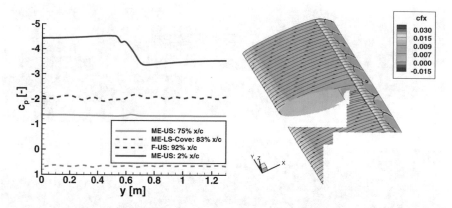

Fig. 5 Left: spanwise pressure coefficient distribution at four x-positions of the DLR-F15: solid lines: main element upper side; dashed line red: flap upper side; dashed orange line: main element lower side in the cove; right: streamlines on the main element and cf_x distribution

coefficient levels than the other side. This results from a higher angle of attack, which is caused at this side by the rotation of the streamwise vortex and induces a higher circulation of the main element. The flap is less loaded, through the increased main element circulation, see [9]. At the trailing edge of the main element and the flap the differences are negligibly small.

The vortex convects downstream above the DLR-F15, visualized by the stream-lines in Fig. 5 (right). The streamlines in the middle on the upper side of the main element are slightly deflected to the right side (viewing direction: x-direction), as a result of the lower side of the counter rotating vortex. This effect is limited to a small middle area. Most of the airfoil's main element does not show any cross flow.

Overall, the influence of the vortex on the high lift airfoil is small and restricted locally. The expected three-dimensional effects of the inflow conditions are rather small and limited to a small strip in the middle region. Both sides of the main element, considered individually, show a behaviour without strong gradients. The flap region is mainly dominated by three-dimensional effects which can not be separated into left and right section.

3.3 Pressure Distribution with Dynamic Inflow Disturbance

In this section the gust generator performs a single pitch motion of $\Delta\alpha = 10°$ to introduce a dynamic inflow distortion. At the trailing edge of the gust generator, a quasi two-dimensional transversal vortex develops, which convects downstream the test section, see Fig. 6. The rotation direction of the gust is anti-clockwise (viewing direction: x-direction), which first induces a positive w-velocity travelling along

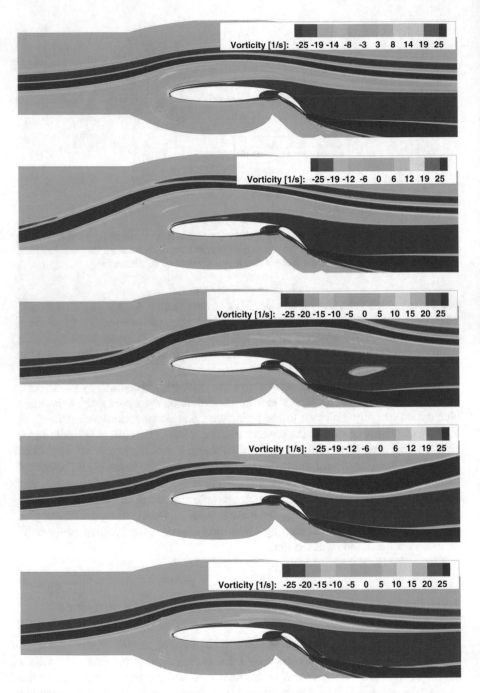

Fig. 6 Figure depicts the vortex path at $y/c_t = 0.44$ and $t = t_0$, $t = 0.0190$ s, $t = 0.0317$ s, $t = 0.0461$, $t = t_{end}$ (top down)

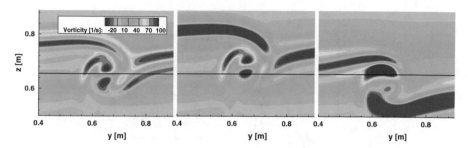

Fig. 7 Three different points in time showing the position of the transversal vortex during the simulation including the pitching of the gust generator. The black line represents the height of the main element's leading edge. The figures are extracted at the position $x/c_l = 0.5$

the high lift airfoil, followed by a negative one. Investigations concerning the two-dimensional simulation case can be found in [10].

While travelling downstream the gust interacts with the vortex generator followed by the high lift airfoil. The gust influences the high lift airfoil as well as the position of the transversal vortex. Figure 7 shows the motion of the vortex core which is caused by the gust. First the vortex core is shifted upwards followed by a downwards motion. The vortex core movement by itself influences the high lift airfoil. This effect on the DLR-F15 can't be investigated isolated from the influence of the gust. However this influence is assumed to be small and locally restricted to the midsection of the airfoil.

Figure 8 depicts the temporal development of the pressure coefficient difference at several positions at the DLR-F15. The pressure coefficient delta is given by the difference of the actual pressure coefficient to the basis pressure coefficient before the pitching motion started at the same position. This method almost eliminates the influence of the static inflow effects and focuses on the gust influence. The dashed lines represent the results for $y/c_t < 0.5$, the solid lines for $y/c_t > 0.5$. The particular results on each side of the test section behave like it is described in [10] and [4] In this paper the focus is on the comparison between both sides. The temporal development of the dashed lines shows that the reaction peaks of the vortex flanks are slightly delayed. There are two possible explanations, which are not mutually exclusive. First, the gust part for $y/c_t < 0.5$ experiences a reduced u-transport velocity, due to the circulation of the vortex generator, see Fig. 3 (left). Second, the gust follows different paths, depending on the spanwise position. The distance of the gust part for $y/c_t < 0.5$ to the high lift airfoil is larger than the distance of the side $y/c_t > 0.5$. This results from the streamwise vortex rotation of the tip vortex in the y-z-plane. On the side $y/c_t < 0.5$ the vortex induces a positive w-velocity, which moves the gust upwards. The contrary is true in this case, see Figs. 7, 3 (right) and 9. This can cause different x-positions of the vortex core on each side at the same moment in time. Unfortunately, it is not possible to detect the position of the vortex core in the simulation data. The vortex structure is more like a shear layer of varying strength than a consolidated vortex. Neither the velocity

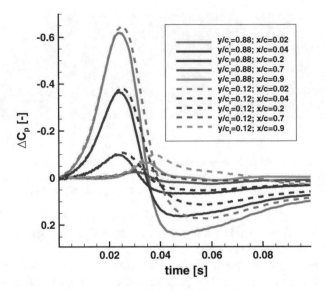

Fig. 8 Pressure coefficient differences at several x-positions on the main element at two spanwise positions. Dashed lines: $y/c_t = 0.12$; solid lines: $y/c_t = 0.88$

difference method, used in [10], nor the vorticity nor the λ_2-criterion detect the position of the vortex core.

Beside the offset in the temporal development, there are differences recognisable in both peak values of the pressure coefficient difference. The reaction of the front vortex flank provides a weaker influence for $y/c_t > 0.5$, although the vortex position is closer to the high lift airfoil and should induce higher positive w-velocities. In a two-dimensional case, a larger distance between airfoil and gust causes a weaker reaction, see [10]. As it is described in [10] and [4] the position of the vortex is slightly upstream of the high lift airfoil, when the first peak of the pressure coefficient difference appears. These results are transferred to this three-dimensional test case. In case of $y/c_t > 0.5$ the vortex is slightly below the leading edge before it finally barely travels along the upper side of the high lift airfoil. At this short moment the leading edge experiences a positive w-velocity but also a negative x-velocity component from the upper part of the gust which might reduce the first peak reaction. A second reason might be found in the different base level of the high lift airfoil. In [10] the reaction of the test case with an angle of attack of $\alpha = 0°$ shows lower peak values for both vortex flanks than the test case with an angle of attack of $\alpha = 6°$. It might be a combination of different effects because only the reaction of the first vortex flank is stronger for $y/c_t < 0.5$ but not the second, see Fig. 8. Most lines in Fig. 8 don't return to the initial starting level. This results from the varied set-up. After the pitch movement the gust generator remains in the deflected position. This influences the flow simulation in the test section. In Fig. 9 different end positions are depicted. Whereas the gust generator wake clearly travels along the upper side of the high lift airfoil at all spanwise positions at the starting

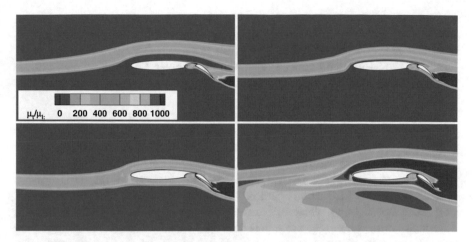

Fig. 9 Figure depicts different gust wake positions with μ_t/μ_l. Four spanwise position after the gust interaction took place. Top left: $y/c_t = 0.12$; bottom left: $y/c_t = 0.61$; top right: $y/c_t = 0.88$; bottom right: $y/c_t = 0.46$

point, the gust generator wake partly impinges on the leading edge or travels along the lower side of the high lift airfoil.

Overall the three-dimensional effects are a complex system of different influencing parameters. Considering each side individually, the effects resemble the two-dimensional results of other investigations. Comparing both sides, additional effects can be recognised.

4 Scaling Test

Simulations of the type presented in the previous sections are very demanding, especially regarding the required computational effort. Resolving all aspects of complex geometries and small turbulence scales requires a mesh with a considerably large number of cells. Additionally, the representation of low frequencies demands long recording times with small physical time steps, which results in numerically expensive simulations. Modern supercomputers such as the Cray XC-40 with a large number of compute nodes, fast cores and modern architecture progressively enable more expensive simulations and accordingly a wider range of more accurate simulation results. In order to use this potential the flow solver must provide good scaling capabilities. Therefore, a scaling test on the CRAY XC-40 system Hazel Hen was performed.

The system provides 7712 compute nodes. Each node consists of two sockets with twelve cores each. All together 185,088 cores are available. The run-time for a job is limited to 24 h.

The chosen mesh for the scaling test is a hybrid grid with additional Chimera overset meshes as it is common for most simulations with complex geometries and highly resolved areas. The mesh consists of about 60 million cells, which equals 54 million points. The chosen flow solver, DLR TAU-Code, version taudir_release_2015.2.0, was compiled using the GNU compiler, MPICH version 7.3.0 and NETCDF version 4.3.2.

For an automated partitioning of the original grid into subgrids two codes are available: the TAU native partitioner called private and the external graph partitioner Chaco, developed by Sandia National Labs. Unfortunately, a direct coupling of the Chaco partitioner and the current TAU version is no longer possible since TAU version taudir_release_2013.1.0. For this reason the TAU version taudir_release_2012.1.0 in combination with Chaco was used to subdivide the initial grid. The preprocessing, which includes the preparation of the multigrid levels, and the actual simulation is done using the latest TAU version. The code's strong scaling behaviour was investigated due to its major importance to the user, compared to the weak scaling. In this case the same initial mesh is divided into different numbers of subgrids, which enables performing the simulation on the respective number of cores. By increasing the number of subgrid domains the number of cells per domain is reduced. Thus, the computing costs per core are reduced. On the other hand the communication among the cores is increased, which leads to an increase in overall simulation time. At some point the communication effort exceeds the benefit of the reduced compute costs per core and further partitioning of the initial grid is unfavorable. For this scaling test the chosen test cases range from 24 to 9216 subgrids, which corresponds to 743,000 to 6500 cells per core. Each test case's subgrid number is a multiple of 24, so all allocated nodes during the simulation are entirely utilized.

The partitioning runs sequentially, so only one core performs the partitioning independent of the number of required subgrids. As consequence the internal core memory must be sufficient to handle the entire mesh size. No issues were observed in this respect for the given mesh size. Hence, the memory was sufficient in all cases. The wallclock time of each run is analyzed and serves as basis for comparison. Figure 10a shows the results of the partitioning. Both partitioners have similar runtimes for the grid subdivision into 24 domains. Increasing the number of requested subgrids increases the process duration for both partitioners. The external partitioner Chaco operates more efficiently. The difference between the partitioners increases with increased subgrid domains. Chaco is the superior partitioner from a runtime point of view, with the gap between the two increasing the more subgrids need to be created.

Opposite to the partitioning, the preprocessing operates in parallel, so the number of required cores equals the number of subgrid domains. The TAU code performs the preprocessing with the subgrids previously created by the different partitioners. The preprocessing generates five levels of multigrids for each test case, which further reduces the number of cells per core, depending on the selected multigrid level. In Fig. 10b the results for the preprocessing are depicted. Increasing the number of cores reduces the time for conducting all preprocessing steps. The additional time-

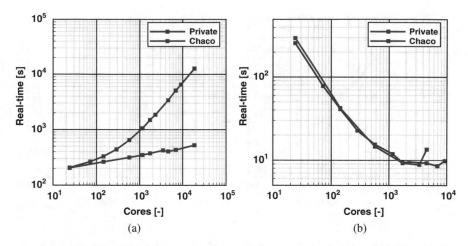

Fig. 10 Scaling test results of the partitioning (**a**) and the preprocessing (**b**) process

savings decrease with an increasing number of cores. Above 1200 cores, which equals roughly 50,000 cells per core, no further time savings can be achieved by increasing the number of cores. There are almost no differences between the two different subgrid versions. Small deviations are most likely attributable to different core selections in both runs. The deviations in the last Chaco test case can probably be ascribed to the core selection as well.

In order to compare the different simulation runs, the time from the first to the last iteration was logged excluding the duration of initialisation and output process. The simulation performs 10,000 iterations per run and the multigrid level three is activated. Test cases smaller than 72 subgrids would need more than 24h to perform the given iterations, which would require a second run, and are therefore omitted. To reduce the influence of the core selection the solver test cases are repeated 3 times. The results can be seen in Fig. 11a. Between 100 and 1000 cores the results differ slightly. Differences can be found above 1000 cores. In that range, the improvement achieved by increasing the core count seems to decrease, with some noticeable variations. Above 4600 cores a linear behavior can be seen in the double logarithmic plotted figure, which nearly equals the linear slope below 1000 cores. It seems that even further partitioning, which results in subgrids containing less than 6500 cells, still provides good scaling results, however the benefit is slightly reduced.

Figure 11b shows the comparison of the solver runtimes for different partitioned test cases. There is almost no difference in the simulation time, indicating that each partitioner provides equal workload for the cores.

The TAU code provides excellent scaling qualities on the supercomputer CRAY XC-40. Both presented partitioners achieve similar results, solely during the partitioning Chaco shows superior characteristics regarding the run-time. However the time consumption of the partitioning process in comparison to the time consumption of the simulation itself is negligible. Therefore, the choice of the partitioner will be

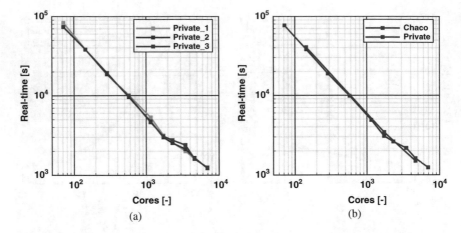

Fig. 11 Solver real-time results for TAU version 2015.2.0. (**a**) Results for 3 test runs. (**b**) Results for differently partitioned test cases

made regarding other aspects such as numerical stability, positions and shapes of subgrid domain boundaries, etc.

5 Conclusion

Numerical studies on the influence of static and dynamic inflow disturbances on a two-element high lift airfoil were presented in this paper. The static three-dimensional effects of this set-up turned out to cause only small three-dimensional effects. The impact area at the high lift airfoil can clearly be separated into three domains. Neither the left side nor the right side show three-dimensional flow characteristics. Only a small streak in the middle experiences cross flow. The additional superposition of a gust leads to a complex system of mutual interaction among all involved parameters. It is not possible to analyse each influence separately due to the dependence among themselves. Both sides of the high lift airfoil, individually considered, show the same behaviour like it is found in the two-dimensional test cases. However, the comparison between both sides reveals new aspects which can't be fully explained yet. It is not clear if each side acts for itself and is not involved in the reaction of the other or if they interact. As these numerical investigations are very demanding, a scaling test of the TAU code on the CRAY XC-40 system was performed to ensure adequate scaling qualities.

Acknowledgements The authors would like to acknowledge the members of the FOR1066 research unit and the "Deutsche Forschungsgemeinschaft DFG" (German Research Foundation) for funding this research and the authors appreciate the cooperation with the *High Performance Computing Center Stuttgart* to enable highly resolved simulations of inflow disturbances and high lift airfoils.

References

1. L. Gilling, Airfoils in Turbulent Inflow, PhD thesis, Aalborg University, 2009
2. D. Hahn, P. Scholz, R. Radespiel, Vortex generation in a low speed wind tunnel and vortex interactions with a high-lift airfoil, in *30th AIAA Applied Aerodynamics Conference* (2012), p. 3024
3. P. Kelleners, R. Heinrich, Simulation of interaction of aircraft with gust and resolved LES-simulated atmospheric turbulence, in *Advances in Simulation of Wing and Nacelle Stall*. Notes on Numerical Fluid Mechanics and Multidisciplinary Design, vol. 131 (Springer, Berlin, 2016), pp. 203–221
4. S. Klein, D.G. Hahn, P. Scholz, R. Radespiel, Vortex interactions with a high-lift airfoil in a low speed wind tunnel. American Institute of Aeronautics and Astronautics, June 2013
5. S. Klein, P. Scholz, R. Radespiel, Transient, three-dimensional disturbances interacting with a high-lift airfoil - wind tunnel experiments. American Institute of Aeronautics and Astronautics, 2016
6. S. Reuß, A. Probst, T. Knopp, Numerical investigation of the DLR F15 two-element airfoil using a Reynolds stress model, in *Third Symposium "Simulation of Wing and Nacelle Stall"* (2012)
7. S. Reuß, A. Probst, T. Knopp, K. Wawrzinek, Hybrid RANS/LES study of development of an airfoil-generated vortex, in *Notes on Numerical Fluid Mechanics and Multidisciplinary Design*, vol. 131 (Springer, Berlin, 2016), pp. 41–54
8. D. Schwamborn, T. Gerhold, R. Heinrich, The DLR TAU-code: recent applications in research and industry, in *ECCOMAS CFD 2006: Proceedings of the European Conference on Computational Fluid Dynamics, Egmond aan Zee, September 5–8, 2006*. Delft University of Technology; European Community on Computational Methods in Applied Sciences (ECCOMAS) (2006)
9. A.M.O. Smith, High-lift aerodynamics. J. Aircraft **12**(6), 501–530 (1975)
10. K. Wawrzinek, T. Lutz, E. Krämer, Numerical studies of turbulent flow influence on a two-element airfoil, in *Notes on Numerical Fluid Mechanics and Multidisciplinary Design*, vol. 131 (Springer, Berlin, 2016), pp. 111–135

About the Influence of Wind Tunnel Walls, Tower and Nozzle on the Performance of a Model Wind Turbine

Annette Klein (née Fischer), Sven Zabel, Thorsten Lutz, and Ewald Krämer

Abstract The influence of wind tunnel walls, tower and nozzle on the performance of a model wind turbine is investigated in the present paper using Computational Fluid Dynamics (CFD). The model wind turbine has a radius of 1.5 m and is located in a wind tunnel with a cross section of 4.2 m × 4.2 m. Global loads, angle of attack distributions as well as flow fields are compared to each other to evaluate the influence of the different configurations.

1 Introduction

In order to increase the competitiveness of wind energy against conventional sources of energy, wind turbines have to be improved and further developed. However, in an early state of development, new applications can't be tested on real wind turbines but have to be investigated on model wind turbines. Such experiments are cheaper than experiments on real wind turbines and in addition, the same inflow conditions can be repeated in a wind tunnel as often as necessary. Cheaper and faster than experiments on turbines are CFD (computational fluid dynamics) simulations of wind turbines. Thereby, the same inflow conditions can be repeated as well, but moreover, numerical models of full size turbines can be simulated and the inflow conditions can be changed easier than in a wind tunnel.

In advance of such simulations, the suitability of the CFD model to display the loads on and the flow around wind turbines has to be validated by experiments. Therefore, simulations of the experimental setup are necessary. As not all components, which are present in the wind tunnel (probe rigs, bottom plate, steps, etc.), can be considered in the simulation and in order to save computational time, the numerical setups are often simplified. As wind turbine rotors are rotationally symmetric, it can be sufficient to simulate only one blade and use rotationally periodic boundary conditions. Thereby, the tower has to be neglected. If the blockage ratio of a wind

A. Klein (née Fischer) • S. Zabel • T. Lutz (✉) • E. Krämer
Institute of Aerodynamics and Gas Dynamics (IAG), University of Stuttgart, Pfaffenwaldring 21, 70569 Stuttgart, Germany
e-mail: fischer@iag.uni-stuttgart.de; lutz@iag.uni-stuttgart.de

© Springer International Publishing AG 2018
W.E. Nagel et al. (eds.), *High Performance Computing in Science and Engineering '17*, https://doi.org/10.1007/978-3-319-68394-2_20

tunnel, which is defined as the rotor swept area divided by the wind tunnel cross section area, is smaller than 10%, no wind tunnel effects should be experienced, according to Schümann [22]. In such a case, the wind tunnel walls could be neglected, too. However, for each experimental setup it should be checked, which components can be neglected in the simulation, in order to get a good accordance between simulation and experiment with as little computational time as possible.

There are several publications about the simulation of model wind turbines. In the *MEXICO* project [19], as well as in the *INNWIND.EU* project [11], model wind turbines in an open jet section are investigated experimentally and numerically. The *NREL PHASE-VI* model wind turbine was investigated, amongst other, by Sørensen et al. [23] numerically. The closed test section in this investigation had a blockage ratio of 8.8% and only small blockage effects occur. Krogstad and Lund [13] expected also only small blockage effects in their experimental and numerical investigation of a model wind turbine in a wind tunnel with a blockage ratio of 11.8%. Schümann [22], however, experienced an effect of the wind tunnel walls for his investigations of a model wind turbine with 14% blockage ratio.

The model wind turbine, which will be investigated in the present paper, is the Berlin Research Turbine (*BeRT*), which was designed and built by the Technical University of Berlin in cooperation with the *SMART BLADE GmbH*. It is investigated in the course of the *DFG PAK 780* project [16], where six partners from five different Universities (TU Berlin, University of Stuttgart, RWTH Aachen, Technical University of Darmstadt and Carl von Ossietzky University Oldenburg) are working together on the field of wind turbine load control under realistic turbulent inflow conditions. Thereby, different load alleviation systems are investigated numerically and experimentally, amongst other, on the *BeRT* turbine.

A one third model of the *BeRT* turbine was already simulated under uniform free stream condition with a large-eddy (LES) approach [6]. As the turbine is located in the closed 4.2 m × 4.2 m test section of the great wind tunnel (*GroWiKa*) of the TU Berlin, where a blockage ratio of over 40% is achieved, the wind tunnel walls must not be neglected in the simulations. In order to estimate the influence of the wind tunnel environment, Reynolds-averaged Navier-Stokes simulations (*RANS*) under uniform free stream condition, but also with a wind tunnel with a slightly higher blockage ratio then in the *GroWiKa*, were performed by Fischer et al. [3] and the results were compared to each other. This approximated wind tunnel had a blockage ratio of around 50%, was realized on the one hand with a slip wall and on the other hand with a no-slip wall and had, due to the one third model of the turbine, a cylindrical cross section instead of an angular one. The minimal distance between the blade tip and the wind tunnel wall was the same as in reality. As expected for such a high blockage ratio, the wind tunnel walls had a huge influence on the performance of the wind turbine, which did not behave like a free stream turbine any more, as even the Betz limit [5] was exceeded. In the present investigation, the influence of wind tunnel walls, tower and a nozzle, which is located $2.5R$ downstream the turbine, on the loads, the angle of attack (AoA) distribution and on the flow field is investigated and compared to each other in order to estimate which simplifications in the numerical setup are acceptable.

2 Model Wind Turbine

The *BeRT* turbine has a radius of $R = 1.5$ m and a hub height of $h = 2.1$ m. In the present investigations, the inflow velocity is set to $v_{inflow} = 6.5$ m/s and the turbine rotates with $180 rpm$. This leads to a tip speed ratio of $\lambda = 4.35$. The blades of the turbine consist of only one cross section, which is based on the *CLARK-Y*-airfoil, and are exchangeable. The choice fell on this airfoil as it can provide attached flow for low Reynolds numbers and it has a good effectiveness with leading and trailing edge flaps. Those flaps can be used for active and passive load alleviation, which is investigated on this turbine in the course of the *DFG PAK 780* project [16]. More information about this turbine can be found in Pechlivanoglou et al. [17]. The turbine is placed in the settling chamber of the *GroWiKa* of the TU Berlin, which was converted to a 4.2 m × 4.2 m test section. As a consequence, the nozzle, which is usually placed in front of a test section, is now located behind the turbine.

3 Numerical and Computational Details

3.1 Numerical Methods

The *RANS* simulations of the *BeRT* turbine are performed using the block structured code *FLOWer*, which was developed by the *German Aerospace Centre* (DLR) in the course of the *MEGAFLOW* project [14]. It uses the finite volume method to solve the unsteady Reynold-averaged Navier-Stokes equations (*URANS*) on block-structured grids. For the spatial discretisation, a second order central discretisation scheme JST [8] is used and the time is discretised with an implicit dual time stepping scheme [7]. Several state of the art turbulence models are implemented, but for the present case, the Menter SST turbulence model was used. All components are meshed separately with a fully resolved boundary layer, ensuring $y^+ \approx 1$ of the first cell, and are overlapped using the *CHIMERA* technique [1]. For the numerical simulation of wind turbines, a process chain, which was developed at the Institute of Aerodynamics and Gas Dynamics (*IAG*) [15], and which was also used in several other wind energy projects [20, 21, 24], was used for the present investigations, too.

3.2 Performance of the Solver

For the present investigations, a *FLOWer* version from 2016 was used, which had only minor changes concerning the performance of the version compared to the version which was used for the investigations in [4]. Therefore, the strong and weak scaling test, as shown in [4], was over taken for the present paper (Fig. 1).

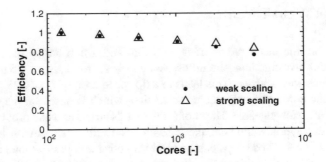

Fig. 1 Efficiency of *FLOWer* on *Cray XC40* using *ifort* fortran compiler and a constant cell loading of 32^3 for each MPI process in case of weak scaling and 4096 times 32^3 cells in case of strong scaling. Taken from [4]

Table 1 Characteristics of the *BeRT* turbine and the wind tunnel

Rotor radius (m)	1.5
Tower height (m)	2.1
Cone angle (°)	0
RPM (–)	180
Inflow velocity (m/s)	6.5
Wind tunnel cross section (m × m)	4.2 × 4.2

For up to 1024 cores, the weak and strong scaling showed the same efficiency. For more cores, the efficiency of the strong scaling is slightly better than for the weak scaling. With a total usage of 4096 cores, *FLOWer* has an efficiency of 0.77 for the weak scaling and of 0.83 for the strong scaling. As an example for an average simulation, the case of the full wind turbine without tower in the far field is presented. The total amount of approximately 36 million cells was computed on 1296 cpus. A simulation of 73 revolutions with 120 to 240 time steps per revolution and 30 inner iterations, as shown in the present investigations, consumes approximately 140 h (wall clock time) and consequently 181,000 cpuh.

3.3 Numerical Setup

Table 1 gives an overview of the turbine and its operating condition used for the investigations in this paper.

Four different cases are investigated in the present paper. One case is simulated under far field condition (hereinafter designated as *FF*), two cases include wind tunnel walls with constant cross section (hereinafter designated as *WT*1) and another case with wind tunnel walls and a nozzle 2.5*R* behind the rotor plane is designated as *WT*2. Simulations of the pure rotor are denominated with the affix $_{rot}$, if the tower is included, the affix is $_{tow}$. The numerical setup of the model wind turbine consists of nine ($_{rot}$) respectively eleven ($_{tow}$) independent meshes which are overlapped using

Table 2 Characteristics of the different setups

	FF_{rot}	$WT1_{rot}$	$WT1_{tow}$	$WT2_{tow}$
Cell number	35,992,576	31,678,976	33,529,856	34,884,608
No. of grids	9	9	11	11
Background	FF	WT	WT	WT
Tower	–	–	x	x
Nozzle	–	–	–	x

the *CHIMERA* technique [1]. Table 2 gives an overview of the different setups used for the investigations in this paper.

The blade mesh is of CH-topology and was created through a script, which was developed at the *IAG*. It has a fully resolved boundary layer (37 cell layers), ensuring $y^+ < 1$ for the first grid layer. In radial direction, 101 cells are used, around the airfoil 181, leading to an amount of approximately 5.5 million cells per blade. The mesh for the blade connection, hub, nacelle, tower connection and tower are created manually. The background grid for the far field case was generated with an automated script [12] and hanging grid nodes are used for the refinement. Thereby, refinements can be realised only were it is needed, whereas usual refinement in a H-topology is leading to refinements at unnecessary spots. The grid is approximately $20.5R$ long ($8R$ upstream and $12.5R$ downstream of the rotor plane), approximately $24.6R$ wide and has a height of approximately $14R$. According to Sayed [18], this extension is big enough to prevent any influence of the boundary conditions on the turbine. The bottom is a slip wall, all other boundaries are realised as far field. The cells around the turbine are $0.025\,\text{m} \times 0.025\,\text{m} \times 0.025\,\text{m}$ and at the outflow $0.1\,\text{m} \times 0.1\,\text{m} \times 0.1\,\text{m}$. The far field mesh has an overall amount of 13.7 million cells. In the experiment, the *BeRT* turbine is located in the settling chamber of the *GroWiKa* and the rotor plane is located 1.245 m behind the beginning of the chamber. This chamber has a cross section of $4.2\,\text{m} \times 4.2\,\text{m}$ and is 5 m long. However, in order to prevent disturbances of the boundary condition to convect to the turbine, the wind tunnel was extended in the numerical setup. It is approximately $16.5R$ long, whereby the rotor plane is located approximately $7.5R$ downstream of the inflow boundary condition and thus far away enough according to Sayed [18]. Like for the far field case, the cells around the turbine measure $0.025\,\text{m} \times 0.025\,\text{m} \times 0.025\,\text{m}$. At the inflow boundary, the cells have a size of $0.4\,\text{m} \times 0.025\,\text{m} \times 0.025\,\text{m}$ and at the outflow of $0.2\,\text{m} \times 0.025\,\text{m} \times 0.025\,\text{m}$. As no hanging grid nodes were used here, only a coarsening in streamwise direction was possible. The wind tunnel walls are realized as slip walls, whereby a calculated displacement thickness is added on the real walls, leading to a continuous reduction of the cross section for the settling chamber. In front and behind the chamber, the cross section is constant. For the setup *WT2*, a nozzle with a total length of 3.0 m and a tapering of 2.2 follows the settling chamber. The inflow boundary is realized as far field and for *WT1*, the outflow boundary, too. For *WT2*, the outflow boundary had to be changed to constant pressure in order to maintain mass continuity. An

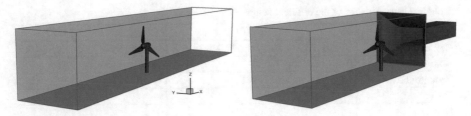

Fig. 2 Setup for $WT1_{tow}$ (left) and $WT2_{tow}$ (right)

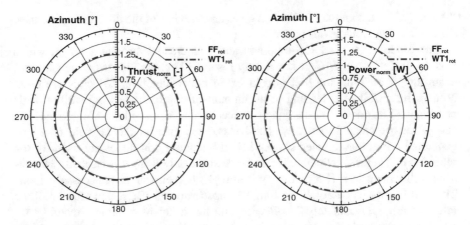

Fig. 3 Thrust (left) and power (right) over one revolution for FF_{rot} and $WT1_{rot}$, normalised with the averaged value of FF_{rot}. Azimuth indicates the position of the first blade

investigation of the grid convergence index according to Celik [2] was already performed for a one third model by Fischer et al. [3] and the grids for the full turbine were created according to the results of this study. Figure 2 shows the setup for $WT1_{tow}$ and $WT2_{tow}$.

4 Results

4.1 Influence of the Wind Tunnel Walls

As already reported by Fischer et al. [3], the wind tunnel walls have a large influence on the flow around the turbine. Figure 3 shows the normalized thrust and power of the whole turbine for FF_{rot} and $WT1_{rot}$. The azimuth indicates the position of the first blade, whereby an azimuth of 0° means, that the blade is pointing upward.

Due to the presence of the wind tunnel environment, thrust is increased by approximately 25%, power is increased by even 50%. This is in good accordance to the results from Fischer et al. [3] who found an increase by 25–30% respectively

67–78% at an inflow velocity of $v_{inflow} = 8\,m/s$ and a blockage ratio of approximately 50%. This increase is due to the limited space caused by the wind tunnel walls. The wind turbine doesn't work like a free stream turbine any more as the wake can not expand like under far field condition, leading to higher velocities in the rotor plane and consequently to higher loads. More information about this topic can be found in Fischer et al. [3].

4.2 Influence of the Tower

As the wind tunnel walls have a large impact on the performance of the turbine, the walls will be considered in the following subsections. Figure 4 shows the normalized thrust and power of the whole turbine for $WT1_{rot}$ and $WT1_{tow}$.

Whereby thrust and power of the pure rotor are almost constant over one revolution, the curve for the setup including tower shows higher variations caused by the displacement effect of the tower. There are three drops in thrust and power at 60°, 180° and 300° azimuth with a phase shift of 120°, which is characteristic for a three bladed turbine. Moreover, three maxima at approximately 0°, 120° and 240° are visible.

Table 3 shows the mean values and the amplitude for thrust and power for one revolution.

It can be seen, that the tower shadow results in fluctuations as the lower velocity in front of the tower, due to the blockage effect, leads to a short-term load reduction. However, for the case under investigation, the influence of the tower does not only lead to a decrease of the load in front of the tower, but also to a higher maximum. This is due to the fact of the limited space around the turbine. In the wind tunnel, the tower increases the blockage even more, leading to higher velocities around the

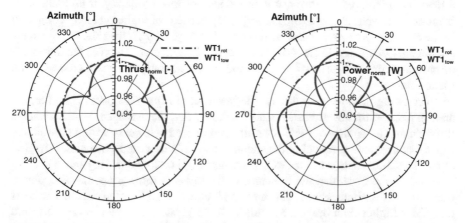

Fig. 4 Thrust (left) and power (right) over one revolution for $WT1_{rot}$ and $WT1_{tow}$, normalised with the averaged value of $WT1_{rot}$. Azimuth indicates the position of the first blade

Table 3 Thrust and power over one revolution, Δ value with regard to the mean value of $WT1_{rot}$

	Thrust (N)		Power (W)	
	$WT1_{rot}$	$WT1_{tow}$	$WT1_{rot}$	$WT1_{tow}$
Mean value	163.1	162.4	730.1	725.7
Δ Mean	–	−0.43%	–	−0.60%
Amplitude	–	2.8	–	18.3

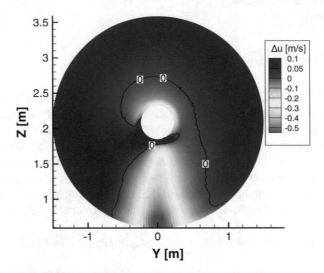

Fig. 5 Difference in streamwise velocity (averaged over one revolution) between $WT1_{tow}$ and $WT1_{rot}$ ($\Delta u = u_{WT1_{tow}} - u_{WT1_{rot}}$) in the rotor plane, viewing direction is upstream

turbine compared to the case $WT1_{rot}$. This can be seen in Fig. 5, where the difference of the averaged velocity between $WT1_{tow}$ and $WT1_{rot}$ in the rotor plane is plotted. Except in front of the tower, the velocity is higher for the case including tower. If a blade is in front of the tower, it is in the area of lower velocity. If no blade is in front of the tower, all blades are consequently in areas of higher velocity compared to the case without tower. These differences lead to the higher maxima. As the loads are not only decreased in front of the tower but also increased next to the tower, the minima and maxima partly offset one another, leading only to a small change of the mean value.

Even though the velocity around the tower and in the rotor plane is higher, the area in front of the tower has such an influence, that the averaged AoA is reduced throughout the majority of the blade radius, compared to the case without tower (see Fig. 6), consequently reducing the mean values of thrust and power. Information about the determination of the AoA distribution can be found in [10] and [9].

Another aspect is, that the influence of the tower is more pronounced in power than in thrust for both, mean value and amplitude. As the lift of an airfoil is at least one order higher than drag, the resulting force is more oriented towards lift and therefore to the normal of the rotor plane. And as power originates from the driving

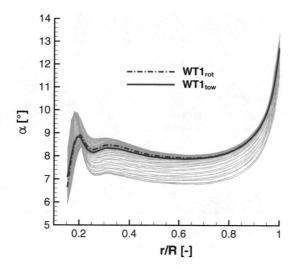

Fig. 6 Angle of attack distribution over the blade, averaged over one revolution. Light curves show the solutions of $WT1_{tow}$ every 1.5° azimuth

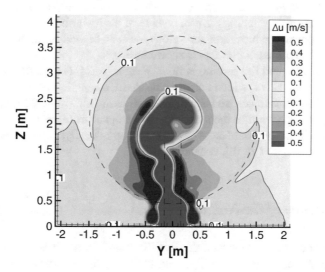

Fig. 7 Difference in streamwise velocity between $WT1_{tow}$ and $WT1_{rot}$, averaged over one revolution ($\Delta u = u_{WT1_{tow}} - u_{WT1_{rot}}$). $1R$ downstream the rotor, viewing direction is upstream

force, which lies in the rotor plane, it is more sensitive to AoA variations than thrust, which originates from the force perpendicular to the rotor plane.

Figure 7 shows the velocity difference between $WT1_{tow}$ and $WT1_{rot}$ $1R$ behind the rotor for the complete wind tunnel cross section. In addition, the rotor and the tower are indicated by dashed lines.

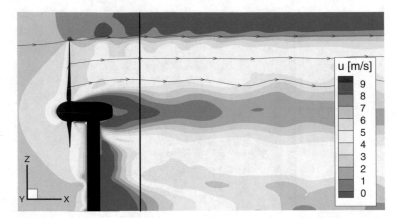

Fig. 8 Snapshot of the streamwise velocity for $WT1_{tow}$ in the middle of the wind tunnel ($y = 0m$) including three streamtraces. Wind turbine illustrated in black, black line indicates the position of Fig. 7

The velocity in the wake of the turbine is higher for the case with tower, indicated by the red colours dominating the plot outside of the tower blockage. Left and right of the tower, due to the blockage, the velocity is even higher. A cut through the flow field of the wind tunnel at $y = 0$ m is shown in Fig. 8.

It is a snapshot and the turbine is illustrated in black. The vertical black line is positioned at $x = 1.5$ m, which corresponds to the evaluation position of Fig. 7. In addition, three streamtraces are shown. These lines are almost horizontal, indicating the only slight expansion of the wake. The wake space of the nacelle, as well as of the tower, can be seen, too. Moreover, the speed-up in the bypassing flow, as already described by Fischer et al. [3], is clearly visible.

4.3 Influence of the Nozzle

As the influence of the tower on the loads is quite strong, it is considered in the following by comparing case $WT1_{tow}$ and $WT2_{tow}$. Figure 9 shows the normalized thrust and power of the whole turbine for these two cases.

Due to the nozzle, the mean values of thrust and power are increased.

Table 4 shows the mean values and the amplitude for thrust and power for one revolution for $WT1_{tow}$ and $WT2_{tow}$ with regard to $WT1_{tow}$ in order to estimate the influence of the nozzle.

As already seen in the load distribution, the mean values are increased due to the nozzle. And again, the influence on the mean value is more pronounced for power than for thrust. The amplitudes, however, are stronger in the case without nozzle. In Fig. 10, the velocity difference between $WT2_{tow}$ and $WT1_{tow}$ are plotted in the rotor plane.

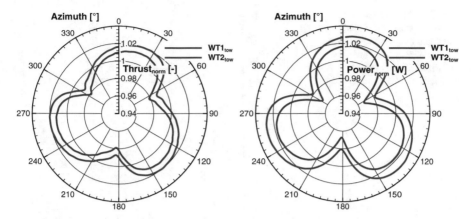

Fig. 9 Thrust (left) and power (right) over one revolution for $WT1_{tow}$ and $WT2_{tow}$, normalised with the averaged value of $WT1_{tow}$. Azimuth indicates the position of the first blade

Table 4 Thrust and power over one revolution, Δ values with regard to the corresponding values of $WT1_{tow}$

	Thrust (N)		Power (W)	
	$WT1_{tow}$	$WT2_{tow}$	$WT1_{tow}$	$WT2_{tow}$
Mean value	162.4	163.7	725.7	734.6
Δ Mean	—	0.80%	—	1.23%
Amplitude	2.8	2.6	18.3	17.3
Δ Amplitude	—	−7.14%	—	−5.46%

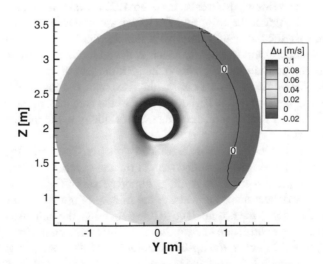

Fig. 10 Difference in streamwise velocity (averaged over one revolution) between $WT2_{tow}$ and $WT1_{tow}$ ($\Delta u = u_{WT2_{tow}} - u_{WT1_{tow}}$) in the rotor plane, viewing direction is upstream

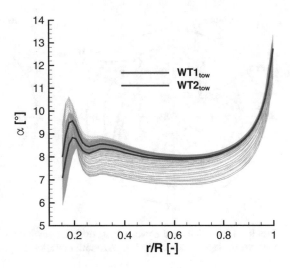

Fig. 11 Angle of attack distribution over the blade, averaged over one revolution. Light curves show the solutions for every 1.5° azimuth

It can be seen that, especially in the inner part, the velocity is higher for the $WT2_{tow}$ case. Consequently, the level of the AoA distribution is higher, too, which can be seen in Fig. 11. The averaged AoA distribution over one revolution shows, due to the smaller impact of the tower, higher values for the $WT2_{tow}$ case, leading to higher mean values of thrust and power, which was already seen in Table 4. The biggest difference between the AoA distributions can be seen again in the inner part of the rotor.

In Fig. 12, the velocity difference between $WT2_{tow}$ and $WT1_{tow}$ is plotted $1R$ behind the rotor. Again, the rotor and the tower are indicated by dashed lines. It can be seen, that for the wind tunnel with nozzle, the velocity in the outer part of the rotor is slightly increased, whereas it is decreased in the corner of the wind tunnel and around the half radius position. In the inner part of the rotor, however, the velocity is higher compared to the case without nozzle. The tower blockage is less pronounced, as indicated by the higher velocity in the area behind the tower.

Figure 13 shows a snapshot of the streamwise velocity in the middle of the wind tunnel ($y = 0$ m).

The increase of speed in the nozzle is clearly visible. Consequently, the wake is less elongated and even less expanded than in the case without nozzle. The higher velocities in the inner part of the rotor, as already seen in Figs. 10 and 12, might be a result of the deflection of the wake to the middle of the wind tunnel as the tip vortices are sucked into the nozzle. Due to this deflection, the induction of the wake changes, which has in turn, due to Biot-Savart, an influence on the induced velocity in the rotor plane. Moreover, the upstream influence of the nozzle might lead to an acceleration in the middle of the wind tunnel, too. A comparison of the stream traces between Figs. 8 and 13 shows, that due to the nozzle, the stream traces are curved, following the shape of the nozzle as they are sucked in.

In order to differentiate between the influence of the tower and the influence of the nozzle, Table 5 shows the mean values and the amplitude for thrust and power for one revolution for $WT1_{rot}$ and $WT2_{tow}$ with regard to $WT1_{rot}$.

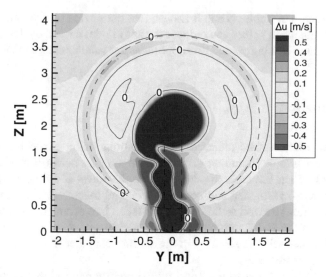

Fig. 12 Difference in streamwise velocity between $WT2_{tow}$ and $WT1_{tow}$, averaged over one revolution ($\Delta u = u_{WT2_{tow}} - u_{WT1_{tow}}$). $1R$ downstream of the rotor, viewing direction is upstream

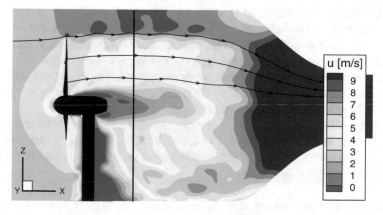

Fig. 13 Snapshot of the streamwise velocity for $WT2_{tow}$ in the middle of the wind tunnel ($y = 0$ m) including three streamtraces. Wind turbine illustrated in black, black line indicates the position of Fig. 12

Table 5 Thrust and power over one revolution, Δ values with regard to the mean value of $WT1_{rot}$

	Thrust (N)		Power (W)	
	$WT1_{rot}$	$WT2_{tow}$	$WT1_{rot}$	$WT2_{tow}$
Mean value	163.1	163.7	730.1	734.6
Δ Mean	—	0.37%	—	0.62%
Amplitude	—	2.6	—	17.3

A comparison of Tables 5 and 3 shows, that the nozzle partly compensates the reduction of the mean values and reduces the amplitudes caused by the tower, as the Δ Mean in Table 5 is bigger and the amplitudes are smaller compared to Table 3.

5 Conclusion

The present article shows numerical investigations of an experimental setup of a model wind turbine including different components, performed with the CFD solver *FLOWer*. The influence of wind tunnel walls, tower and a nozzle downstream of the rotor are investigated and compared. Global loads, velocities and angle of attack distributions as well as flow fields are taken into account. The turbine under investigation is the *BeRT* turbine which has a radius of $R = 1.5$ m, a hub height of $h = 2.1$ m and which is located in the settling chamber of the *GroWiKa* of the Technical University of Berlin.

It turned out, that due to the wind tunnel environment, thrust is increased by approximately 25% and power by 50% for the same inflow velocity and pitch. This strong influence was expected, as the blockage ratio is higher than 40%. Compared to the influence of the walls, the impact of tower and nozzle on the mean values are small (approximately 1%). However, the tower leads to an increase of fluctuations as the blades pass through areas with lower velocity upstream of the tower and areas with higher velocities next to the tower. These load variations should not be neglected in the simulation. The reduction of the mean values of thrust and power, which are also a result of the tower, is partly compensated by the nozzle, which also reduces the amplitudes. This leads to the conclusion that if the tower is taken into account, the nozzle should be considered, too.

To sum up, for such a high blockage ratio, a direct transfer from the wind tunnel results to far field condition is not reasonable. A comparison to the numerical solutions, however, is feasible, but the modelling of the wind tunnel walls, the tower and the nozzle is thereby mandatory.

Acknowledgements The authors gratefully acknowledge the *High Performance Computing Center Stuttgart* for providing computational resources within the project *WEALoads* and the TU Berlin for the provision of the geometric data of the *BeRT* turbine. The studies presented in this article have been funded by the *German Research Foundation* (DFG).

References

1. J.A. Benek, J.L. Steger, F.C. Dougherty, P.G. Buning, *Chimera. A Grid-Embedding Technique* (Arnold Engineering Development Center Arnold Air Force Station, Tennessee Air Force Systems Command United States Air Force, 1986)
2. I.B. Celik, U. Ghia, P.J. Roache et al., Procedure for estimation and reporting of uncertainty due to discretization in {CFD} applications. J. Fluids Eng.-Trans. ASME **130**(7), 078001-078004 (2008)

3. A. Fischer, A. Flamm, E. Jost, T. Lutz, E. Krämer, Numerical investigation of a model wind turbine, in *Contributions to the 21st STAB Symposium Braunschweig, Germany 2016*. Notes on Numerical Fluid Mechanics and Multidisciplinary Design. STAB (Springer, Berlin, 2016)
4. A. Fischer, L. Klein, T. Lutz, E. Krämer, Simulations of unsteady aerodynamic effects on innovative wind turbine concepts, in *High Performance Computing in Science and Engineering'16* (Springer, Berlin, 2016), pp. 529–543
5. R. Gasch, J. Twele, *Windkraftanlagen: Grundlagen, Entwurf, Planung und Betrieb* (Springer, Berlin, 2010)
6. X. Huang, S. Vey, M. Meinke, W. Schroeder, G. Pechlivanoglou, C. Nayeri, C.O. Paschereit, Numerical and experimental investigation of wind turbine wakes, in *45th AIAA Fluid Dynamics Conference* (2015), p. 2310
7. A. Jameson, Time dependent calculations using multigrid, with applications to unsteady flows past airfoils and wings. AIAA Paper 1596:1991 (1991)
8. A. Jameson, W. Schmidt, E. Turkel et al., Numerical solutions of the euler equations by finite volume methods using Runge-Kutta time-stepping schemes. AIAA Paper 1259:1981 (1981)
9. E. Jost, A. Fischer, T. Lutz, E. Krämer, An investigation of unsteady 3d effects on trailing edge flaps. J. Phys.: Conf. Ser. **753**, 022009 (2016)
10. L. Klein, T. Lutz, E. Krämer, CFD analysis of a 2-bladed multi-megawatt turbine, in *10th PhD Seminar on Wind Energy in Europe*, EAWE, 2014, 28–31 October 2014, Orléans, pp. 47–50
11. L. Klein, C. Schulz, T. Lutz, E. Krämer et al., Influence of jet flow on the aerodynamics of a floating model wind turbine, in *The 26th International Ocean and Polar Engineering Conference* (International Society of Offshore and Polar Engineers, 2016)
12. U. Kowarsch, C. Öhrle, M. Keßler, E. Krämer, Aeroacoustic simulation of a complete h145 helicopter in descent flight. J. Am. Helicopter Soc. **61**(4), 1–13 (2016)
13. P.-Å. Krogstad, J. Lund, An experimental and numerical study of the performance of a model turbine. Wind Energy **15**(3), 443–457 (2012)
14. N. Kroll, C.-C. Rossow, K. Becker, F. Thiele, The megaflow project. Aerosp. Sci. Technol. **4**(4), 223–237 (2000)
15. K. Meister, *Numerische Untersuchung zum aerodynamischen und aeroelastischen Verhalten einer Windenergieanlage bei turbulenter atmosphärischer Zuströmung* (Shaker Verlag, Herzogenrath, 2015)
16. C.N. Nayeri, S. Vey, D. Marten, G. Pechlivanoglou, C.O. Paschereit, X. Huang, M. Meinke, W. Schöder, G. Kampers, M. Hölling, J. Peinke, A. Fischer, T. Lutz, E. Krämer, U. Cordes, K. Hufnagel, K. Schiffmann, H. Spiegelberg, C. Tropea, Collaborative research on wind turbine load control under realistic turbulent inflow conditions, in *DEWEK* (2015)
17. G. Pechlivanoglou, J. Fischer, O. Eisele, S. Vey, C.N. Nayeri, C.O. Paschereit, Development of a medium scale research hawt for inflow and aerodynamic research in the TU Berlin wind tunnel. *DEWEK* (2015)
18. M. Sayed, T. Lutz, E. Krämer, Aerodynamic investigation of flow over a multi-megawatt slender bladed horizontal-axis wind turbine, in *Renewable Energies Offshore* (2015)
19. J. Schepers, H. Snel, Model experiments in controlled conditions. ECN Report (2007)
20. C. Schulz, L. Klein, P. Weihing, T. Lutz et al., CFD studies on wind turbines in complex terrain under atmospheric inflow conditions. J. Phys.: Conf. Ser. **524**, 012134 (2014)
21. C. Schulz, K. Meister, T. Lutz, E. Krämer, Investigations on the wake development of the Mexico rotor considering different inflow conditions, in *Contributions to the 19th STAB/DGLR Symposium Munich, Germany 2014*. Notes on Numerical Fluid Mechanics and Multidisciplinary Design. STAB (Springer, Berlin, 2014)
22. H. Schümann, F. Pierella, L. Sætran, Experimental investigation of wind turbine wakes in the wind tunnel. Energy Procedia **35**, 285–296 (2013)
23. N. Sorensen, J. Michelsen, S. Schreck, Navier-stokes predictions of the NREL phase vi rotor in the NASA ames 80-by-120 wind tunnel, in *ASME 2002 Wind Energy Symposium* (American Society of Mechanical Engineers, 2002), pp. 94–105
24. P. Weihing, K. Meister, C. Schulz, T. Lutz et al., CFD simulations on interference effects between offshore wind turbines. J. Phys.: Conf. Ser. **524**, 012143 (2014)

Numerical Analysis of a Propeller Turbine Operated in Part Load Conditions

Bernd Junginger and Stefan Riedelbauch

Abstract A part load operating point of a hydraulic propeller turbine is numerically analyzed. Transient flow simulations without geometry simplifications are performed on meshes up to 100 million nodes using RANS and hybrid RANS-LES turbulence models. The weaknesses of steady state and transient RANS computations are evaluated against transient approaches with hybrid RANS-LES turbulence models. Due to the dominating phenomena of the vortex rope the focus of the results presented is on the draft tube flow field. Therefore, velocity profiles at 5 evaluation lines located in the draft tube are compared as well as the capabilities of the turbulence models to resolve the turbulent flow structures and developing vorticies for different mesh densities. Additionally, hydraulic losses in the machine components are compared.

1 Introduction

In the year 2016, hydro power is the leading source for renewable electric power generation [27]. About 71% of the global renewable energy is generated by hydro power which correlates to an installed capacity of 1064 GW. In total hydro power supplies about 16.4% of the world's electricity. In Germany, the installed hydropower capacity is about 5.6 GW which correlates to 18.6% of the total generated electric renewable energy or respectively 3.5% of the overall electric energy generation [4]. Hydro power in Germany is well developed, however there is still undeveloped technical potential available. According to a potential analysis carried out for the German federal ministry of economics and energy in 2010, there is a potential of about 0.4 TWh or additional 450 small hydro power plants [3]. Due to promotions given by the European Union, in line with the Water Framework Directive, potential small hydro locations are attractive for energy providers [6].

B. Junginger (✉) • S. Riedelbauch
Institute of Fluid Mechanics and Hydraulic Machinery, University of Stuttgart,
Pfaffenwaldring 10, 70550 Stuttgart, Germany
e-mail: junginger@ihs.uni-stuttgart.de

© Springer International Publishing AG 2018
W.E. Nagel et al. (eds.), *High Performance Computing in Science and Engineering '17*, https://doi.org/10.1007/978-3-319-68394-2_21

All member states of the EU are advised to achieve a good ecological status for all flowing water. River power plants, which are typically equipped with bulb or Kaplan turbines are suitable to revitalize unused dams and weirs and hence fulfill the requirements of the Water Framework Directive.

Another huge advantage of hydro power is the fast adjustability of the energy output which is a quality well needed in times of the energy revolution. Thus, it is possible to guarantee the electric grid stability in times of fluctuating energy production of other renewable energy sources like wind or photovoltaic. This leads to an increase of the operating range of hydraulic machines to ensure the necessary adjustability to balance the grid. An enlarged operating range increases the risk of transient phenomena like vortex ropes, cavitation and pressure oscillations. A part load operating point is a typical example for an off design operating point. The strong swirling flow at the runner outlet leads to the development of a vortex rope resembling a corkscrew [5, 13, 23]. The vortex rope induces pressure oscillations which can lead to considerable damage of the turbine.

For low head turbines the correct prediction of the losses of the machine in particular in the draft tube is essential. In particular geometry simplifications like the negligence of gaps can lead to a falsification of the computation results. An accurate recreation of the geometry is required since especially in these regions strong gradients of essential quantities occur. Furthermore, the complex three dimensional flow field generated by the rotating vortex rope is difficult to resolve correctly with classical RANS (Reynolds Averaged Navier-Stokes) approaches. Hence, different turbulence models are applied to various grid densities to investigate their effects on the results. Hybrid turbulence models like the SAS (Scale Adaptive Simulation) turbulence model provide promising results in the field of hydraulic turbomachinery and are also investigated in this paper [11, 12, 14]. Additionally, the SAS turbulence model is compared to the new implemented SBES (Stress Blended Eddy Simulation) turbulence model which provides a faster transition from RANS to LES (Large Eddy Simulation) than the SAS turbulence model [16, 17].

2 Investigated Operating Point

The investigated operating point possesses the same head as the best efficiency operating point (BEP) with a reduced discharge and a smaller guide vane opening (Fig. 1). The dimensionless terms of the part load operating point are a specific speed $n_{ED} = n_{ED_{BEP}}$ and specific discharge factor $Q_{ED} = 0.78\, Q_{ED_{BEP}}$. The guide vane opening $\Delta\gamma$ is about $0.53\, \Delta\gamma_{BEP}$.

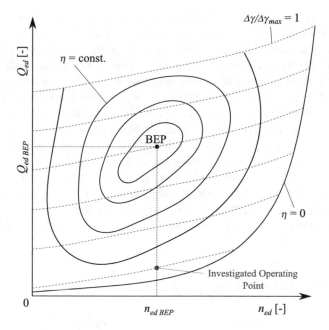

Fig. 1 Schematic hill chart of a turbine

3 Numerical Methods and Turbulence Modeling

All simulations are performed with the commercial CFD (Computational Fluid Dynamics) code ANSYS CFX version 16.0, 17.2 and 18.0 [1]. For the discretization a finite volume method is used on an implicit pressure formulation. Coupling of different meshes as well as the rotation of the meshes, which is important for the turbine runner, can be handled by the CFD code. A coupled algebraic multigrid solver with an ILU based algorithm is used [22].

Three turbulence models are applied in this paper. A classical RANS model, the k-ω-SST model (Shear Stress Transport), which represents the standard turbulence model in the development process of hydraulic turbomachinery [15]. Additionally, two hybrid turbulence models, namely the SAS-SST [7, 8, 17, 19] and the SBES [16, 18, 20, 24] model are applied in this work. The SAS model is a combination of an unsteady RANS SST model and a SRS framework (Scale Resolving Simulation), like in a LES [21]. An additional source term Q_{SAS} is introduced in the transport equation of the turbulence eddy frequency ω of the SST turbulence model which leads to a reduction of the turbulent eddy viscosity. A high wave number limit is set, to avoid the reduction of the eddy viscosity below the level of the LES eddy viscosity based on the WALE model (Wall-Adapting Local Eddy-Viscosity). Hence, an overestimation of the eddy viscosity in fine meshes can be prevented. Similar to the SAS model the SBES model is a combination of a RANS and a LES model. For the investigation the SBES blends between the unsteady RANS

SST model and the WALE model, which is the same function as used in the SDES (Shielded Delayed Eddy Simulation) but with an explicit switch to an algebraic LES model. The crux of the DES (Detached Eddy Simulation) formulation is the protection of the wall against influences of the LES. An improved definition of the shielding function especially of the grid length scale provides a faster transition from RANS to LES and a clear distinguishing between RANS and LES. A premature switching could otherwise cause strong deterioration of the RANS capabilities in the attached boundary layer [25]. A second order Euler backward scheme is used for the temporal discretization, whereas the spatial discretization is depending on the applied turbulence model. For the spatial discretization of the RANS simulations a high-resolution (HR) scheme is used [2], whereas a less dissipative bounded central differencing (BCD) is applied for all hybrid RANS-LES turbulence models [9]. The BCD discretization scheme is second order. For the spatial discretization of the convection term of the turbulence model a first order scheme is applied [16].

4 Computational Setup

The geometry of the propeller turbine is illustrated in Fig. 2. The fluid flows from the inflow through the guide vanes to the runner followed by the draft tube. An expansion tank downstream the draft tube is also part of the computational domain of the draft tube to ensure that the outlet boundary condition does not effect the simulation results. The computational model is divided in three domains which are coupled in the transient simulations with transient rotor-stator-interfaces and in the steady state computation with mixing plane interfaces. A circumferential averaging at the interface is performed when using a mixing plane interface, whereas no simplification takes place when using the transient rotor stator interface. The entire geometry is built exactly like the model turbine installed in the laboratory of the Institute. The relevant gaps of the machine, at the trailing edges of the guide vanes and the tip clearance of the runner, are modeled for all transient simulations. The gap at the trailing edge of the guide vane is neglected in the steady state approach. The

Fig. 2 Hydraulic contour of the propeller turbine with evaluation lines (red lines) 1–5 (in flow direction) for flow quantities

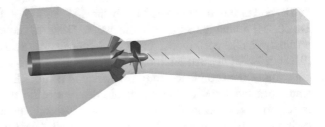

Table 1 Number of elements for the meshes with a normalized runner gap of $\tau = 0.3$ in million nodes, corresponding time steps Δt_{deg} in $°/\Delta t$ and $\Delta t_{ru_{rev}}$ in Δt/revolution

	Steady state	Transient			
Turbine part	5M	15M	30M	50M	100M
Guide vanes	1.8M	3.2M	5.4M	8.0M	13.3M
Runner	2.0M	5.9M	10.0M	16.5M	27.8M
Draft tube	1.5M	5.0M	14.9M	24.6M	59.8M
Total	5.2M	14.1M	30.3M	49.2M	100.8M
Nodes in runner gap	12	20	30	40	50
Δt_{deg} [$°/\Delta t$]	–	0.5	0.4	0.36	0.25
$\Delta t_{ru_{rev}}$ [Δt/revolution]	–	720	900	1000	1440

runner gap, in radial direction, has a normalized size of $\tau = 0.3$ which is defined as:

$$\tau = \frac{s_{gap}}{s_{gap_{max}}} \tag{1}$$

with the actual gap size s_{gap} and the maximum investigated gap size $s_{gap_{max}}$.

A steady state mass flow boundary condition with velocity profiles and turbulence quantities is set at the inlet of the guide vane domain. At the outlet of the draft tube domain a static averaged pressure is set as boundary condition. A listing of the investigated meshes with used time step in degree Δt_{deg} and the corresponding revolution of the runner per time step $\Delta t_{ru_{rev}}$ is listed in Table 1. The time step for the respective grid densities is chosen so the CFL number is below 1. Except for the gaps this criterion is fulfilled in the entire computational domain. The dimensionless wall distance y^+ for the 5 million (M) grid is between 7–10, for the 15M mesh the near wall resolution $y^+ = 1$–7 and for all other mesh densities $y^+ \leq 1$.

5   Results

All computations have run a sufficient number of time steps to achieve a periodic behavior. Additionally, selected quantities are time averaged over 50 runner revolutions.

5.1   Machine Data and Flow Analysis

The hydraulic losses of the turbine are plotted in Fig. 3. All losses are normalized to the measured head of the turbine. The highest losses are predicted in the steady state and the transient SST simulation. However, the reason for the losses is differing in

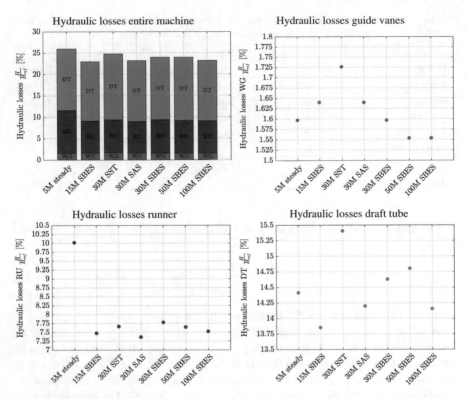

Fig. 3 Hydraulic losses of each machine component for the different numerical approaches, total machine (*top left*), guide vanes (WG) (*top right*), runner (RU) (*bottom left*), draft tube (DT) (*bottom right*)

both cases. In the steady state simulation the losses in the runner component are strongly overestimated, compared to all other computations, whereas the hydraulic losses in the other machine components are on an equal level similar to the losses when applying hybrid turbulence models. The transient RANS simulation predicts the highest losses in the guide vane and the draft tube domain. An explanation for the high losses in particular in the draft tube, can be the dissipative character of the RANS turbulence model and the lack to resolve the turbulent flow structures. Hence, the prediction of flow detachment can be wrong.

The losses in the guide vane domain reduce with an increasing mesh density when applying a hybrid turbulence model. The hydraulic losses computed for the 15M SBES case are on the same level than the losses of the 30M SAS simulation. This might be explained with the transition to LES and hence a faster blending to a less dissipative turbulence model. Computations with finer meshes and the SBES model predict higher losses in the draft tube. The differences can arise from the different runner outlet velocity profiles.

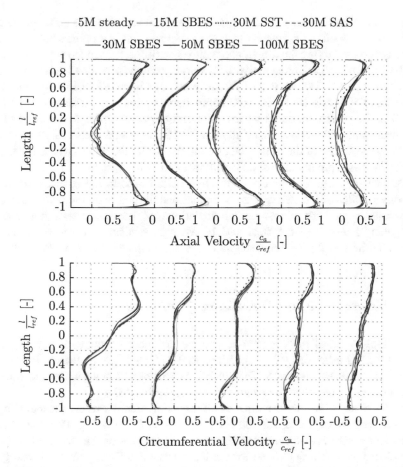

Fig. 4 Velocity profiles in streamwise direction at 5 evaluation lines in the draft tube, axial velocity (*top*), circumferential velocity (*bottom*)

Velocity profiles in streamwise direction for five evaluation lines, which are presented in Fig. 2, are shown in Fig. 4. The evaluation lines are located about $\frac{D}{2}$, D, $2D$, $3D$ and $4D$ downstream the runner trailing edge. The length plotted on the ordinate represents the radius for line 1–3 which are located in the conical part of the draft tube. Line 4 and 5 are located in the transition part of the draft tube where the shape transfers from a cylindrical to a rectangular shape.

Only computations of the cases 50M SBES and 100M SBES predict a negative axial velocity component at the first evaluation line behind the runner. Additionally, it can be observed that the tangential velocity component in the stagnation region for these two cases is reduced. The circumferential velocity component shows no trend in the second half of draft tube. At a length of $l/l_{ref} = 0.8$ at line 1 and 2 the 50M and 100M cases show a higher axial and circumferential velocity component than all other computations. Closer to the shroud of the machine the effect appears vice versa. This effect may develop due to the runner gap, which leads to an increase of the axial velocity component close to the shroud and is a point which is further investigated by varying the runner gap width. The differences in the velocity profiles can be noted until line 3. A negative axial velocity component for all calculated cases develops at line 3. Hence, two different trends can be noted. While for fine mesh densities of 50M and 100M the axial component continues to increase starting from line 3 to the draft tube outlet an inverse trend appears for the coarse meshes and the steady state simulation. Especially the 30M SST case predicts a large stagnation region in the center of the draft tube close to the draft tube outlet, whereas the 50M and 100M cases predict almost no backflow at all.

Despite the large differences of the hydraulic losses in the runner domain the velocity profiles of the steady state computations fit quite well the results achieved with coarser mesh densities. Larger deviations of the steady state computation compared to results of the 50M and 100M cases can be observed at the end of the draft tube and the predicted stagnation region in the center of the draft tube.

5.2   Vortex Rope and Turbulence Quantities

The visualization of the vortex rope and the turbulent structures is performed by usage of the Q-criterion [10]. Q is defined as $0.5\,(\Omega^2 - S^2)$, with the vorticity tensor Ω and the strain rate tensor S. Thus it defines vorticies as areas where the vorticity magnitude is greater than the magnitude of the strain rate.

In Fig. 5 the turbulent structures of all transient simulations are presented by an iso-surface $Q = 1$ colored with the viscosity ratio $v_t / v = 0$–250. The visualization of the vortex rope is also carried out by using the Q-criterion. However, the value of the velocity invariant Q utilized for the iso-surface is differing for the investigated cases to be able to visualize the vortex rope.

The largest turbulent structures in the stagnation region occur behind the runner hub, where the vortex rope is situated. Towards the outlet of the draft tube the dominant effect of the vortex rope is declining which is going with a decay of the large flow structures in streamwise direction. The lack of predicting turbulent flow structures can be observed for the RANS simulation, which is only capable to resolve the largest flow structures. The smallest turbulent flow structures, on an equal mesh, can be resolved by the SBES turbulence model. The turbulent eddy viscosity ratio for all transient simulations for line 1 and 5 is shown in Fig. 6. The viscosity ratio of the RANS simulation is about one magnitude larger than in the SAS case with the same mesh density and the SBES model using a 15M grid since the SST model is not able to switch to a SRS mode to reduce the turbulent eddy viscosity ratio. Rates of the turbulent eddy viscosity ratio are decreasing with an increasing mesh density. The distribution of the viscosity ratio over the evaluation lines are comparatively smooth when applying the SST or the SAS turbulence model, whereas in the SBES cases fluctuations can be observed in the time averaged

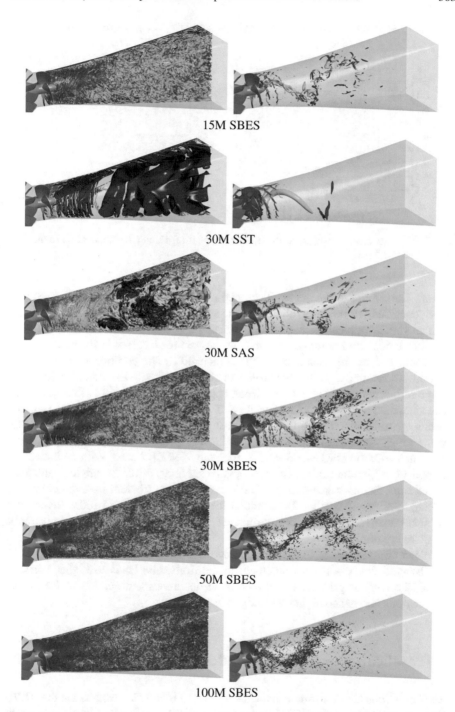

Fig. 5 Flow structures in the draft tube with an iso-surface of the velocity invariant Q colored with the viscosity ratio $\nu_t/\nu = 0-250$ (*left*) and visualization of the vortex rope with an iso-surface of the velocity invariant Q colored with the vorticity $z = 0-250$ (*right*)

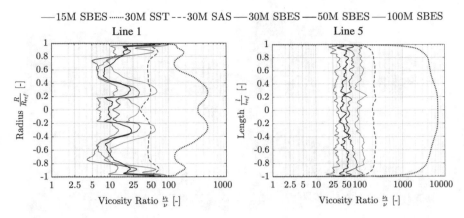

Fig. 6 Viscosity ratio v_t/v close to the inlet of the draft tube line 1 (*left*) and close to the draft tube outlet line 5 (*right*)

signal. The effect seen in the axial velocity profile at line 1 (see Fig. 3) between the mesh densities can be observed in the time averaged viscosity ratio.

The vortex rope can be captured by using the Q-criterion. Even when applying the SST model it is possible to visualize the part load vortex in the center of the draft tube and the tip clearance vorticies induced by the gap between runner and shroud. Additionally, the blade wakes can be captured with the RANS model. The shape of the vorticies is changing when applying hybrid RANS-LES approaches. The smoothed shape visualized for the SST turbulence model is burst into smaller turbulent structures. This effect is increasing with increasing mesh density. Due to the faster transition from RANS to LES, it is possible to capture even smaller structures with the 15M SBES case as with the 30M SAS case. With an increasing number of mesh nodes it is getting harder and harder to differ the single phenomena vortex rope, blade wakes and gap vorticies since the turbulent flow structures are getting smaller and smaller. In particular for the 50M SBES and 100M SBES case the small resolved turbulent flow structures of the vortex rope and the tip clearance vorticies are mingling.

When comparing the different mesh densities for the SBES approaches it can be observed that the mesh refinement in the draft tube leads to higher vorticity rates. The smaller grid cells lead to a more accurate identification of the vortex core regions and hence higher vorticity rates.

6 Parallelization and Computational Resources

All simulations are performed on the CRAY XC40 Hazel Hen installed at the HLRS Stuttgart, which consists of 7712 compute nodes with Intel® Xeon® processors with 128 GB memory interconnected with Arias. A compute node has 2 sockets each with

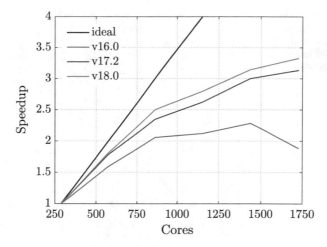

Fig. 7 Speed up test of a transient flow simulation of a part load operating point using the SBES turbulence model on a 50M mesh

12 cores. For the comparison of the parallel performance all cases are partitioned upfront.

A comparison of the 50M and the 100M mesh of the propeller turbine, with two rotor stator interfaces, is carried out for the version v16.0, v17.2 and v18.0 of ANSYS CFX. The speedup results of 50M mesh are shown in Fig. 7. A significant improvement of the parallel performance can be noticed between v16.0 and v17.2 in particular for larger core number where an decrease of the parallel performance can be observed when using v16.0. Between v17.2 and v18.0, which is released in spring 2017, the parallel performance is slightly improved. The efficiency of 91% for v18.0 compared to 80% for v16.0 is reached when using 576 cores An even better efficiency can be achieved in cases where no rotor stator interfaces are required [26].

The results of the speedup with 100M mesh are illustrated in Fig. 8. Similar to the 50M mesh, a significant improvement of the parallel efficiency can be observed between v16.0 and the newer versions v17.2 and v18.0. For the 100M mesh a stagnation of the parallel performance occurs for larger core numbers when using v16.0, whereas a decrease of the needed computation time can be observed for the newer versions, v17.2 and v18.0.

In total about 65 computation days are required for the 50M grid on 576 cores This time can be divided in 20 days to achieve a periodic flow behavior and 45 days for the time averaging of the evaluation quantities. About 120 computation days are necessary to achieve results for the 100M mesh on 1200 cores. The required time to a achieve a periodic flow behavior increases to 30 days. A time of 90 days is needed for the time averaging of the flow quantities. A minimum of 70 runner revolutions is required for all computations, including 50 runner revolutions for averaging the the flow quantities. Depending on the operating point and the resultant flow in the

Fig. 8 Speed up test of a
transient flow simulation of a
part load operating point
using the SBES turbulence
model on a 100M mesh

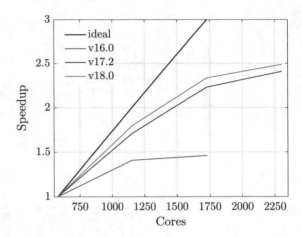

draft tube up to 80 runner revolution of time averaging can be necessary to obtain
satisfying results close to the draft tube outlet.

7   Conclusions and Outlook

Flow simulations of a propeller turbine operated in part load conditions were
performed at the HLRS by using the commercial CFD code Ansys CFX. Hybrid
RANS-LES turbulence models, SAS and SBES, are compared to a RANS SST
turbulence model and a steady state approach. Additionally, the influence of the
mesh density up to 100 million nodes is investigated when applying the SBES
turbulence model.

The RANS model and the steady state computation show the highest losses in
the entire machine, whereas the machine component where the deviation to the
other computation occurs is varying. For the steady state approach the losses in the
runner domain are significantly overestimated, while for the transient SST model
the losses in the draft tube are overrated because of the too dissipative character of
the RANS model. Approaches with hybrid RANS-LES turbulence models lead to
smaller losses. The smallest losses are predicted by the SBES model, independent
of the mesh density.

Only the 50 million and 100 million mesh with the SBES turbulence model
predict a stagnation region at the first evaluation line downstream the runner, while
all other approaches still predict an positive axial velocity component. This effect
can also be seen in the circumferential velocity profiles. Close to the draft tube outlet
the effect is vice versa, the coarser approaches (15M and 30M) predict the smallest
axial component compared to the finer mesh densities (50M and 100M).

The effect of the influence of the mesh density as seen in the velocity profiles can
also be observed in the time averaged turbulent eddy viscosity ratio. The SBES

turbulence model leads to the lowest viscosity ratios and resolves the smallest turbulent structures in the draft tube of the investigated approaches. Compared with the smoothed shape of the developing part load vortex for the RANS model all hybrid turbulence models resolve a more detailed vortex rope, which consists of vortex streaks. All approaches are able to resolve the developing tip clearance vorticies and the blade wakes. With increasing mesh densities the resolved turbulence flow structures are getting smaller and hence the differing between the developing phenomena is getting harder due to mingling.

A detailed experimental validation of the CFD data is planned to achieve an extensive set of data. Besides measurements of the velocity profiles at several positions in the draft tube, pressure pulsations in the draft tube are measured for different operating points. Furthermore the runner gap size is varied in the experiment to validate the CFD data.

Acknowledgements The authors gratefully acknowledge the High Performance Computing Center Stuttgart for providing computational resources.

References

1. ANSYS Inc.: ANSYS CFX Version 17.2, 2017
2. T.J. Barth, D.C. Jesperson, The design and application of upwind schemes on unstructured meshes, in *27th AerospaceSciences Meeting* (1989)
3. Bundesministerium für Wirtschaft und Energie, Potentialermittlung für den Ausbau der Wasserkraftnutzung in Deutschland als Grundlage für die Entwicklung einer geeigneten Ausbaustrategie (2010)
4. Bundesministerium für Wirtschaft und Energie, Arbeitsgruppe Erneuerbare Energien: Zeitreihen zur Entwicklung der erneucrbaren Energien in Deutschland (2017)
5. G.D. Ciocan, M.S. Iliescu, T.C. Vu, B. Nennemann, F. Avellan, Experimental study and numerical simulation of the FLINDT draft tube rotating vortex. J. Fluids Eng. **129**(2), 146–158 (2007)
6. Commission of the European Communities: Towards sustainable water management in the European Union - first stage in the implementation of the Water Framework Directive. 2000/60/ECCOM 2007;128:1–13 (2000)
7. Y. Egorov, F.R. Menter, Development and application of SST-SAS turbulence model in the DESIDER project, in *Advances in Hybrid RANS-LES Modelling: Papers contributed to the 2007 Symposium of Hybrid RANS-LES Methods*, Corfu, 17–18 June 2007 (Springer, Berlin, 2008), pp. 261–270
8. Y. Egorov, F.R. Menter, D. Cokljat, The scale-adaptive simulation method for unsteady turbulent flow predictions. Part 2: Application to aerodynamic flows. J. Flow Turbul. Combust. **85**(1), 139–165 (2010)
9. H. Jasak, H.G. Weller, A.D. Gosman, High resolution NVD differencing scheme for arbitrarily unstructured meshes. Int. J. Numer. Methods Fluids **31**, 431–449 (1999)
10. J. Jeong, F. Hussain, On the identification of a vortex. J. Fluid Mech. **285**, 69–94 (1995)
11. D. Jošt, A. Škerlavaj, Efficiency prediction for a low head bulb turbine with SAS SST and zonal LES turbulence models, in *27th IAHR Symposium on Hydraulic Machinery and Systems*, Montreal (2014)
12. D. Jošt, A. Škerlavaj, A. Lipej, Numerical ow simulation and efficiency prediction for axial turbines by advanced turbulence models, in *6th IAHR Symposium on Hydraulic Machinery and Systems*, Beijing (2012)

13. O. Kirschner, Experimentelle Untersuchung des Wirbelzopfes im geraden Saugrohr einer Modellpumpturbine, PhD thesis, IHS-Mitteilung 32, University of Stuttgart, 2011
14. T. Krappel, S. Riedelbauch, R. Jester-Zuerker, A. Jung, B. Flurl, F. Unger, P. Galpin, Turbulence resolving flow simulations of a francis turbine in part load using highly parallel CFD simulations, in *28th IAHR Symposium on Hydraulic Machines and Systems*, Grenoble (2016)
15. F.R. Menter, Two-equation eddy-viscosity turbulence models for engineering applications. AIAA J. **32**(8), 269–289 (1994)
16. F.R. Menter, Best Practise: Scale-Resolving Simulations in ANSYS CFD Version 2.0, ANSYS Germany, 2015
17. F. Menter, Scale-adaptive simulation, in *ERCOFTAC Hybrid RANS-LES Methods for Industrial CFD*, Otterfing (2015)
18. F. Menter, The stress blended Eddy simulation model - SBES, in *ERCOFTAC Hybrid RANS-LES Methods for Industrial CFD*, Otterfing (2015)
19. F.R. Menter, Y. Egorov, The scale-adaptive simulation method for unsteady turbulent flow predictions. Part 1: Theory and model description. J. Flow Turbul. Combust. **85**(1), 113–138 (2010)
20. F. Menter, M. Strelets, Detached Eddy Simulation (DES), in *ERCOFTAC Hybrid RANS-LES Methods for Industrial CFD*, Otterfing (2015)
21. F.R. Menter, J. Schütze, M. Gritskevich, *Global vs. Zonal Approaches in Hybrid RANS-LES Turbulence Modelling Progress in Hybrid RANS-LES Modelling*. Notes on Numerical Fluid Mechanics and Multidisciplinary Design, vol. 117 (Springer, Berlin, 2012), pp. 15–28
22. M. Raw, Robustness of coupled algebraic multigrid for the Navier-Stokes equations, in *34th Aerospace Sciences Meeting and Exhibit, Aerospace Sciences Meetings*, AIAA 96-0297, Reno, NV (1996)
23. A. Ruprecht, T. Helmrich, T. Aschenbrenner, T. Scherer, Simulation of vortex rope in a turbine draft tubine, in *21st IAHR Symposium on Hydraulic Machinery and Systems*, Lausanne (2002)
24. M.L. Shur, P.R. Spalart, M.K. Strelets, A.K. Travin, A hybrid RANS-LES approach with delayed-DES and wall-modelled LES capabilities. Int. J. Heat Fluid Flow **29**, 1638–1649 (2008)
25. P.R. Spalart, S. Deck, M.L. Shur, K.D. Squires, M.K. Strelets, A. Travin, A new version of detached-eddy simulation, resistant to ambiguous grid densities. Theor. Comput. Fluid Dyn. **20**, 181–195 (2006)
26. J. Wack, B. Junginger, Untersuchung des Netzeinfluss auf den Spaltwirbel im Kaviationskanal, Internal report (2017)
27. World Energy Council: World Energy Resources 2016. ISBN:978 0 946121 58 8 (2016)

Mesoscale Simulations of Janus Particles and Deformable Capsules in Flow

Othmane Aouane, Qingguang Xie, Andrea Scagliarini, and Jens Harting

Abstract Complex fluids are common in our daily life and play an important role in many industrial applications. The understanding of the dynamical properties of these fluids and interfacial effects is still lacking. Computer simulations pose an attractive way to gain insight into the underlying physics. In this report we restrict ourselves to two examples of complex fluids and their simulation by means of numerical schemes coupled to the lattice Boltzmann method as a solver for the hydrodynamics of the problem. First, we study Janus particles at a fluid-fluid interface using the Shan-Chen pseudopotential approach for multicomponent fluids in combination with a discrete element algorithm. Second, we study the dense suspension of deformable capsules in a Kolmogorov flow by combining the lattice Boltzmann method with the immersed boundary method.

1 Introduction

Complex fluids are ubiquitous in soft matter systems, such as colloidal suspensions and biofluids. Understanding the dynamics of such systems is important for example for industrial production or disease detection. However, such problems are hard to solve analytically since the systems follow non-linear dynamics and might involve complex interfaces. These circumstances call for numerical methods and computer simulations which can also provide access to observables not traceable in

---

O. Aouane • A. Scagliarini
Helmholtz Institute Erlangen-Nürnberg for Renewable Energy (IEK-11), Forschungszentrum Jülich GmbH, Fürther Straße 248, 90429 Nürnberg, Germany

Q. Xie
Department of Applied Physics, Eindhoven University of Technology, P.O. Box 513, NL-5600MB, Eindhoven, The Netherlands

J. Harting (✉)
Helmholtz Institute Erlangen-Nürnberg for Renewable Energy (IEK-11), Forschungszentrum Jülich GmbH, Fürther Straße 248, 90429 Nürnberg, Germany

Department of Applied Physics, Eindhoven University of Technology, P.O. Box 513, NL-5600MB Eindhoven, The Netherlands
e-mail: j.harting@fz-juelich.de

© Springer International Publishing AG 2018
W.E. Nagel et al. (eds.), *High Performance Computing in Science and Engineering '17*, https://doi.org/10.1007/978-3-319-68394-2_22

experiments such as the local interface curvature or local fluid properties. Therefore, computer simulations can be used to complement experiments. In this report, we focus on computational simulations of colloidal particles at fluid interfaces and dense suspensions of soft particles under flow.

Colloidal particles can strongly adsorb at fluid-fluid interfaces [1]. Due to the particle properties (e.g. weight/buoyancy, roughness or anisotropic shape), or external fields (e.g. electric), particles deform the interfaces. When interface deformation caused by neighbouring particles overlaps, capillary interactions arise causing the migration and assembly of particles [2]. The capillary interactions are characterized by the way how interfaces deform. The modes of interface deformation can be represented analytically in a multipole expansion of the interface height [3]. The assembled structures of particles are dependent on the dominant mode and strength of capillary interactions, providing the possibility of dynamic control. Xie et al. [4] recently showed how to create tunable dipolar capillary deformations using a spherical magnetic Janus particle adsorbed at a fluid-fluid interface. When the particle is influenced by an external magnetic field directed parallel to the interface, the field causes the particle to experience a magnetic torque and to tilt with respect to the interface. When tilted, the particle deforms the interface in a dipolar fashion. This deformation can be dynamically controlled by adjusting the external magnetic field.

Our study on magnetic Janus particles considers the particle as a solid object. However, deformable particles are common in complex fluids [5], such as for example gel particles [6], red blood cells [7, 8], vesicles [9–11] or polymer capsules [12]. While systems like foams, emulsions, colloidal and polymer gels have been widely studied in the last decades [13, 14], still little is known about the flow behavior of dense suspensions (volume fraction above 60%) of drops encapsulated by an inextensible and impermeable surface membrane, i.e. capsules [15]. Their dynamics are governed by shear elastic and bending forces making them a suitable model to catch the complexity of soft matter in flow.

In this report, we firstly present our numerical study on the capillary interactions between spherical magnetic Janus particles at fluid-fluid interfaces [16]. We measure the pair lateral capillary force and find the minimal free energy configuration of a pair of particles. Driven by capillary interactions, multiple Janus particles assemble into chain-like structures, which are tunable by applying an external magnetic field. Then, we investigate the rheology of dense suspensions of capsules under a Kolmogorov flow. We use the shear-stress/shear-rate relationship to study the effect of the volume fraction and the mechanical properties of the individual capsules (stiffness/softness) on the fluid-jammed phase transition.

This current report is organized as follows. We present our method in Sect. 2 and discuss the performance and scaling of our code on Hazel Hen. In Sect. 3 we present two different applications of our LB code, namely colloidal Janus particles at fluid interfaces and dense suspensions of deformable particles (capsules) subject to a Kolmogorov flow. Finally, we conclude in Sect. 4.

2 Numerical Method

2.1 The Multicomponent Lattice Boltzmann Method

The lattice Boltzmann method (LBM) is a local mesoscopic algorithm, allowing for efficient parallel implementations. It has demonstrated itself as a powerful tool for numerical simulations of fluid flows [17, 18] and has been extended to allow the simulation of, for example, multiphase/multicomponent fluids [19, 20] and suspensions of particles of arbitrary shape and wettability [8, 21–24]. We implement the pseudopotential multicomponent LBM of Shan and Chen [19] with a D3Q19 lattice and review some relevant details in the following [23–27]. Two fluid components are modelled by following the evolution of two distribution functions discretized in space and time according to the lattice Boltzmann equation:

$$f_i^c(\mathbf{x} + \mathbf{c}_i \Delta t, t + \Delta t) = f_i^c(\mathbf{x}, t) - \frac{\Delta t}{\tau^c} [f_i^c(\mathbf{x}, t) - f_i^{\text{eq}}(\rho^c(\mathbf{x}, t), \mathbf{u}^c(\mathbf{x}, t))], \tag{1}$$

where $i = 1, \ldots, 19, f_i^c(\mathbf{x}, t)$ are the single-particle distribution functions for fluid component $c = 1$ or 2, $\mathbf{c}_i$ is the discrete velocity in ith direction, and τ^c is the relaxation time for component c. The macroscopic densities and velocities are defined as $\rho^c(\mathbf{x}, t) = \rho_0 \sum_i f_i^c(\mathbf{x}, t)$, where ρ_0 is a reference density, and $\mathbf{u}^c(\mathbf{x}, t) = \sum_i f_i^c(\mathbf{x}, t) \mathbf{c}_i / \rho^c(\mathbf{x}, t)$, respectively. Here, $f_i^{\text{eq}}(\rho^c(\mathbf{x}, t), \mathbf{u}^c(\mathbf{x}, t))$ is a third-order equilibrium distribution function. When sufficient lattice symmetry is guaranteed, the Navier-Stokes equations can be recovered from Eq. (1) on appropriate length and time scales [17]. For convenience we choose the lattice constant Δx, the timestep Δt, the unit mass ρ_0 and the relaxation time τ^c to be unity. The latter leads to a kinematic viscosity $\nu^c = \frac{1}{6}$ in lattice units. In the multicomponent model a mean-field interaction force

$$\mathbf{F}_C^c(\mathbf{x}, t) = -\Psi^c(\mathbf{x}, t) \sum_{c'} g_{cc'} \sum_{\mathbf{x}'} \Psi^{c'}(\mathbf{x}', t)(\mathbf{x}' - \mathbf{x}) \tag{2}$$

is introduced between fluid components c and c' [19], in which $\mathbf{x}'$ denote the nearest neighbours of lattice site $\mathbf{x}$ and $g_{cc'}$ is a coupling constant determining the surface tension. $\Psi^c(\mathbf{x}, t)$ is an "effective mass", which is chosen as

$$\Psi^c(\mathbf{x}, t) \equiv \Psi(\rho^c(\mathbf{x}, t)) = 1 - e^{-\rho^c(\mathbf{x}, t)}. \tag{3}$$

This force is then applied to the component c by adding a shift $\Delta \mathbf{u}^c(\mathbf{x}, t) = \frac{\tau^c \mathbf{F}_C^c(\mathbf{x}, t)}{\rho^c(\mathbf{x}, t)}$ to the velocity $\mathbf{u}^c(\mathbf{x}, t)$ in the equilibrium distribution. The Shan-Chen LB method is a diffuse interface method with typical interface widths of $\approx 5\Delta x$ [25].

2.2 Colloidal Particles

The trajectories of the colloidal particles are updated using a leap-frog integrator. Every particle is discretized on the fluid lattice and coupled to the fluid species by means of a modified bounce-back boundary condition as pioneered by Ladd and Aidun [21, 28] and extended to multicomponent flows by Jansen and Harting [23]. Hydrodynamics leads to a lubrication force between the particles. This force is reproduced automatically by the simulation for sufficiently large particle separations. If the distance between the particles is so small that no free lattice point exists between them this reproduction fails. We apply a lubrication correction when the distance of the two particles is less than one lattice constant, where the force resulting from lubrication interaction is not correctly reproduced in the simulation [29]. The lubrication force (including the correction) already reduces the probability that the particles come closely together and overlap. For the few cases where the particles still would overlap we introduce the direct potential between the particles which is assumed to be a Hertz potential [30]. The outer shell of the particle is filled with a "virtual" fluid with the density

$$\rho_{\text{virt}}^1(\mathbf{x}, t) = \overline{\rho}^1(\mathbf{x}, t) + |\Delta\rho|, \tag{4}$$

$$\rho_{\text{virt}}^2(\mathbf{x}, t) = \overline{\rho}^2(\mathbf{x}, t) - |\Delta\rho|, \tag{5}$$

where $\overline{\rho}^1(\mathbf{x}, t)$ and $\overline{\rho}^2(\mathbf{x}, t)$ are the average of the density of neighbouring fluid nodes for component 1 and 2, respectively. The parameter $\Delta\rho$ is called the "particle colour" and dictates the contact angle of the particle. A particle colour $\Delta\rho = 0$ corresponds to a contact angle of $\theta = 90°$, i.e. a neutrally wetting particle. In order to simulate a Janus particle, we set different particle colours in well defined surface areas corresponding to the different hemispheres of the particle.

2.3 The Immersed Boundary Method

In the immersed boundary method (IBM) [10, 12, 31], interfaces are considered as sharp and are represented by marker points constituting a moving Lagrangian mesh. A bi-directional coupling of the lattice fluid and the moving Lagrangian mesh is applied. The interface is moving along with the ambient fluid velocity. A deformation of the interface generally leads to stresses reacting back onto the fluid via local forces. The stresses depend on the chosen constitutive behaviour of the interface and are not predicted by the IBM itself. The discretized form of the IBM equations is given by

$$\mathbf{f}(\mathbf{x}) = \sum_i \mathbf{F}_i \Delta(\mathbf{x} - \mathbf{r}_i) \tag{6}$$

$$\dot{\mathbf{r}}_i = \sum_x \mathbf{u}(\mathbf{x}) \Delta(\mathbf{x} - \mathbf{r}_i), \tag{7}$$

where $\Delta(\mathbf{x} - \mathbf{r}_i)$ is the appropriate discretized approximation of Dirac's delta function [32] and $\mathbf{F}_i$ is the force exerted on node i located at position $\mathbf{r}_i$ with velocity $\dot{\mathbf{r}}_i$ (see Fig. 1). Equations (6) and (7) describe the spreading of the interfacial forces to the fluid nodes and the interpolation of the fluid velocity to the deformable interface. The update of the interface node positions is performed using a forward Euler approach. In what follows, we consider our deformable interfaces to be capsules, so to say deformable drops enclosed by a surface membrane. Their membrane is endowed with resistance to in-plane (shear elasticity) and out-of-plane (bending) deformations. Shear elastic and area dilatation effects are modeled using the Skalak constitutive law [33] and the out-of-plane deformations are calculated using the Helfrich free energy [34]. A discrete differential geometry operators approach [35] is used to evaluate the curvatures and the Laplace-Beltrami operator. The constraints of constant area and volume are prescribed using penalty functions [36]. To avoid overlap of particles at high volume fractions, a short range repulsion force that vanishes at node-to-node distance larger than one lattice unit is added [15]. The surface of the capsule is modelled using a triangular mesh and the forces are evaluated using a finite element method (FEM) [36]. The spherical capsule is generated from an icosahedron (Fig. 2a) that is refined recursively N_r times (Fig. 2b–e), so to say each triangular face is subdivided into 4 smaller triangles then the new

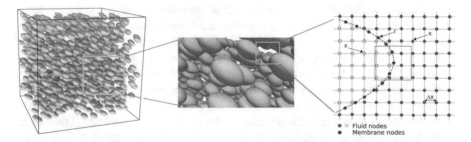

Fig. 1 A schematic depicting the coupling between the fluid and the interface nodes using the immersed boundary method. First, the membrane forces $\mathbf{F}_i$ are distributed to the neighboring fluid nodes within the stencil's range (represented here by the square box). Then, the velocities of the neighboring fluid nodes $\mathbf{u}(\mathbf{x})$ are interpolated back to the membrane nodes following the same scheme. Δx in the sketch refers to the lattice constant

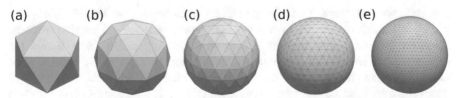

Fig. 2 Discretization of the surface of a spherical capsule using triangular elements. (**a**) $N_r = 0$, $N_n = 12$ and $N_f = 20$, (**b**) $N_r = 1$, $N_n = 42$ and $N_f = 80$, (**c**) $N_r = 2$, $N_n = 162$ and $N_f = 320$, (**d**) $N_r = 3$, $N_n = 642$ and $N_f = 1280$ and (**e**) $N_r = 4$, $N_n = 2562$ and $N_f = 5120$

nodes are shifted radially to ensure that they lie on the unit sphere. The total number of triangular faces (N_f) is defined based on the total number of nodes (N_n) and the number of recursive refinements (N_r) such as $N_f = 2N_n - 4$ and $N_n = 2 + 10 \cdot 4^{N_r}$.

2.4 Performance and Scalability

The core of the code is a Lattice-Boltzmann solver (LB), written for use on multi-CPU architectures. The parallel codes are written in standard FORTRAN90, and make use of a number of features of that language that are object-oriented in spirit. The code utilizes the message passing interface (MPI) for synchronization and communication between processors. It can be used in single data multiple processors (SDMP) mode, where the load of one large task is split across processors. This mode is used to perform large-scale calculations whose time and memory requirements are prohibitive on a single processor. Parallelization of the code in this mode is performed by means of a Cartesian domain decomposition strategy. The underlying 3D lattice is partitioned into sub-domains (boxes) and each box is assigned to one processor. Each processor is responsible for the particles within its sub-domain and performs exactly the same operations on these particles. Two rounds of communication between neighboring sub-domains are required: at the propagation step, where particles on a border node can move to a lattice point in the sub-domain of a neighboring processor, and in evaluating the forces. By using a ghost layer of lattice points around each sub-domain (halo), the propagation and collision steps can be isolated from the communication step. Before the propagation step is carried out the values at the border grid points are sent to the ghost layers of the neighboring processor and after the propagation step an additional round of communication is performed to update the ghost layers. This additional round of communication is required because of the presence of non-local interactions in the model whose computation requires (the updated) single-particle distribution functions at neighboring sites. Parallel HDF5 format is used for I/O operations.

2.4.1 Colloidal Particles

Our code was the work horse for several previous projects at HLRS [5, 18, 24–26, 37, 38] and its performance has been documented in several previous reports. Therefore, we only give a short update on its strong scaling behavior on the supercomputer Hazel Hen. We study a system containing $1024 \times 1200 \times 1024$ lattice Boltzmann nodes: a pure LB system containing only one fluid species (Fig. 3 (left)) and a more expensive multicomponent system with an additional 0.5 million colloidal particles (Fig. 3(right)). Due to the larger memory requirements, the code required to run on at least 192 cores for the single component case and on 384 cores for the latter one. Figure 3(left) shows that the nearly linear scaling was achieved on up to 98,304 cores or 56% of the whole machine. Figure 3(right) shows

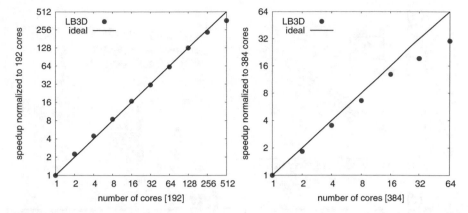

Fig. 3 Strong scaling of LB3D on Hazel Hen when simulating a single component fluid on up to 98,304 cores (left) and two fluid components with 0.5 million particles on up to 24,576 cores (right). The solid line represents ideal scaling relative to the performance of 192 cores (left) and 384 cores (right)

that the linear scalability breaks down at an earlier stage for colloidal suspension simulations. Here, the scaling is limited by the size of the suspended particles compared to the size of a single processor's domain. If the latter becomes too small, next-nearest neighbor interactions become common for the discrete element part resulting in a breakdown of the linear scalability for large processor counts.

2.4.2 The Immersed Boundary Method

We report here on the scaling and performance measurements of our LB code coupled to the IBM module on Hazel Hen. The number of fluid lattice site updates per second (FLUPS) is used as unit for performance measurement. The domain size is fixed to $512^3 \Delta x^3$ in all the IBM-LB benchmarks and a shear flow using velocity boundary conditions is imposed. The particle radius is fixed to $R = 7\Delta x$. Figure 4a shows the strong scaling when simulating a suspension of deformable particles in a shear flow on up to 8192 cores for volume fractions of 1% (upper triangles) and 10% (lower triangles), and a single component fluid without particles on up to 2048 cores (squares). The number of fluid lattice site updates per second as function of the volume fraction of particles and as function of the number of particle mesh points for a fixed volume fraction are reported in Fig. 4b, c. The benchmark in Fig. 4d shows the FLUPS as a function of the number of mesh nodes per particle on 2048 cores for different implementations of the bending forces: (1) a discrete model introduced by Kantor and Nelson [39] where the bending force is estimated using the angle between two neighboring faces, (2) a continuum model based on the Helfrich force density where the mean and Gaussian curvatures and the Laplace-Beltrami operator are approximated using discrete differential geometry

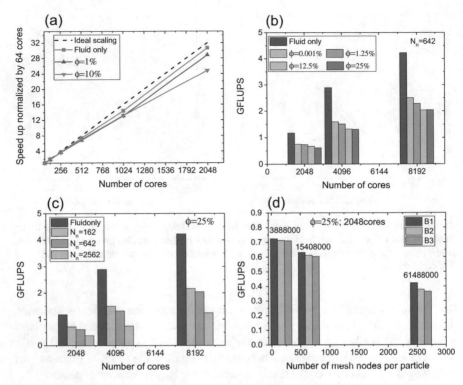

Fig. 4 Strong scaling and performance of our LB code coupled with the IBM module. (**a**) Strong scaling for a suspension of deformable particles in a shear flow on up to 8192 cores and volume fractions $\Phi = 1\%$ (upper triangles) and $\Phi = 10\%$ (lower triangles), as well as a single component fluid (bare fluid) on up to 2048 cores (squares), (**b**) Billion fluid lattice site updates per second (GFLUPS) as function of the number of cores for different volume fractions and a fixed number of mesh nodes per particle $N_n = 642$, (**c**) GFLUPS as a function of the number of cores for a fixed $\Phi = 25\%$ and different N_n, and (**d**) GFLUPS as a function of the number of mesh nodes per particle for $\Phi = 25\%$ on 2048 cores for three different implementations of the bending forces described in Sect. 2.4.2. The numbers on top of the bar plots refer to the total number of particle mesh nodes

operators approach [35], and (3) same as (2) but using a paraboloid fitting approach of the surface around the node of interest to estimate the mean and Gaussian curvatures [40]. Hereafter, we denote the three bending algorithms as B1, B2 and B3 respectively. In terms of numerical cost and accuracy, the continuum model B2 is a good compromise. Detailed benchmarks on the computational efficiency and accuracy of these algorithms can be found in [41–43].

The IBM module is a relatively recent addition to our simulation package and we are currently busy with improving the scaling behavior and the single core performance of this part of the code. However, due to the inherent data structures required for the description of the soft particle surfaces and the limitation that for efficient computation the size of a CPU domain should not be smaller than a particle, we do not expect to be able to reach the excellent scaling properties of the solid particle and LB parts of code.

3 Results

3.1 Janus Particles at Fluid Interfaces

We report on the behavior of magnetic Janus particles at a fluid-fluid interface as an example of the coupled lattice Boltzmann-discrete element method explained in Sect. 2.1. We follow Ref. [16] and present the results on capillary interactions and self-assembly of magnetic Janus particles adsorbed at fluid-fluid interfaces interacting with an external magnetic field.

We consider two spherical Janus particles, each composed of apolar and polar hemispheres, adsorbed at a fluid-fluid interface, as illustrated in Fig. 5. The two hemispheres have opposite wettability, represented by the three-phase contact angles $\theta_a = 90° + \beta$ and $\theta_p = 90° - \beta$, respectively, where β represents the amphiphilicity of the particle. Interacting with a horizontal magnetic field, $\mathbf{H}$, the particles experience a torque $\tau = \mathbf{m} \times \mathbf{H}$ and take tilted orientations with respect to the interface for a given dipole-field strength $B = |\mathbf{m}||\mathbf{H}|$. The tilt angle ϕ is defined as the angle between the particle dipole-moment and the undeformed interface normal. φ_A, φ_B are defined as the angles between the projection of the orientation of the magnetic dipole on the undeformed interface and the center-to-center vector of the particles. In this report we limit ourselves to study Janus particles with equal bond angles. A tilted Janus particle causes the interface to deform. The deformed interface around a tilted Janus particle forms a symmetric rise and depression on opposite sides of the particle [4], which is a dipolar deformation. Capillary interactions arise when interface deformations induced by neighboring

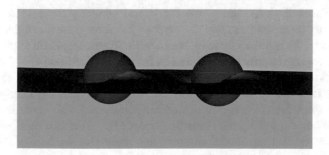

Fig. 5 Snapshot of two tilted Janus particles adsorbed at a fluid-fluid interface as obtained from our simulations. The interface deforms in a dipolar fashion resulting in dipolar capillary interactions. The Janus particles have a center to center distance $L_{AB}/R = 4$, bond angles $\varphi = 0°$, tilt angle $\phi = 60°$, and amphiphilicity $\beta = 39°$

particles overlap. We have derived an expression for the interaction energy between two Janus particles interacting via capillary interactions of the kind described above. In this model, we assume that (1) the leading order deformation mode is dipolar (2) the superposition approximation is valid (3) interface deformations are small [3]. The dipolar interaction energy for two Janus particles ΔE using cylindrical coordinates is

$$\Delta E = 2\pi \zeta^2 \gamma_{12} R^2 L_{AB}^{-2} + \frac{8\zeta R^2 \gamma_{12} \sin \beta}{L_{AB}} \left(\tan^{-1} \frac{L_{AB} + R}{L_{AB} - R} - \frac{\pi}{4} \right) \tag{8}$$

where γ_{12} is the fluid-fluid interface tension, R is the particle radius (both particles have the same radii), and L_{AB} is the centre-centre separation of the particles. To reiterate, β is the amphiphilicity of the particles and ζ is the height of the maximal interface deformation caused by the particles. The lateral capillary force $\Delta F = \frac{\partial(\Delta E)}{\partial L_{AB}}$ is therefore

$$\Delta F = -4\pi \zeta^2 \gamma_{12} R^2 L_{AB}^{-3}$$
$$- \frac{8\zeta R^2 \gamma_{12} \sin \beta}{L_{AB}^2} \left(\frac{L_{AB} R}{R^2 + L_{AB}^2} + \tan^{-1} \frac{L_{AB} + R}{L_{AB} - R} - \frac{\pi}{4} \right) \tag{9}$$

For a detailed derivation, we refer the reader to Ref. [16]. We compare the theoretical lateral capillary force Eq. (9) (solid lines) to the measured lateral capillary force from our simulations (circles) in Fig. 6 (left) for two different particle amphiphilicities $\beta = 14°$ and $\beta = 21°$. We place two particles of radius $R = 14$ at a distance $L_{AB} = 60\Delta x$ apart along the x-axis with a total system size $S = 1536 \times 384 \times 512\Delta x^3$. We fix the bond angle $\varphi = 0°$ between the particles and measure the lateral force on the particles as the tilt angle varies. Such large systems are mandatory due to the inherent long-range capillary interactions. This application shows that even relatively simple systems containing only two particles become true supercomputing applications once high precision measurements are required. Figure 6 (left) shows that the lateral capillary force increases as the amphiphilicity increases from $\beta = 14°$ to $\beta = 21°$ for a given tilt angle. For a given amphiphilicity, the capillary force increases with tilt angle up to tilt angles $\phi \approx 30°$. This is because the interface area increases for small tilt angles [4] which increases the interaction energy. As the tilt angle increases further $\phi > 30°$, the capillary force tends to a nearly constant value, due to the fact that the maximal contact line height (and therefore the deformed interface area) also tends to a constant value [4]. When comparing our theoretical model (solid lines) with simulation data (circles), we see that our model captures the qualitative features of the capillary interaction well, and quantitatively agrees with the numerical results for small tilt angles $\phi < 25°$ and small amphiphilicities $\beta = 14°$. The quantitative deviations at large tilt angles in the $\beta = 21°$ case are due to the breakdown of various assumptions in the theoretical model, namely the assumption of small interface slopes, and of finite-size effects in our simulations. The important predictions of our model are

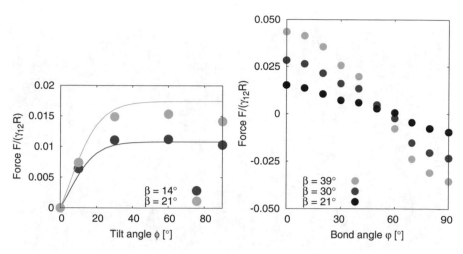

Fig. 6 Left: Lateral capillary force as a function of tilt angle for particles with amphiphilicities $\beta = 14°$ (red) and $\beta = 21°$ (green). The particles have equal bond angles $\varphi_A = \varphi_B = 0$. The solid line represents values from our theoretical model (Eq. (9)), and the symbols are simulation data. The theoretical analysis agrees well with our simulation results in the limit of small interface deformations and for large system sizes (see text). Right: Lateral capillary force as a function of bond angles of particles with amphiphilicity $\beta = 21°$ (red), $\beta = 30°$ (green) and $\beta = 39°$ (blue). The particles have a tilt angle of $90°$. The capillary force is repulsive for small bond angles, and becomes attractive for large bond angles. There is maximal attractive force between two particles for bond angles $\varphi = 90°$ (Reproduced from Xie et al. [16])

that the capillary force between particles can be tuned by increasing the particle amphiphilicity and/or the particle tilt angle. Since the external field strength controls the tilt angle, this allows the tuning of capillary interactions using an external field (Fig. 7).

In order to understand the self-assembled structures of many-particles, it is required to consider the minimum energy orientation between two particles for a given tilt angle and separation. In the current case of equal bond angles $\varphi_A = \varphi_B = \varphi$, minimising the total interaction energy with respect to the bond angle using our theoretical model indicates that the interaction energy decreases as the bond angle increases from $\varphi = 0°$ to $\varphi = 90°$. This theoretical analysis predicts that bond angles $\varphi_A = \varphi_B = 90°$ minimise the interaction energy, and that there is no energy barrier stopping the particles arranging into this configuration. In order to test the predictions of our model, we performed simulations of two particles of radius $R = 10$ separated by a distance $L_{AB} = 40\Delta x$ along the x-axis with a total system size $S = 512 \times 96 \times 512\Delta x^3$. We fix the tilt angles $\phi = 90°$ and measure the lateral force on the particles as the bond angle varies. Figure 6 (right) shows that the capillary force is repulsive for bond angles $\varphi < 50°$ and attractive for bond angles $\varphi > 50°$. There is a maximal attractive force between two particles for $\varphi_A = \varphi_B = 90°$. The simulation results show that for these parameters, two Janus particles with equal bond angles $\varphi_A = \varphi_B = \varphi = 90°$ minimise the

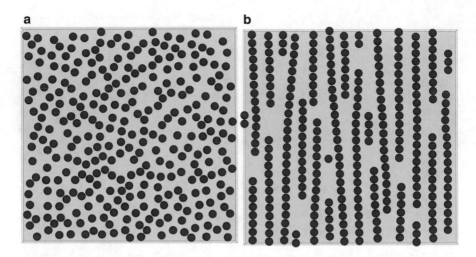

Fig. 7 Simulation snapshots of 300 magnetic Janus particles at fluid-fluid interface. (**a**) Initially the particles are distributed randomly at the interface. (**b**) Driven by capillary interactions, the particles form chain-like structures

interaction energy, agreeing with our theoretical predictions. Moreover, the lateral force decreases monotonically with increasing the bond angle indicating that there is no energy barrier stopping the particles from achieving the minimum energy state. Therefore, two Janus particles of the kind investigated in this paper interacting as capillary dipoles should rearrange into a configuration with $\varphi = 90°$ bond angle. Then, we study the alignment of multiple Janus particles at a flat fluid interface in simulations. We perform simulations with a system size $S = 512 \times 96 \times 512 \Delta x^3$. The particles have a radius $R = 10 \Delta x$. At first, we randomly distribute the particles at the interface. We equilibrate the system and then apply an external magnetic field parallel to the interface (Fig. 7a). The Janus particles experience a magnetic torque and align in similar tilted orientation with the interface. The magnetic torque is dominated by a strong external magnetic field and a weak dipole moment so that the dipole-dipole interactions are negligible. Therefore, the particles are purely driven by capillary interactions to assemble towards their minimal interaction energy configuration. The particles form chain-like structures (Fig. 7b), which demonstrates that the straight chain alignment (i.e. bond angle $\varphi = 90°$) is the minimum interaction energy configuration, which is consistent with our conclusion for two particles. Moreover, by controlling the external field, we are able to tune the assembled structures, which is promising to fabricate controlled reconfigurable colloidal materials.

3.2 Dense Suspension of Deformable Capsules Under a Kolmogorov Flow

This study is motivated by the work of Benzi et al. [14] on the rheology of dense emulsions. The focus is whether there are some similarities between the rheological response of emulsions and capsules under external stresses. The dynamical behavior of capsules under flow shows more complexity than drops and foams which makes them an interesting model system to study soft glassy materials. The rheology of dense suspensions of polymeric coated core-shells, i.e. capsules is investigated numerically under a Kolmogorov flow for volume fractions of up to 90% using the lattice Boltzmann method. The fluid-interface coupling is achieved using the immersed boundary method and forces are computed on the surface of the capsules using a finite element scheme. The total force exerted by the particle on the fluid is decomposed to three parts: (1) bending forces in the form of the functional derivative of the Helfrich energy [34]; (2) shear forces obtained from the derivative of the Skalak energy with respect to the nodes' position [33]; and (3) area and volume penalty forces to prescribe globally the conditions of inextensibility and impermeability of the capsule. The details of the implementation of these forces can be found in [12, 32, 35]. This study includes the effect of the mechanical properties of the individual capsules (stiffness/softness) on the local rheological behavior.

Local Rheology The system is $128^3 \Delta x^3$, athermal, and 3D periodic. The suspension is monodisperse with spherical unstressed shapes. The radius of the particles is fixed to $R = 8\Delta x$. The surface of each capsule is discretized with 642 mesh nodes and the total number of mesh nodes is between 385,200 and 577,800 for $60\% \leq \phi \leq 90\%$. The corresponding GFLUPS are 8.137, 7.398, 6.751, and 5.255 for $\phi = 60\%, 70\%, 80\%$ and 90%. It should be noticed that this work serves as a reference case to determine the relevant parameters of our system for a further, more detailed study. Indeed the Kolmogorov flow requires a large domain size with a large number of particle layers. All parameters and variables are given in lattice units. The capsules are distributed randomly in the domain with an initial rescaled radius. The size of the particles is increased linearly in time until reaching the desired radius. A friction force is added during the growing stage to counterbalance the increase of kinetic energy in absence of hydrodynamic interactions. Figure 8 shows the equilibrium structures obtained for volume fractions ranging from 60% to 90% after switching on hydrodynamic interactions. The suspension of capsules is subject to a Kolmogorov flow with an external force along the z-axis in the form $f_z = A_f \sin(k\{x + 1/2\})$, where A_f is the amplitude of the forcing and k is the wave number. The choice of a Kolmogorov flow is motivated by the absence of heterogeneities introduced by walls. At steady state the gradient of the shear stress has to counterbalance the external forcing, i.e. $\partial_x \sigma_{xz} = f_z$. It follows that the expression of the shear stress reads as $\sigma_{xz} = -\frac{A_f}{k} \cos(k\{x + 1/2\})$ and the local shear rate along the cross-flow axis, i.e. x-axis is obtained by evaluating numerically $\gamma(x) = \langle \partial_x u_z(x) \rangle_{yz}$. The rheological behavior of the suspension and the existence of

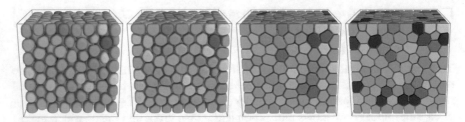

Fig. 8 Structures of densely packed suspension of capsules: 60%, 70%, 80% and 90%. The color scheme denotes the index of the particles

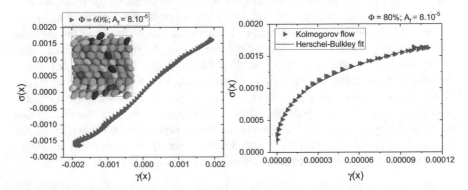

Fig. 9 Shear stress versus shear rate for a suspension of capsules at $\Phi = 60\%$ (left) and for a suspension of capsules at $\Phi = 80\%$ (triangles) together with the corresponding Herschel-Bulkley fit (solid line) (right)

yield stresses can be extracted from the shear-rate/shear-stress curve and described by the Herschel-Bulkley model [44]. In what follows, the shear-stress is averaged over time. Figure 9 (left) shows a linear behavior of the shear-rate/shear-stress relation at a volume fraction of 60%. This is a proof that the suspension is still in a fluid phase. At a volume fraction of 80% the suspension exhibits the properties of a yield fluid and the flow curve can be fitted using the Herschel-Bulkley model (see Fig. 9, right). But due to the limited size of our system, the values of the local yield points cannot be extracted with high enough accuracy. Currently, simulations of larger systems are ongoing on Hazel Hen.

Above $\Phi = 80\%$, the suspension of capsules undergoes a transition to a fully "jammed" state (Fig. 10-left). The question arises as to whether or not the "jamming transition" depends on the stiffness/softness of the particles. We increase the stiffness of the individual particles (shear elasticity and rigidity) while fixing the volume fraction to $\Phi \leq 80\%$. Figure 10-right shows a shift of the "jamming" point to $\Phi = 80\%$. The same trend is observed for $\Phi = 60\%$ (not shown here).

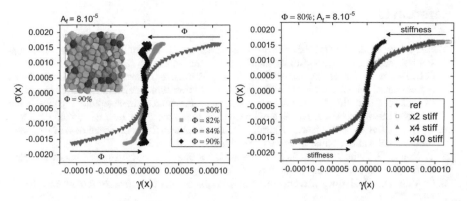

Fig. 10 Shear stress versus shear rate for a suspension of capsules showing transition to a fully "jammed" state as function of: the volume fraction (left), and the stiffness of the particles at a fixed volume fraction (right)

4 Conclusions

We presented two examples of our recent applications of lattice Boltzmann based simulations to complex fluid and interface problems. With the multicomponent lattice Boltzmann method coupled to a discrete element algorithm for the description of suspended (colloidal) particles, we studied the behavior of magnetic Janus particles at interfaces. With our method, we are able to resolve individual particles and their interactions. Moreover, we studied the dynamics of dense suspensions of capsules in Kolmogorov flow with the combined lattice-Boltzmann and immersed boundary method. The immersed boundary method offers a flexible and simple approach to model deformable particles immersed in suspending fluids. Our implementation is capable to simulate hundreds of thousands of suspended solid or deformable particles flowing in complex geometries on current supercomputers such as Hazel Hen. This is due to the highly efficient implementation and a consistent parallelization based on MPI and HDF5 parallel IO as demonstrated by the very good strong scaling of our code to almost 100,000 CPU cores for the pure fluid case.

Acknowledgements We thank M. Zellhöfer for fruitful discussions and technical support. Q. Xie and J. Harting acknowledge financial support from NWO/STW (STW project 13291). We thank the High Performance Computing Center Stuttgart for the allocation of computing time on Hornet and Hazel Hen.

References

1. B.P. Binks, P.D.I. Fletcher, Particles adsorbed at the oil-water interface: a theoretical comparison between spheres of uniform wettability and "Janus" particles. Langmuir **17**, 4708 (2001)
2. G.B. Davies, T. Krüger, P.V. Coveney, J. Harting, F. Bresme, Assembling ellipsoidal particles at fluid interfaces using switchable dipolar capillary interactions. Adv. Mater. **26**, 6715 (2014)
3. D. Stamou, C. Duschl, D. Johannsmann, Long-range attraction between colloidal spheres at the air-water interface: the consequence of an irregular meniscus. Phys. Rev. E **62**, 5263 (2000)
4. Q. Xie, G. Davies, F. Günther, J. Harting, Tunable dipolar capillary deformations for magnetic Janus particles at fluid-fluid interfaces. Soft Matter **11**, 3581 (2015)
5. T. Krüger, S. Frijters, F. Günther, B. Kaoui, J. Harting, Numerical simulations of complex fluid-fluid interface dynamics. Eur. Phys. J. Spec. Top. **222**, 177 (2013)
6. H. Mehrabian, J. Harting, J.H. Snoeijer, Soft particles at a fluid interface. Soft Matter **12**, 1062–1073 (2016)
7. M. Thiébaud, Z. Shen, J. Harting, C. Misbah, Prediction of anomalous blood viscosity in confined shear flow. Phys. Rev. Lett. **112**, 238304 (2014)
8. F. Janoschek, F. Toschi, J. Harting, Simplified particulate model for coarse-grained hemodynamics simulations. Phys. Rev. E **82**, 056710 (2010)
9. C. Misbah, Vacillating breathing and tumbling of vesicles under shear flow. Phys. Rev. Lett. **96**, 28104 (2006)
10. B. Kaoui, T. Krüger, J. Harting, How does confinement affect the dynamics of viscous vesicles and red blood cells? Soft Matter **8**, 9246 (2012)
11. R. Kusters, T. van der Heijden, B. Kaoui, J. Harting, C. Storm, Forced transport of deformable containers through narrow constrictions. Phys. Rev. E **90**, 033006 (2014)
12. T. Krüger, B. Kaoui, J. Harting, Interplay of inertia and deformability on rheological properties of a suspension of capsules. J. Fluid Mech. **751**, 725 (2014)
13. A.J. Liu, S.R. Nagel, Nonlinear dynamics: jamming is not just cool any more. Nature **396**, 21–22 (1998)
14. R. Benzi, M. Bernaschi, M. Sbragaglia, S. Succi, Rheological properties of soft-glassy flows from hydro-kinetic simulations. Europhys. Lett. **104**, 48006 (2013)
15. M. Gross, T. Krüger, F. Varnik, Rheology of dense suspensions of elastic capsules: normal stresses, yield stress, jamming and confinement effects. Soft Matter **10**, 4360–4372 (2014)
16. Q. Xie, G.B. Davies, J. Harting, Controlled capillary assembly of magnetic Janus particles at fluid-fluid interfaces. Soft Matter **12**, 6566–6574 (2016)
17. S. Succi, *The Lattice Boltzmann Equation* (Oxford University Press, Oxford, 2001)
18. J. Harting, M. Harvey, J. Chin, M. Venturoli, P.V. Coveney, Large-scale lattice Boltzmann simulations of complex fluids: advances through the advent of computational grids. Philos. Trans. R. Soc. Lond. A **363**, 1895 (2005)
19. X. Shan, H. Chen, Lattice Boltzmann model for simulating flows with multiple phases and components. Phys. Rev. E **47**, 1815 (1993)
20. S. Cappelli, Q. Xie, J. Harting, A.M. Jong, M.W.J. Prins, Dynamic wetting: status and prospective of single particle based experiments and simulations. New Biotechnol. **32**, 420–432 (2015)
21. A.J.C. Ladd, R. Verberg, Lattice-Boltzmann simulations of particle-fluid suspensions. J. Stat. Phys. **104**, 1191 (2001)
22. A. Komnik, J. Harting, H.J. Herrmann, Transport phenomena and structuring in shear flow of suspensions near solid walls. J. Stat. Mech: Theory Exp. **2004**, P12003 (2004)
23. F. Jansen, J. Harting, From bijels to Pickering emulsions: a lattice Boltzmann study. Phys. Rev. E **83**, 046707 (2011)
24. F. Günther, F. Janoschek, S. Frijters, J. Harting, Lattice Boltzmann simulations of anisotropic particles at liquid interfaces. Comput. Fluids **80**, 184 (2013)
25. S. Frijters, F. Günther, J. Harting, Effects of nanoparticles and surfactant on droplets in shear flow. Soft Matter **8**, 6542 (2012)

26. F. Günther, S. Frijters, J. Harting, Timescales of emulsion formation caused by anisotropic particles. Soft Matter **10**, 4977 (2014)
27. S. Frijters, F. Günther, J. Harting, Domain and droplet sizes in emulsions stabilized by colloidal particles. Phys. Rev. E **90**, 042307 (2014)
28. C.K. Aidun, Y. Lu, E.-J. Ding, Direct analysis of particulate suspensions with inertia using the discrete Boltzmann equation. J. Fluid Mech. **373**, 287 (1998)
29. F. Janoschek, J. Harting, F. Toschi, Accurate lubrication corrections for spherical and non-spherical particles in discretized fluid simulations (2016). arXiv:1308.6482
30. H. Hertz, Über die Berührung fester elastischer Körper. J. Reine Angew. Math. **92**, 156 (1881)
31. C.S. Peskin, The immersed boundary method. Acta Numer. **11**, 479 (2002)
32. T. Krüger, *Computer Simulation Study of Collective Phenomena in Dense Suspensions of Red Blood Cells Under Shear* (Springer, Berlin, 2012)
33. R. Skalak, Modelling the mechanical behavior of red blood cells. Biorheology **10**(2), 229–238 (1973)
34. W. Helfrich, Elastic properties of lipid bilayers: theory and possible experiments. Z. Naturforsch. C **28**, 693–703 (1973)
35. M. Meyer, M. Desbrun, P. Schröder, A.H. Barr et al., Discrete differential-geometry operators for triangulated 2-manifolds. Vis. Math. **3**, 52–58 (2002)
36. T. Krüger, F. Varnik, D. Raabe, Efficient and accurate simulations of deformable particles immersed in a fluid using a combined immersed boundary lattice Boltzmann finite element method. Comput. Math. Appl. **61**, 3485–3505 (2011)
37. S. Schmieschek, A. Narváez Salazar, J. Harting, Multi relaxation time lattice Boltzmann simulations of multiple component fluid flows in porous media, in *High Performance Computing in Science and Engineering '12*, ed. by M. Resch W. Nagel, D. Kröner (Springer, Berlin, 2013), p. 39
38. Q. Xie, F. Günther, J. Harting, Mesoscale simulations of anisotropic particles at fluid-fluid interfaces, in *High Performance Computing in Science and Engineering '15*, ed. by E.W. Nagel, H.D. Kröner, M.M. Resch (Springer, Berlin, 2016), pp. 565–577
39. Y. Kantor, D.R. Nelson, Phase transitions in flexible polymeric surfaces. Phys. Rev. A **36**, 4020 (1987)
40. A. Farutin, T. Biben, C. Misbah, 3d numerical simulations of vesicle and inextensible capsule dynamics. J. Comput. Phys. **275**, 539–568 (2014)
41. T. Surazhsky, E. Magid, O. Soldea, G. Elber, E. Rivlin, A comparison of gaussian and mean curvatures estimation methods on triangular meshes, in *IEEE International Conference on Robotics and Automation, 2003. Proceedings. ICRA'03*, vol. 1 (IEEE, New York, 2003), pp. 1021–1026
42. K. Tsubota, Short note on the bending models for a membrane in capsule mechanics: comparison between continuum and discrete models. J. Comput. Phys. **277**, 320–328 (2014)
43. A. Guckenberger, S. Gekle, Theory and algorithms to compute Helfrich bending forces: a review. J. Phys.: Condens. Matter **29**(20), 203001 (2017)
44. W.H. Herschel, R. Bulkley, Measurement of consistency as applied to rubber-benzene solutions, in *American Society of Test Proceedings*, vol. 26 (1926), pp. 621–633

Application and Development of the High Order Discontinuous Galerkin Spectral Element Method for Compressible Multiscale Flows

Andrea Beck, Thomas Bolemann, David Flad, Hannes Frank, Nico Krais, Kristina Kukuschkin, Matthias Sonntag, and Claus-Dieter Munz

Abstract This paper summarizes our progress in the application of a high-order discontinuous Galerkin (DG) method for scale resolving fluid dynamics simulations on the Cray XC40 Hazel Hen cluster at HLRS. We present the large eddy simulation (LES) of flow around a wall mounted cylinder, a LES of flow around an airfoil at realistic Reynolds number using a recently introduced kinetic energy preserving flux formulation and a simulation of transitional flow in a low pressure turbine. Furthermore, it provides an overview over the parallel efficiency reached by our code when using up to 49,152 CPUs and the latest developments of our DG framework.

1 Introduction

Most engineering problems in fluid dynamics contain turbulent flow. While in reality, these flows are highly unsteady and comprise a broad range of dynamic length scales, in engineering applications they are mostly treated by statistical models based on the Reynolds averaged Navier Stokes equations (RANS). This approach leads to steady partial differential equations that can be approximated with a comparatively low computational effort. Today it is widely recognized that for a broad range of engineering applications, scale-resolving simulations such as direct numerical simulation (DNS) or large eddy simulation (LES) are necessary to reliably predict anisotropy effects in turbulence, off-design conditions, bluff-body flows, transitional flows or acoustic emissions [13].

A. Beck • T. Bolemann • D. Flad • H. Frank • N. Krais (✉) • M. Sonntag • C.-D. Munz
Institute of Aerodynamics and Gasdynamics, University of Stuttgart, Pfaffenwaldring 21, 70569 Stuttgart, Germany
e-mail: beck@iag.uni-stuttgart.de; bolemann@iag.uni-stuttgart.de; flad@iag.uni-stuttgart.de; frank@iag.uni-stuttgart.de; krais@iag.uni-stuttgart.de; sonntag@iag.uni-stuttgart.de; munz@iag.uni-stuttgart.de

K. Kukuschkin
Robert Bosch GmbH, Wernerstr. 51, 70469 Stuttgart, Germany
e-mail: Kristina.Kukuschkin@de.bosch.com

© Springer International Publishing AG 2018
W.E. Nagel et al. (eds.), *High Performance Computing in Science and Engineering '17*, https://doi.org/10.1007/978-3-319-68394-2_23

However, compared to RANS, for both of the above approaches, the computational cost is significantly higher. This stems from (1) the wide range of physical scales that have to be resolved and (2) the long time intervals needed to collect converged flow statistics. Thus, DNS or LES of scientifically relevant flow problems often require HPC capacity.

Our group is active in the development of an efficient approach for and the application of LES and DNS for engineering flows. In order to efficiently meet the resolution requirement (1), it is essential to employ a simulation approach with high accuracy and computational efficiency in a massively parallel environment. We pursue this target using a high order discontinuous Galerkin spectral element method (DGSEM) in our CFD simulation framework FLEXI. Its applicability to LES and DNS has been demonstrated in past contributions, such that our focus in the last grant period was twofold:

- development of LES closure as well as stabilization approaches that synergize with the strengths of the DGSEM method,
- application of the framework for DNS and LES of complex flow problems.

In the current paper, we present simulation examples from both fields. In Chap. 2, the numerical method is briefly introduced. Furthermore, we describe current developments enhancing the high efficiency in all components of the software, maintaining the suitability of our framework for ever-enlarging computations. In Chap. 4, we present the LES of flow around a wall mounted cylinder at Reynolds number 32,000 that is used to validate RANS computations in an industrial setting. Chapter 5 presents LES results of flow around an NACA 0012 airfoil employing a recently introduced split formulation of the DG method. We show the advantages of this formulation compared to other methods to achieve non-linear stability. As a final application of our code, we present a LES of transitional flow around a low pressure turbine in Chap. 6. In this case we apply our previously presented adaptive filtering approach to a setting that is close to industrial applications. In Chap. 3 we also present scaling results that have been obtained using the current version of our simulation framework with up to 49,152 CPUs. These results include the use of a shock capturing mechanism employing finite volume sub-cells.

2   Simulation Framework

Our code FLEXI is a simulation software consequently designed for DNS and LES in an HPC context. Ongoing development ensures an optimal exploitation of the capacities of modern large-scale computer clusters. At present, typical production simulation runs are conducted using 1000–20,000 processors.

The major building block of the framework is the underlying discontinuous Galerkin spectral element method (DGSEM) [9]. In each grid element, a local high order polynomial tensor-product basis is employed to represent the solution. Numerical flux functions adopted from finite volume schemes provide the coupling between the elements. Thus, beside delivering high order accuracy and geometrical flexibility, the method is inherently parallel because only direct neighbor element data must be transferred.

Our implementation of the method in FLEXI proved that this important property of the method can directly be translated to large-scale, massively parallel simulations [1]. A recently released open-source version of our simulation framework includes a mechanism for shock-capturing that is based on finite volume sub-cells. In Chap. 3 we show that the combination of these two discretization methods does not significantly affect our scaling abilities.

Beside the parallel efficiency of the pure operator other components of the flow solver must be taken into account in order to achieve an overall optimal performance. With increasing computational power, problem sizes and the required disk space grow dramatically. Hence, a critical element during runtime is the parallel I/O, simulation results have to be written on the hard drives by all processors simultaneously. In the past, we employed the collective I/O operations of the HDF5 library. This procedure was found to be sufficient up to about 10,000 processors, but not for larger numbers of processors. For instance, the collective write operation of a 50 TB file took about 30 min on 93,840 processors.

Therefore, our parallel output routine was extended by implementing dedicated MPI groups for I/O. While in the original routine each processor had to write its data to the file, in the new version the data is collected on so-called I/O masters, which collectively write the file. The I/O groups are also used for the parallel read-in in the beginning of the computation. Different group sizes of the MPI I/O groups were investigated. While an optimal group size of 12 or 24 processors might be expected, we determined an optimal group size of about 4–6 processors. For a large number of processors (10,000 and higher), the novel I/O routine reduces the time required by I/O by more than one order of magnitude. This leaves us with an efficient parallel I/O method that is sufficient for our current calculation sizes.

In the next development step, MPI3 functionality will be integrated in FLEXI. Shared memory within the compute node potentially saves time for inner-node communication. This could enable larger I/O groups.

3 Parallel Efficiency

In this section we present the results of the investigations on the parallel efficiency of the open source version of our FLEXI code. In contrast to previous scaling test [2, 7], where only the pure discontinuous Galerkin method was analyzed, now a shock capturing with finite volume sub-cells is included. In this hybrid shock capturing approach, the DG operator is replaced by the finite volume method in all troubled elements. To prevent a large loss in resolution, the finite volume method is performed on logical sub-cells of the original elements, where one FV sub-cell is associated with each degree of freedom of the DG element. Thereby the number of DOFs is exactly the same and the data structures/variables can be used for both methods. This specific approach is used with the aim to achieve well balanced loads per element, regardless of which method is used as space operator. The finite volume operator is used in shock regions which are detected by indicator functions. Since the numerical example used for the investigations of the parallel efficiency does not require any shock capturing, we apply different synthetic distributions of DG and FV sub-cell elements.

All simulations for this scaling test where performed in March 2017 on the super-computer "Hazel Hen", a Cray XC40-system of the High-Performance Computing Center (HLRS) in Stuttgart. The FLEXI code is built with the GNU Fortran (GCC) 6.3.0 compiler and the Cray MPI library 7.5.2 with the default compiler options of the FLEXI code, except for the switch on of the FV sub-cells. The full set of compiler options is:

```
-DEQNSYSNR=2 -DFV_ENABLED=1 -DFV_RECONSTRUCT=1
-DH5DIFF=\"/opt/cray/hdf5/1.10.0.1/bin/h5diff\" -DLUSTRE
-DPARABOLIC=1 -DPP_Lifting=1 -DPP_N=N -DPP_NodeType=1
-DPP_VISC=0 -DPP_nVar=5 -DPP_nVarPrim=6 -DUSE_MPI=1
-fdefault-real-8 -fdefault-double-8 -fbackslash
-ffree-line-length-0 -DGNU -O3 -march=core-avx2
-finline-functions -fstack-arrays -Jinclude -xf95-cpp-input -fPIC
```

We have investigated the parallel efficiency of our code for a wide range of problem sizes. The numerical setup for all cases is the same except for the number of elements in the mesh. The computational domain is always a cuboid, which is discretized with 6^3 elements for the smallest case. This number of elements is increased for each successive case by doubling the number of elements in one direction. All 12 test cases are summarized in Table 1. This table also includes the minimal and maximal number of nodes (with 24 cores each) used for the scaling test of every single case. The simulation for each case is performed on all powers of 2 between the minimum and maximum number of nodes. For the largest case (case 12), the minimum number of nodes is 2, instead of 1, since the memory requirement for this case is not satisfied by a single node. The maximal number of nodes is, for all cases, chosen such that only nine elements are computed by every core.

Table 1 The parallel efficiency is investigated for different problem sizes

Case	1	2	3	4	5	6	7	8	9	10	11	12
#elementsX	6	12	12	12	24	24	24	48	48	48	96	96
#elementsY	6	6	12	12	12	24	24	24	48	48	48	96
#elementsZ	6	6	6	12	12	12	24	24	24	48	48	48
Min #nodes	1	1	1	1	1	1	1	1	1	1	1	2
Max #nodes	1	2	4	8	16	32	64	128	256	512	1024	2048

Each case has twice the number of elements than the previous one. The maximal number of nodes, a case is run on, also doubles for each case, which corresponds to only nine elements per core for all cases

Each simulation is executed for exactly 100 time steps and the performance index (PID) is measured for the pure computational time without any file input/output or initialization. The performance index is the time that is required to update one DOF on one core and can be calculated as

$$PID = \frac{\text{wall-clock-time} \cdot \#\text{cores}}{\#\text{DOF} \cdot \#\text{time steps} \cdot \#\text{RK-stages}}.$$ (1)

To investigate the influence of the different space operators, either the discontinuous Galerkin method or the finite volume sub-cells scheme, four different synthetic distributions of the DG and FV sub-cell elements are used. A pure discontinuous Galerkin computation and a pure finite volume computation, where all elements are updated with the respective operator, are performed to compare the parallel performance of these operators. Furthermore we use a checkerboard type indicator where every second element is marked for the FV method, which leads to mixed DG/FV interfaces in the whole domain. This is the worst scenario in terms of operation counts, since the mixed interfaces require additional computations. Nevertheless the distribution of DG and FV elements among the processors is more or less even. In a real world example, where the shock capturing is only required locally, this cannot be expected. Therefore, we consider a further synthetic distribution where the domain is halved in x-direction into a discontinuous Galerkin part and a finite volume sub-cells part. This leads to extra work only at the mixed interfaces, which are located at $x = 0$.

Since our investigations will always be influenced by other jobs running at the same time on the supercomputer, we used a single job to perform the simulations for the four different distributions of DG and FV elements. This makes the results comparable and to gather some statistics we furthermore repeated every single run five times. The parallel efficiency of those runs is than defined as

$$\frac{\widetilde{PID}_1}{\widetilde{PID}_P} \cdot 100\%,$$ (2)

where $\widetilde{PID}_1$ is the median of the performance indices on a single node and $\widetilde{PID}_P$ the respective value for the computations on P nodes. This parallel efficiency for all four distributions of DG and FV sub-cell elements is plotted in Fig. 1.

Each problem size is represented by a colored line, where the symbols mark the median value and the error bars show the worst/best run. These lines all start at 100% for a baseline run on a single node (except for the largest case, which uses two nodes for the baseline run). The right endpoints of every line always corresponds to a setting, where exactly nine elements reside on a single core. In the top left plot of Fig. 1 the pure discontinuous Galerkin computations shows a superlinear scaling, which can be either explained with caching/memory effects of the processors or not so good baseline runs. But there are also runs where the efficiency substantially drops below 100% and also shows a large variability. This is especially visible for runs with a high number of cores, where the amount of communication compared to local work is large. Nevertheless the upper bounds of the error bars (the best runs)

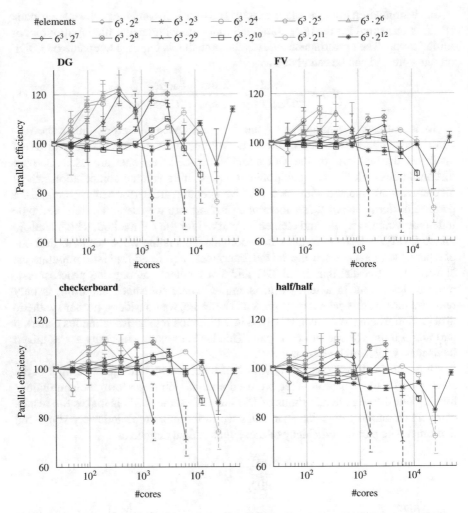

Fig. 1 Parallel efficiency of a strong scaling for different problem sizes. Each color/line corresponds to a specific mesh, for which the number of elements is given in the legend. The baseline simulation is performed on a single node (24 cores) except for the largest cases, which does not fit on a single node due to memory restrictions. For all meshes, the number of cores is doubled until each core has only nine elements. To reduce statistical effects every single simulation is repeated five times and the variability is visualized with the error bars

are quite near the 100%. Therefore, it is most likely that the interaction with other jobs running at the same time on the supercomputer affects the parallel efficiency. The other distributions of DG and FV sub-cell elements show a similar behavior, but the superlinear scaling effects are less pronounced. In the bottom right of Fig. 1 the load imbalance of the half/half distribution is visible. The parallel efficiency, compared to the other three plots, is slightly diminished. Nevertheless the results

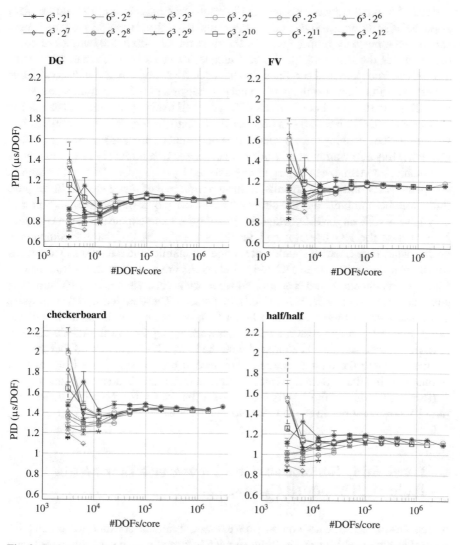

Fig. 2 Performance index of a strong scaling for different problem sizes and distributions of DG and FV elements. A good performance can be achieved for about 10,000 DOFs and more per core. Lower numbers of DOFs per core can lead to better results but the variability, caused by the sharing of network resources with other jobs, may reduce the parallel performance

are still remarkably good, as the efficiency is over 90% in most of the computations. However, these plots of the parallel efficiency present relative data only and are not comparable among the different distributions of the DG and FV elements and even not among the different problem sizes in each of the four plots.

Therefore, we present in Fig. 2 the same data from another point of view. Now the performance index is plotted as absolute values over the number of DOFs per core instead of relative values. This makes the results comparable for different problem sizes and distributions of DG and FV elements. In contrast to Fig. 1 the data is now shown the other way round, going to the right on the x-axis the number of cores decrease and the runs with the highest number of cores (nine elements per core) are the left end points of the lines. In the top left plot of Fig. 2 the discontinuous Galerkin results show that the performance is independent of the problem size, since the PID is for all lines about 1 μs/DOF. For a small load per core the PID can fall to about 0.8 μs/DOF, but also can become significantly larger. The better results can be explained by caching effects where most of the data a processor works on fits into its cache. But a low amount of work per core leads to a relatively large amount of communication which can be negatively affected by other jobs. This can be seen in the greater variability of the results. In the top right plot of Fig. 2 the results for pure finite volume computations are shown. They show the same behavior, except that they have a 15% larger PID, which is caused by a more expensive space operator. Nevertheless the DG operator and the FV sub-cells method have a comparable performance index, which should not require a dynamic load balancing. The results for distributions with mixed DG and FV elements are plotted in the bottom row of Fig. 2. In the checkerboard scenario, where every second element is a FV sub-cells element, only mixed DG/FV interfaces are present. These mixed interfaces require additional work, which results in a larger PID, but the overall behavior is still the same. However, in this case the additional work is required in the whole domain. A more realistic scenario is emulated with the half/half distribution of DG and FV elements, where the mixed interfaces are needed only locally. This leads to load imbalances, but the bottom right plot of Fig. 2 shows that this does not influence the performance index as the results are between the values for the pure DG and the pure FV computations in the top row plots.

4   Large Eddy Simulation of the Flow over a Low Aspect Ratio Wall Mounted Cylinder

In the design of modern combustion engines, Lambda sensors are crucial for controlling the fuel-injection as they provide information whether the combustion is rich or lean. To further optimize the engines Lambda sensors must be as accurate as possible. The sensors are wall-mounted and are directly located in the exhaust stream, with their height being usually similar to the local boundary layer thickness. While the internal geometry of the sensors can vary, its basic outer shape mostly resembles a surface-mounted finite-length cylinder, which are well investigated in literature, a comprehensive overview of wall-mounted cylinder flow phenomena is given by Sumner [14]. Assuming ambient conditions and low Mach numbers, the flow characteristics for these cylinders are mainly determined by two parameters: the aspect or slenderness ratio $AR = H/D$ and the relative thickness of the boundary layer δ/H, where H is the height, D the diameter and δ the boundary layer thickness. The flow field in the wake is characterized by a horseshoe vortex, emitted at the cylinder-wall junction and the tip vortex and downwash at the free end. Above a critical aspect ratio, Kármán-like vortex shedding occurs in the wake, interacting with the horse-shoe vortex and the downwash. Below the critical aspect ratio, Kármán vortex shedding is suppressed by the downwash and is replaced by symmetric arch vortex shedding, which is depicted in Fig. 3. The flow field on the free end surface is characterized by a large symmetric separation bubble forming at the leading edge and subsequent reattachment near the trailing edge, both interacting with the downwash in the wake, all these effects being illustrated in Fig. 4.

The Lambda sensor geometries are usually in a range of $0.7 < AR < 1.15$ and $1 < \delta/H < 1.48$, for the sake of simplicity we thus chose a cylinder with $AR = \delta/H = 1$ for the present numerical investigation. Despite the aforementioned extensive literature, only few published results exist which consider both, a very-

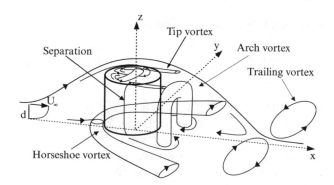

Fig. 3 Schematic of the overall flow field for a low-aspect-ratio surface-mounted cylinder [11]

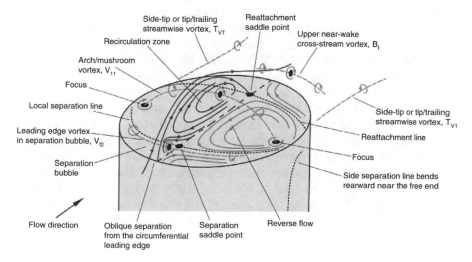

Fig. 4 Schematic of the local flow field phenomena on the free end of a surface-mounted cylinder [14]

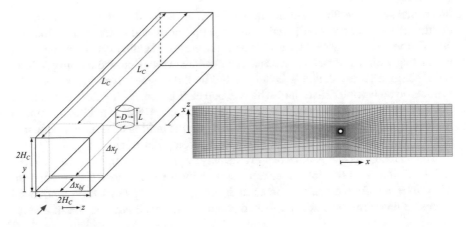

Fig. 5 *Left:* Schematics of geometrical setup. *Right:* Surface mesh used for computation

Table 2 Geometry and flow parameters of the simulation

$2H_C$	L_C^*	H_f	Δx_{bf}	Δx_f	D	D/L	u_∞	Ma_∞	$t^* = D/u_\infty$	Re_D
0.3 m	1.5 m	0.001 m	0.025 m	0.85 m	0.02 m	1	24 m/s	0.07	$8.333 \cdot 10^{-4}$ s	32,000

low-aspect-ratio cylinder with $AR \leq 1$ and a high relative boundary layer thickness $\delta/H \geq 1$ at the same time. Experimental results for this setup include the work by Kawamura et al. [8] and Tsutsui [15]. Since the experiments conducted by Kawamura are close to the intended use-case, the simulation has been designed to accurately reproduce the wind-tunnel experiments, the setup being depicted in Fig. 5 and the corresponding parameters listed in Table 2.

The wire turbulators used in the experiment to generate the turbulent boundary layer have been modeled by a zero-dimensional inner-domain pseudo boundary condition located at Δx_{bf} from the inflow, where a block inflow-profile is used. At the left, right and upper domain boundaries symmetry boundary conditions have been employed to incorporate potential blockage effects present in the experiment, without explicitly resolving the walls. At the outflow a pressure boundary is used. The flat-plate and the cylinder itself are adiabatic no-slip walls.

To save computing time the mesh has been designed to be rather fine near the inflow, where the turbulator is located, becoming coarser again in the flat-plate region and finer when approaching the cylinder. Prior to the actual cylinder simulations, calibration runs for the pure flat-plate region have been conducted using the same setup, to ensure that the chosen resolution is sufficient, for the correct boundary layer properties to develop, where especially the thickness was of major importance. Figure 6 shows the distribution of the non-dimensional wall spacings

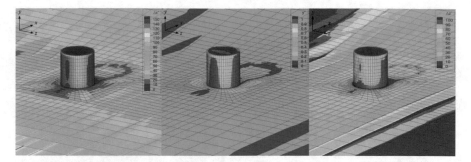

Fig. 6 Distribution of the x^+, y^+, z^+ values in the near cylinder region. x^+, z^+ computed assuming equidistant inner-cell interpolation point distribution and y^+ using exact wall-normal distance of first interpolation point

Table 3 Simulation parameters and computational cost ($t^* = D/u_\infty = 8.33 \cdot 10^{-4}$)

Case	Cores	DOF [10^6]	DOF/Core	t_{wall} [h]	t_{sim} [s]	t_{wall}/t^* [10^3 h]
N7	16,440	50.41	3066	214.5	0.2675	60.5
N5	8976	21.27	2370	82.7	0.2675	25.5
N3	1296	6.30	4861	16.4	0.2675	7.6
URANS	48	2.48	51,666	51.2	0.2881	3.0

For comparison key parameters of URANS simulation have been added

x^+, y^+, z^+ in the near-cylinder region. The block-structured mesh has been designed to guarantee $y^+ < 1$ throughout the domain and $x^+ < 120$, $z^+ < 50$ approaching the cylinder. For all simulations the mesh contains 98,472 elements, at the curved cylinder boundaries a geometry polynomial degree of $N_{geo} = 4$ has been used.

For the simulations a p-refinement strategy has been chosen, using the polynomial degrees $N = \{3, 5, 7\}$. Polynomial de-aliasing using the filter approach as described in [3] has been applied with a baseline polynomial degree of $M = \{4, 7, 9\}$. Parallel to the LES, RANS computations have been performed for validation purposes and to benchmark both performance and solution quality. For the actual design of the Lambda sensors RANS will remain the primary design tool. As a time marching scheme an explicit 4th-order low-storage Runge-Kutta method by Carpenter et al. has been employed. Roe's approximate Riemann solver has been used for computing the numerical fluxes. Time averaging has been started at $t/t^* = 215$ and data has been collected for $\Delta t/t^* \geq 100$ convective times based on the cylinder diameter, which should be sufficient for accurate statistics. The computations have been run on up to 16,440 cores with a targeted load of few 1000 DOF/core. Relevant parameters for the simulation and the associated computational cost are listed in Table 3.

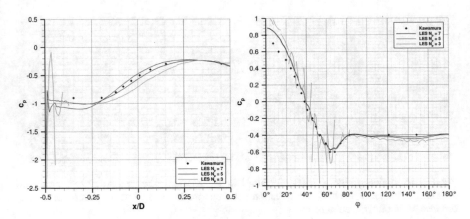

Fig. 7 Time-averaged distribution of the pressure coefficient c_p for the LES simulation. *Left:* on free-end $z/D = 0$. *Right:* on barrel $y/D = 0.5$

Figure 7 plots the distribution of the pressure coefficient on the free-end and the barrel. The $N = 5$ and $N = 7$ cases are in well agreement to the experimental results, the $N = 7$ result being closest to the reference. The oscillations which are visible for the $N = 3$ case are mostly artifacts due to the under-resolution occurring at the element boundaries. Despite these artifacts and the low resolution the simulation can reproduce the principal slope of the c_p distribution quite well.

Figure 8 depicts the instantaneous streamlines around the cylinder. As expected, a separation can be observed at the free-end of the cylinder and subsequent reattachment. The size of the recirculation region and the position of the reattachment line are in good agreement to the experimental results [8]. As predicted by the theory Kármán vortex shedding is not present, but the arch vortex in the cylinder wake as well as the horseshoe vortex are well visible. The separation line on the cylinder side bends towards the front at approximately half the cylinder height, then shifts towards the back near the free-end, as observed in the experiments. When directly comparing the LES and RANS results depicted in Fig. 9, it is clearly visible that RANS fails to reproduce the position and size of the separation bubble on the free-end. The reattachment line is strongly shifted towards the trailing edge, causing the pressure coefficient c_p to be much higher near the leading and much lower near the trailing edge.

Overall the results show good agreement with the experimental results. While the $N = 3$ case is quite under-resolved, it still captures many relevant flow features. A clear convergence towards the experimental results can be observed with increased resolution. While RANS is an order of magnitude faster than the LES, it has major

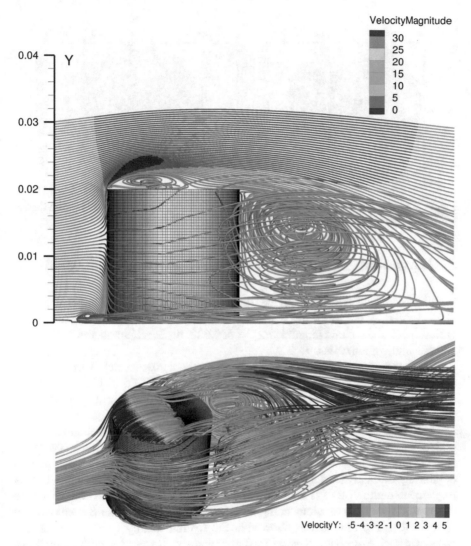

Fig. 8 Streamlines showing flow-structures around the wall-mounted cylinder and in the cylinder wake. *Top:* colored by velocity magnitude. *Bottom:* colored by vertical velocity component

shortcomings for the separated flow and fails to reproduce most of the relevant flow features on the free-end. Especially the $N = 7$ case could accurately reproduce the present experimental data, providing a further validation to these results. It can also be regarded as a viable benchmark solution for subsequent optimizations to the RANS strategy.

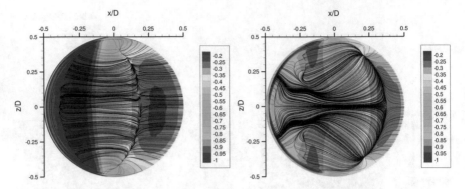

Fig. 9 Time-averaged distribution of the pressure coefficient c_p and time-averaged stream lines at the cylinder free-end. *Left:* LES $N = 7$. *Right:* RANS

5 Large Eddy Simulation of an Airfoil at Reynolds 660,000

In this project we have computed the flow around a NACA 0012 airfoil. The flow is characterized by the following non-dimensional quantities. The Reynolds number with respect to the chord length is $Re_c = 660,000$, the Mach number is $Ma = 0.12$ and the angle of attack is $\alpha = 2°$.

For the computation we use the LES methodology, which was originally developed and described in [5]. The key for successful LES at higher Reynolds numbers is to use kinetic energy preserving flux formulations. Such split fluxes are widely used in finite difference CFD codes, but were only recently introduced for DG methods. The key property of these flux formulations is their inherent stability, making additional numerical viscosity for stable computations unnecessary. There is also a significant gain in computational efficiency compared to using polynomial de-aliasing normally used to obtain non-linear stability. The split flux approach was implemented into FLEXI at the end of this report period, and here we compare some of the first results to previous computations obtained with the LES methodology from in [6]. The simulations show that the overall quality is better, while the computational cost could even be further decreased. So far, for both simulations an implicit approach for the subgrid-scale (SGS) stress is used, relying solely on the dissipation of the added inter-cell Riemann solvers, in this case Roe's approximate matrix dissipation. We will explore the effect of explicit SGS stress terms in the following period of the project, also further increasing the Reynolds number. The computation is run on an unstructured hexahedral grid with 138,950 cells and a polynomial degree of $N = 7$, in total 71,142,400 degrees of freedom. The flow is tripped at 5% chord using a geometrical step on the airfoil surface. For a qualitative impression of the flow Fig. 10 shows iso-surfaces of the Q-criterion within the boundary layer colored by velocity magnitude. To validate the simulation we compare the time-averaged wall friction and pressure coefficient to those obtained from a 2D panel method computation using the program XFOIL [4]. We observe a

Fig. 10 ISO-surface of "Q-criterion" (Q= 1000) colored by velocity magnitude: LES NACA 0012, $Re = 660{,}000$

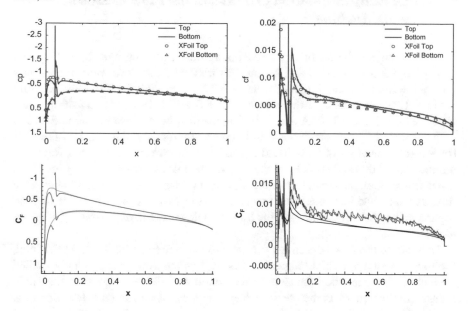

Fig. 11 Dimensionless pressure (left) and wall-friction (right). XFOIL: black, LES: suction sideblue, pressure side red

good agreement with the XFOIL (Fig. 11) computation for the pressure coefficient, with differences in the front part of the wing, as the step is not modeled in the 2D calculations. Compared to the previous results we use smaller cells in the boundary layer, which significantly reduces the unsteady shape of the wall friction results. The novel method can make use of the full polynomial discretization, whereas previous

Table 4 Computational
details for the LES of NACA
0012 aeroacoustic case

N	n_{DOF}	Procs	PID [μs]	Runtime/T^*
11	$3.2 \cdot 10^7$	9423	1.4	3.2 h
7	$7.1 \cdot 10^7$	8688	1.18	2.63 h

methods had to sacrifice parts of the resolution for the filter, to obtain non-linear stability. This remarkable property directly results in higher accuracy per effective DOF. The integral dimensionless pressure is less sensitive to discretization errors and is in good agreement for either method. In Table 4 the computational details are listed. For the current simulation the total number of degrees of freedom is about twice as large compared to the former one, yet the simulation is by a factor of approximately 1.3 less computational expensive. This gain in efficiency can predominantly be attributed to the larger time step obtained with the new method.

6 Large Eddy Simulation of Transitional Flow in a Low Pressure Turbine

In this subproject, the adaptive de-aliasing approach presented in [6] and applied in [1] has been extended from canonical turbulent flows to a more industrially relevant setting, the simulation of transitional flow through the T106 low pressure turbine (LPT) cascades. In this test case, DNS and LES of the transitional and separated flow on the T106A and T106C high-lift subsonic turbine cascades are to be investigated. This cascade is a well-known setup for assessing transition models for Reynolds numbers of 50,000 and beyond. For the current simulations, Reynolds numbers of 60,000 and 80,000 are chosen. As the inlet turbulence is very low, both flows feature laminar separation and a slow natural transition. Figure 12 shows the time-averaged and instantaneous flow field for the T106A case, where the laminar separation on the suction side and the transition downstream of the trailing edge are visible.

In Table 5, the flow conditions for the two cases are summarized. At the inlet, the total pressure and total temperature as well as the flow direction are specified through a characteristic-based boundary condition [12]. By choosing the static pressure at the outlet (i.e. the pressure drop over the cascade) and the flow viscosity, the corresponding isentropic flow conditions can be set. This particular set of boundary conditions mimics the experimental setup in a blow-down wind tunnel.

Figure 13 shows the mesh used for the T106A case (left) and the inlet and outlet Mach numbers (right). The simulation is conducted on the computational grid (shown in black), with periodic boundary conditions chosen in the spanwise and crossflow direction. The mesh consists of 5706 hexahedral elements (4359 for T106C), using a polynomial of degree $N_{geo} = 3$ to approximate the geometry. Since

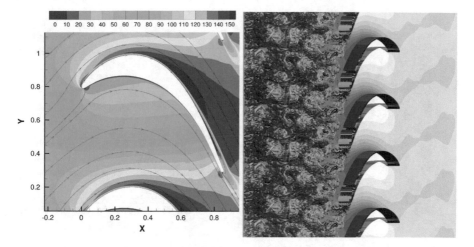

Fig. 12 *Left:* Time- and spanwise averaged velocity magnitude and streamlines for the T106A case. *Right:* Instantaneous isocontours of vorticity magnitude

Table 5 Flow conditions for the LPT

Geometry	T106A	T106C
$Re_{2,s}$	60,000	80,000
$M_{2,s}$	0.405	0.65
α_1	45.5	32.7

$M_{2,s}$ denotes the isentropic exit Mach number and $Re_{2,s}$ the Reynolds number based upon the velocity obtained by isentropic expansion and the airfoil chord

the boundary conditions in the flow direction are chosen as described above, the prescribed quantities according to Table 5 are not enforced directly, but instead establish themselves once the flow has reached a quasi-steady state (neglecting non-isentropic processes along the flow path). To check if this is the case, the inlet and outlet Mach numbers are plotted over approx. 40 characteristic time units in Fig. 13 (right). The instantaneous quantities have been averaged over the inflow and outlet faces for each time step. For both cases, the resulting Mach numbers at inflow and outflow are in very close agreement with the specifications.

A number of simulations for the T106A/C cases were run. A relevant selection of cases is shown in Table 5. One focus of this investigation was to extend the approach described in [6], in which the de-aliasing operator is coupled to a cell-local indicator, which triggers the de-aliasing only when necessary. In this investigation,

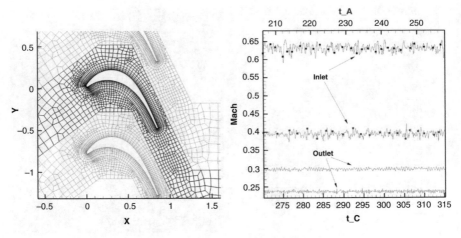

Fig. 13 *Left:* Close-up of the computational grid for the T106A case. The actual grid is shown in black, the blue grids show the periodic continuation of the domain in the *y*-direction. *Right:* Temporal evolution of Mach number (averaged over inlet and outlet faces). Blue lines denote the T106A case, black lines T106C case. T_a and T_c denote characteristic flow times per chord length c/u_2

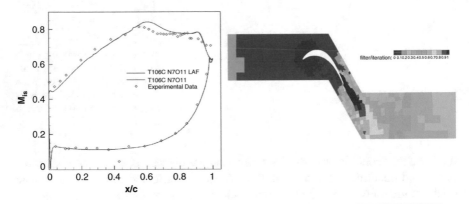

Fig. 14 *Left:* Comparison of isentropic blade Mach number $M_{is} = \sqrt{\frac{2}{\gamma-1}\left(\frac{p_{t1}}{p}^{\frac{\gamma}{\gamma-1}} - 1\right)}$ for T106C with experimental data from [10]. *Right:* Ratio R of de-aliasing operations to operator evaluation: $R = 0$: no de-aliasing during computation, $R = 1$: de-aliasing at every iteration step

this approach, which is labeled "local adaptive filtering" (LAF) is compared to the standard de-aliasing, which is applied at every iteration and every element (standard over-integration).

In Fig. 14 (left), the isentropic Mach number along the turbine blade (a measure of the pressure losses) is compared against experimental data from [10]. The overall agreement on the pressure side is excellent, while some discrepancies occur on the

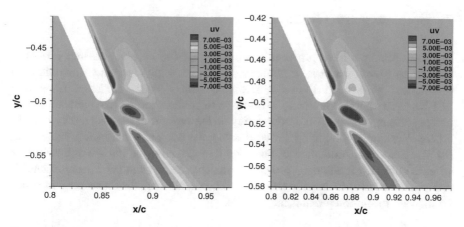

Fig. 15 Close up of $u'v'$ stresses at the trailing edge. *Left:* De-Aliasing at every element and operator iteration (Case N7O11) *Right:* Locally adaptive de-aliasing (Case N7O11 LAF)

Table 6 Computational details for the LES of T106A/C LPT case

Case	n_{DOF}	Procs	PID [μs]	CPUh/T^*
T106A-N7O11	$9.9 \cdot 10^6$	5706	1.5	450
T106A-N7O11 LAF	$9.9 \cdot 10^6$	5706	1.5	720
T106C-N7O11	$7.5 \cdot 10^6$	4359	1.4	310
T106C-N7O11 LAF	$7.5 \cdot 10^6$	4359	1.4	534

suction side near the point of separation. There is also a perceivable difference between the LAF and over-integration results in this region. This is supported by Fig. 14 (right), in which the ratio of filtering/de-aliasing operations per element to the iteration number is plotted for LAF. Note that for over-integration, this ratio would be equal to 1 everywhere. Here, the indicator triggers the de-aliasing in a physically motivated sense: Upstream of the blade and along the laminar suction side, no de-aliasing occurs, while in the separation region and the wake, full de-aliasing is active. This selective approach also influences the structure of the near wake and is less dissipative on the small scale structures, as shown in Fig. 15.

Table 6 summarizes the computational details for these investigations. The difference in computing time between the LAF and over-integration cases is due to the stricter timestep requirement of the LAF and the increase in computational costs due to the evaluation of the indicator.

This investigation has shown that the local adaptive de-aliasing approach is generally applicable and robust for non-trivial cases. Further investigations into its efficiency with regards to higher time step penalty are necessary. Also, for achieving LES of higher Reynolds numbers, additional modeling will be required.

7 Summary and Outlook

In this report, we have provided a summary of the current research efforts in our group geared towards accurate and efficient scale-resolving simulations of turbulence and acoustics. In the last grant period, our simulation framework FLEXI has matured sufficiently to allow us to conduct large-scale simulations and postprocessing that push the boundaries of the current available hardware at the HLRS. We aim at continuing our work along the two paths outlined in this report: (a) The numerical simulation of unsteady flow phenomena that are at the limit of what is currently feasible, to provide a framework for simulation-based scientific exploration and (b) to develop new modelling capabilities for those cases in which a full scale-resolution remains impractical.

Acknowledgements The research presented in this paper was supported in parts by the Deutsche Forschungsgemeinschaft (DFG) and the Boysen Stiftung. We truly appreciate the ongoing kind support by HLRS and Cray in Stuttgart.

References

1. M. Atak, A. Beck, T. Bolemann, D. Flad, H. Frank, F. Hindenlang, C.-D. Munz, Discontinuous Galerkin for high performance computational fluid dynamics, in *High Performance Computing in Science and Engineering '14* (Springer International Publishing, Berlin/Heidelberg, 2015), pp. 499–518
2. M. Atak, A. Beck, T. Bolemann, D. Flad, H. Frank, C.-D. Munz, High fidelity scale-resolving computational fluid dynamics using the high order discontinuous Galerkin spectral element method, in *High Performance Computing in Science and Engineering '15* (Springer International Publishing, Cham, 2016), pp. 511–530
3. A.D. Beck, T. Bolemann, D. Flad, H. Frank, G.J. Gassner, F. Hindenlang, C.-D. Munz, High-order discontinuous Galerkin spectral element methods for transitional and turbulent flow simulations. Int. J. Numer. Methods Fluids **76**(8), 522–548 (2014)
4. M. Drela, Xfoil: an analysis and design system for low Reynolds number airfoils, in *Low Reynolds Number Aerodynamics* (Springer, Berlin/Heidelberg, 1989), pp. 1–12
5. D. Flad, G. Gassner, On the use of kinetic energy preserving DG-schemes for large eddy simulation. J. Comput. Phys. **350**, 782–795 (2017)
6. D. Flad, A. Beck, C.-D. Munz, Simulation of underresolved turbulent flows by adaptive filtering using the high order discontinuous Galerkin spectral element method. J. Comput. Phys. **313**, 1–12 (2016)
7. F. Hindenlang, Mesh curving techniques for high order parallel simulations on unstructured meshes, PhD thesis, Universität Stuttgart, 2014
8. T. Kawamura, M. Hiwada, T. Hibino, I. Mabuchi, M. Kimada, Flow around a finite circular cylinder on a flat plate: cylinder height greater than turbulent boundary layer thickness. Bull. JSME **27**(232), 2142–2151 (1984)
9. D.A. Kopriva, G. Gassner, On the quadrature and weak form choices in collocation type discontinuous Galerkin spectral element methods. J. Sci. Comput. **44**, 136–155 (2010)
10. J. Michálek, M. Monaldi, T. Arts, Aerodynamic performance of a very high lift low pressure turbine airfoil (T106C) at low Reynolds and high Mach number with effect of free stream turbulence intensity. J. Turbomach. **134**(6), 061009–0610010 (2012)

11. R.J. Pattenden, N.W. Bressloff, S.R. Turnock, X. Zhang, Unsteady simulations of the flow around a short surface-mounted cylinder. Int. J. Numer. Methods Fluids **53**(6), 895–914 (2007)
12. J.W. Slater, Verification assessment of flow boundary conditions for CFD analysis of supersonic inlet flows, in *37th Joint Propulsion Conference and Exhibit* (2002), p. 3882
13. J. Slotnick, A. Khodadoust, J. Alonso, D. Darmofal, W. Gropp, E. Lurie, D. Mavriplis, CFD vision 2030 study: a path to revolutionary computational aerosciences. Technical report, NASA Langley Research Center, Hampton, VA (2014)
14. D. Sumner, Flow above the free end of a surface-mounted finite-height circular cylinder: a review. J. Fluids Struct. **43**, 41–63 (2013)
15. T. Tsutsui, Flow around a cylindrical structure mounted in a plane turbulent boundary layer. J. Wind Eng. Ind. Aerodyn. **104–106**, 239–247 (2012). 13th International Conference on Wind Engineering

Part V
Transport and Climate

Markus Uhlmann

The simulations in the category "Transport and Climate" feature a broad spectrum of studies in numerical weather forecast and climate modelling. In this granting period there have been five projects running on HazelHen (HLRS) and two on ForHLR II (SCC), consuming nearly 80 million core-hours in total. Community modelling is clearly the paradigm here, and all projects have been realized with the aid of only three numerical model families (COSMO, WRF and AWI-CM).

In the first of the following contributions (Schädler et al.) the DWD model "COSMO" is used in climate mode in order to investigate trends in past and future regional climate in the broad area of central Europe. The authors address various aspects, such as energy considerations in urban scenarios, paleoclimate and the interaction between land surface and the atmosphere. Their high resolution simulations (with grid width of 2.8 km) show a significantly better agreement with precipitation measurements.

The following two contributions were realized with the aid of the model "WRF". The authors of the second project (Warrach-Sagi et al.) have investigated (global) climate change scenarios for a period until the end of the twenty-first century, employing spatial resolutions down to 3 km grid sizes. This effort is embedded in two project consortia, and when completed it will provide highly sought-after data to the community of climate impact researchers.

Finally, the project by Schwitalla et al. is focused upon a faithful reproduction/prediction of extreme events in climate research by using grids at convection permitting scales of a few kilometers. They have performed highly-resolved

M. Uhlmann (✉)
Institute for Hydromechanics, Karlsruhe Institute of Technology, Kaiserstr. 12, 76131 Karlsruhe, Germany
e-mail: markus.uhlmann@kit.edu

simulations of the entire tropical region over a seasonal time scale. The preliminary results suggest a very good correspondence with the available observation data. This project demonstrates once more the benefit of increasing availability of HPC resources upon the predictive capabilities in the context of climate research.

Regional Climate Simulations with COSMO-CLM: Ensembles, Very High Resolution and Paleoclimate

G. Schädler, H.-J. Panitz, E. Christner, H. Feldmann, M. Karremann, and N. Laube

Abstract The IMK-TRO (KIT) presents in the HLRS annual report for 2016–2017 projects and their results using the CRAY XC40 "Hazel Hen". The research focuses on the very high resolution regional climate simulations including the modeling of land surface processes and urban climate, the generation of ensemble projections, and regional paleoclimate (PALMOD). The simulations are performed with the regional climate model (RCM) COSMO-CLM (CCLM) and cover spatial resolutions from 50 to 2.8 km. Within the projects, the standard CCLM is enhanced; for the analysis of the impact of different soil-vegetation transfer schemes (SVATs) VEG3D is coupled via OASIS3-MCT to CCLM. For the PALMOD project, a special isotope-enabled version CCLMiso is used. To highlight the added value, the results of the higher resolution climate predictions are compared to those of simulations with coarser resolutions. In addition, the impact of different global driving data sets is investigated. Climate projections are performed for two future time slices, 2021–2050 and 2071–2100. The urban climate and its change are also investigated using very high resolution simulations to enable the energetic optimisation of buildings. The required Wall-Clock-Times (WCT) range from 9 to 2000 node-hours per simulated year.

1 Overview

The working group "Regional Climate and Water Cycle" of the Institute of Meteorology and Climate Research—Tropospheric Research (IMK-TRO) at the Karlsruhe Institute of Technology (KIT) (www.imk-tro.kit.edu) uses the climate version of the COSMO model (COSMO-CLM) on the CRAY XC40 'Hazel Hen' at the HLRS high performance computing facilities to investigate past, present and

G. Schädler (✉) • H.-J. Panitz • E. Christner • H. Feldmann • M. Karremann • N. Laube
Institut für Meteorologie und Klimaforschung Forschungsbereich Troposphäre (IMK-TRO),
Karlsruher Institut für Technologie (KIT), Karlsruhe, Germany
e-mail: gerd.schaedler@kit.edu

W.E. Nagel et al. (eds.), *High Performance Computing in Science
and Engineering '17*, https://doi.org/10.1007/978-3-319-68394-2_24

future regional climate with a focus on subregions of Central Europe and Germany. Topics include

- Very high resolution regional climate prognoses and projections
- Urban climate and energetic optimisation of buildings
- Analysis of climate extremes (floods, draughts)
- Assessment of uncertainty via ensemble simulations
- Interaction between land surfaces and the atmosphere
- Simulation of regional paleoclimates

In this report, we describe in some detail three of these topics which are interesting for both their practical relevance and their numerical/computational interest. First, the very high resolution regional climate projections and the simulations concerning urban climate and energetic optimization of buildings; these projects are third-party funded (KLIMOPASS, KLIWA and Baden Württemberg Stiftung). Second, the PALMOD Project, funded by the BMBF, will be presented and the structure of the regional paleo climate simulations illustrated. Third, first results and details to using stable water isotopes to evaluate temperature in paleo simulations are given. Additionally, a description and latest results of the CORDEX project are included.

The text is structured as follows: Sect. 2 describes the model used and Sect. 3 contains the CORDEX (Sect. 3.1) and the PALMOD project (Sect. 3.2), and also the very high resolution simulations (Sect. 3.3).

2 The CCLM Model

The regional climate model (RCM) COSMO-CLM (CCLM) is the climate version of the operational weather forecast model COSMO (Consortium for Small-scale Modeling) of DWD. It is a three-dimensional, non-hydrostatic, fully compressible numerical model for the atmosphere. The model solves prognostic equations for wind, pressure, air temperature, different phases of atmospheric water, soil temperature, and soil water content.

Further details on COSMO and its application as a RCM can be found in [1, 2] and on the web-page of the COSMO consortium (www.cosmo-model.org).

3 Regional Climate Simulations Using the HLRS Facilities

3.1 CORDEX (Coordinated Regional Downscaling Experiment) Project

Since the United Nations climate change conference in Paris in December 2015 (COP21) the optimistic Representative Concentration Pathway 2.6 (RCP2.6, [3])

Fig. 1 The EURO-CORDEX
domain; see also
www.cordex.org

came into the focus as the future scenario for which the global mean temperature increase can be kept below +2 °C [4]. RCP2.6 assumes that global annual Green-House-Gas GHG emissions (measured in CO2-equivalents) peak between 2010–2020, with emissions declining substantially thereafter. In order not only to consider the global aspect, there is also an increasing demand to study the response of regional climate to RCP2.6. Up to now, the matrix of regional climate simulation under RCP2.6 is rather sparse, however. Therefore, we regionalized results of the global model EC-EARTH, realization 12, with RCP2.6 for the EURO-CORDEX region (Fig. 1). The spatial resolution of these CCLM simulations is 0.11° (about 12 km) and spans the period from 2006 until 2100. The GCM EC-EARTH and its realization 12 have been used because the same global model in its realization 12 has already been downscaled with CCLM on the same 0.11° domain, but assuming the future scenarios RCP4.5 and RCP8.5 [5].

The impact of the different emission scenarios is most pronounced in the future trends of the near-surface temperature in 2 m height (Fig. 2). For the future scenario RCP 8.5 and averaged over the whole EURO-CORDEX domain (Fig. 2, red curve), the mean yearly temperature in 2 m height increases up to about 4.5 °C at the end of the twenty-first century, relative to a 30 years reference period from 1971 until 2000. For RCP4.5, the temperature is also rising, but seems to stabilize at the end of the current century. However, applying RCP2.6, the temperature increases slightly until around 2060, and then seems to decrease again.

Note that the example shown in Fig. 2 represents temporal and spatial averages, temporal averages over each year, and spatial ones over the whole model domain. Considering different seasons and different regions in Europe, the results will be

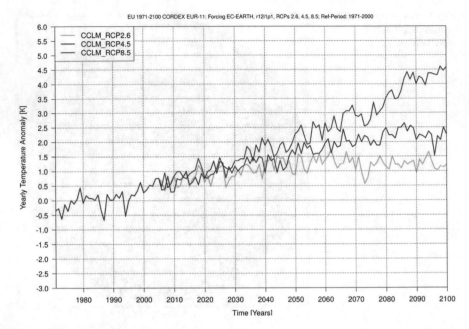

Fig. 2 Yearly 2 m temperature anomaly under different future emission scenarios. Reference period: 1971–2000, regional climate model: CCLM, Version 4-8-17; driving GCM: EC-EARTH, realization 12. Green line: result assuming RCP2.6, Blue line: result assuming RCP4.5, red line: result assuming RCP8.5

qualitatively similar to those shown in Fig. 2, but quantitatively they might differ considerably from season to season and from region to region. A more detailed analysis of the results is still ongoing. Note too, that the basis of the results discussed so far is the regionalization of a single GCM by only one RCM. To investigate the robustness of the results it is necessary to consider a multi member ensemble consisting of several GCMs that have been downscaled by several RCMs. The climate simulations described here will be part of such a multi member ensemble. A final analysis of this ensemble is pending. The HPC resources needed for the CORDEX simulations were in the order of 32,000 node-hours for the computations, and 90 TB for the total data storage. It should also be mentioned that these simulations had been used by members of the CRAY staff at HLRS to analyse the variations of wall-clock-time (WCT) we reported in our HLRS report 2015 [6]. In order to carry out the analysis it was necessary to implement in the code special routines for time-measurements, which sent real-time information to the CRAY staff. We do not know whether any causes for the WCT variations could be detected. However, WCT variations up to 20% and more still exist.

3.2 Regional Paleoclimate Simulations

The Paleo Modeling project PALMOD (www.palmod.de) aims to improve the understanding and modeling of paleoclimate system dynamics and variability. Therefore, fundamental processes determining the Earth's climate trajectory and variability during the last glacial cycle need to be identified and quantified. Our working group contributes to two work packages within the project: "Physical System: Scale Interactions" and "Proxy Data Synthesis/Data-Model Interface". The aim of the first work package is to identify and understand relevant subgrid scale processes, which significantly contribute to the Greenland ice sheet surface mass balance. Such processes can not be resolved with coarse resolution GCMs and, therefore, need to be parameterized. In the second work package, we apply CCLM to validate regional simulations of paleoclimates against water isotope records and provide a combined model-data framework for the interpretation of proxy data.

3.2.1 Modeling of the Greenland Surface Mass Balance

For the identification of subgrid scale key processes, we adapted and set up the regional climate model CCLM for the Greenland region. The model was driven with ERA-Interim reanalysis data for the period of 1992–2015 using the SVAT TERRA-ML. Sensitivity tests concerning e.g. the modeling domain, the time step, the horizontal resolution or the inclusion of sea ice were performed.

Comparisons between simulations performed with CCLM and observation data show a generally good agreement throughout Greenland. However, the quality of simulations depends on the analysed region (Fig. 3a), the season and the variable (precipitation or temperature). Simulations are also sensitive to the modeling domain, the resolution and the impact of sea ice. Best agreement of simulations and observation data (Fig. 3a) is found for the CORDEX-Arctic region (Fig. 3b) as modeling domain on a resolution of 25 km and a time step of 150 s. Moreover an increase of the value of the maximum albedo as well as the implementation of sea ice lead to better agreement with observation data. Fig. 4a, b show a comparison of the CCLM results with three different observational data sets for the southwestern region in February (Fig. 4a) and the northern region in June (Fig. 4b). There is a good agreement in the medians in both cases, but a larger data spread within and between the different data sets.

3.2.2 Modeling of Stable Water Isotopes in the Arctic Region

Stable isotopes of atmospheric water such as $H_2^{16}O$ and $H_2^{18}O$ are fractionated during phase changes. This is measured as $\delta^{18}O = R_{18}O/R_{18O,VSMOW} - 1$ with the isotope ratio $R_{18O} = [H_2^{18}O]/[H_2^{16}O]$ and the ocean water ratio $R_{18O,VSMOW} = 0.002005$.

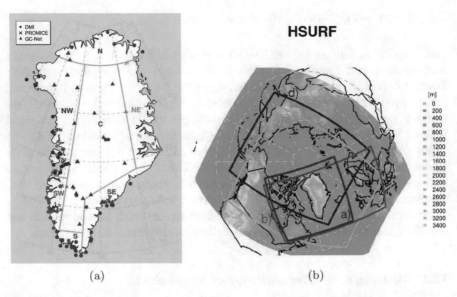

(a)   (b)

Fig. 3 (**a**) Location of measuring stations used for validation of the model: Danish meteorological Institute (DMI; dots), Programme for Monitoring of the Greenland Ice Sheet (PROMICE; x), and of the Greenland Climate network (GC-Net; triangles). Colours show seven analyzed regions of Greenland: north (N; blue), northwest (NW; dark blue), central (C; brown), northeast (NE; orange), southwest (SW, green), south (S; red), and southeast (SE, purple). (**b**) Orography [m] and modeling domains: a) region around Greenland. b) Region covering almost region a) but extended towards Europe. c) Modeling area covering parts of Europe and crossing the north pole. d) CORDEX-Arctic region

In this context, an effect of particular importance is the gradual decrease of $\delta^{18}O$ in cooling and raining air masses, which are depleted in $H_2^{18}O$ because of a preferential fractionation of the heavier $H_2^{18}O$ into the condensate. The resulting general relation between air temperature and the $\delta^{18}O$ of water vapor or precipitation (e.g.[7, 8]), in turn, allows reconstructing variations of air temperature in the past from Arctic ice sheets, which record the $\delta^{18}O$ of precipitation from thousands of centuries [9].

Within the PALMOD project, the relation between $\delta^{18}O$ and temperature is used for validating paleo simulations with general circulation models (GCMs) against the $\delta^{18}O$ in ice core samples from Greenland. To allow a direct comparison between the model and the $\delta^{18}O$ in ice cores, isotope physics was implemented into the employed GCMs. To account for the relatively low horizontal resolution of the GCMs, KIT uses an isotope-enabled version [10] of the regional CCLM model (CCLMiso) for a dynamical downscaling of the global paleo simulations.

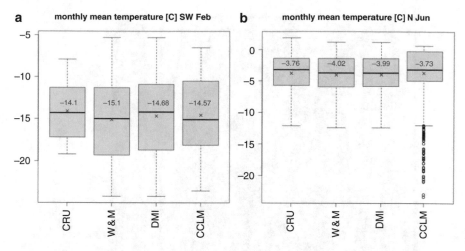

Fig. 4 Box-Whisker-plots for monthly mean temperature in 2 m for "Climate Research Unit" (CRU), "Willmott and Matsuura" (W & M), "Danish meteorological Institute" (DMI) and CCLM simulations on 25 km resolution using the CORDEX-Arctic region as modeling area. Blue: monthly mean of all grid points (**a**) south west Greenland in February 2001. (**b**) North in June 2001

In a first step, we validated present-day CCLMiso simulations for the Arctic region against the $\delta^{18}O$ of top core samples from ice cores and snow pits (Fig. 5). CCLMiso is capable of capturing the observed spatial distribution of $\delta^{18}O$. The RMSE between the modeled and the observed $\delta^{18}O$ is only 1.6‰. For the arctic region, both findings confirm that the most important isotpe physics is reliably implemented. In a next step, the validated CCLMiso may therefore be used for paleo simulations of the Arctic region and a comprehensive translation of the $\delta^{18}O$ in Greenland ice cores into paleo temperatures.

Table 1 summarizes the computing demands on CRAY XC40 "Hazel Hen" at HLRS for the simulations. Due to the increase of prognostic equations, the computational costs of isotope modeling are considerably higher than comparable simulations that do not consider isotopes.

3.3 Very High Resolution Simulations

3.3.1 Added Value of Very High Resolution Simulations

As already done operationally for weather forecasts, climate simulations at very high spatial horizontal resolution in the order of 1–3 km are becoming now feasible. Apart from the increasing demand for such simulations—they reduce the scale gap between climate models and impact models (urban climate, hydrological and agricultural models often run at resolutions of a few hundred meters)—there

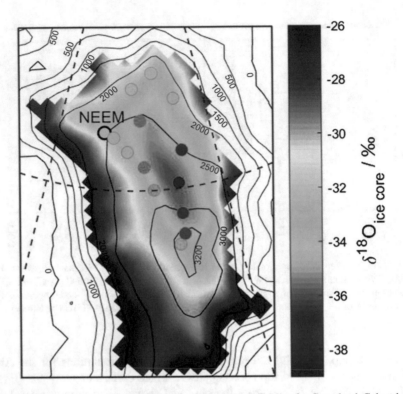

Fig. 5 Regional present-day (2000–2014) simulation with CCLMiso for Greenland. Colored area: accumulation-weighted annual model values of $\delta^{18}O$; colored dots: respective annual $\delta^{18}O$ in top core samples from ice cores and snow pits of [11]

Table 1 Project PALMOD: Summary of computing demands on CRAY XC40 "Hazel Hen" at HLRS

Project	Domain size	Grid resolution (km)	Time-step (s)	Computing time (node-h/ year)	No. of sim. years	Storage needs (Gbyte/ year)	Total computing time (node-h)	Total storage needs (Tbyte)
PALMOD	246 * 120 * 50	50	120	190	350	300	66,500	11

are also physical and numerical pros for very high resolution: a considerable reduction, if not an elimination of the model bias, i.e. the systematic difference between model results and observations, can be obtained. This is due to a better representation of orography and land cover as well as an explicit calculation of deep convective precipitation instead of a parameterisation. Moreover, a high spatial resolution induces a more realistic small scale temporal and spatial variability of meteorological variables due to a better representation of the orography and land use patterns, which is especially true in complex terrain. This small scale variability is also important to capture weather extremes more realistically and to better represent persistent and therefore climatically relevant small scale features like local/regional wind, precipitation systems, and temperature regimes.

In [12], it had already been demonstrated that the quality of the model results could be improved due to a refinement of the resolution of the climate model. Altogether, model simulations at 2.8 km horizontal resolution provide an added value compared to coarser resolutions (for example, 7 km) for temperature, precipitation, relative humidity, global radiation, and regional wind systems, even though the magnitude of the added value varies with variable, season, region, altitude and statistics [12]. This makes the high resolution data more suitable for applications and impact studies on regional or local scales where high resolution is required, e.g. in cities or small river catchments. It can also be expected that higher variability improves the tails of probability distributions, i.e. the representation of extremes.

After having shown that there is an added value of very high resolutions, we started to work on two projects involving very high resolution simulations.

3.3.2 Simulations of Present and Future Climate

In the frame of the project KLIWA (Klimaveränderung und Konsequenzen für die Wasserwirtschaft) we started to generate very high resolution (about 2.8 km) regional climate projection ensembles focusing on an area that comprises the basins of major rivers of the southern part of Germany (e.g. Rhine, Moselle, Danube, Inn). The first GCM that we downscaled in a three-stage nesting approach was ECHAM6, realisation 1. The three-nest approach regionalized the GCM data down to a resolution of 50 km in the first step. The second step uses the RCM result of the first step and performs the downscaling to 7 km, and in the third step the final resolution of 2.8 km is reached. The historical control period ranges from 1968 to 2005, for the future we consider the periods 2018 until 2050 and 2068 until 2100 using emission scenario RCP8.5 [3]. For all three periods, the first 3 years are considered as spin-up time, and they are not used in the analyses of the results. In order to establish a small ensemble, two other GCMs, namely HadGEM2-ES, realization 1, and EC-EARTH, realization 12, have been downscaled, again applying the three-nest approach. In its finest resolution of 2.8 km the downscaling of HadGEM2-ES is already finished for the historical control period. The EC-EARTH driven simulation for the historical period is close to its end.

Table 2 summarizes the computing demands on CRAY XC40 "Hazel Hen" at HLRS for the three-nest downscaling of one GCM, taking into account the number of simulation years of all three periods considered.

The model domain for the very high resolution (2.8 km) simulations covers large parts of Germany and neighboring regions, having a size of 322 * 328 horizontal grid-points. The analyses of results are performed in the region marked by the red box in Fig. 6.

Table 2 Project KLIWA2.8: Summary of computing demands on CRAY XC40 "Hazel Hen" at HLRS for the three-nest downscaling of one GCM

Project	Domain size	Grid resolution (km)	Time-step (s)	Computing time (node-h/year)	No. of sim. years	Storage needs (Gbyte/year)	Total computing time (node-h)	Total storage needs (Tbyte)
Very high resolution	118 * 110 * 40	50	360	9	203	115	1827	24
KLIWA 2.8 km Ensemble in three nest approach	165 * 200 * 40	7	60	33	203	300	6699	61
	322 * 328 * 49	2.8	25	2000	104	500	208,000	52
Total amount				2042	510	915	216,526	137

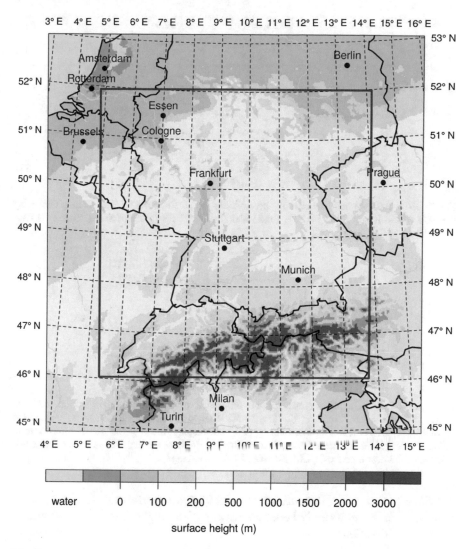

Fig. 6 KLIWA project. Actual model domain for very high resolution simulations and its orography. The red box encloses the region where results are analysed

Figure 7 displays the annual mean daily precipitation sums for 1971–2005, resulting from CCLM simulations with MPI-ESM-LR, EC-EARTH and HadGEM2-ES forcings, and compares them with the gridded EOBS V.11 observations [13]. This comparison reveals a general wet bias of the model results, whose amount strongly depends on the driving GCM.

In Fig. 8, a comparison between the CCLM_MPI-ESM-LR third nest (2.8 km) and the CCLM_MPI-ESM-LR second nest (7 km) to HYRAS observations [14] is displayed in terms of whisker box plots indicating the 25, 50 and 75 percentiles.

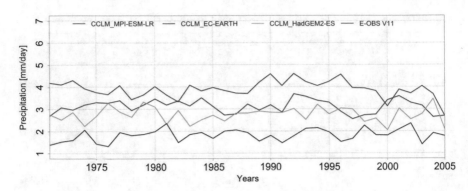

Fig. 7 1971–2005 Annual mean daily precipitation sums for three CCLM simulations using the MPI-ESM-LR (red), EC-EARTH (blue) and HadGEM2-ES (green) forcing for the KLIWA evaluation area

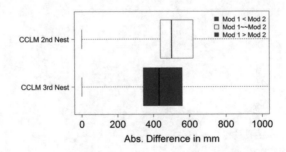

Fig. 8 1971–2000 Boxplot comparing the difference of the annual precipitation sums of CCLM_MPI-ESM-LR third nest (2.8 km) to HYRAS to the difference CCLM_MPI-ESM-LR second nest (7 km) to HYRAS. The coloring of the box displays the 95 percentile significance level (Wilcoxon test). Red indicates the significantly lower differences to HYRAS of the CCLM_MPI-ESM-LR third nest than the CCLM_MPI-ESM-LR second nest

A Wilcoxon test with a significance level of 95% was performed. In Fig. 8 the red box indicates a significantly better agreement of the higher resolved third nest to HYRAS observations.

3.3.3 Urban Climate: Performance Criteria for Indoor Comfort Wall-Building Materials Under the Influence of the Climate Change in Baden-Württemberg ("Room/Climate/Plaster")

Under changing environmental conditions—which are a consequence of global climate change—living comfort should be maintained. Expected changes of temperature and humidity, which affect the living comfort of people, are analyzed in the project. The study is performed in the framework of a project, funded by the Baden-Württemberg Stiftung, that couples the outdoor and the indoor climate as well as the thermal-hygric behavior of walls by thermal-energetic building simulations driven with regional climate model data. The intention is to avoid too wet and sultry indoor climate by passive plaster systems.

The project is divided in three parts, which correspond essentially to the spatial scales weather and climate ("climate"), room climate and user behavior ("room") as well as building envelope and material behavior ("plaster"). The part "climate" is processed by IMK-TRO and serves as basis for the other two parts, as the resulting climate data output are the boundary data for the impact simulations (thermal-energetic building simulations and moisture transport simulations in walls and multi-layer building components). For these simulations hourly weather data (temperature, humidity, pressure, radiation, precipitation and wind) in the epw (EnergyPlus Weather Data)-format was built and provided by IMK-TRO.

The high resolution regional climate simulations were performed with CCLM, which was driven by data from the GCM ECHAM6 for projection, and ERA Interim reanalysis for validation. The global data are dynamically downscaled with CCLM up to a convection permitting mesh size of 2.8 km. The dynamical downscaling of ECHAM6 data, which have a resolution of about 180 km, is done via two nests (55 and 11 km) to the target resolution of convection permitting 2.8 km. ERA-Interim data, having already a grid-spacing of about 75 km, are scaled down via an 11 km nest to 2.8 km mesh size. The 11 km runs cover the whole of Europe, the 2.8 km runs cover Baden-Württemberg and surrounding areas; past (1981–2010) and future (2021–2050) periods are considered. To estimate the range of possible future developments, an ensemble is created by:

- the use of two emission scenarios, RCP4.5 and RCP8.5,
- coupling COSMO-CLM with the SVAT VEG3D as a more complex alternative scheme to the reference SVAT TERRA-ML,
- and climate simulations with different GCMs as forcing models, which were already available at IMK-TRO.

Table 3 summarizes the computing demands on CRAY XC40 "Hazel Hen" at HLRS for these simulations for a single driving GCM.

The focus of the evaluation of the data lies on sultry conditions in summer in Baden-Württemberg. To identify for the large-scale atmospheric conditions of sultry weather in Baden-Württemberg, the objective weather type classification of the DWD [15] was used. By viewing the results of the validation run, warm and humid conditions in Baden–Württemberg show a strong dependency on the large scale flow conditions. Using the dew point temperature as an indicator for warm and humid conditions, most of these conditions result from the southwesterly weather type SW, followed by undefined directions XX (see Fig. 9). Relative to the total occurrence of weather types of one direction (sector), in summer even 2/3 of the southeasterly SE and also 1/4 of the southwesterly SW weather patterns produce sultry conditions.

Table 3 Summary of computing demands on CRAY XC40 "Hazel Hen" at HLRS for the project room/climate/plaster of one GCM

Project	Domain size	Grid resolution (km)	Time-step (s)	Computing time (node-h/year)	No. of sim. years	Storage needs (Gbyte/year)	Total computing time (node-h)	Total storage needs (Tbyte)
Room/climate/plaster	160 × 120 * 40	55	400	12	170	75	2040	13
Ensemble three nest approach	320 × 360 * 40	11	80	260	175	2500	66,300	437
	140 × 150 * 50	2.8	25	300	265	800	93,000	212
Total amount				572	735	3375	161,340	662

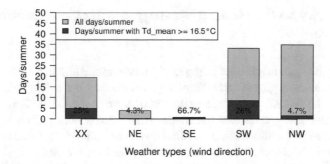

Fig. 9 Weather types in summer (mean of 1981–2010) sorted with respect to wind directions. XX is undefined direction (<2/3 of all viewed grid points not in one wind direction sector); NE = North East, SE = South East, SW = South West, NW = North. The grey bars represent the total number of directions, the red bars the total number of directions, which cause a daily mean of the dew point temperature ≥ 16.5 °C at a grid point near Karlsruhe. The numbers show the height of the red bars relative to the grey bars

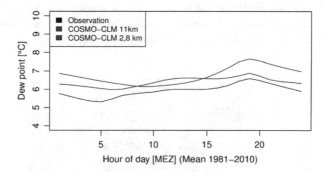

Fig. 10 Comparison between observation data (black line; meteorological tower at KIT near Karlsruhe) and model data with horizontal resolutions of 11 km (red line) and 2.8 km (blue line). The daily cycles of the dew point temperature averaged from 1981 to 2010 are shown

The evaluation of the finer grid runs is currently in progress. An example for a benefit of a combined humidity and temperature measure is shown in Fig. 10. It shows the daily cycle of the dew point temperature averaged from 1981–2010. The observation data are from the 200 m meteorological tower at KIT (near Karlsruhe). From the models (forcing data: ERA-interim, first nest 11 km, second nest 2.8 km) the nearest grid point to the tower location was chosen. The 2.8 km run (blue line) shows better agreement with the observations (black line) in the statistics and in the shape of the curve than the 11 km run (red color). Further evaluations related to sultry conditions like the influence of urban surfaces or local wind systems are in progress.

4   Remarks on the Nesting Strategy and Its Computational Aspects

Dynamical downscaling is used in regional climate modelling to transfer large scale information to the regional scale. Basically, this method is a nesting of the regional climate model (RCM) into large-scale global climate model (GCM) projections or reanalyses. This means that the model is initialized once with a state derived from the large scale information and that this information is updated at the lateral boundaries of the regional model domain at regular time intervals.

The horizontal grid sizes of the global data sets are generally considerably larger than 100 km. To avoid too large jumps in the grid sizes, especially if the ultimate goal are regional climate simulations at very high spatial horizontal resolution in the order of 1–3 km, a multiple nesting technique is used for the dynamical downscaling. The advantages of multiple nests and of regional simulations at very high spatial horizontal resolutions are discussed in [12], for example. The technique of multiple nesting is illustrated in Fig. 11. It has to be pointed out that, with respect to the application of CCLM, the multiple nesting requires consecutive simulations, one for each nest. Local grid refinement is not possible in CCLM.

Fig. 11 From global resolution to very high regional resolution: illustration of the multiple nesting strategy (here two nests) used in CCLM applications (courtesy of Süddeutsches Klimabüro: http://www.sueddeutsches-klimabuero.de/)

In a first nesting step the global scale data are used to drive a CCLM simulation with the coarsest horizontal grid size, here 0.44° (about 50 km in mid-latitudes, the "European" scale in Fig. 11). In the second step the results of this regional coarse grid simulation are used to drive CCLM simulations with finer grid spacing (here 0.0625°), in general with more specific regional or even local focus (for example, the "German" scale in Fig. 11). This procedure is continued down the final very high resolution scale (here 0.025°).

Of course, going from coarser grid to finer ones does not necessarily imply a reduction of the geographical size of the model domain. However, one has to keep in mind that a refinement of the horizontal grid spacing by a factor of two and keeping the domain of the coarse grid would increase the computational needs considerably. The required storage capacity would increase by a factor of four. Roughly, the computing time would increase by a factor of eight, because not only the total number of horizontal grid points would increase by a factor of four, but also the numerical time step has to be reduced by factor of two at least. Of course, the real increase of computing time would be less than the "linear" estimate because more horizontal grid points would allow the usage of more computational nodes. However, the number of cores and thus the number of nodes that sensibly can be used is limited by the number of grid-points in each horizontal coordinate direction and by the numerical advection scheme used. Experience with the CCLM model shows that the ratio between the number of grid points and the number of computational cores in each horizontal direction should be in the order of ten in order to achieve a good balance between the times needed for the pure computations and the communication between parallel cores.

As already said, a nest with refined grid spacing often focuses on regions that have a smaller geographical size than the region represented in the nest with coarser resolution. However, this does not necessarily imply that the number of grid points considered also becomes smaller. On the contrary, this number might even increase, as demonstrated, for example, in Tables 2 and 3. Together with the smaller numerical time steps, the computational needs for the simulations in the nests with higher resolution are considerably larger than for the first nest with 50 km resolution.

Thus, the consideration of multiple nests with consecutive refinements of the horizontal grid spacing has not necessarily a computational benefit. But, and more important from the viewpoint of a climate researcher, there is certainly a scientific benefit as demonstrated by Hackenbruch et al. [12].

The benefit (the speedup) of using the nesting compared with running a whole domain in finer resolution can also roughly be deduced from the first lines in Table 2 (and Table 3), since both simulations presented there only differ in the spatial resolution, the number of horizontal grid points, and the numerical time step. Performing the simulation of the first nest with a resolution of 7 km instead of 50 km would increase the number of horizontal grid points by about a factor of 49, and the time-step would have to be reduced by factor of six. Assuming that the number of cores used does not change, the computational need would roughly increase by a factor of 294, leading to a computational time of 2646 node-hours per simulation year. Theoretically, due to the larger number of grid-points, about 266 nodes could be used for the higher resolution case (7 km) compared to six for the 50 km case. Thus, the computational time would be in the order of 60 ($= 2646/44$) node-hours per simulation year. However, this is still larger than the sum of the computing times, which is 41 node-hours per simulation year, for the two consecutive nesting steps.

References

1. M. Baldauf, A. Seifert, J. Förstner, D. Majewski, M. Raschendorfer, T. Reinhardt, Operational convective-scale numerical weather prediction with the cosmo model: description and sensitivities. Mon. Weather Rev. **139**(12), 3887–3905 (2011)
2. B. Rockel, A. Will, A. Hense, The regional climate model cosmo-clm (cclm). Meteorol. Z. **17**(4), 347–348 (2008)
3. R.H. Moss, J.A. Edmonds, K.A. Hibbard, M.R. Manning, S.K. Rose, D.P. van Vuuren, T.R. Carter, S. Emori, M. Kainuma, T. Kram, G.A. Meehl, J.F.B. Mitchell, N. Nakicenovic, K. Riahi, S.J. Smith, R.J. Stouffer, A.M. Thomson, J.P. Weyant, T.J. Wilbanks, The next generation of scenarios for climate change research and assessment. Nature **463**(7282), 747–756 (2010)
4. D.P. van Vuuren, E. Stehfest, M. den Elzen, T. Kram, J. van Vliet, S. Deetman, M. Isaac, K. Klien Goldewijk, A. Hof, A. Mendoza Beltran, R. Oostenrijk, B. van Ruijven, Rcp2.6: exploring the possibility to keep global mean temperature increase below 2°C. Clim. Chang. **109**, 95–116 (2011)
5. K. Keuler, K. Radtke, S. Kotlarski, D. Lüthi, Regional climate change over Europe in cosmo-clm: influence of emission scenario and driving global model. Meteorol. Z. **25**(2), 121–136 (2016)
6. H.-J. Panitz, G. Schädler, M. Breil, S. Mieruch, H. Feldmann, K. Sedlmeier, N. Laube, M. Uhlig, Application of the regional climate model CCLM for studies on urban climate change in Stuttgart and decadal climate prediction in Europe and Africa, in *High Performance Computing in Science and Engineering '15* (Springer, Cham, 2016), pp. 593–606
7. L. Araguas, P. Danesi, K. Froehlich, K. Rozanski, Global monitoring of the isotopic composition of precipitation. J. Radioanal. Nucl. Chem. **205**(2), 189–200 (1996)
8. K. Yoshimura, Stable water isotopes in climatology, meteorology, and hydrology: a review. J. Meteorol. Soc. Jpn. **93**, 513–533 (2015)
9. W. Dansgaard, S.J. Johnsen, J. Moller, C.C. Langway Jr., One thousand centuries of climatic record from camp century on the greenland ice sheet. Science **166**(3903), 377–380 (1969)
10. S. Pfahl, H. Wernli, K. Yoshimura, The isotopic composition of precipitation from a winter storm - a case study with the limited-area model $cosmo_{iso}$. Atmos. Chem. Phys. **12**, 1629–1648 (2012)
11. S. Weißbach, A. Wegner, T. Opel, H. Oerter, B.M. Vinther, S. Kipfstuhl, Accumulation rate and stable oxygen isotope ratios of the ice cores from the North Greenland Traverse (2016). Supplement to: S. Weißbach et al., Spatial and temporal oxygen isotope variability in northern Greenland - implications for a new climate record over the past millennium. Clim. Past **12**(2), 171–188 (2016). https://doi.org/10.5194/cp-12-171-2016
12. J. Hackenbruch, G. Schädler, J.W. Schipper, Added value of high-resolution regional climate simulations for regional impact studies. Meteorol. Z. **25**, 291–304 (2015)
13. M.R. Haylock, N. Hofstra, A.M.G. Klein Tank, E.J. Klok, P.D. Jones, M. New, A European daily high-resolution gridded dataset of surface temperature and precipitation. J. Geophys. Res. **113**, D20119 (2008)

14. M. Rauthe, H. Steiner, U. Riediger, A. Mazurkiewicz, A. Gratzki, A central European precipitation climatology? Part i: generation and validation of a high-resolution gridded daily data set (hyras). Meteorol. Z. **22**(3), 235–256 (2013)
15. P. Bissolli, E. Dittmann, The objective weather type classification of the German weather service and its possibilities of application to environmental and meteorological investigations. Meteorol. Z. **10**(4), 253–260 (2001)

High Resolution WRF Simulations for Climate Change Studies in Germany

Kirsten Warrach-Sagi, Viktoria Mohr, and Volker Wulfmeyer

Abstract The scope of WRFCLIM is to produce high resolution regional climate simulations with WRF from 1958 to 2100 in the framework of the BMBF funded ReKliEs-De (Regionales Klimaensemble für Deutschland) and the DFG funded research unit on regional climate change (FOR 1695) on a 0.11° (approx. 12 km) grid and further downscale one projection to 0.0275° (approx. 3 km) resolution within the FOR 1695 project from 2000 to 2040.

The overall goal of ReKliEs-De is to derive robust climate change information on high spatial resolution for Germany and the large river catchments draining into Germany from this unique model ensemble. The University of Hohenheim (UHOH) contributes five simulations using WRF to the joint Global Climate Model-Regional Climate Model-matrix, including pre- and post-processing of model input and output data, evaluation of the single model results and delivery of the resulting data to the partners for further post-processing, distribution and storage. The scientific focus of the analyses performed at UHOH is the estimation of the bandwidth of the ensemble results in close cooperation with the project partners.

It is the joint objective of the Research Unit on regional climate change to investigate the effects of global climate change on structure and functions of agricultural landscapes on a regional scale and to work out projections for their development until 2040 taking into consideration various possible socio-economic conditions and adaptation processes. In order to achieve these objectives, the WRF model is coupled with land surface and crop models as well as with multi agent systems in a new integrated land-system model system, and structure and parameterisation of various model components are optimised. Within WRFCLIM verification runs of WRF with and without the crop model GECROS, which includes recent improvements with respect to the representation of agricultural land cover, over Central Europe with convection-permitting (CP) resolution will be simulated downscaling ERA-Interim forced WRFCLIM simulation data from 0.11° are currently ongoing.

K. Warrach-Sagi (✉) · V. Mohr · V. Wulfmeyer
Institute of Physics and Meteorology, University of Hohenheim, Garbenstrasse 30, 70599 Stuttgart, Germany
e-mail: kirsten.warrach-sagi@uni-hohenheim.de; viktoria.mohr@uni-hohenheim.de; volker.wulfmeyer@uni-hohenheim.de

© Springer International Publishing AG 2018
W.E. Nagel et al. (eds.), *High Performance Computing in Science and Engineering '17*, https://doi.org/10.1007/978-3-319-68394-2_25

1 Introduction

WRFCLIM is an ongoing project at the HLRS performing high resolution climate simulations with the Weather Research and Forecast (WRF) model [12] for Europe focusing on Germany (e.g. [7–9, 17]). While climate change is a fundamentally global phenomenon, it will unfold its effects on the regional scale. There will be manifold impacts of climate change on the regional energy and water cycles; of particular concern are changes of the intensity and frequency of extreme events such as droughts or extreme precipitation. Of special concern is the impact of climate change on yield and yield quality and crop rotations, water regime, agricultural production systems and land use. End users like federal agencies and climate impact and adaptation researchers require high resolution climate information.

Within the world wide coordinated effort of the Coupled Model Intercomparison Project Phase 5 (CMIP5, http://cmip-pcmdi.llnl.gov/cmip5/), the impact of the new emission scenarios based on Representative Concentration Pathways (RCPs) [15] on climate is simulated with global models of the climate system. A sample of the global simulations is dynamically downscaled for Europe to 0.44° and 0.11° resolution within the frame of EURO-CORDEX (EUROpe—COordinated Regional climate Downscaling EXperiment (http://www.euro-cordex.net)). Within the BMBF funded German national coordinated effort ReKliEs-De (Regionale Klimaprojektionen Ensemble für Deutschland) (http://reklies.hlnug.de/) this new set of simulations for Germany is evaluated and systematically complemented by further simulations with both dynamical and statistical downscaling methods. The overall goal of ReKliEs-De is to derive robust climate change information on high spatial resolution for Germany and the large river catchments draining into Germany from this unique model ensemble to drive climate impact research and adaptation planning as well as policy advice. A table of the matrix of the applied models and forcing GCMs forming the model ensemble can be found e.g. at http://reklies.hlnug. de/fileadmin/tmpl/reklies/dokumente/ReKliEs-newsletter-nr2.pdf. The University of Hohenheim (Institute of Physics and Meteorology) contributes to ReKliEs-De to this ensemble with five climate simulations with the Weather Research and Forecast (WRF) model within WRFCLIM (see Table 1). All ReKliEs-De partners analyse the results of the ensemble to provide information on climate change in Germany (see Chap. 3 and [4]).

Soil vegetation atmosphere (SVA) feedback is a key controlling factor of weather and climate from local to global scales [6, 11, 13]. The strength of SVA feedback varies across seasons, and the domains where SVA feedback plays an important role will likely increase in a changing climate [3, 14]. Currently, in non-convection permitting model simulations (i.e. grid cells larger than 3 km, convection parameterization) three major systematic errors remain: (1) the windward-lee effect (i.e. too much precipitation at the windward side of mountain ranges and too little precipitation on their lee side), (2) phase errors in the diurnal cycle of precipitation, (3) precipitation return periods (e.g. [1, 10, 16]). An objective of the DFG funded Research Unit 'Agricultural Landscapes under Climate Change Processes and

Table 1 Downscaled Global Climate Model data with WRF simulations by the University of Hohenheim in WRFCLIM to 0.44° and 0.11° resolution

GCM name	Scenarios	Simulation period	Grid resolution
MPI-ESM-LR (Max Planck Institute-Earth System Model-Low Resolution)	Historical	1958–2005	1.8653° × 1.875°
	RCP 8.5	2006–2100	
	RCP 2.6	2006–2100	
MIROC5 (Model for Interdisciplinary Research on Climate)	Historical	1958–2005	1.4008° × 1.40625°
	RCP 8.5	2006–2100	
HadGEM2-ES (Hadley Global Environment Model 2—Earth System)	Historical	1958–2005	1.25° × 1.875°
	RCP 8.5	2006–2100	
EC-EARTH (European Earth System Model)	Historical	1958–2005	1.1215° × 1.125°
	RCP 8.5	2006–2100	

Feedbacks on a Regional Scale (FOR 1695)' is the development and verification of a convection-permitting regional climate model including an advanced representation of SVA feedback processes with emphasis on the water and energy cycling between croplands, atmospheric boundary layer and the free atmosphere. Since croplands show a specific impact on the SVA feedback depending on winter or summer crops, the WRF model in this project is coupled with the crop model GECROS (http://models.pps.wur.nl/node/942). For this new model development an evaluation climate simulation downscaling the reanalysis data ERA-Interim [2] to the convection-permitting scale is under way. After this evaluation run is completed one of the ReKliEs-De climate projections will be downscaled until 2040 to study the climate change in agricultural landscapes in Germany.

2 WRF Simulations at HLRS

2.1 Model Setup and Forcing Data

For the simulations the WRF model is applied, using the CRAY XC40 System of the HLRS. Within WRF, several possibilities of the use of physical parameterizations exist, to carry out the climate simulations. In the WRF simulations by the University of Hohenheim the land surface model NOAH, the Morrison two-moment microphysics scheme, the Yonsei University (YSU) planetary boundary layer parameterization, the Kain-Fritsch-Eta convection scheme and the radiation parameterization scheme CAM for longwave and shortwave radiation are chosen for the regional climate simulations at 50 and 12 km are used (within EURO-CORDEX

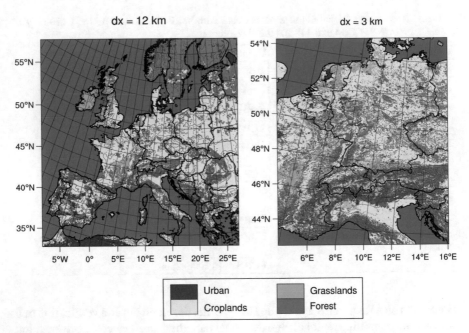

Fig. 1 Land use in the applied model domains for 12 and 3 km horizontal resolution

other institutions provide different combinations so a multi-physics ensemble can be studied as e.g. in [5]). This model configuration is enhanced by the crop model GECROS to the land surface model for the convection permitting simulations in Germany. The model domains for the 12 and 3 km resolving simulations are displayed in Fig. 1.

Evaluation climate simulation are performed downscaling the reanalysis data ERA-Interim [2]. For the future climate projections, four different GCMs and two different RCP scenarios of the CMIP5 project, are applied as boundary forcing with the WRF model. The historical runs of the GCMs cover the period from 1850 to 2005. This period is forced by observed atmospheric composition changes of anthropogenic and natural sources. The RCP scenarios of the GCM's cover the period from 2006 to 2100. They represent mitigation scenarios that assume policy actions will be taken into account to achieve certain emission targets [14]. The numbers of the RCPs give a rough estimate of the range in the change of the radiative forcing by the year 2100 relative to the pre-industrial values. The forcing data we applied, the resolution of the GCMs, its scenarios and the chosen simulation period is presented in Table 1.

2.2 Status of WRFCLIM Simulations

The coarse grid (100–200 km resolution) climate data from five global climate model simulations (GCMs) were retrieved from the CMIP5 archive and pre-

Table 2 Summary of WRF simulation status in April 2017

GCM name	Forcing, WRF grid	Total simulation period	Simulation status	Archiving status (HPSS)
MPI-ESM-LR	Historical, 0.44° and 0.11°	1958–2005	Completed	Completed
MPI-ESM-LR	RCP 8.5, 0.44° and 0.11°	2006–2100	Completed	Ongoing
MPI-ESM-LR	RCP 2.6, 0.44° and 0.11°	2006–2100	2006–2085	Ongoing
MIROC5	Historical, 0.44° and 0.11°	1958–2005	Completed	Completed
MIROC5	RCP 8.5, 0.44° and 0.11°	2006–2100	2006–2076	Ongoing
HadGEM2-ES	Historical, 0.44° and 0.11°	1958–2005	Completed	Completed
HadGEM2-ES	RCP 8.5, 0.44° and 0.11°	2006–2100	2006–2080	Ongoing
EC-EARTH	Historical, 0.44° and 0.11°	1958–2005	Completed	Completed
EC-EARTH	RCP 8.5, 0.44° and 0.11°	2006–2100	2006–2082	Ongoing
ERA-Interim	Reanalysis, 0.11°	1987–2014	1987–2009	Ongoing
ERA-Interim	Reanalysis, 0.0275°	2000–2010	2000–2001	Ongoing
ERA-Interim	Reanalysis, 0.0275°, with GECROS	2000–2010	2000–2001	Ongoing

processed for their downscaling with WRF. Due to their coarse resolution numerical stability requires an intermediate downscaling simulation with WRF to 0.44° resolution prior to the required resolution of 0.11°. Further evaluation simulations are required from 1989 to 2012 downscaling the ECMWF reanalyses data ERA-Interim with the same set-up. Within WRFCLIM to date these evaluation simulations are finished for the 0.44° grid and performed for 1989–2002 for the 0.11° grid. The convection permitting simulation with WRF-NOAHMP-GECROS has started in September 2016.

The first step of the nesting to the 0.44° horizontal resolution applying all five forcing GCMs to WRF were carried out for the simulation period of 1958–2100 on hazelhen until June 2016. Simultaneously, when the first simulated years were available on the 0.44°, the nesting step to the 0.11° grid resolution could be started. By the end of August 2016 all historical simulations for the simulation time of 1958–2005 were finalized on the 0.11° grid and are currently subject to analyses. Table 2 summarizes the status of the simulations including the ERA-Interim driven evaluation runs in April 2017. All data is post-processed on-the-fly extracting the three-hourly time-series of the climate variables for analyses, impact modelling and storage in the archive (see Sect. 3). A first analysis was presented by Viktoria Mohr et al. at the 16th European Meteorological Society Annual Meeting in Trieste in September 2016, the 19th Results and Review Workshop of the HLRS at 13–14th October 2016 in Stuttgart.

3 Technical Description

For simulations at 12 km resolution, the open source WRF model version 3.6.1 is applied (http://www.wrf-model.org). The scalability of the WRF model is depending on the problem size. Starting from 240 cores (10 nodes), the scaling is linear up to 1200 cores with a slight decrease and recovering when applying 1680 cores. The total number of MPI tasks is limited by the fact that each task should not contain less than 15*15 grid points to avoid spending too much time with the communication between adjacent cores. For 3 km simulations the version 3.7.1 is applied with the more sophisticated land surface model NOAHMP.

The WRF model can also be run in hybrid mode (MPI+OpenMP). The scaling performance in the speedup is not proportionally to the used Cores as the code is not preferably written for OpenMP applications. However, due to a strict time limitation for the climate runs any speedup is necessary in order to reach the goal of a fast as possible simulation completion. Using the hybrid mode starting from 1680 cores including also the number of applied OpenMP Threads. Longterm simulations show that the best speedup for our problem size we get when applying 5400 cores.

The WRF Model system requires NetCDF and PnetCDF as external libraries. The I/O performance strongly depends on the current load on the file system. Previous projects revealed a large variation between 1 and 8 GB/s on the WS7 file system of Hazelhen. Analyses are performed with a variety of software packages including CDO (climate Data Operator), NCO (netCDF Operator) and the NCAR Command Language (NCL). WRF is compiled at HLRS with PGI 14.7 and applied in a hybrid configuration using MPI and OpenMP to optimize the speed of the simulation. Within 24 h walltime, it was possible to simulate approximately 1 year of the 12 km domain and 2.5 months of the 3 km domain respectively. So far (status April 2017) we were able to downscale the majority of the GCM's years to the grid size of 12 km. These simulations are about to be finalized within the next weeks. However, all data need to be post processed for archiving and analysis. This work will be ongoing until late autumn. Simulations on the 3 km grid started in December 2016. Details for the simulations are given in Table 3. The storage required on

Table 3 Simulations and approximated computational needs for the WRFCLIM project June 2016–June 2017

GCM forcing	Num. simul.	$\Delta x, \Delta y$	Grid size	Storage interval in h	Wall-time per steps in s	Cores per run	Mio. core-h
ERA-interim	1 (10 years)	0.11°	$460 \times 425 \times 50$	3	0.15	5400	1
RCPs (projections)	5 (each 110 years)	0.11°	$460 \times 425 \times 50$	3	0.15	5400	64
Convection permitting ERA-interim	2 (each 5 years)	0.03°	$413 \times 457 \times 50$	3	0.15	5400	1
Post-processing							0.01
Total							66

HPSS for the model results from the WRFCLIM is currently 1.3 PB. The workspace of 400 TB is needed to ensure that the simulations can be performed in time and parallel (i.e. 1 job is up to assessing 27,000 cores). Simulated data needs to be hold in the workspace until the next year is completed and the data is postprocessed.

4 Results and Outlook

4.1 ReKliEs-De

At a climate model data user workshop in May 2016 in Potsdam the concept and climate indices were discussed between the ReKliEs-De modellers and potential users in Germany. Hüebener et al. [4] give an overview of this workshop. The historical simulations are completed and currently analyzed within ReKliEs-De. The projections are ongoing until the end of May 2017 and then will be post processed (e.g. climate indices will be calculated) and the climate change signals in the ensemble of all regional model simulations will be studied. On the 6th and 7th December the results of ReKliEs-De will be presented at a final workshop in Wiesbaden and a user handbook will be published. Preliminary results can be found e.g. in Mohr et al. [8, 9]. Figure 2 exemplarily shows a joint ReKliEs-De model ensemble analyses in form of a Normalized Taylor Diagram for precipitation and temperature for the historical runs in comparison with the gridded observational data HYRAS from the German Weather Service (DWD). The normalized Taylor diagram is a method to evaluate a model against observations discarding the systematic bias by normalizing the quantities. The distance from the origin is the standard deviation of the field. If the standard deviation of the model is same as that of the observation, then the radius is 1. The distance from the reference point to the plotted point gives the root mean square difference (RMSE). The nearer the plotted point is to the reference point, the lesser will be the RMSE. The correlation between the model and the climatology is the cosine of the polar angle (if the correlation between the model and observation is 1, then the point will lie on the horizontal axis). Thus the model which has largest correlation coefficient, smaller RMSE and comparable variance will be close to the reference point (i.e., the observation) is considered to be the best among all.

4.2 FOR 1695

The simulations with and without a crop model coupled to WRF are ongoing. First results are expected in autumn 2017. Afterwards the performance of a first CP-resolution climate projection until 2040 downscaling the MPI-ESM-LR RCP 8.5 forced WRFCLIM simulation in order to reduce the current deficiencies of global

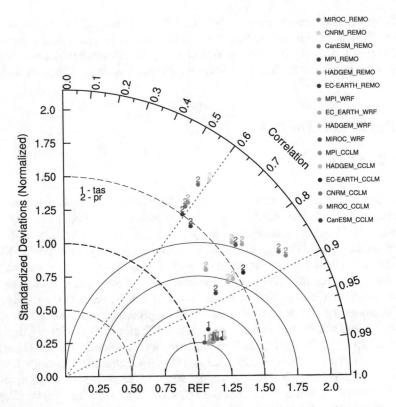

Fig. 2 Normalized Taylor Diagram for mean annual precipitation (pr) and temperature (tas) between 1971 and 2000 for the ReKliEs-De Ensemble in comparison with the HYRAS data for Germany

and regional climate projections such as errors introduced by the parameterisation of deep convection will be investigated. This will also permit to study the respective importance of land cover changes and greenhouse forcing on the evolution of regional climate within the next decades.

Acknowledgements This work is part of the ReKliEs-De project funded by the BMBF (Federal Ministry for Education and Research) and the Research Unit 1695 funded by the DFG (Deutsche Forschungsgemeinschaft). We are thankful for the support from the staff of the DKRZ (Deutsches Klimarechenzentrum), to be able to access GCM data. Computational Resources for the model simulations on the CRAY XE6 and XC40 within WRFCLIM were kindly provided by HLRS, we thank staff for their great support.

References

1. H.S. Bauer, T. Weusthoff, M. Dorninger, V. Wulfmeyer, T. Schwitalla, T. Gorgas, M. Arpagaus, K. Warrach-Sagi, Predictive skill of a subset of models participating in D-PHASE in the COPS region. Q. J. R. Meteorol. Soc. **137**(S1), 287–305 (2011)
2. D.P. Dee, S.M. Uppala, A.J. Simmons, P. Berrisford, P. Poli, S. Kobayashi, U. Andrae, M.A. Balmaseda, G. Balsamo, P. Bauer, P. Bechtold, A.C.M. Beljaars, L. van de Berg, J. Bidlot, N. Bormann, C. Delsol, R. Dragani, M. Fuentes, A.J. Geer, L. Haimberger, S.B. Healy, H. Hersbach, E.V. Hólm, L. Isaksen, P. Kållberg, M. Köhler, M. Matricardi, A.P. McNally, B.M. Monge-Sanz, J.J. Morcrette, B.K. Park, C. Peubey, P. de Rosnay, C. Tavolato, J.N. Thépaut, F. Vitart, The ERA-Interim reanalysis: configuration and performance of the data assimilation system. Q. J. R. Meteorol. Soc. **137**(656), 553–597 (2011). https://doi.org/10.1002/qj.828
3. P.A. Dirmeyer, B.A. Cash, J.L. Kinter III, C. Stan, T. Jung, L. Marx, P. Towers, N. Wedi, J.M. Adams, E.L. Altshuler, B. Huang, E.K. Jin, J. Manganello, Evidence for enhanced land–atmosphere feedback in a warming climate. J. Hydrometeorol. **13**(3), 981–995 (2012). https://doi.org/10.1175/JHM-D-11-0104.1
4. H. Hübener, P. Hoffmann, K. Keuler, S. Pfeifer, H. Ramthun, A. Spekat, C. Steger, K. Warrach-Sagi, Deriving user-informed climate information from climate model ensemble results. Adv. Sci. Res. **14**, 261–269 (2017). https://doi.org/10.5194/asr-14-261-2017
5. S. Kotlarski, K. Keuler, O.B. Christensen, A. Colette, M. Déqué, A. Gobiet, K. Goergen, D. Jacob, D. Lüthi, E. van Meijgaard, G. Nikulin, C. Schär, C. Teichmann, R. Vautard, K. Warrach-Sagi, V. Wulfmeyer, Regional climate modeling on European scales: a joint standard evaluation of the EURO-CORDEX RCM ensemble. Geosci. Model Dev. **7**(4), 1297–1333 (2014). https://doi.org/10.5194/gmd-7-1297-2014
6. R. Mahmood, R.A. Pielke, K.G. Hubbard, D. Niyogi, P.A. Dirmeyer, C. McAlpine, A.M. Carleton, R. Hale, S. Gameda, A. Beltrán-Przekurat, B. Baker, R. McNider, D.R. Legates, M. Shepherd, J. Du, P.D. Blanken, O.W. Frauenfeld, U. Nair, S. Fall, Land cover changes and their biogeophysical effects on climate. Int. J. Climatol. **34**(4), 929–953 (2014). https://doi.org/10.1002/joc.3736
7. J. Milovac, O. Branch, H.S. Bauer, T. Schwitalla, K. Warrach-Sagi, V. Wulfmeyer, High-resolution WRF model simulations of critical land surface-atmosphere interactions within arid and temperate climates (WRFCLIM), in *High Performance Computing in Science and Engineering '15*, ed. by W. Nagel, D. Kroener, M. Resch (Springer International Publishing, Cham, 2016), pp. 607–622. https://doi.org/10.1007/978-3-319-47066-5
8. V. Mohr, T. Schwitalla, V. Wulfmeyer, K.Warrach-Sagi, High-resolution climate projections using the WRF model on the HLRS, in *Sustained Simulation Performance 2016*, ed. by M. Resch, W. Bez, E. Focht, N. Patel, H. Kobayashi (Springer International Publishing, Cham, 2016), pp. 173–184. https://doi.org/l10.1007/978-3-319-46735-1_14
9. V. Mohr, K. Warrach-Sagi, T. Schwitalla, H.S. Bauer, V. Wulfmeyer, High-resolution climate projections using the WRF model on the HLRS, in *High Performance Computing in Science and Engineering '16*, ed. by W. Nagel, D. Kroener, M. Resch (Springer International Publishing, Cham, 2016), pp. 577–587. https://doi.org/10.1007/978-3-319-47066-5
10. A.F. Prein, A. Gobiet, M. Suklitsch, H. Truhetz, N.K. Awan, K. Keuler, G. Georgievski, Added value of convection permitting seasonal simulations. Clim. Dyn. **41**(9), 2655–2677 (2013). https://doi.org/10.1007/s00382-013-1744-6
11. S.I. Seneviratne, T. Corti, E.L. Davin, M. Hirschi, E.B. Jaeger, I. Lehner, B. Orlowsky, A.J. Teuling, Investigating soil moisture–climate interactions in a changing climate: a review. Earth Sci. Rev. **99**(3), 125–161 (2010). https://doi.org/10.1016/j.earscirev.2010.02.004
12. W.C. Skamarock, J.B. Klemp, J. Dudhia, D. Gill, D.O. Barker, M.G. Duda, W. Wang, J.G. Powers, A description of the advanced research WRF Version 3. NCAR Technical Note TN-475+STR, NCAR, Boulder/CO (2008). http://www.mmm.ucar.edu/wrf/users/docs/arw_v3.pdf

13. M. Stéfanon, P. Drobinski, F. D'Andrea, C. Lebeaupin-Brossier, S. Bastin, Soil moisture-temperature feedbacks at meso-scale during summer heat waves over Western Europe. Clim. Dyn. **42**(5), 1309–1324 (2014). https://doi.org/10.1007/s00382-013-1794-9

14. K.E. Taylor, R.J. Stouffer, G.A. Meehl, An overview of CMIP5 and the experiment design. Bull. Am. Meteorol. Soc. **93**(4), 485–498 (2012). https://doi.org/10.1175/BAMS-D-11-00094.1

15. D.P. van Vuuren, J.A. Edmonds, M. Kainuma, K. Riahi, J. Weyant, A special issue on the RCPs. Clim. Chang. **109**, 1 (2011)

16. K. Warrach-Sagi, T. Schwitalla, V. Wulfmeyer, H.S. Bauer, Evaluation of a climate simulation in Europe based on the WRF-NOAH Model System: precipitation in Germany. Clim. Dyn. **41**(3–4), 755–774 (2013). https://doi.org/10.1007/S00382-013-1727-7

17. K. Warrach-Sagi, H.S. Bauer, T. Schwitalla, J. Milovac, O. Branch, V. Wulfmeyer, High-resolution climate predictions and short-range forecasts to improve the process understanding and the representation of land-surface interactions in the WRF model in Southwest Germany (WRFCLIM), in *High Performance Computing in Science and Engineering '14*, ed. by W. Nagel, D. Kroener, M. Resch (Springer International Publishing, Berlin/Heidelberg, 2015), pp. 575–592

Seasonal Simulation of Weather Extremes

Thomas Schwitalla, Volker Wulfmeyer, and Kirsten Warrach-Sagi

Abstract To date, seasonal forecasts are often performed by applying horizontal resolutions of 75–150 km due to lack of computational resources and associated operational constraints. As this resolution is too coarse to represent fine scale structures impacting the large scale circulation, a convection permitting (CP) resolution of less than 4 km horizontal resolution is required. Most of the simulations are carried out as limited area model (LAMs) and thus require boundary conditions at all four domain boundaries. In this study, the Weather Research and Forecasting (WRF) model is applied in a latitude belt set-up in order to avoid the application of zonal boundaries. The horizontal resolution is 0.03° spanning a belt between 65° N and 57° S encompassing 12000 ∗ 4060 ∗ 57 grid boxes. The simulation is forced by ECMWF analysis data and high resolution SST data from multiple sources. The simulation period is February to beginning of July 2015 which is a strong El Niño period.

1 Introduction

To date, climate simulations and long-term weather forecasts are extremely challenging. This is due to the complex interaction between large-scale and small-scale meteorological processes such as the monsoon circulation and mesoscale circulations induced by complex terrain and heterogeneous land surface properties. Furthermore, e.g. the African continent is strongly affected by teleconnections triggered by El Niño Southern Oscillation (ENSO) events. During an El Niño period, e.g. South East Africa and Australia can suffer from extreme aridness while countries like Kenya and Ethiopia can be influenced by additional precipitation during extreme events. Thus, in principle, forecasting phenomena like ENSO, which can influence weather patterns over the entire globe, would improve predictability of extreme events for up to several months in advance. Consequently, it should

---

T. Schwitalla (✉) • V. Wulfmeyer • K. Warrach-Sagi
Institute of Physics and Meteorology, University of Hohenheim, Garbenstrasse 30,
70599 Stuttgart, Germany
e-mail: thomas.schwitalla@uni-hohenheim.de; volker.wulfmeyer@uni-hohenheim.de;
kirsten.warrach-sagi@uni-hohenheim.de

© Springer International Publishing AG 2018
W.E. Nagel et al. (eds.), *High Performance Computing in Science
and Engineering '17*, https://doi.org/10.1007/978-3-319-68394-2_26

be possible to predict extreme events statistics up to the seasonal scale for the protection of the environment and humankind.

Unfortunately, still to date, it was not possible to exploit this potential of predictability in seasonal simulations and weather forecasts. Particularly, current model systems are inadequate to simulate droughts and extreme precipitation with acceptable predictive skill. This is due to the limited grid resolution of current global general circulation models (GCMs), limitations in model physics, poor model physics, particularly with respect to the simulation of the atmospheric boundary layer (ABL), the land-ABL interaction, clouds, and precipitation, as well as an inaccurate initialization of the state of the ocean-land-atmosphere system. For instance, current climate projections performed in the CORDEX project (www. cordex.org) for the African continent are performed with a horizontal resolution of 50 km, which is not sufficient to resolve mesoscale and orographically-induced circulations.

To enhance the prediction of climate change and extreme events, it is necessary to perform simulations on the convection permitting (CP) scale of the order of 1–3 km. Rotach et al. [3], Wulfmeyer et al. [7], and Bauer et al. [1] clearly demonstrated that this resolution is leading to a considerable improvement of the predictive skill with respect to the simulation of surface temperatures, clouds, and precipitation.

The commonly applied limited area models (LAM) are dependent on the lateral boundary forcing, usually provided by coarser scale global general circulation models (GCM). As these models often apply different physics schemes compared to LAMs, this can lead to considerable spin-up effects at the domain boundaries. To avoid these inconsistencies, latitude-belt simulations are performed where the external forcing is only required at the northern and southern domain boundaries. The experiment described here is a successive simulation based on the experience of Schwitalla et al. [5].

2 Domain Setup

Since the recent system upgrade at HLRS from Hermit to Hazel Hen, additional computing power became available for the performance of new seasonal simulations of the ENSO year 2015 along a latitude belt, which would then cover the entire tropical region with a convection permitting horizontal resolution of 0.03°. The final model domain for this simulation is shown in Fig. 1 covering an area between 56° S and 65° N by applying 12000 ∗ 4060 grid cells and 57 levels up to 10 hPa.

The applied model is the WRF-NOAH-MP model system [6]. The simulation will be performed for the period from February to July 2015 and the selected domain will also allow for investigating extratropical transitions, such as tropical cyclones transforming to storm systems affecting Europe. This simulation is considered as single realization of a potential seasonal forecasting ensemble on the convection permitting scale.

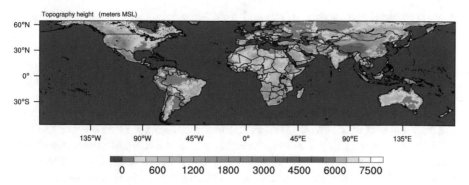

Fig. 1 Domain configuration for the new latitude belt simulation

3 Technical Aspects

Compared to the previous study of Schwitalla et al. [4, 5], the applied version of the WRF model system is now 3.8.1 and the code is compiled with the Intel compiler suite version 16.0.3. In addition, Cray NetCDF 4.3.3.2 and Cray Parallel NetCDF (PNetCDF) 1.7.0 is applied. For this large domain a few changes to the I/O routines have been made in order to be able to follow the CDF5 convention. This allows using variable arrays larger than 2^{32}-4 bytes which is necessary to run such large domains.

All necessary input fields have been processed by the WRF preprocessing system WPS. This initially required the processing of approx. 2 TB of input data from the European Centre for Medium Range Weather Forecasting (ECMWF). As the WRF preprocessing system WPS does not support PNetCDF, the preprocessing was performed in a way that each of 840 MPI tasks (35 nodes) writes out its single file with about 105 MB size. Unfortunately this resulted in 504,000 files occupying 50 TB which had to be temporally stored on the WS9 file system before being further processed. A minimum Number of 35 nodes was required due to high memory demands during this step.

As initial tests revealed that using Sea Surface Temperature (SST) data from ECMWF during the winter period is not sufficient at this high resolution, additional effort was made in order to combine the coarser resolution SST data from ECMWF and other sources providing higher horizontal resolutions. Initially, the 0.011° resolution JPL SST data set (https://podaac.jpl.nasa.gov/dataset/JPL-L4UHfnd-GLOB-MUR) was applied. Tests revealed that although the resolution is considerably higher over the sea compared to ECMWF, problems with unresolved inland lakes like the Aral Sea or the lakes in Kazakhstan, and Mongolia are not well resolved. This leads to an unrealistic simulation of surface fluxes and eventually the model stopped with a segmentation fault.

In the next step the combination of the high-resolution OSTIA-SST data set (www.ghrsst-pp.metoffice.com/pages/latest_analysis/ostia.html) together with the operational SST data from ECMWF has been tested. This combination gave a much more realistic representation of all the frozen lakes over Kazakhstan and Mongolia so that this combination will be applied for the seasonal simulation. In order to make use of both data sets concurrently in the WPS system, adjustments to the interpolations methods have been made. The interpolation option has been changed from a four-point average to the nearest neighbor method in order to avoid any erroneous values as best as possible especially along the coast.

Further tests with the initialization program REAL still showed deficiencies for the SST data in case inland lakes are neither resolved by ECMWF nor OSTIA. Therefore a new SST field was introduced according to the ingested data from WPS. The source code was enhanced in such a way that the reasonable range for SSTs is between 271 K (frozen saline water) and 307.15 K which is the highest recorded SST data so far. Also some minor changes were necessary in case sea ice and snow were present over inland lakes. In order to prevent negative water vapor mixing ratios in very dry areas at high altitudes associated with temperatures below $-30\,°C$, a minor change was introduced into the sea ice subroutine of the NOAH-MP land-surface model concerning the latent heat release of snow.

Compared to the previous study of Schwitalla et al. [5], additional aerosol optical depth data (AOD) at 550 nm from the operational ECMWF Monitoring atmospheric composition&climate (MACC) reanalysis (http://apps.ecmwf.int/datasets/data/macc-reanalysis/levtype=sfc/, [2]) have been included. These data interact with the shortwave radiation thus having a beneficial influence on surface/skin temperatures which can impact the development of the boundary layer and thus enhance precipitation. This can become especially important in desert areas where lots of dust is available.

As some of the diagnostic variables from the Air Force Weather Agency (AFWA) package cannot be derived when operating the model in hybrid mode (MPI + OpenMP), the code was modified in a way that diagnostic variables like mean sea level pressure (MSLP), integrated water vapor, or cloud liquid water path are calculated during the model integration. As this is fully parallelized, this procedure saves a lot of time compared to a retrospective calculation of these variables.

In order to avoid a tremendous amount of data for a statistical evaluation, the output design is the following:

1. Restart files are written every 24 h simulation time. File size is about 1150 GB per file
2. 3-dimensional fields are omitted. If desired, they can be reproduced by starting from restart files for a limited time period if desired.
3. Data on 11 meteorological standard pressure levels are stored every 6 h in accordance with ECMWF data for verification. The file size is approx. 12 GB.
4. A separate output stream was defined to store only diagnostic variables like e.g. 2-m temperatures, surface fluxes, and integrated water vapor. Currently these files are stored at 30 min frequency. The file size is approx. 6 GB.

The simulation will be performed in hybrid mode (MPI + OpenMP) by applying 4096 nodes on the current Hazel Hen system. This is a similar set-up as applied in the study of [4]. Tests revealed that the model integration time reduces by increasing the number of MPI tasks, but then the I/O rate drastically drops down. By using 4096 nodes with pure MPI, the I/O speeds drop down to few hundred Megabytes/s. In case using 6 OpenMP threads per node, the I/O rate stays around 4- GB/s using all 54 OSTs of WS9.

The necessary run time is expected to be 192 h when using 4096 nodes. The expected real time is about 10 weeks as the average waiting time for such big jobs is around 7 days. The total data amount is expected to be 172 TB for the restart files, 7.2 TB for the standard pressure level data, and 46 TB for the surface variables resulting in a total data amount of 215 TB. These data will be stored in the HPSS archive.

Recently it was figured out how to make use of combining parallel NetCDF together with the so-called quilting procedure. This procedure dedicates a specific number of cores which are responsible for I/O. With this procedure, data I/O rates of 60 GB/s can be achieved. However this procedure currently does not work for the very large restart files (1.2 TB each) and thus needs further investigation.

4 Preliminary Results

To date, 43 out of 150 simulation days have been accomplished. Figure 2 shows an example output of the outgoing long wave radiation at the model top after 21 days of simulation time. The convective clouds, indicated by the dark blue areas over the southern hemisphere, are clearly visible.

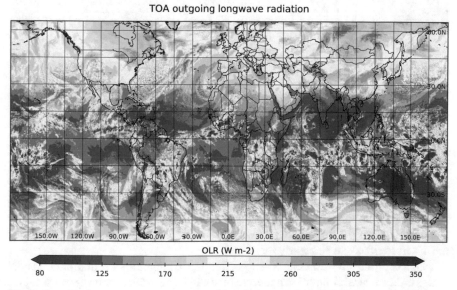

Fig. 2 Outgoing long wave radiation [W m^{-2}] at the model top valid at February 21, 2015 at 00 UTC

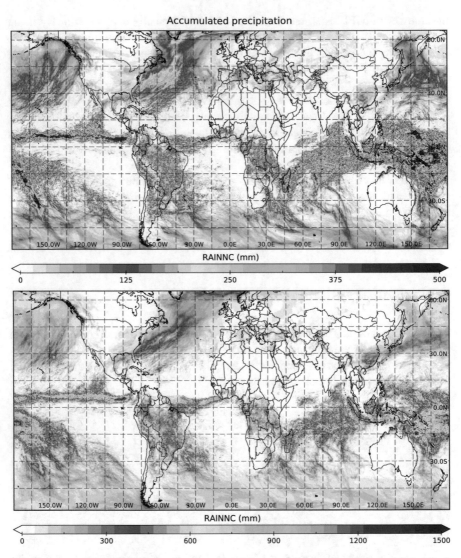

Fig. 3 Accumulated precipitation [mm] on 21st February 2015 (upper panel) and 13th March 2015 (lower panel)

Figure 3 shows an example of the 21 day accumulated precipitation (upper panel) and 43 day accumulated precipitation since the start of simulation on 1st February 2015. The higher precipitation amounts over the tropics and the very dry areas over the desert regions can be clearly identified.

Another interesting feature when performing simulations on this large spatial scales is the general circulation pattern which can be described by the mean sea level

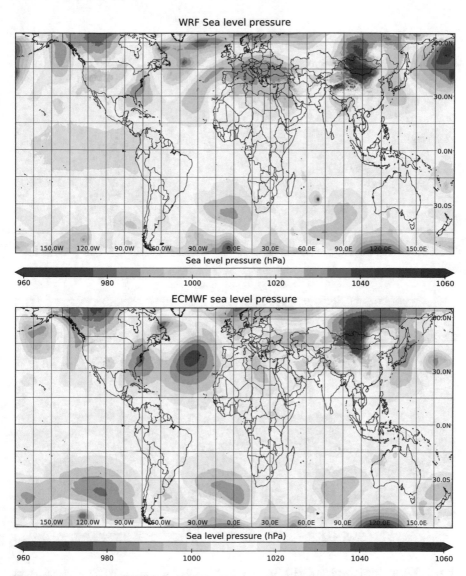

Fig. 4 Upper panel: MSLP [hPa] of WRF at February 21, 2015 00 UTC. Lower panel: MSLP of the ECMWF analysis

pressure (MSLP) and the Jet Stream. The upper panel of Fig. 4 exemplarily displays MSLP at February 21, 2015 00 UTC. The very high pressure exceeding 1045 hPa over Tibet, Mongolia, and Canada is clearly visible associated with temperatures below −35 °C. The lower panel of Fig. 4 shows the MSLP field of the ECMWF analysis for the same time step. Larger differences are visible in the northern part of

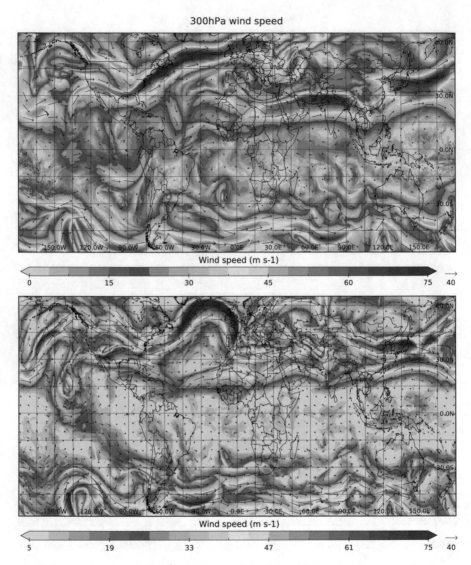

Fig. 5 300 hPa wind speed [m s^{-1}] at February 21, 2015 00 UTC from the WRF simulation (upper panel) and ECMWF analysis (lower panel)

the model domain as this is the most sensitive region. On the southern hemisphere, the simulation agrees much better with the ECMWF analysis field apart from the tropical storm in the South-West Indian Ocean.

The reddish colors in Fig. 5 nicely show the location of the Jetstream of the WRF simulation (upper panel) and the ECMWF analysis for the same time step

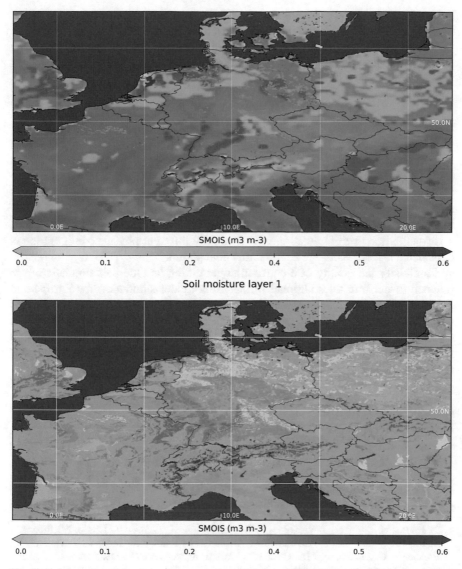

Fig. 6 Soil moisture of the first soil layer at the initial time step (1st February 2015, upper panel) and at 13th March 2015

as mentioned above. The location of the Jet Stream maximum is strongly correlated with the pressure gradients shown in the previous Figure.

Another important point of such simulations is the soil moisture spin-up phase. The following Fig. 6 shows a comparison of the soil moisture between 0 and 10 cm below the ground over Central Europe. The upper panel shows the status at the beginning of the forecast and the lower panel the situation after 43 days.

The soil moisture adjustment is clearly visible in the lower panel of Fig. 6. More details about that can be found in the study of Schwitalla et al. [5].

5   Summary

A convection permitting latitude belt simulation has been set up as a successor of a latitude belt simulation performed in 2014. The domain covers an area between 56°S and 65°N. The simulation will be performed during a El Niñjo year and was started at February 1st, 2015 and will be continued at least until 1st July, 2015. In order to get a more detailed insight about the model behavior on these scales, a second latitude-belt simulation with a slightly different physics configuration is considered if it is feasible in a reasonable time period.

To classify the quality of a convection permitting latitude-belt simulation, it is planned to perform an additional limited area model simulation for Europe and the Horn of Africa at the same horizontal and vertical resolution. All simulations will be verified against the operational ECMWF analysis and other data sets like Tropical Rainfall Measuring Mission (TRMM) data, the Global Precipitation Measurement data and the EUMETSAT land surface analysis (LANDSAF; https://landsaf.ipma.pt/).

References

1. H.S. Bauer, T. Weusthoff, M. Dorninger, V. Wulfmeyer, T. Schwitalla, T. Gorgas, M. Arpagaus, K. Warrach-Sagi, Predictive skill of a subset of models participating in D-PHASE in the COPS region. Q. J. R. Meteorol. Soc. **137**(S1), 287–305 (2011)
2. A. Inness, F. Baier, A. Benedetti, I. Bouarar, S. Chabrillat, H. Clark, C. Clerbaux, P. Coheur, R.J. Engelen, Q. Errera, J. Flemming, M. George, C. Granier, J. Hadji-Lazaro, V. Huijnen, D. Hurtmans, L. Jones, J.W. Kaiser, J. Kapsomenakis, K. Lefever, J. Leitão, M. Razinger, A. Richter, M.G. Schultz, A.J. Simmons, M. Suttie, O. Stein, J.N. Thépaut, V. Thouret, M. Vrekoussis, C. Zerefos, the MACC team, The MACC reanalysis: an 8 yr data set of atmospheric composition. Atmos. Chem. Phys. **13**(8), 4073–4109 (2013). https://doi.org/10.5194/acp-13-4073-2013, http://www.atmos-chem-phys.net/13/4073/2013/
3. M.W. Rotach, M. Arpagaus, M. Dorninger, C. Hegg, A. Montani, R. Ranzi, F. Bouttier, A. Buzzi, G. Frustaci, K. Mylne, E. Richard, A. Rossa, C. Schär, M. Staudinger, H. Volkert, V. Wulfmeyer, P. Ambrosetti, F. Ament, C. Appenzeller, H.S. Bauer, S. Davolio, M. Denhard, L. Fontannaz, J. Frick, F. Fundel, U. Germann, A. Hering, C. Keil, M. Liniger, C. Marsigli, Y. Seity, M. Stoll, A. Walser, M. Zappa, MAP D-PHASE: real-time demonstration of weather forecast quality in the alpine region. Bull. Am. Meteorol. Soc. **90**, 1321–1336 (2009)
4. T. Schwitalla, K. Warrach-Sagi, V. Wulfmeyer, High-resolution latitude belt simulation with the weather research and forecasting model, in *Sustained Simulation Performance 2015*, ed. by M.M. Resch, W. Bez, E. Focht, H. Kobayashi, J. Qi, S. Roller (Springer International Publishing, Cham, 2015), pp. 185–194. ISBN 978-3-319-20339-3

5. T. Schwitalla, H.S. Bauer, V. Wulfmeyer, K. Warrach-Sagi, Continuous high-resolution midlatitude-belt simulations for july–august 2013 with wrf. Geosci. Model Dev. **10**(5), 2031–2055 (2017). https://doi.org/10.5194/gmd-10-2031-2017, https://www.geosci-model-dev.net/10/2031/2017/

6. W.C. Skamarock, J.B. Klemp, J. Dudhia, D. Gill, D.O. Barker, M.G. Duda, W. Wang, J.G. Powers, A description of the advanced research WRF version 3. NCAR Technical Note TN-475+STR, NCAR, Boulder/CO (2008), http://www.mmm.ucar.edu/wrf/users/docs/arw_v3.pdf

7. V. Wulfmeyer, A. Behrendt, C. Kottmeier, U. Corsmeier, C. Barthlott, G.C. Craig, M. Hagen, D. Althausen, F. Aoshima, M. Arpagaus, H.S. Bauer, L. Bennett, A. Blyth, C. Brandau, C. Champollion, S. Crewell, G. Dick, P. DiGirolamo, M. Dorninger, Y. Dufournet, R. Eigenmann, R. Engelmann, C. Flamant, T. Foken, T. Gorgas, M. Grzeschik, J. Handwerker, C. Hauck, H. Höller, W. Junkermann, N. Kalthoff, C. Kiemle, S. Klink, M. König, L. Krauss, C.N. Long, F. Madonna, S. Mobbs, B. Neininger, S. Pal, G. Peters, G. Pigeon, E. Richard, M.W. Rotach, H. Russchenberg, T. Schwitalla, V. Smith, R. Steinacker, J. Trentmann, D.D. Turner, J. van Baelen, S. Vogt, H. Volker, T. Weckwerth, H. Wernli, A. Wieser, M. Wirth, The convective and orographically induced precipitation study (cops): the scientific strategy, the field phase, and research highlights. Q. J. R. Meteorol. Soc. **137**(S1), 3–30 (2011)

Part VI
Miscellaneous Topics

Wolfgang Schröder

In the previous chapters, topics such as fluid mechanics, structural mechanics, aerodynamics, thermodynamics, chemistry, combustion, and so forth have been addressed. In the following, another degree of interdisciplinary research is emphasized. The articles clearly show the link between applied mathematics, fundamental physics, computer science, and the ability to develop certain models such that a closed mathematical description can be achieved which can be solved by massively parallel algorithms on up-to-date high performance computers. In other words, it is the collaboration of several scientific fields which defines the area of numerical simulations and determines the progress in fundamental and applied research. The subsequent papers will confirm simulations to be used to corroborate physical models and to develop new theories.

In the contribution of the Goethe Center for Scientific Computing of the Goethe University Frankfurt mathematical modeling is considered to enhance the understanding of permeation processes in the skin. For the epidermis, a morphology model consisting of agglomerated tetrakaidecahedra was proposed. A tetrakaidecahedron (TKD) is a polygon with 14 faces providing a dense spatial packing. This property has been discovered by Lord Kelvin studying foam cells. Based on TKD shaped cells a geometry model for the stratum corneum (SC), i.e., the outermost layer of the epidermis was introduced. This region is believed to be responsible for the barrier property of the skin. Nowadays, computing interactions in large networks of cells became feasible. To fully resolve the highly differing scales between corneocytes and the surrounding lipid layer, very fine grids are required. Considering the complex TKD based geometric model and the unstructured nature of the associated grids, massively parallel computers are required. In the article

W. Schröder (✉)
Institute of Aerodynamics, RWTH Aachen University, Wüllnerstraße 5a, 52062 Aachen, Germany
e-mail: office@aia.rwth-aachen.de

of the Goethe University Frankfurt the parallel meshing involved in preparing distributed grid hierarchies for the application of massively parallel multigrid solvers on TKD based grids is discussed.

Seismic applications of full waveform investigation are discussed by the Geophysical Institute of the Karlsruhe Institute of Technology. Since natural resources are precious, the number of underground constructions increases. It is important to map earth's geological structures accurately by collecting seismic data and transforming them into subsurface images. Therefore, full waveform inversion (FWI) that accounts for the full information content of seismic recordings is developed. FWI retrieves multiparameter models of the subsurface by solving the full wave equation. It allows to map structures on sub-wavelength scales. Thus, FWI helps to improve both petrophysical interpretation and geotechnical characterization of the subsurface. The focus of this study is on the implementation of the time-domain FWI and its application to seismic field-data problems. It comprises two- and three-dimensional modeling of viscoelastic wavefields and exploits straightforward and efficient parallelization by domain decomposition and source parallelization leading to a significant speedup on parallel computers.

The contribution on the impact of pores on microstructure conduction is from the Institute of Applied Materials, Karlsruhe Institut of Technology and the Institute of Materials and Processes, Hochschule Karlsruhe Technik und Wirtschaft. Several of the most important problems in the research of materials science deal with ceramic materials. Besides ceramics, many materials with unique electrical, chemical, mechanical and optical properties can be found. However, in most cases, processing ceramics and fabricating ceramic parts involve powder techniques and, in particular, sintering to form a functional microstructure. During sintering, a body of compacted powder redistributes its volume to fill inner cavities or pores. This process is driven by minimizing the surface energy and involves diffusion of material and simultaneous migration of interfaces. According to a qualitative model, the sintering process can be divided in three stages. In the initial stage, particles are in contact with each other forming a grain boundary and a sintering neck. In the intermediate stage, the surface energy drives the coalescent network of particles to local shape changes and enhanced shrinking. In the final stage of sintering, the remaining porosity is decreased to 10% and grain growth starts to occur in addition to shrinkage and diffusion. This grain growth decreases the driving force for further sintering and is undesirable for most applications, but hard to avoid. In this contribution, a model to treat pore coalescence and to describe more complex pore interactions is introduced.

The interior dynamics of the terrestrial planets is considered in the contribution of the Institute of Planetary Research of the German Aerospace Center and the University of Applied Sciences in Berlin. Numerous data from various space missions over the past decades have revealed that all other terrestrial bodies apart from the Earth operate in the so-called stagnant lid regime. While the surface of the Earth is broken into seven major plates, the surface of all other terrestrial bodies is covered by an immobile layer. These two different convection styles have important implications for the interior evolution and result in different evolutionary paths for the volcanic outgassing and the core cooling. The atmosphere and the magnetic field are fundamental in maintaining habitable conditions at the surface by regulating the temperatures to allow for liquid water and shielding against solar radiations. In order to constrain the conditions necessary for habitability it is important to understand the thermo-chemical evolution of the interior of the Earth and other terrestrial bodies. To this end, numerical simulations of planetary interiors have become one of the most important tools to tackle complex fluid dynamics problems. In this study recent improvements and results using the mantle convection code Gaia will be presented.

The contribution by Mario Heene, Alfrdeo Parra Hinojosa, Michael Obersteiner, and Dirk Pflüger presents results from the EXAHD project. EXAHD is on of the sixteen cross-disciplinary consortia in DFG's Priority Programme 1648 "SPPEXA—Software for Exascale Computing". The overall research topic of EXAHD is to demonstrate that higher dimensional simulation problems—here the 5D gyrokinetic equations in plasma physics—can be successfully, efficiently, and scalably tackled via minimal-invasive extensions to existing software (here GENE) with the help of the sparse grid combination technique. The concrete topic of the paper presented here, however, has a different flavor. It explores the elegant fault-tolerance features the combination technique offers through its multi-level characteristics. This means an algorithmic or application-driven fault tolerance that can do without checkpointing or process replication—which will be an asset in exascale times.

Massively Parallel Multigrid for the Simulation of Skin Permeation on Anisotropic Tetrakaidecahedral Cell Geometries

Sebastian Reiter, Arne Nägel, Andreas Vogel, and Gabriel Wittum

Abstract Numerical simulation based on mathematical models is an important pillar for enhancing the understanding of permeation processes in the skin. To adequately resolve the complex geometrical structure of the skin, special models based on tetrakaidecahedral cells have been suggested. While these models preserve many of the desirable properties of the underlying geometry, they impose challenges regarding mesh generation and solver robustness.

To improve robustness of the used multigrid solver, we propose a new mesh and hierarchy structure with good aspect ratios and angle conditions. Furthermore, we show how those meshes can be used in scalable massively parallel multigrid based computations of permeation processes in the skin.

1 Introduction

Mathematical modeling is one important pillar for enhancing the understanding of permeation processes in the skin [1, 2]. The sub-class of microscopic models consider conservation laws and morphology on the cellular scale and combine both to improve the qualitative and quantitative understanding.

For the epidermis Allen and Potten [3] suggested a morphology model consisting of agglomerated tetrakaidecahedra. A tetrakaidecahedron (*TKD*), depicted in the rightmost figure in Fig. 1, is a polyhedron with 14 faces providing a dense spatial packing. This property has been discovered by Lord Kelvin [4] studying foam cells and makes it an attractive cell template for the construction of idealized biological tissues.

Based on *TKD* shaped cells Feuchter et al. [5] suggested a geometry model for the stratum corneum (SC), i.e., the outermost layer of the epidermis. This region

S. Reiter (✉) • A. Nägel • A. Vogel
G-CSC, Goethe-Universität Frankfurt, Kettenhofweg 139, 60325 Frankfurt (M.), Germany
e-mail: sebastian.reiter@gcsc.uni-frankfurt.de; arne.naegel@gcsc.uni-frankfurt.de;
andreas.vogel@gcsc.uni-frankfurt.de

G. Wittum
Goethe Center for Scientific Computing (G-CSC), Simulation and Modelling Department,
Goethe-Universität Frankfurt am Main, Kettenhofweg 139, 60325 Frankfurt am Main, Germany
e-mail: wittum@gcsc.uni-frankfurt.de

© Springer International Publishing AG 2018   457
W.E. Nagel et al. (eds.), *High Performance Computing in Science
and Engineering '17*, https://doi.org/10.1007/978-3-319-68394-2_27

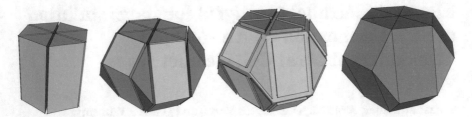

Fig. 1 New simplified meshing of tetrakaidecahedral elements. *Left:* Inner prism structure, *middle left:* full *TKD* mesh with additional prism and tetrahedral elements, *middle right:* Surrounding lipid elements in an explosion visualization, *right:* Surface of a complete building block consisting of the inner *TKD* and the surrounding lipid layer

is believed to be responsible for the barrier property of the skin and has been studied by modeling in [6, 7]. Nowadays, computing interactions in large networks of cells becomes feasible. At the same time, the models do not only focus on the SC only, but extend to the living epidermis [8, 9]. From a computational perspective this is important as the increase in size should be accommodated by a scaling of computational resources. Multigrid solvers were employed before [10], but weak scalability has not been the primary focus of the research in this area.

To fully resolve the highly differing scales between corneocytes and the surrounding lipid layer, very fine grids are required. Considering the complex *TKD* based geometric model and the unstructured nature of the associated grids, massively parallel computers are required to handle the large number of elements and unknowns involved.

In [11] we presented a massively parallel solver for the problem of drug diffusion through the skin on a much simpler geometric model. In this article, we consider the parallel meshing involved in preparing distributed grid hierarchies for the application of massively parallel multigrid solvers on *TKD* based grids. To this end, we develop an alternative meshing which leads to a very robust and still highly scalable problem setup.

This work is organized as follows: The mathematical model is formulated in Sect. 2, underlying geometries and new meshing approaches are described in Sect. 3. The description of the solver setup is given in Sect. 4 and numerical results are presented in Sect. 5.

2   Problem Description

Permeation can be described by the diffusion equation

$$\frac{\partial Ku}{\partial t} + \nabla \cdot [-K\mathbb{D}\nabla u] = 0 \qquad (1)$$

Fig. 2 New meshing of flat tetrakaidecahedral cells with surrounding lipid layer. *Left:* Surface of the new *TKD* meshing, *middle:* Explosion visualization of the new *TKD* meshing, *right:* Element structure near small lipid angles

with suitable initial and boundary conditions. The unknown u is a normalized reference concentration, and $K = K(x)$ and $\mathbb{D} = \mathbb{D}(x)$ are the partition coefficient and diffusion tensor respectively.

Both parameters are position dependent. In general, K is a function of the chemical potential μ (or alternatively the Gibbs energy $\triangle G := \mu - \mu_0$) [12, 13]:

$$K(x) := \exp\left(-\frac{\mu(x) - \mu_0}{RT}\right)$$

Similarly, diffusion depends also on the spatial position $x \in \mathbb{R}^3$.

The focus of this work is to demonstrate experimentally, how an efficient multigrid solver can be constructed for this problem. This should be exemplified for the *stratum corneum*, which is the outermost layer of the epidermis, consisting of keratinized cells embedded in a matrix of lipid bilayers. As a consequence the coefficient $\epsilon(x) := K(x)\mathbb{D}(x)$ is highly variable and jumps between corneocytes and the lipids occur. Morcover, the geometry features a large degree of anisotropy. Both properties enter the discretization and must be resolved by the solver as well.

As a model problem, we consider the equilibrium case with the simplifying assumptions $K = 1$ and

$$\epsilon(x) = \begin{cases} 1, & x \in \Omega_{LIP} \\ \epsilon_{COR}, & x \in \Omega_{COR} \end{cases}.$$

Here, Ω_{COR} and Ω_{LIP} correspond to the regions of the corneocytes and the lipids respectively. In Fig. 2 these are shown in light red and light blue respectively.

3 Meshing

To create a geometry model of the stratum corneum, we consider so called building blocks consisting of one tetrakaidecahedral cell with one layer of surrounding lipid, following [5]. Stacking those building blocks in the different space directions leads to a mesh which closely resembles the structure of the stratum corneum.

3.1 Previous Mesh Structure

Feuchter et al. [5, 14] introduced a meshing of a tetrakaidecahedral cell with 3 hexahedral, 36 prism, 6 pyramidal, and 18 tetrahedral elements. Together with the surrounding lipid layer, which was meshed with 18 hexahedral and 36 prism elements, a total of 117 elements per building block was required. The final mesh was then created by shifting copies of this building-block in the different space dimensions.

Besides the large number of elements, the proposed structure led to issues when the cells were flattened, as required to resemble the flat structure of the stratum corneum. When flattened, the elements in the surrounding lipid layer exposed highly obtuse angles. Solvers applied to linearized systems on such grids tend to require large iteration numbers or even diverge completely. Multigrid methods, which are required for scalable parallel computations, need complex smoothers and base solvers to address those issues. Such base solvers and smoothers may not scale well or may even fail completely in massively parallel environments.

We thus developed an alternative mesh structure which still allows to stack the *TKD* based building blocks in all space directions, yet provides a grid with much better element qualities than previous meshing attempts.

3.2 New Mesh Structure

We created the alternative meshing in two steps. First, we created a mesh for a tetrakaidecahedral cell using only 18 elements (12 prisms and 6 tetrahedra) for the corneocyte, and 36 elements (24 prisms and 12 hexahedra) for the surrounding lipid layer, totaling in 54 elements for one building block (cf. Fig. 1).

However, while consisting of less than half of the number of original elements, the new meshing still exhibits the same highly obtuse angles when flattened. We thus further adjusted the mesh structure by subdividing certain elements to remove the obtuse angles. Hereby it was important to subdivide in a way which allows to stack the building blocks in all space directions while guaranteeing mesh consistency. The resulting mesh structure is depicted in Fig. 2. A comparison between the meshing at the outer rim of the old and new mesh structures is given in Fig. 3.

The new mesh structure for a complete building block consists of 6 tetrahedra, 96 prisms, and 72 hexahedra (174 elements in total). No highly obtuse angles are contained in the new mesh anymore. A cut through a mesh resulting from stacking those building blocks is given in Fig. 4. Meshing was performed using the software *ProMesh* (cf. [15]).

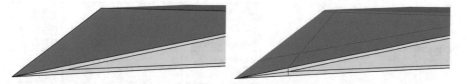

Fig. 3 Comparison of the meshing at the outer rim of the *TKDs*. *Left:* The old meshing exposed highly obtuse angles where now angles close to 90° can be found. *Right:* New meshing with vastly improved angles

Fig. 4 Cuts through a mesh resulting from stacking *TKDs* with the proposed new mesh structure. *Left:* Overview, *right:* Closeup

3.3 Refinement

We are using an anisotropic refinement scheme analogous to [11], where we only refine those edges which are longer than a certain threshold. After each refinement step, we halve the threshold and consequently more edges are considered during refinement. The initial threshold can for example be chosen as $0.75 \times maxEdgeLen$, where *maxEdgeLen* is the length of the longest edge of the given mesh.

We perform five anisotropic refinements followed by further isotropic refinements, depending on the number of intended multigrid levels. With each refinement we introduce a new level in the multigrid hierarchy. During anisotropic refinement element ratios of refined elements are improved compared to their respective parent elements in each consecutive level. Elements which are not refined, since none of their edges are longer than the given threshold, are copied to the next level.

Refinement is performed in parallel and interweaved with parallel redistribution, using the techniques described in [11]. Using this technique, we can efficiently generate large meshes spanning thousands or hundred thousands of processes.

Table 1 Solver components: Solvers on lower levels serve as base solvers for the preconditioners on higher levels, (cf. [11])

Levels	2	2–4	4–top
Solver	LU (exact)	BiCGStab	BiCGStab
Preconditioner	–	GMG	GMG
Smoother	–	ILU	Jacobi
Cycle	–	V (3,3)	V (3,3)

4 Solver Setup

We are using the same solver setup as described in [11]. It consists of two stacked multigrid preconditioned BiCGStab solvers, one serving as base solver for the other. The base-multigrid thereby only spans up to 64 processes and uses an ILU smoother, which works very efficiently on anisotropic geometries on small process numbers. Since the ILU smoother is not required for the isotropic elements encountered in higher levels, we are using a perfectly scalable damped Jacobi smoother for the upper multigrid. The solver setup is shown in Table 1. For details on the highly scalable MPI based parallelization of the underlying multigrid solver for hundreds of thousands of processes, please refer to [16].

5 Results

We performed weak scaling studies of the described model problem on the proposed *TKD* based mesh. The computations were executed on the Cray XC40 super computer *Hazel Hen* at the HLRS. *Hazel Hen* has a peak performance of 7420 TFlops. It features 7712 compute nodes which provide 128 GB of memory each. On each node 24 cores are available (up to 48 through hyperthreading).

Runs were performed starting at 8 processes and scaling up to 32,768 processes, increasing the number of processes by a factor of 8 between each run. With each run we also increased the number of elements by refining the grid. However, due to the presence of prism elements, the number of elements grows slightly slower than the number of processes. Between runs 1 and 2 the number of elements grows by a factor of 7.5, between runs 4 and 5 by a factor of 7.93. The workload per process is thus only approximately constant over all runs.

In each run we computed the solution of the steady state of the described model problem. To investigate the robustness of the solver regarding jumps in diffusion coefficients, we performed two studies with different coefficients $\epsilon_{COR} = 1$ (study A) and $\epsilon_{COR} = 0.001$ (study B). The distribution of the underlying multigrid hierarchy and the general solver setup was identical in both studies, only the number of iterations required to reach a given defect varied slightly.

Table 2 Each line corresponds to an individual run. Recorded are the number of processes (*PEs*), the number of levels (*Levels*), the number of unknowns (*DoFs*), the run times of assembly (T_{ass}), solver initialization (T_{ini}), and solving in study A (T_{sol}^A), and study B (T_{sol}^B) in seconds

PEs	Levels	DoFs	T_{ass}	T_{ini}	T_{sol}^A	T_{sol}^B
8	5	374,223	0.30	1.44	3.72	7.88
64	6	2,807,351	0.58	2.21	4.95	9.36
512	7	21,703,869	0.71	2.40	5.31	9.58
4096	8	170,592,761	0.80	2.51	5.69	10.4
32,768	9	1,352,552,433	0.83	2.53	5.95	10.6

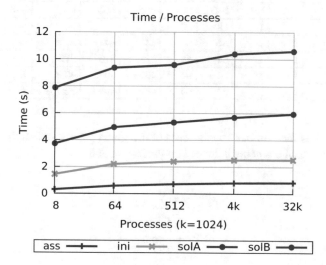

Fig. 5 Scaling of the run times of assembly (*ass*), solver initialization (*ini*), and solving of study A (*solA*), and study B (*solB*)

The number of CG-iterations were fixed to 10 in each run of study *A* and to 14 in each run of study *B* to allow for a good comparability of the timings. In each run we achieved a reduction of the initial defect by a factor of 10^{-5}.

The number of unknowns and the timings of the assembly and solution process in each run are given in Table 2. Since assembly and solver initialization are mostly identical in studies *A* and *B*, we only show the different timings in the actual solving stage T_{sol}. The scaling behavior of assembly, solver initialization, and solving is also shown in Fig. 5.

In Table 3 the levelwise distribution qualities for each run are shown. The distribution quality q_l of a level *l* of the hierarchy is computed as

$$q_l := \frac{n_l^{total} - n_l^{max}}{n_l^{max} \cdot (P_l - 1)},$$

Table 3 Distribution qualities for each level of the multigrid hierarchy for the different runs

PE	lv-0	lv-1	lv-2	lv-3	lv-4	lv-5	lv-6	lv-7	lv-8	lv-9
8	1	1	1	0.96	0.97	0.99	–	–	–	–
64	1	1	1	0.59	0.81	0.89	0.88	–	–	–
512	1	1	1	0.71	0.96	0.86	0.85	0.84	–	–
4096	1	1	1	0.71	0.96	0.85	0.84	0.82	0.80	–
32,768	1	1	1	0.71	0.96	0.87	0.89	0.85	0.84	0.82

Table 4 Number of processes used on each level for the individual runs

PE	lv-0	lv-1	lv-2	lv-3	lv-4	lv-5	lv-6	lv-7	lv-8	lv-9
8	1	1	1	8	8	8	–	–	–	–
64	1	1	1	64	64	64	64	–	–	–
512	1	1	1	64	64	512	512	512	–	–
4096	1	1	1	64	64	4096	4096	4096	4096	–
32,768	1	1	1	64	64	4096	4096	32,768	32,768	32,768

where $P_l > 1$ is the number of processes of the given process-hierarchy on level l, n_l^p is the number of elements in level l on process p, and

$$n_l^{total} := \sum_{p=1}^{P_l} n_l^p,$$

$$n_l^{max} := \max_{p=1,\dots,P_l} n_l^p.$$

For $P_l = 1$ numerator and denominator both vanish and we define $q_l = 1$. q_l is thus in the range $[0, 1]$, where $q_l = 0$ means that all elements of level l are contained on one process only and $q_l = 1$ reflects an equal share of elements among all processes.

Since we are using a hierarchical distribution scheme (cf. [16]), not all processes are contributing on all levels. Table 4 gives the number of involved processes on each level of the multigrid hierarchy for each run.

Runtimes of the solver are higher in study B compared to study A. Besides the fact that more solver iterations are performed in study B, the problem is now more difficult to solve due to the jumping coefficients. Due to this, the iteration numbers of the base solver increase, which affects the overall performance. Nevertheless, runtimes and scalability can be considered good for both studies.

While the first run (8 PEs) is contained on one node only, subsequent runs on 64 and more PEs are distributed onto several nodes. The overhead introduced through message passing between nodes is already seen in the jump in runtimes between the first two runs.

Given the highly scalable solver setup (cf. [16] and [11]), the biggest issues regarding scalability are to be expected from load imbalances. Indeed, correlating Table 3 and timings in Table 2, we see that imbalances on higher levels affect the efficiency on large process numbers. Still, overall efficiency is very good considering the unstructured nature of the problem.

While improvements regarding load balancing of unstructured multigrid hierarchies on massively parallel architectures could further improve scalability of the presented simulation, the high efficiency and robustness of the current approach is more than sufficient to perform large scale computations of complex biological problems on massively parallel computers.

Acknowledgements This work has been supported by the DFG in the *German Priority Programme 1648—Software for Exascale Computing* in the project *Exasolvers* (WI 1037/24-2) and by the *German Ministry of Economics and Technology (BMWi)* (02E11476B). We thank the HLRS for the opportunity to use *Hazel Hen* and their kind support.

References

1. S. Mitragotri, Y.G. Anissimov, A.L. Bunge, H.F. Frasch, J.Hadgraft, R.H. Guy, G.B. Kasting, M.E. Lane, M.S. Roberts, Mathematical models of skin permeability: an overview. Int. J. Pharm. **418**, 115–129 (2011)
2. A. Naegel, M. Heisig, G. Wittum, Detailed modeling of skin penetration - an overview. Adv. Drug Deliv. Rev. **65**(2), 191–207 (2013)
3. T.D. Allen, C.S. Potten, Significance of cell shape in tissue architecture. Nature **264**(5586), 545–547 (1976)
4. W. Thomson, On the division of space with minimum partitional area. Philos. Mag. **24**, 503 (1887)
5. D. Feuchter, M. Heisig, G. Wittum, A geometry model for the simulation of drug diffusion through the stratum corneum. Comput. Vis. Sci. **9**, 117–130 (2006)
6. A. Nägel, M. Heisig, G. Wittum, A comparison of two-and three-dimensional models for the simulation of the permeability of human stratum corneum. Eur. J. Pharm. Biopharm. **72**(2), 332–338 (2009)
7. I. Muha, A. Naegel, S. Stichel, A. Grillo, M. Heisig, G. Wittum, Effective diffusivity in membranes with tetrakaidekahedral cells and implications for the permeability of human stratum corneum. J. Membr. Sci. **368**, 18–25 (2010)
8. M. Yokouchi, T. Atsugi, M. Van Logtestijn, R.J. Tanaka, M. Kajimura, M. Suematsu, M. Furuse, M. Amagai, A. Kubo, Epidermal cell turnover across tight junctions based on Kelvin's tetrakaidecahedron cell shape. eLife **5**:e19593 (2016). https://doi.org/10.7554/eLife.19593
9. R. Wittum, A. Naegel, M. Heisig, G. Wittum, Mathematical modelling of the viable epidermis: impact of cell shape and vertical arrangement (2017, submitted)
10. A. Nägel, Schnelle Löser für große Gleichungssysteme mit Anwendungen in der Biophysik und den Lebenswissenschaften, PhD thesis, Universität Heidelberg, 2010
11. S. Reiter, A. Vogel, A. Nägel, G. Wittum, A massively parallel multigrid method with level dependent smoothers for problems with high anisotropies, in *High Performance Computing in Science and Engineering '16* (Springer, Cham, 2016), pp. 667–675
12. Y.G. Anissimov, M.S. Roberts, Diffusion modeling of percutaneous absorption kinetics: 3. Variable diffusion and partition coefficients, consequences for stratum corneum depth profiles and desorption kinetics. J. Pharm. Sci. **93**(2), 470–487 (2004)
13. R. Schulz, K. Yamamoto, A. Klossek, R. Flesch, S. Hönzke, F. Rancan, A. Vogt, U. Blume-Peytavi, S. Hedtrich, M. Schäfer-Korting et al., Data-based modeling of drug penetration relates human skin barrier function to the interplay of diffusivity and free-energy profiles. Proc. Natl. Acad. Sci. **114**(14), 3631–3636 (2017)

14. D. Feuchter, Geometrie- und Gittererzeugung für anisotrope Schichtengebiete, PhD thesis, Universität Heidelberg, 2008
15. Promesh, http://www.promesh3d.com
16. S. Reiter, A. Vogel, I. Heppner, M. Rupp, G. Wittum, A massively parallel geometric multigrid solver on hierarchically distributed grids. Comput. Vis. Sci. **16**(4), 151–164 (2013)

Real Data Applications of Seismic Full Waveform Inversion

A. Kurzmann, L. Gaßner, R. Shigapov, N. Thiel, N. Athanasopoulos, T. Bohlen, and T. Steinweg

Abstract Full waveform inversion (FWI) is a powerful imaging technique which exploits the richness of seismic waveforms. We further developed FWI to obtain multi-parameter images at high resolution. Here, we involve physical parameters, such as velocities and attenuation of seismic waves as well as mass density, which are essential for a reliable petrophysical characterization of subsurface structures in hydrocarbon exploration, geotechnical applications and underground constructions. Referring to this, we successfully applied FWI to field datasets recorded in the Black Sea and in the shallow-water area of a river delta in the Atlantic Ocean. We obtained detailed subsurface images containing rock formations which might be potential gas deposits. Additionally, we performed synthetic studies as preparatory steps to verify methodological improvements for further field-data applications. Here, we demonstrate resolution capabilities of FWI for imaging geological structures beneath salt bodies, investigate strategies to recover attenuation information from seismic data and perform a joint inversion of surface waves to image the very shallow subsurface.

1 Introduction

In a time where natural resources are precious, the number of underground constructions is increasing. It is important to map earth's geological structures accurately by collecting seismic data and transforming them into subsurface images. We develop full waveform inversion (FWI) as a cutting-edge seismic inversion technique that accounts for the full information content of seismic recordings. Each echo from geological discontinuities is used to unscramble the subsurface. FWI retrieves multi-parameter models of the subsurface by solving the full wave equation. It allows to map structures on sub-wavelength scales. Thus, FWI helps to improve both petrophysical interpretation and geotechnical characterization of the subsurface.

A. Kurzmann • L. Gaßner • R. Shigapov • N. Thiel • N. Athanasopoulos • T. Bohlen •
T. Steinweg (✉)
Karlsruhe Institute of Technology (Geophysical Institute), Karlsruhe, Germany
e-mail: andre.kurzmann@kit.edu; thomas.bohlen@kit.edu; tilman.steinweg@kit.edu

© Springer International Publishing AG 2018
W.E. Nagel et al. (eds.), *High Performance Computing in Science and Engineering '17*, https://doi.org/10.1007/978-3-319-68394-2_28

First implementations of FWI were conducted in the 1980s in the time domain by Tarantola [17] and Mora [13] as well as in the frequency-domain in the 1990s by Pratt [15] (see [19] for a general FWI overview). Particularly due to huge improvements in high-performance computing, these FWI strategies have emerged as an efficient imaging tool. In our work we concentrate on the implementation of the time-domain FWI and its application to seismic field-data problems. It comprises two- and three-dimensional (e. g., [4]) modeling of viscoelastic wavefields and exploits—as a main advantage—straightforward and efficient parallelization by domain decomposition [2] and source parallelization [12] leading to a significant speedup on parallel computers.

Within the scope of the HPC project *KITFWT*, we present applications of FWI to field-data (Sects. 3.1–3.4)

2 Methodology

2.1 Full Waveform Inversion

FWI aims to find the optimal subsurface model by iteratively minimizing the misfit function between recorded and synthetic seismic data. That is, by solving the "forward problem" this model has to explain the recorded seismic data. The iterative optimization scheme of FWI—combining "forward problem" and "inverse problem"—comprises several steps shown in Fig. 1.

In detail, the method is initialized by two main inputs. First, we choose 2D or 3D initial parameter models of the subsurface, such as seismic velocities v_P of compressional wave (P-wave) and v_S of shear wave (S-wave), mass density ρ as well as attenuation represented by quality factors Q_P and Q_S for both wave types. They are assigned to the starting model at the first FWI iteration. The initial model can be estimated from a-priori information or computed by conventional seismic imaging methods. Second, the recorded data is obtained from seismic measurements involving many source locations and receivers. Typical acquisitions are performed offshore by utilizing air guns as sources and hydrophones located at sea surface/sea floor or onshore with hammerblow sources and geophones.

Within the FWI framework, for each source of the acquisition geometry, seismic modeling is applied (solution of "forward problem", see [2, 18]). That is, using the initial source wavelet the wavefield is emitted by the source and forward-propagates across the medium. A time series of spatial wavefield volumes has to be stored in memory. Synthetic seismic data is obtained at the receivers and the difference of synthetic and recorded data is calculated—resulting in residuals. In order to improve the minimization of the misfit function, a comprehensive multi-stage workflow focusses on both different model scales by applying frequency filtering to the data (e. g., [3, 16]) or choosing subsets of the data.

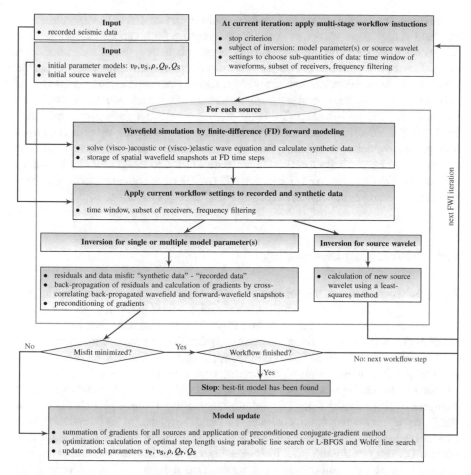

Fig. 1 General FWI scheme used for iterative improvement of physical model parameters of the subsurface by minimizing the misfit between modeled and recorded seismic data

For each source, the residual wavefield is back-propagated from the receivers to the source position. The cross-correlation of forward- and back-propagated wavefields yields source-specific steepest-descent gradients (solution of "inverse problem", see [10, 13, 14, 17]). The computation of the global gradient for the entire acquisition geometry is given by the summation of all source-specific gradients. Subsequent optimization methods, such as preconditioned conjugate-gradient method and L-BFGS method are applied. The update of the model parameter(s) is the final step of a FWI iteration. The gradient has to be scaled by an optimal step length to get a proper model update. The estimation of the step length might require a significant amount of additional forward modelings. Furthermore, regarding the source wavelet, an equivalent inverse problem is given. The true wavelet is not known. The initial wavelet (a rough estimation or synthetic signal) is subject to FWI, too, and optimized during inversion by a least-squares method [6, 15].

Seismic modeling represents the fundamental part of FWI. In dependence of the field of application, the wave-propagation physics for an underlying subsurface model has to be described by an appropriate wave equation. On the one hand, that comprises the utilization of (visco-)acoustic or (visco-)elastic wave equation. On the other hand, the problem has to be solved for two-dimensional or three-dimensional subsurface models. The numerical implementation of the wave equations consists of a time-domain finite-difference (FD) time-stepping method in cartesian coordinates. In detail, the FD-scheme solves the stress-velocity formulation by utilizing stress and particle-velocity wavefields. Due to finite model sizes, the wave equations are expanded by perfectly matched layer terms (PML) to avoid artificial boundary reflections. Finally, at each FWI iteration a 2D or 3D wave equation has to be solved for a certain number of sources in forward- and back-propagation.

2.2 Parallel Implementation

Apart from other factors, the success of a FWI depends on a sufficient illumination of the model area. Thus, several source and receiver positions are necessary (reasonable numbers may vary between 20 and more than 500). For each source, modelings have to be performed separately requiring most of the entire computation time of FWI. That results in huge computational efforts, which can be handled by a massive parallelization. Our FWI implementation offers two types of parallelization. On the one hand, the model area can be decomposed into subdomains, which are assigned to all available cores [2]. Additional padding layers with half the size of the spatial differential operator are located around the model. At each time step these model boundaries are exchanged by Message-Passing-Interface communication (*MPI*). On the other hand, due to increasing communication, modelings cannot benefit from the decomposition of a model into a high number of very small subdomains. Hence, it should be supplemented with parallelized modelings with respect to the sources [11, 12]. The combination of domain decomposition and source parallelization results in nearly perfect speedup on supercomputers.

3 Results Obtained on FORHLR Phase I

3.1 Application of FWI to Image Submarine Gas Hydrate Deposits

3.1.1 Motivation

We want to characterize gas hydrate deposits in the Western Black Sea from data collected by ocean-bottom seismometers (OBS). Gas hydrate deposits can be detected with seismic methods through the observation of so called bottom-simulating reflectors (BSR). They result from a velocity increase in sediments where hydrate accumulates and a strong decrease in velocity where gas is trapped beneath the hydrated sediment layer, i.e., the gas hydrate stability zone. Stability conditions of hydrate are closely related to the thickness of the sediment column. Consequently, the BSR is often parallel to the seafloor and crosscuts geological reflectors. Nevertheless, seismic interpretation can not provide detailed information on seismic velocities, which are relatable to hydrate and gas saturation. Detailed velocity models are therefore obtained by FWI.

3.1.2 Field Data Application

OBS data utilized for FWI was collected together with streamer data along two parallel profiles of 14 km length. Both profiles are situated 1 km apart along a channel axis where a BSR horizon of varying distinctness has been mapped. Five stations were deployed in the central part of each profile with 1 km spacing. Data for a total of 1300–1700 shots has been acquired for profile P1 and profile P2, respectively.

Synthetic studies showed the sufficiency of the field geometry to resolve an extended hydrate and gas layer in a v_P-model with a typical BSR signature. Under field data conditions, i.e., noise at the required low frequencies and insufficient starting models, no resolution in multiparameter FWI for v_S and density was observed. For the moderate quality field data available, we therefore concentrate on acoustic inversion, inverting simultaneously for v_P and density. Attenuation is considered as a passive inversion parameter, with a constant value of $Q = 100$ in the sediment column.

Field data preparation includes data transformation to correct for 3D-geometrical spreading effects not considered in the 2D-FWI scheme. During the inversion process filtering is applied to increase the frequency content gradually. We use time windowing to exclude the direct wave arrivals from the inversion process, as they exhibit strong ringing which can not be explained by the applied approach. We therefore only invert refracted wave signals and multiple reflections. The comparability of field and synthetic data is ensured by the normalization of each trace with its rms-amplitude.

To suppress undesired model updates, e.g., in the watercolumn and close to the OBS positions we apply spatial preconditioning to the gradients. Furthermore, we enhance updates in the deeper parts of the model were coverage is reduced. To suppress horizontal parameter fluctuations which can arise due to the inadequate consideration of S-wave propagation we apply a smoothing filter.

3.1.3 Results

For the inverted v_P-models a satisfactory match of input and modeled data is achieved and a high similarity of the recovered STF is obtained for both profiles. Inverted v_P-models show significantly more detail than the starting models, with more pronounced layers and smaller scale velocity structures. On both profiles, we observe a velocity reduction in the first 30 m below the seafloor, and shallow low-velocity zones at around 100 mbsf. On profile P1 we resolve an extended low velocity zone together with increased velocities above of a length of 4.5 km and approximately 100 m vertical extent. The discovered velocity behaviour strongly indicates the occurrence of hydrated sediments and gas accumulation. No such behaviour can be observed at profile P2 where only a slight velocity increase is detected at BSR depth.

Velocity structures obtained by FWI can also be identified as reflection patterns in the migrated seismic streamer data (compare Fig. 2). Shallow low velocity zones as well as the observed signature at BSR depth in profile P1 are coincident with increased reflectivity amplitude areas.

3.1.4 Summary

We applied acoustic 2D FWI to ocean-bottom seismometer data to identify hydrated sediments and gas accumulations. A bottom simulating reflector (BSR) which was tracked in high resolution reflection seismic data indicates their occurrence. The observation of a sharp BSR coincides with the recovered velocity signature. Low velocity zones recovered by FWI are aligned with areas of increased reflectivity amplitudes.

3.2 Application of FWI to Shallow-Water OBC-Data with Minimal Data Preprocessing and Simple Inversion Strategy

3.2.1 Motivation

FWI with minimal data preprocessing and simple inversion strategy is still a dream, rather a routine method. The level of noise in the real data and modelization errors are often so high that theoretically sound FWI becomes an ugly and complicated algorithm. In this work we present application of a 2D acoustic FWI with a minimal data preprocessing and simple inversion strategy to the real marine data acquired in a river delta using ocean bottom cables (OBC) with relatively high signal-to-noise ratio.

3.2.2 FWI of Shallow-Water OBC-Data with Minimal Data Preprocessing and Simple Inversion Strategy

We used 57 shot gathers acquired in a shallow-water marine experiment. Each shot gather corresponds to the air gun placed at 6 m depth and contains 240 traces

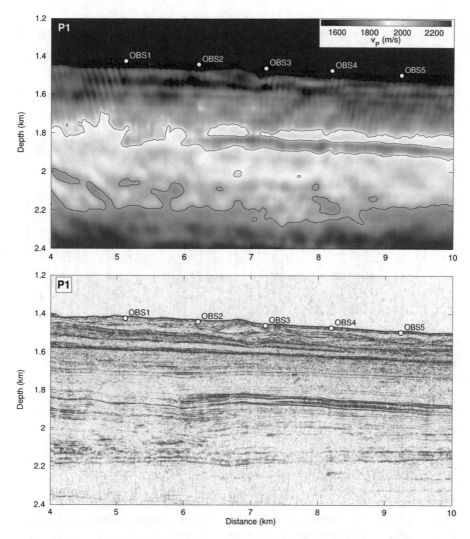

Fig. 2 Top: Inverted v_P-model with OBS locations. A typical BSR signature is visible beneath OBS2 to OBS5. Bottom: Migrated seismic section of profile P1 with increased reflection amplitudes correlating with the low velocity zone of the inverted v_P-model

registered by hydrophones placed at the sea floor (120–140 m depth). As a data preprocessing we used only 3D-to-2D conversion, based on the exact formula strictly valid for a homogeneous acoustic medium, and data interpolation with a proper time sampling dictated by the stability and dispersion limits of the 2–8 order finite-difference scheme used to solve the acoustic wave equations.

Our simple inversion strategy includes only a standard multiscale approach in the frequency band 2–32 Hz. We do not use any time- and offset-windowing and invert full seismograms. We use the pseudo-Hessian as a preconditioner for the gradient of

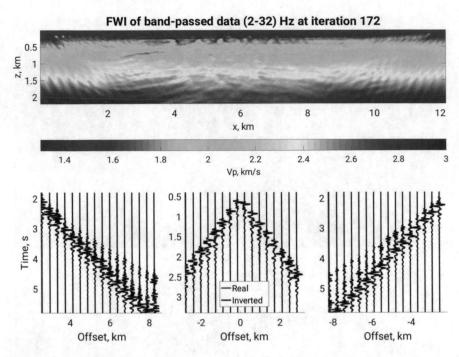

Fig. 3 The results of 2D FWI and data fit at different offsets

a standard L_2 misfit between real and modelled data. The initial model was provided by a third party and corresponds to a travel-time tomography result.

3.2.3 Results

The inverted model, shown in Fig. 3, provides a very good data fit at all offsets. The inverted model, overlaid by the 3D migrated image provided by a third party, is shown in Fig. 4. The main high-contrast zones in the inverted model correlate well with the high-reflectivity zones in the migrated image.

3.2.4 Summary

We presented application of a 2D acoustic FWI with a minimal data preprocessing and simple multiscale strategy to the real shallow-water marine data acquired in a river delta using ocean bottom cables (OBC) with relatively high signal-to-noise ratio. The inverted model provides a very good data fit at all offsets and correlates well with a 3D migrated image provided by a third party.

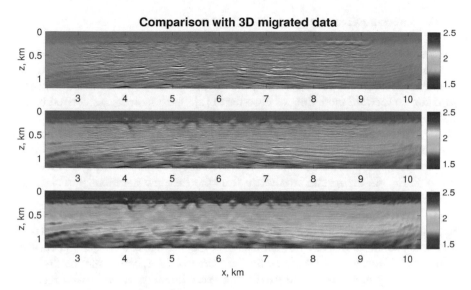

Fig. 4 The results of 2D FWI overlaid by the 3D migrated image for different levels of image transparency

3.3 Subsalt Imaging with Acoustic and Elastic FWI

3.3.1 Motivation

Salt bodies proved to be promising sites for the search for hydro carbonates. For classical imaging techniques the reconstruction of structures beneath or near salt bodies is challenging. One reason for this is the high reflection coefficient at the salt-sediment-interface that results in only weak scattered energy returning from the subsalt regions. Additional reasons are the complex shape of the salt bodies, trapped sediments in the salt body and a rugose surface. These characteristics lead to complex wavefields and regions with poor illumination. The solution for these problems can be FWI that is capable to use weak scattered waves travelling in complex velocity models. In this work we explore the performance of 2D acoustic and elastic FWI in time domain for subsalt imaging.

For this we use field data provided by PGS (marine 2D line). The profile is 265 km long with a total number of 5300 shots. The model is 15 km deep and for a better handling, the model was divided into three subpart. Only the right subpart will be shown in the following. The subpart has a size of 88.5 × 12 km, a grid distance of 12.5 m and a record length of 12 s. for each of the 99 source points 804 receivers are used (moving streamer geometry).

3.3.2 Inversion Strategy

In this work we use the two dimensional FWI code IFOS2D. In order to minimize the computational costs acoustic inversion is performed (e.g. [9, 17]). The code uses the time domain stress-velocity finite-difference formulation on a standard staggered grid for the forward modelling as describe in [2]. The gradients are derived by the adjoint state method [17].

In order to quantify the misfit between the real and modelled data the normalised L2-norm is used. It proofed to be more robust for field data than the L2-norm [5]. The inversion starts at 5 Hz and increases the frequency content during the inversion up to 12 Hz. As the results are mostly from the first few iterations most inverted models were inverted with a frequency content of 5 Hz maximum.

To decrease the influence of noise on the inversion result and therefore the appearance of artefacts the field seismograms are muted before the first arrival. Important for a successful inversion is also the wavefield used for the forward simulation. Thus, we need to invert for the source wavelet, before starting the FWI.

The density model is not inverted. After every flooding a new density model is calculated from the velocity model by using the Gardner relation [7].

3.3.3 Results of Field Data Inversion

Despite of the availability of a interpreted model from PGS we started from zero and used the provided reference model only as quality control for our inversion and picking result. Therefore, the first step was to reproduce the sea floor. For this we used a homogeneous starting model with the water velocity of 1490 m/s. From the inversion result we picked the water bottom and flooded sediments below the picked line (Fig. 5a). The also plotted sea floor line of the 'true' model shows the high accuracy of the picking. In addition this test verifies the correctness of the acquisition geometry for the forward modelling.

The inversion was started again with the new velocity model. After one iteration the salt surface was picked and the model flooded with salt velocity. In Fig. 5b is the contour of our reference model plotted on top. The inverted salt surface is comparable with the on in the reference model. Due to some internal velocity layers in the top salt region and therefore, some difficulties in the picking, we get a slightly smaller depth of the surface. Especially in the middle part of the model between $x = 25\text{–}40$ km. The greatest differences are about 400 m, which is less than one wavelength in the sediments at 5 Hz. Also the deep valleys cannot be picked in the inversion result due to the lack of high frequencies. Furthermore, structures at a depth of about 4 km and between profile length 70–90 km are inverted, not visible in the starting model.

After starting the inversion again and picking the bottom salt line after 22 iterations, the model was flooded below the picked line with a sediment-velocity gradient. The comparison of the inverted salt bottom with the line of the reference model show high similarities. The rough structure is the same. The combination of

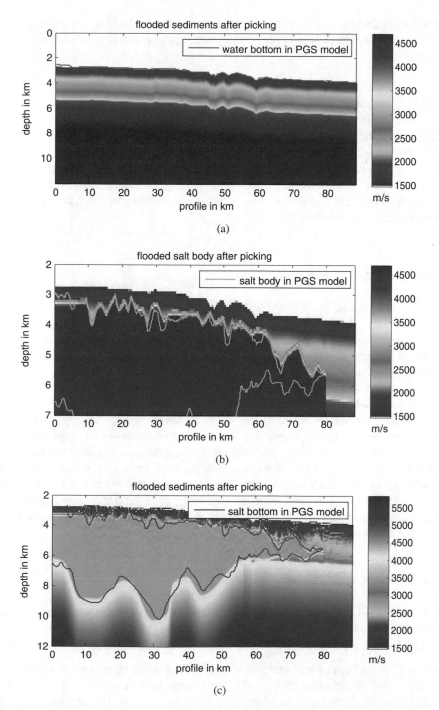

Fig. 5 V_p models in different inversion and flooding stages (field data). (**a**) Flooded sediments before inversion. (**b**) Model flooded with salt velocity below picked top salt. (**c**) Model flooded with sediments velocity gradient below picked salt bottom

flooding technique and FWI seems to be very effective in imaging the salt body in only few iterations. Moving to higher frequencies should improve the resolution considerably.

3.3.4 Summary

Our results indicate that FWI is able to image the sea bottom as well as the top salt interface. The latter is required for the application of the Flooding Technique allowing also the inversion of subsalt structures. Further tests with a better imaging of the shallow area below the seabed and the usage of higher frequencies will be the next step to get a higher resolution of the overburden which will in turn also improve the subsalt reconstruction.

3.4 Sequential FWI

3.4.1 Motivation

Shallow seismic applications are important for geotechnical site characterization. Vertical hammer blows allow the excitation of Rayleigh waves and compressional P-waves. However, in most cases the amplitudes of the Rayleigh waves are much higher than the amplitudes of the compressional P-waves. In these particular cases the full-waveform inversion (FWI, [17]) fails to properly incorporate the contribution of the P-waves, as the misfit is dominated by the high amplitude Rayleigh waves. To compensate for this effect we suggest a new sequential FWI (SFWI) strategy, in which we initially invert using the low-amplitude refracted P-waves and then the full wavefield including the Rayleigh waves, subsequently. We further apply the SFWI to field data and compare it with the conventional multi-parameter FWI. The results reveal an improvement of the data fitting and sharper reconstructed models, although the overall improvements did not reach the level of the synthetic cases.

3.4.2 Method and Theory

The new SFWI consists of a two-stage strategy. In the first stage we invert only the refracted P-waves to obtain an initial P-wave velocity model. In the second stage, we apply a three-parameter FWI using the full wavefield, including both wave types.

Stage 1 In the mono-parameter inversion of the P-waves, we separate the waveform of the refracted waves by time-windowing around the first arrivals. Outside of this window we apply an exponential damping of the amplitudes. The very near-offset is removed because a clear distinction between the direct, refracted and surface waves is not possible. After obtaining a satisfactory decrease in the misfit function by using this mono-parameter FWI, we move on to the second stage.

Stage 2 At this stage, we apply a multi-parameter FWI of the full wavefield in order to reconstruct the P-wave velocity (v_p), S-wave velocity (v_s) and density (ρ). As initial model for the P-wave velocity we use the model from stage 1. The inversion of the three parameters is performed simultaneously, using again the multi-scale approach (frequency filtering).

3.4.3 Field Data

The location of the survey is on a glider airfield in Rheinstetten near Karlsruhe (Germany). Previous studies also performed in the area [8] revealing a predominantly depth-dependent 1D subsurface. Also located in the area is a shallow low-velocity anomaly, remnant of a man-made trench served as a defensive line in the early eighteenth century. Wittkamp and Bohlen [20] extensively studied the same area and applied FWI of Rayleigh and Love waves (individual and joint inversion).

3.4.4 Results: Mono-Parameter Inversion (Stage 1)

In this section, we present the results of the reconstructed v_p model from only the refracted waves. In Fig. 6 (left) we compare the time-windowed (only refracted

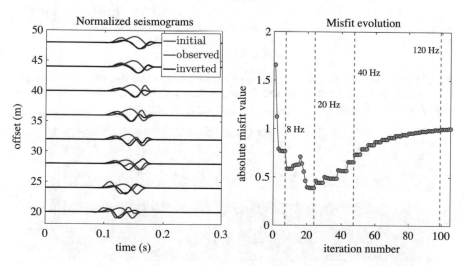

Fig. 6 Left: normalized displacement seismograms for 8 different offsets from 5 up to 120 Hz, calculated from the observed (blue), the inverted (black) and the initial (red) data for the first shot. Right: evolution of the misfit value with the number of iterations. The jumps in the misfit values correspond to updates on the frequency bandwidth

waves) displacement seismograms for eight different offsets of the observed data (blue) with the data obtained by the mono-parameter inversion (black) as well as the initial model (red). We observe that mainly phases but also the amplitudes are fitted exceptionally well. The misfit function plotted over the number of iterations (Fig. 6, right) reveals the increased fitting of the seismograms. This good fitting of the first arrivals is reflected in the model space as an increase of the velocity in the upper 4 m and the presence of a local discontinuity. The model obtained from this inversion is then used as initial v_p model for SFWI.

3.4.5 Results: FWI and SFWI

We present and compare the results of the conventional multi-parameter FWI (stage 2, without an initially updated v_p model) with the SFWI (stage 1 and 2). Figures 7 and 8 show the reconstructed v_s and v_p models obtained from SFWI (top) and FWI (bottom), respectively.

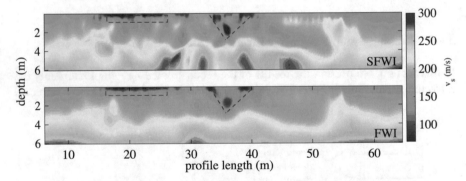

Fig. 7 S-wave velocity models reconstructed from SFWI and FWI, respectively. The location of the trench (red-dotted triangular) and the near surface low-velocity artifacts (black-dashed rectangular) are indicated

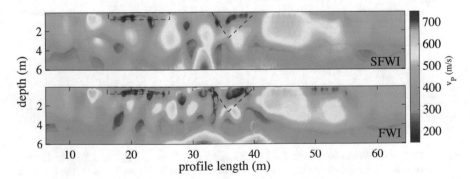

Fig. 8 P-wave velocity models reconstructed from SFWI and FWI, respectively. The location of the trench (red-dotted triangular) and the near surface low-velocity artifacts (black-dashed rectangular) are indicated

Both strategies produce a similar reconstruction of the v_s models (Fig. 7). The 2D variations seem to be superimposed to a predominantly depth dependent homogeneous background model. In the middle of the profile a triangular shaped low-velocity anomaly (Fig. 7, red-dashed) is presented, corresponding to the trench, up to a depth of about 2.2 m and horizontal length of 8 m. The velocity values vary between the two inversions. SFWI (Fig. 7, top) reveals a sharper structure, that agrees better with the joint FWI of Rayleigh and Love waves presented by Wittkamp and Bohlen [20]. They found that the trench holds higher velocities in the shallow part than in the lower part and bears a triangular shape. A second shallow low-velocity anomaly is also visible at the top left of the profile Fig. 7 (black-dashed) which could be related to increased saturation of the shallow soil in this area. This anomaly appears sharper in SFWI. In general, S-wave velocity model variations obtained by SFWI seem to be more pronounced since information from the refracted waves are accounted for in the inversion [1]. The v_s models suffer in all cases (synthetic and field data) from vertically orientated artifacts underneath the source positions.

The structure of the P-wave velocity above the estimated groundwater table revealed an intermediate layer at around 4 m depth in the case of SFWI. The low-velocity anomaly is also visible in v_p at offsets of 30 m and up to a depth of around 2 m. During our synthetic studies we found that there was no indication of significant cross-talk leaking from v_s to v_p models. The information from the updated v_p model is reflected also in the v_s model. SFWI incorporates better the information obtained from the refracted P-waves. The reconstructed density models did not provide valuable information, apart from the fact that higher density values are revealed compared to our initial model. However, from previous studies we know that the reconstruction of density can lead to significant cross-talk leaking from the v_s and v_p models.

To evaluate the resulting models we compare the displacement seismograms for four different offsets between the observed data, the conventional FWI and SFWI (Fig. 9, left) for frequencies up to 65 Hz. The inversion fitted the fundamental Rayleigh mode in both inversion types sufficiently well. In the case of SFWI the first arrivals are fitted better, especially their phase. The differences in seismogram fitting of this field data set do not reveal a big advantage of SFWI, opposite to our previous synthetic study. However, SFWI reaches a lower misfit value than the conventional FWI at most frequency stages (Fig. 9, right).

3.4.6 Conclusions

The conventional multi-parameter elastic FWI fails to reconstruct accurately the P-wave velocity model due to its low sensitivity to the Rayleigh waves which dominate the calculation of the misfit. We tested a two-stage sequential approach. In a first stage, only the information from the refracted waves was used to reconstruct the v_p model. Then a multi-parameter FWI of total wavefield is applied. Several 2D structures in the v_s models along with the low-velocity shallow anomaly

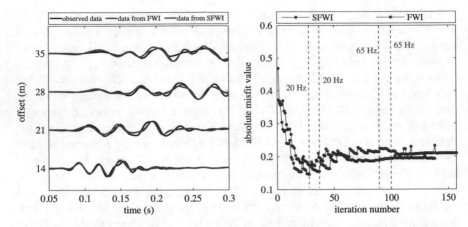

Fig. 9 Left: normalized displacement seismograms for 4 different offsets, calculated from the observed data (black), the inverted data from FWI (red) and SFWI (blue) for the first shot. Right: evolution of the misfit value with the number of iterations for both FWI and SFWI

were reconstructed. The SFWI provided a higher resolution and more pronounced variations in the v_s model. The v_p model revealed a better correlation to the v_s model in the case of SFWI. Although more accurate models were reconstructed, source-related artifacts near the surface were not eliminated, such as in the case of the synthetic study. In conclusion, the new sequential approach shows the improved capability to recover P- and S-wave velocities in the shallow subsurface. Further improvements are necessary to use this strategy in routine shallow seismic investigations.

4 Computational Efforts of FWI on FORHLR Phase I

Based on the field-data example shown in Sect. 3.3, we estimated the resource consumption of one FWI application as follows:

- finite-difference discretization and 2D wave simulations:

  - spatial discretization: 960×7080 grid points (**6.8 million grid points**)
  - 12,000 time steps for each simulation
  - number of seismic sources: 155
  - total amount of **28,750 simulations** within whole FWI

- parallelization:

  - domain decomposition: 2D model is divided into 24×40 subdomains
  - allocation of **960 CPU cores**

- resource consumption and computational performance of FWI framework:
  - number of iterations: 64
  - computation time for whole FWI job: 45 h (**43,200 core h**)
  - memory consumption for wavefield storage: 766 MB/core (**total: 0.74 TB**)

Acknowledgements The scientific projects in this work are kindly supported by the sponsors of the Wave Inversion Technology (WIT) Consortium and funded by BMWi (grant number 03SX381C). We also gratefully acknowledge financial support by the Deutsche Forschungsgemeinschaft (DFG) through CRC 1173. The computations were performed on the computational resource "ForHLR Phase I" funded by the Ministry of Science, Research and the Arts Baden-Württemberg and DFG ("Deutsche Forschungsgemeinschaft").

References

1. N. Athanasopoulos, T. Bohlen, Sequential full-waveform inversion of refracted and Rayleigh waves, in *Near Surface Geoscience 2016-22nd European Meeting of Environmental and Engineering Geophysics* (2016)
2. T. Bohlen, Parallel 3-D viscoelastic finite difference seismic modeling. Comput. Geosci. **28**, 887–899 (2002)
3. C. Bunks, F.M. Saleck, S. Zaleski, G. Chavent, Multiscale seismic waveform inversion. Geophysics **60**(5), 1457–1473 (1995)
4. S. Butzer, A. Kurzmann, T. Bohlen, 3D elastic full-waveform inversion of small-scale heterogeneities in transmission geometry. Geophys. Prospect. **61**(6), 1238–1251 (2013)
5. Y. Choi, T. Alkhalifah, Application of multi-source waveform inversion to marine streamer data using the global correlation norm. Geophys. Prospect. **60**(4), 748–758 (2012)
6. T. Forbriger, Inversion flachseismischer Wellenfeldspektren. Dissertation, Stuttgart, University of Stuttgart, 2001
7. G.H.F. Gardner, L.W. Gardner, A.R. Gregory, Formation velocity and density – the diagnostic basics for stratigraphic traps. Geophysics **39**(6), 770–780 (1974)
8. L. Groos, M. Schäfer, T. Forbriger, T. Bohlen, Application of a complete workflow for 2D elastic full-waveform inversion to recorded shallow-seismic Rayleigh waves. Geophysics **82**(2), R109–R117 (2017)
9. D. Köhn, Time domain 2D elastic full waveform tomography. Dissertation, Kiel, Christian-Albrechts-Universität zu Kiel, 2011
10. D. Köhn, D. De Nil, A. Kurzmann, A. Przebindowska, T. Bohlen, On the influence of model parametrization in elastic full waveform tomography. Geophys. J. Int. **191**(1), 325–345 (2012)
11. A. Kurzmann, Applications of 2D and 3D full waveform tomography in acoustic and viscoacoustic complex media. Dissertation, Karlsruhe, Karlsruhe Institute of Technology, 2012
12. A. Kurzmann, D. Köhn, A. Przebindowska, N. Nguyen, T. Bohlen, 2D acoustic full waveform tomography: performance and optimization, in *71st EAGE Conference and Technical Exhibition* (2009)
13. P. Mora, Nonlinear two-dimensional elastic inversion of multioffset seismic data. Geophysics **52**, 1211–1228 (1987)
14. R.-E. Plessix, A review of the adjoint-state method for computing the gradient of a functional with geophysical applications. Geophys. J. Int. **167**(2), 495–503 (2006)
15. R. Pratt, Seismic waveform inversion in the frequency domain, part 1: theory and verification in a physical scale model. Geophysics **64**, 888–901 (1999)
16. L. Sirgue, R.G. Pratt, Efficient waveform inversion and imaging: a strategy for selecting temporal frequencies. Geophysics **69**(1), 231–248 (2004)

17. A. Tarantola, Inversion of seismic reflection data in the acoustic approximation. Geophysics **49**, 1259–1266 (1984)
18. J. Virieux, P-SV wave propagation in heterogeneous media: velocity-stress finite-difference method. Geophysics **51**(4), 889–901 (1986)
19. J. Virieux, S. Operto, An overview of full-waveform inversion in exploration geophysics. Geophysics **74**, WCC1–WCC26 (2009)
20. F. Wittkamp, T. Bohlen, Field data application of individual and joint 2-D elastic full waveform inversion of Rayleigh and love waves, in *Near Surface Geoscience 2016 – 22nd European Meeting of Environmental and Engineering Geophysics* (2016)

The Impact of Pores on Microstructure Evolution: A Phase-Field Study of Pore-Grain Boundary Interaction

V. Rehn, J. Hötzer, M. Kellner, M. Seiz, C. Serr, W. Rheinheimer, M.J. Hoffmann, and B. Nestler

Abstract Among the most important issues of today's materials research ceramic materials play a key role as e.g. in Lithium batteries, in fuel cells or in photovoltaics. For all these applications a tailored microstructure is needed, which usually requires sintering: A pressed body of compacted powder redistributes its material and shrinks to a compact body without pores. In a very porous polycrystal, pores constrain the motion of interfaces (pore drag) and no grain growth occurs. During further sintering the number and size of pores decreases and the pore drag effect fades away. Accordingly, in the final stage of sintering grain growth emerges. This grain growth decreases the driving force for sintering and is undesirable, but hard to avoid. Since application of ceramic materials usually requires a dense and fine-grained microstructure, it is of high interest to control the interplay of remaining pores and interface migration during sintering.

Unfortunately, the present modeling of sintering does not allow for predicting microstructural evolution in an adequate way. In this study, a previously developed phase-field model of pore-grain boundary interaction during final stage sintering is extended by a pore-interaction module to improve the modeling of final stage

V. Rehn and J. Hötzer contributed equally.

V. Rehn • J. Hötzer (✉) • M. Seiz • W. Rheinheimer • M.J. Hoffmann
Institute of Applied Materials (IAM), Karlsruhe Institute of Technology (KIT),
Straße am Forum 7, 76131 Karlsruhe, Germany
e-mail: johannes.hoetzer@kit.edu

M. Kellner • B. Nestler
Institute of Applied Materials (IAM), Karlsruhe Institute of Technology (KIT),
Straße am Forum 7, 76131 Karlsruhe, Germany

Institute of Materials and Processes, Hochschule Karlsruhe Technik und Wirtschaft,
Moltkestr. 30, Karlsruhe, Germany

C. Serr
Institute of Materials and Processes, Hochschule Karlsruhe Technik und Wirtschaft,
Moltkestr. 30, Karlsruhe, Germany

© Springer International Publishing AG 2018
W.E. Nagel et al. (eds.), *High Performance Computing in Science and Engineering '17*, https://doi.org/10.1007/978-3-319-68394-2_29

sintering. This module handles the growth of pores that come into contact during grain growth.

The model with the extensions is used to simulate interface migration in a well-defined model setup. The results show that the present model is appropriate to describe grain growth during the final stage of sintering. However, the need of large scale simulations becomes evident: pore drag depends critically on the local geometry (i.e. position of the pores at grain boundaries, triple lines or quadruple points). The microstructure evolution during final stage sintering in polycrystalline ceramics underlies strong statistical variations in the local geometry. Accordingly, if grain growth in polycrystals in the presence of remaining pores from sintering is considered in detail, large scale simulations are needed to picture the local statistic variation of pore drag in an adequate way.

1 Introduction

Several of the most important problems of today's research in materials science deal with ceramic materials, as e.g. in the field of Lithium batteries, solar cells, lasers, catalysts, biomaterials and fuel cells. Amongst ceramics, many materials with unique electrical, chemical, mechanical and optical properties can be found. However, in most cases, processing ceramics and fabricating ceramic parts involve powder techniques and, in particular, sintering to form a functional microstructure. During sintering, a body of compacted powder redistributes its volume to fill inner cavities or pores. This process is driven by a minimization of surface energy and involves diffusion of material and simultaneous migration of interfaces. According to the qualitative model by Coble [6], the sintering process can be divided in three stages. In the initial stage, particles in contact with each other form a grain boundary and a sintering neck. Mass is transported to these contact zones from other areas of the particles. In the intermediate stage, the surface energy drives the coalescent network of particles to local shape changes and, thus, enhances shrinkage. In the final stage of sintering, the remaining porosity is decreased to 10% and grain growth starts to occur in addition to shrinkage and diffusion. This grain growth decreases the driving force for further sintering and is undesirable for most applications, but hard to avoid.

In general, grain growth occurs in all polycrystals and is driven by a minimization of grain boundary energy. The sudden occurrence of grain growth during final stage sintering is caused by an interaction of pores and grain boundary migration known as pore drag [3]. If a migrating interface hits a pore, its further movement is dragged by the movement of the pore itself as shown in Fig. 1a–c. In this case, as depicted in Fig. 2, grain growth is strongly suppressed as observed during the initial and intermediate stage of sintering. The details of this drag effect depend on the local driving force and some other parameters (surface and grain boundary energy, diffusion coefficients, mobility or size of pores). However, the density and size of

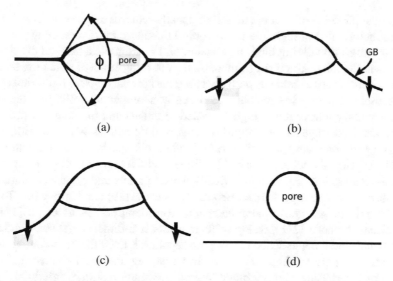

Fig. 1 Interaction of a pore with a migrating grain boundary [5]. (**a**) Initial configuration, (**b**, **c**) dragging of interfacial migration by the pore and (**d**) pore detachment for high local driving forces

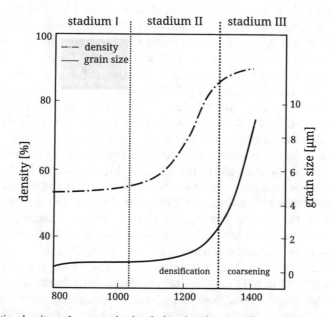

Fig. 2 Relative density and mean grain size during sintering according to [27]

pores in the polycrystal decreases with the progress of sintering. At some point, the density of pores and, thus, the drag effect are low enough, so that interfaces start to migrate and grain growth occurs. Accordingly, the mean grain size strongly increases as shown in Fig. 2.

Since application of ceramic materials usually requires a dense and fine-grained microstructure, it is of high interest to control the interplay of remaining pores and interface migration during final stage sintering. For example a tailored distribution of pores in the microstructure can increase the retarding effect of pore drag on grain growth. This is hard to achieve, partly because the optimal pore configuration cannot be determined from experiments. Accordingly a proper modelling of the pore-grain boundary interaction is highly needed to understand the complex processes during the final stage of sintering. So far very little studies issued the impact of pores on grain boundary migration. A tailored set of experiments along with a very useful analytical model is published by Rödel and Glaeser [22]. On the modelling side, the interactions between spherically isolated pores and grain boundaries are described by simple models from Brook [3], Nichols [19] and Carpay [4]. To take more details of geometric parameters into consideration, Evans et al. [14] and Riedel and Svoboda [21] develop in their works two-dimensional sharp-interface models and find a dependence of the pore separation from the pore configuration. However, the pore configuration underlies a strong statistical scattering in real microstructures. Thus a proper modelling of its evolution during sintering requires the consideration of very large simulation setups in three dimensions. Advances in computer technologies enable the investigation of pore drag effects with complex numerical three-dimensional simulations. The phase-field method has become a powerful tool for describing complex multi-physical systems numerically. A phase-field model for investigating the influence of pores on grain growth in hydrostatic pressure is developed by Jing et al. [15] and a model for solid state sintering which includes grain growth and various diffusion paths is proposed by Wang [26]. Biner [2] studies the pore and grain boundary migration under a temperature gradient in two dimensions. Ahmed et al. [1] introduce a phase-field model using a system of Cahn-Hilliard and Allen-Chan equations with a density field to describe pores. Pore dynamics under consideration of pressure stability using a phase-field model which is found to be in good accordance with analytic equations as well as experiments is investigated by Hötzer et al. [9]. In this work, the model in [9] is extended to treat pore coalescence and thus describes more complex pore interactions. It bases on a well-controllable model geometry of the microstructure using a large grain growing into a polycrystal. However, in order to picture the statistical variations of polycrystals as e.g. grain size distributions, pore sizes and their location in the microstructure in a realistic way, very large simulation domains are needed. Different length scales have to be resolved since pores are about an order of magnitude smaller than grains and significant microstructure evolution needs to be modelled without loss of statistics to completely capture the impact of pore drag on microstructure evolution. These two issues result in the need for large scale three-dimensional simulations on powerful supercomputers. The present paper extends previous work [9] to a larger parameter space. The more thorough results give deep insight into the pore-grain boundary interaction during final stage sintering and allow for an extrapolation towards grain growth in porous polycrystals as evident during final stage sintering in real microstructures.

2 Model

In the following, the phase-field model of [9, 10] to investigate the interaction between pores and grain boundaries during the final sintering stage, is briefly summarized. The model is implemented in the highly parallel multi-physics phase-field framework PACE3D [11, 12, 23–25].

The evolution equation of Allen-Cahn type for the N order-parameters in the vector $\boldsymbol{\phi}$, representing the different grains and pores, is written as:

$$\tau\epsilon\frac{\partial\phi_\alpha}{\partial t} = -\left(\underbrace{\epsilon\left(\frac{\partial a(\boldsymbol{\phi}, \nabla\boldsymbol{\phi})}{\partial\phi_\alpha} - \nabla\cdot\frac{\partial a(\boldsymbol{\phi}, \nabla\boldsymbol{\phi})}{\partial\nabla\phi_\alpha}\right) + \frac{1}{\epsilon}\frac{\partial\omega(\boldsymbol{\phi})}{\partial\phi_\alpha} + \frac{\partial f(\boldsymbol{\phi})}{\partial\phi_\alpha}}_{:=rhs_\alpha} - \lambda\right).$$

(1)

Each order-parameter $\phi_\alpha(\mathbf{x}, t)$ describes the volume fraction of an individual grain at the position $\mathbf{x}$ in the three-dimensional domain Ω at the time t. The parameter ϵ scales the thickness of the diffuse interface between the phase α and β and the parameter τ represents its kinetic coefficient calculated as

$$\tau = \frac{\sum_{\substack{\alpha,\beta=1 \\ (\alpha<\beta)}}^{N,N} \tau_{\alpha\beta}\phi_\alpha\phi_\beta}{\sum_{\substack{\alpha,\beta=1 \\ (\alpha<\beta)}}^{N,N} \phi_\alpha\phi_\beta}$$

(2)

with $\tau_{\alpha\beta}$ as the reciprocal mobility. In the case of pores, $\tau_{\alpha\beta}$ depends on the volume $V_\alpha^t = \int_\Omega \phi_\alpha(\mathbf{x}, t)d\Omega \mid \mathbf{x} \in \Omega_\alpha$ of the phase α with $\tau_{\alpha\beta} = A\pi\left(3/(4\pi)V_\alpha^t\right)^{4/3}$ including a material specific constant $A = 0.0007133$ [9, 10]. The diffuse interface is modeled with the gradient energy density

$$a(\boldsymbol{\phi}, \nabla\boldsymbol{\phi}) = \sum_{\substack{\alpha,\beta=1 \\ (\alpha<\beta)}}^{N,N} \gamma_{\alpha\beta}|\mathbf{q}_{\alpha\beta}|^2$$

(3)

and the interfacial energy density of multi-obstacle type of the form

$$\omega(\boldsymbol{\phi}) = \begin{cases} \dfrac{16}{\pi^2} \displaystyle\sum_{\substack{\alpha,\beta=1 \\ (\alpha<\beta)}}^{N,N} \gamma_{\alpha\beta}\phi_\alpha\phi_\beta + \displaystyle\sum_{\substack{\alpha,\beta,\delta=1 \\ (\alpha<\beta<\delta)}}^{N,N,N} \gamma_{\alpha\beta\delta}\phi_\alpha\phi_\beta\phi_\delta, & \boldsymbol{\phi} \in \Delta^N, \\ \infty, & \text{else.} \end{cases}$$

(4)

The parameter $\gamma_{\alpha\beta}$ describes the interface and surface energy, the parameter $\gamma_{\alpha\beta\delta}$ is a higher order term to reduce the occurrence of third or higher order phases at two phase interface regions [13, 18] and $\mathbf{q}_{\alpha\beta} = \phi_\alpha \nabla \phi_\beta - \phi_\beta \nabla \phi_\alpha$ is the general gradient vector orthogonal to the interface [18]. The driving force with the initial pressure p_α^0 and the current pressure p_α^t is written as

$$\frac{\partial f(\boldsymbol{\phi})}{\partial \phi_\alpha} = \frac{p_\alpha^0 V_\alpha^0}{V_\alpha^t} \frac{\partial h_\alpha(\boldsymbol{\phi})}{\partial \phi_\alpha} \tag{5}$$

with the interpolation function

$$h_\alpha(\boldsymbol{\phi}) = \frac{\phi_\alpha^2}{\displaystyle\sum_{\beta=1}^N \phi_\beta^2} . \tag{6}$$

from [17]. The Lagrange multiplier
$\lambda = 1/N \sum_{\alpha=1}^N rhs_\alpha$ is used in (1) to ensure the constraint $\sum_\alpha^N \partial\phi_\alpha/\partial t = 0$.

To efficiently store and solve an arbitrary number of phases in N component vector $\boldsymbol{\phi}$, the local reduction of the order-parameter (LROP) proposed by Kim et al. [16] is utilized. The implementation in the PACE3D solver is given in [9, 23].

3 Implementation

To describe the interaction between the pores in a physically correct manner, first the implementation of the pore interaction module is described. Afterwards, the parallel implementation of the mesh writer to efficiently store the simulated microstructure is shown.

3.1 Implementation of the Pore Interaction Module in the PACE3D Framework

During the sintering process, pores can merge and split or disappear due to their interaction with the grain boundaries. The different time scales of the pressure change and the grain boundary movement are exploited to describe the pressure in the pores in a computationally efficient way [9]. Therefore, for each pore an initial pressure and volume is assigned. By calculation of the current pore volume and consideration of the relation of the ideal gas law, the pressure for each pore in the next time step can be determined and used as driving force in (5) [9]. The pores in the three-dimensional domain are tracked by using the parallel connected component labeling algorithm of [10, 11]. The interactions of the pores are detected by comparing the connected components of the pores in the current time step with the ones of the previous time step. If the connected components of the pore in the current time step spatially overlap with two pores of the previous time step, the pores are merged by assigning them the same order-parameter and by adjusting the volume and pressure of the merged pore. In the opposite case of splitting, the two new separated pores are assigned with different order-parameters. Based on the pressure and volume of the pore before the split and the two volumes of the new pores after the split, the pressures of the new pores are calculated.

3.2 Mesh Output

To store the simulation data in a memory efficient manner, the PACE3D-framework allows to store the microstructure as meshes. The meshes are generated by marching the isosurface between the phases using the *Cubical Marching Squares algorithm* from [8]. For large domains, the calculations in PACE3D are parallelized using domain decomposition and MPI as indicated exemplarily in Fig. 3 by brown lines. To generate the mesh, each subdomain on the different MPI processes is marched in

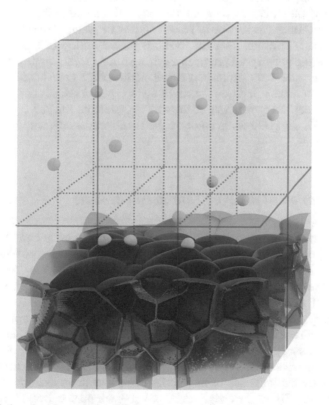

Fig. 3 Simulation domain with the different grouped phase-fields (gray, yellow). The domain decomposition is exemplarily indicated by brown lines

parallel and afterwards collected. Due to linear increase in communication and data with the number of processes, a single master for collecting and writing the data can become a bottleneck. Hence, multiple master processes can be assigned, allowing to efficiently collect the mesh parts of each subdomain and store the combined data in separate files, using the STereoLithography (STL) format. Depending on the post processing steps, the different mesh files containing the parts of the domain need to be combined. However, most (visualization) tools support multiple input files. To reduce the number of individual files for each phase-field, their mesh data can be grouped in one file. The grouping of the different phase-fields is exemplarily shown for the two groups in Fig. 3. The group of pores is highlighted in yellow and the group of grains in gray.

4   Results and Discussion

In the first part of this section, the validation of the pore-interaction module with the Young Laplace equation using a two-dimensional setup is presented. Subsequently, the effect of pore interaction in three-dimensional simulations of realistic microstructures is shown. For all studies the setup described by Rödel et al. [22] is used, in which a single crystal is placed on top of a polycrystal. The capillarity of the polycrystal provides a driving force for the single crystal to grow into the polycrystal. This results in directed motion of the grain boundary. Pores are placed at the boundary between the single and the polycrystal as depicted in Fig. 4. The pores reduce the growth rate of the single crystal into the polycrystal as shown in Fig. 10. This setup allows to compare the simulation results with both, analytics and experiments.

4.1   Validation of the Pore Interaction Module

During sintering, grains grow and pores interact with the grain boundaries. While grains grow, pores migrate and eventually merge. Thereby, the pore radius changes and, thus their impact on grain boundary motion changes as well. To include this effect in the phase-field model, the following two issues need to be considered. First, the effects of the model extension on the pore interactions have to be validated. Second, the pressure of merging pores has to match the Young-Laplace equation. Therefore, the results from [10] are recapitulated. Two pores with the diameter d_p and initial pressure p_α^0 are placed between the single crystal and the polycrystal into a two-dimensional simulation domain consisting of 200×150 cells for each study.

To illustrate the influence of the model extensions in an exemplary way, three different cases I–III are investigated. In case I, each pore is treated as a separate phase without any adjustment of the pores. In case II, the pores can merge and their pressures, volumes and order parameters are adjusted. In case III, additionally the

Fig. 4 Simulation setup based on [9, 10] following the experimental setting described by Rödel et al. [22]

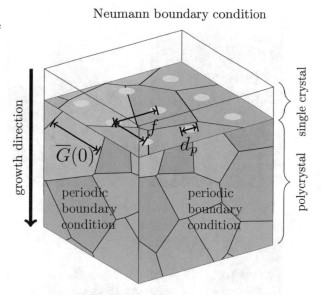

Neumann boundary condition

Neumann boundary condition

mobility is changed when pores are merging according to [9, 10]. In Fig. 5 different time steps for cases I–III are shown. Case I with no pore interaction (Fig. 5a–c) forms a nonphysical interface between the two pores and they stay attached to the grain boundary. In cases II and III Fig. 5e and h, the pores merge to a single pore. For case III as shown in Fig. 5i, the merged pore detaches only if the mobility is dependent on its volume. If pores are interacting with each other, ignoring the mobility dependence leads to a very different behavior in the microstructure evolution. Since the size dependence of the pore mobility influences the overall grain growth kinetics, merging pores require adjusted pore mobilities to capture the separation behavior in a physically correct manner. Therefore, it is shown in the simulations that both pore merging and mobility adjustment are essential for the occurrence of pore separation.

To validate the pressure in the merged pores, the simulation results are compared with the Young-Laplace equation

$$\Delta p = \frac{2\gamma_{\alpha\beta}}{d_p^{equ.}}(D - 1) \tag{7}$$

which describes the pressure difference Δp between the initial (outer) and the equilibrium pressure in the pore. In (7) the parameter $\gamma_{\alpha\beta}$ is the pore-grain surface energy, $d_p^{equ.}$ describes the pore diameter in the equilibrium state and D represents the dimension. For the validation, the diameter of pores and the surface energy are changed systematically using an initial pressure of $p^{initial} = 1.6667$. In the two-dimensional setting, four initial diameters of $d_p = 10, 12, 14, 16$ cells are chosen

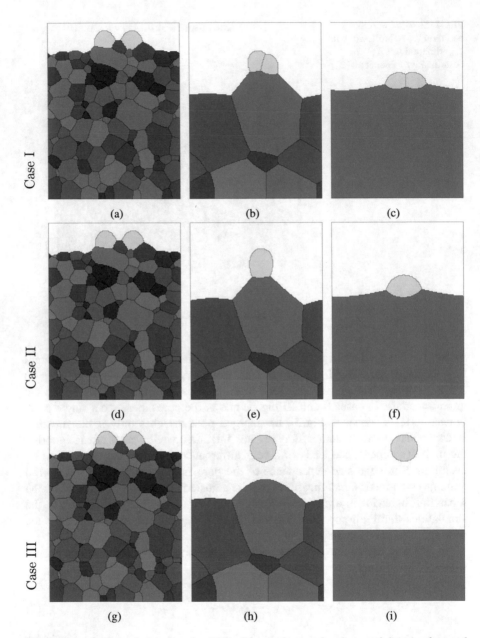

Fig. 5 Phase-field simulations based on [10] of the single crystal grain growth into a polycrystal with two attached pores at the boundary without pore merging (case I), with pore merging (case II) and with pore merging as well as adjusted pore mobilities (case III). (**a**) $t = 95$. (**b**) $t = 1995$. (**c**) $t = 5130$. (**d**) $t = 95$. (**e**) $t = 1995$. (**f**) $t = 5130$. (**g**) $t = 95$. (**h**) $t = 1995$. (**i**) $t = 5130$

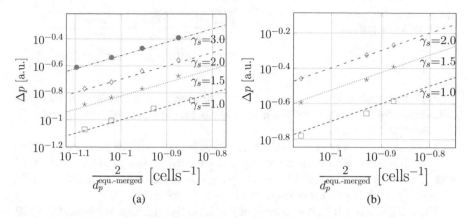

Fig. 6 Validation of the pressure model and the pore interaction module with the Young Laplace equation based on [10]. The pressure difference and the radius are measured for different values of pore-grain surface energy $\gamma_{\alpha\beta} = \gamma_s$ after the pore reaches its equilibrium state. (**a**) 2D. (**b**) 3D

and the surface energy of the pore-grain boundaries is altered as $\gamma_{\alpha\beta} = 1, 1.5, 2, 3$. For the three-dimensional case, three initial diameters of $d_p = 12, 14, 18$ cells are chosen and surface energy values of $\gamma_{\alpha\beta} = 1, 1.5, 2, 3$ are investigated. During the simulation, the pores merge and eventually detach from the grain boundary. After the detachment, the merged pores evolve towards their equilibrium shape of a sphere. Once the equilibrium state is reached, the pressure difference Δp and the pore size $d^{equ.}$ is measured. Similar to [9], the simulation results depicted in Fig. 6 are in good agreement with analytic equations.

4.2 Influence of Interacting Pores at Grain Boundaries

To study pore-grain boundary interactions the setup previously depicted in Fig. 4 is used. A polycrystalline microstructure with a homogeneous grain size distribution and a mean grain diameter $\overline{G}$ is generated by a Voronoi tessellation. A defined pore array with spacing f and pore diameter d_p is placed at the interface between single crystal and polycrystal.

Following Rödel and Glaeser [22], an expression for the critical grain size of the polycrystal for pore detachment is derived as

$$\overline{G}(t) < \frac{6f^2}{\pi d_p} \left(1 - \frac{v_p}{v_b(t)} \right) \qquad (8)$$

with the maximum velocity of the pores v_p and grain boundaries v_b before detachment. A detailed description of the derivation of (8) is given in [9].

Table 1 Simulation parameter for the realistic setup

Parameter		Simulation value
f	[cells]	25, 30, 50, 60
$\overline{G}(0)$	[cells]	20, 30, 35, 40,45, 50, 60, 70, 90

For investigating the merging of pores and their interactions with the grain boundaries, the spacings f and the mean grain diameter $\overline{G}$ are varied systematically as summarized in Table 1.

The initial pore diameter d_p is chosen as 14 cells and the initial pressure as $p^{init.} = 1.5$ for all studies. The grain boundary and the surface energy are chosen as 0.5 and 1, respectively. The mobility of the grain boundaries is kept constant as 0.49, whereas the mobility of the pores depends on their volume.

The domain size for a pore spacing of $f = 25$ cells is chosen as $200 \times 200 \times 320$, for $f = 30$ and $f = 60$ cells as $240 \times 240 \times 320$ and for $f = 50$ cells as $400 \times 400 \times 320$ voxel cells. The simulations are conducted on the Hazel Hen supercomputer [7] at the High Performance Computing Center Stuttgart (HLRS) using 288 cores for 6h for $f = 25$ and $f = 30$ and 512 cores for 8 h for $f = 50$ and $f = 60$.

In addition to two-dimensional experimental micrographs, three-dimensional simulations allow a spatial investigation of pore deformation processes and pore interaction. Since triple lines and quadruple junctions only exist in three dimensions, a two-dimensional simulation is not capable to reflect real microstructure evolution with pore-grain boundary interaction.

An exemplary simulation result in three dimensions is shown for different time steps in Fig. 7a–c with a pore spacing of $f = 30$ and an initial mean grain diameter of $\overline{G}(0) = 50$ cells. In Fig. 7d–f the corresponding top views are presented.

The growth of the single crystal results in a vertical movement of the pores. In horizontal direction the pores follow the movement of the triple lines and quadruple junctions [9], which is the cause of pore merging. The migration of the pores from their position in the previously shown time step is highlighted by black circles in Fig. 7e and f. In Fig. 7b and e, pores merge and subsequently change their size and shape (red and green pores). Due to the increase of the pore size, they become less mobile and detach at lower boundary migration rates from the grain boundary than smaller and more mobile pores. All merged and detached pores are highlighted in green in Fig. 7b, c, e and f. After detachment, the pores are trapped in the single crystal and approach a spherical shape. The detachment of pores leads to a local change of pore density which results in a lower drag force. This increases the local velocity of the grain boundary which can lead to further detachment of smaller pores. These detached but not merged pores are indicated in blue in Fig. 7c, f.

A merging of several pores, their detachment and shape change is shown for different time steps in Fig. 8. Two different cases are highlighted in violet and orange, respectively.

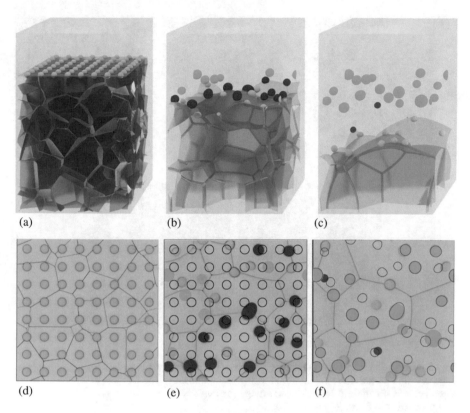

Fig. 7 Simulation results of a single crystal growing into a polycrystal for different time steps in 3D (**a–c**) and as top view (**d–f**). Merged pores are marked in red, merged and detached pores are marked in green and pores that detach at their initial size are marked in blue. In the top view, the black circles indicate the positions of the pores in the previously shown time step. (**a**) $t = 2.1$. (**b**) $t = 7560$. (**c**) $t = 13,020$. (**d**) $t = 2.1$. (**e**) $t = 7560$. (**f**) $t = 13,020$

To quantify the effect of pore detachment, the conducted simulations are summarized in a separation map in Fig. 9. This map continues the work in [9] for smaller pore spacings. The curves give the critical grain size for pore detachment from (8) with respect to the pore spacing for the pore velocities $v_p = 0.02$ (dashed line) and $v_p = 0.03$ (solid line). In the previous study, these two pore velocities are found to represent the upper and lower limit for pore detachment [9]. The filled red circle marks indicate simulations in which all pores stay attached to the grain boundary. The filled green square marks represent simulations in which all pores detach. Simulations with partial detachment are marked with filled blue triangle along with the percentage of detached pores is given. For a spacing of $f = 60$ cells all three cases of pore-grain boundary interaction can be observed. Whereas all simulations with complete pore detachment occur below the predicted critical grain size, simulations in which all pores stay attached are found above the critical grain size. This corresponds well with analytical expectations of (8) and the work

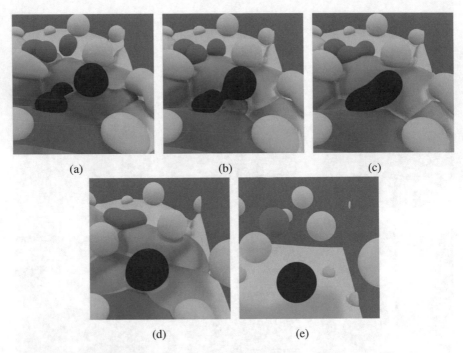

Fig. 8 Merging of several pores and the detachment from the grain boundary for $\overline{G}(0) = 60$ and $f = 25$. Smaller pores are still attached to the grain boundary. The merging and detachment of three pores is highlighted in violet and orange for two cases. (**a**) $t = 5460$. (**b**) $t = 5880$. (**c**) $t = 6300$. (**d**) $t = 8400$. (**e**) $t = 19{,}320$

of Rödel and Glaeser [22]. The simulations with partial detachment of the pores represent a transition from the case of attachment to the case of detachment. This is caused by the statistical variation in the initial position of the pores. As shown in [9], pores tend to migrate from quadruple junctions to triple lines to grain boundary planes before they detach. Accordingly, the initial location of the pores and their distribution impacts the detachment events. Additionally, for smaller pore spacings, merging of pores happens more frequently. Thus more low mobility pores exist, which detach earlier. This results in a stronger statistical variation of the individual detachment events. For example, due to many merging pores, no case of total pore attachment or detachment can be observed for the spacings $f = 25$, $f = 30$ and $f = 50$. It should be noted that the simulations differ from the assumptions in the analytics, where the pore size is assumed to be constant. Therefore a deviation between analytics and simulations is expected. This deviation should increase with increasing fraction of merged pores and, thus, with smaller pore spacing as observed in Fig. 9.

The evolution of the growth of the single crystal under the impact of pore drag is shown in Fig. 10. The blue curve (open blue triangle) gives the single crystal growth without any pores or pore drag. Parabolic growth can be found as expected

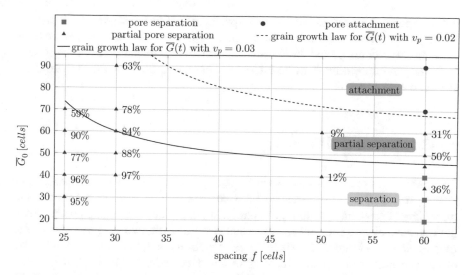

Fig. 9 Separation map for simulation results and comparison with grain growth laws of (8). Simulations with a total attachment of the pores are marked with red filled circle, with a total detachment are marked with green filled square and with partial detachment are marked with blue filled triangle

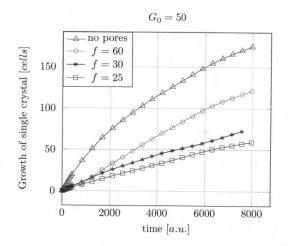

Fig. 10 Grain growth of the single crystal over the time for different pore spacings

for this setup [20]. If pores are added to the initial interface of the polycrystal, the growth is retarded. With a spacing of 60 cells, the impact of pore drag is only visible within the first 3000 time steps (green curve, open green circle). After 3000 time steps the blue and green curve are very close to parallel. This is expected, since a high fraction of pores detaches for this setup (Fig. 9). Accordingly, a strong overall drag effect of pores on the growth of the single crystal is only expected within this period; afterwards the drag effect is negligible and the growth is similar to the case without pores. The setups with a pore spacing of 30 (asterisk) and 25 (open red square) show a similar growth of the single crystal and are significantly retarded in the entire simulation period. In Fig. 9 it is shown that a smaller pore spacing results in fewer pores detaching from the interface. Thus a stronger drag effect is expected as observed in Fig. 10.

5 Conclusion and Outlook

The phase-field model of pore-grain boundary interaction during final stage sintering presented in [9] is extended by a pore-interaction module to improve the modelling of final stage sintering. In particular, the effect of merging pores, if they get into contact with each, other was included. As a result, two or more pores merge to a single much larger pore. Since the pore mobility and, thus, its individual drag effect depends on the pore size, very different behavior is obtained, if merging pores are considered. A comparison to the previous work highlights the need for this additional module. The extension module is validated by observing the inner pressure of merged pores and comparing with the analytic Young-Laplace equation. It is shown that the simulations match the analytic expectations very well. The presented model with the extensions is used to simulate the growth of a single crystal into a polycrystal in the presence of pores. Planar arrays of pores with different spacings are placed at the interface between single crystal and polycrystal. Dependent on the pore spacing, drag and detachment of pores is studied. The main conclusions of these investigations are:

(i) The extended model is capable to investigate the pore merging mechanisms during microstructure evolution.
(ii) Both, the merging process and the change of the mobility of the merged pores have a significant impact on pore drag and detachment of the pores.
(iii) The measured pressure of merged pores agrees with the predictions from the equation of Young-Laplace.
(iv) The application of this model to a well-controllable setup with growing single crystals shows a behavior (pore drag and detachment) which agrees well with expectations from analytic equations.
(v) The simulations show the effect of pore location with respect to microstructure (quadruple point, triple line or grain boundary plane). Accordingly the issue of pore drag needs a large setup to capture the statistical distribution of pores in real microstructure.
(vi) The analytical equations for pore drag and pore detachment cannot resemble the impact of merging pores. Accordingly, the predictions of pore detachment from analytic equations apply to model geometries, but not to realistic microstructure evolution during final stage sintering.

The results show that the present model is appropriate to model grain growth during the final stage of sintering. However, the complexity of this process is highlighted as well. Particularly conclusions (v) and (vi) illustrate the need of large scale simulations: the details of pore drag depend critically on local geometry of the microstructure. The microstructure evolution during final stage sintering in polycrystalline ceramics underlies strong statistical variations of in the local geometry. The size of grains is not uniform, but shows a wide grain size distribution resulting in locally very different driving forces. The size of pores is not uniform as well, but in addition their special distribution is not uniform. Accordingly, if grain growth in polycrystals in the presence of remaining pores from sintering is considered in detail, large scale simulations are needed to picture the local statistic variation of pore drag in an adequate way.

Acknowledgements We are grateful for the provided computational resources at the Höchstleistungsrechenzentrum in Suttgart (HLRS). We further thank the ministry MWK of the state Baden-Wuerttemberg for financial support through the cooperative graduated school "Gefügestrukturanalyse und Prozessbewertung" and the BMBF for funding the project "SKAMPY".

References

1. K. Ahmed, C.A. Yablinsky, A. Schulte, T. Allen, A. El-Azab, Phase field modeling of the effect of porosity on grain growth kinetics in polycrystalline ceramics. Model. Simul. Mater. Sci. Eng. **21**(6), 065005 (2013)
2. S.B. Biner, Pore and grain boundary migration under a temperature gradient: a phase-field model study. Model. Simul. Mater. Sci. Eng. **24**(3), 035019 (2016)
3. R.J. Brook, Pores and grain growth kinetics. J. Am. Ceram. Soc. **52**(6), 339–340 (1969)
4. F.M.A. Carpay, Discontinuous grain growth and pore drag. J. Am. Ceram. Soc. **60**(1), 2–3 (1977)
5. C.B. Carter, M.G. Norton, *Ceramic Materials: Science and Engineering* (Springer, New York, 2007)
6. R.L. Coble, Sintering crystalline solids. I. Intermediate and final state diffusion models. J. Appl. Phys. **32**(5), 787–792 (1961)
7. HLRS, Cray XC40 (Hazel Hen). https://www.hlrs.de/systems/cray-xc40-hazel-hen (2016) [Online; accessed March 2016]
8. C. Ho, F.-C. Wu, B.-Y. Chen, Y.-Y. Chuang, M. Ouhyoung et al., Cubical marching squares: adaptive feature preserving surface extraction from volume data, in *Computer Graphics Forum*, vol. 24. (Wiley Online Library, 2005), pp. 537–545
9. J. Hötzer, V. Rehn, W. Rheinheimer, M.J. Hoffmann, B. Nestler, Phase-field study of pore-grain boundary interaction. J. Ceram. Soc. Jpn. **124**(4), 329–339 (2016)
10. J. Hötzer, Massiv-parallele und großskalige Phasenfeldsimulationen zur Untersuchung der Mikrostrukturentwicklung. PhD thesis (2017). https://doi.org/10.5445/IR/1000069984
11. J. Hötzer, A. Vondrous, J. Ettrich, M. Jainta, A. August, D. Studenvoll, M. Reichardt, M. Selzer, B. Nestler, Phase-field simulations of large-scale microstructures by integrated parallel algorithms, in *High Performance Computing in Science and Engineering '14: Transactions of the High Performance Computing Center, Stuttgart (HLRS) 2014*, 2015 edition, ed. by W.E. Nagel, D.H. Kröner, M.M. Resch (Springer, Cham, 2015)
12. J. Hötzer, M. Jainta, M. Ben Said, P. Steinmetz, M. Berghoff, B. Nestler, Application of large-scale phase-field simulations in the context of high-performance computing, in *High Performance Computing in Science and Engineering '15: Transactions of the High Performance Computing Center, Stuttgart (HLRS) 2015*, ed. by W.E. Nagel, D.H. Kröner, M.M. Resch (Springer, Cham, 2015)

13. J. Hötzer, O. Tschukin, M. Ben Said, M. Berghoff, M. Jainta, G. Barthelemy, N. Smorchkov, D. Schneider, M. Selzer, B. Nestler, Calibration of a multi-phase field model with quantitative angle measurement. J. Mater. Sci. **51**(4), 1788–1797 (2016)

14. C.H. Hsueh, A.G. Evans, R.L. Coble, Microstructure development during finalintermediate stage sintering-I. Pore-grain boundary separation. Acta Metall. **30**, 1269–1279 (1982)

15. X.N. Jing, J.H. Zhao, G. Subhash, X.-L. Gao, Anisotropic grain growth with pore drag under applied loads. Mater. Sci. Eng. A **412**(12), 271–278 (2005)

16. S.G. Kim, D.I. Kim, W.T. Kim, Y.B. Park, Computer simulations of two-dimensional and three-dimensional ideal grain growth. Phys. Rev. E **74**, 061605 (2006)

17. N. Moelans, A quantitative and thermodynamically consistent phase-field interpolation function for multi-phase systems. Acta Mater. **59**(3), 1077–1086 (2011)

18. B. Nestler, H. Garcke, B. Stinner, Multicomponent alloy solidification: phase-field modeling and simulations. Phys. Rev. E **71**, 041609 (2005)

19. F.A. Nichols, Pore migration in ceramic fuel elements. J. Nucl. Mater. **27**, 137–146 (1968)

20. W. Rheinheimer, M. Bäurer, C.A. Handwerker, J.E. Blendell, M.J. Hoffmann, Growth of single crystalline seeds into polycrystalline strontium titanate: anisotropy of the mobility, intrinsic drag effects and kinetic shape of grain boundaries. Acta Mater. **95**, 111–123 (2015)

21. H. Riedel, J. Svoboda, A theoretical study of grain growth in porous solids during sintering. Acta Metall. Mater. **41**(6), 1929–1936 (1993)

22. J. Rödel, A.M. Glaeser, Pore drag and pore-boundary separation in Alumina. J. Am. Ceram. Soc. **73**(11), 3302–3312 (1990)

23. M. Selzer, Mechanische und Strömungsmechanische Topologieoptimierung mit der Phasenfeldmethode. PhD thesis, 2014

24. A. Vondrous, B. Nestler, A. August, E. Wesner, A. Choudhury, J. Hötzer, Metallic foam structures, dendrites and implementation optimizations for phase-field modeling, in *High Performance Computing in Science and Engineering '11: Transactions of the High Performance Computing Center, Stuttgart (HLRS) 2011*, 2015 edition, ed. by W.E. Nagel, D.H. Kröner, M.M. Resch (Springer, Cham, 2011)

25. A. Vondrous, M. Selzer, J. Hötzer, B. Nestler, Parallel computing for phase-field models. Int. J. High Perform. Comput. Appl. **28**(1), 61–72 (2014)

26. Y.U. Wang, Computer modeling and simulation of solid-state sintering: a phase field approach. Acta Mater. **54**(4), 953–961 (2006)

27. M.F. Yan, Microstructural control in the processing of electronic ceramics. Mater. Sci. Eng. **48**, 53–72 (1981)

Modeling the Interior Dynamics of Terrestrial Planets

Ana-Catalina Plesa, Christian Hüttig, and Florian Willich

Abstract Over the past years, large scale numerical simulations of planetary interiors have become an important tool to understand physical processes responsible
for the surface features observed by various space missions visiting the terrestrial
planets of our Solar System. Such large scale applications need to show good
scalability on thousands of computational cores while handling a considerable
amount of data that needs to be read from and stored to a file system. To this
end, we analyzed numerous approaches to write files on the Cray XC40 Hazel
Hen supercomputer. Our study shows that HPC applications parallelized using MPI
highly benefit from utilizing the MPI I/O facilities. By implementing MPI I/O in
Gaia, we improved the I/O performance up to a factor of 100. Additionally, in this
study we present applications of the fluid flow solver Gaia using high resolution
regional spherical shell grids to study the interior dynamics and thermal evolution
of terrestrial bodies of our Solar System.

1 Introduction

Numerous data obtained from various space missions visiting the terrestrial bodies
in our Solar System over the past decades have revealed that apart from Earth, all
other terrestrial bodies operate in the so-called stagnant lid regime e.g., [1, 15].
While the surface of the Earth is broken into seven major plates, with surface
material being recycled at subduction zones and renewed at spreading centers,
the surface of all other terrestrial bodies is covered by an immobile layer, the so-
called stagnant lid through which heat is transported by conduction. These two
different convection styles have important implications for the interior evolution and

A.-C. Plesa (✉) • C. Hüttig
German Aerospace Center, Institute of Planetary Research, Berlin, Germany
e-mail: ana.plesa@dlr.de; christian.huettig@dlr.de

F. Willich
University of Applied Sciences (HTW) Berlin, Berlin, Germany

German Aerospace Center, Institute of Planetary Research, Berlin, Germany
e-mail: florian.willich@student.htw-berlin.de

© Springer International Publishing AG 2018
W.E. Nagel et al. (eds.), *High Performance Computing in Science
and Engineering '17*, https://doi.org/10.1007/978-3-319-68394-2_30

result in different evolutionary paths for the volcanic outgassing which is directly linked to the atmosphere, and the core cooling, which is an important agent for the magnetic field generation. Both the atmosphere and the magnetic field are fundamental in maintaining habitable conditions at the surface by regulating the temperatures to allow for liquid water and shielding against solar radiations [16]. In order to constrain the conditions necessary for habitability it is important to understand the thermo-chemical evolution of the interior of the Earth and other terrestrial bodies. To this end, numerical simulations of planetary interiors have become one of the most important tools to tackle complex fluid dynamics problems. Although analytical solutions exist, they are available only for simple systems while laboratory experiments are restricted to limited parameter ranges. Thus, highly parallelized numerical simulations are often employed to address complex systems in various geometries [13].

The computational time needed in such numerical simulations can be divided in three categories: actual compute time, where usually a system of equations is solved iteratively, the communication time, i.e. the time needed for exchanging data between the computational processes, and I/O time, i.e. the time elapsing while reading from or writing to files on a storage system. An efficiently parallelized application will spend most of the time in the first category, while the time spent in the two other categories should be negligible [9].

In this study, we present recent improvements and results using the mantle convection code Gaia [4–6], a fluid flow solver that can handle high viscosity contrasts as appropriate for the mantle of terrestrial planets. In the next chapter we present the mathematical and physical model used in our simulations, followed by a description of the technical realization with focus on the numerical grid and improved I/O procedures. Further, we show the code performance measured on the Cray XC40 platform at HLRS, Hornet, using the recent implementation and discuss further improvements.

2 Mathematical and Physical Model

Thermal and compositional convection in the mantle of terrestrial bodies is modeled by solving the conservation equations of mass, linear momentum, energy and transport of chemical species e.g., [15]. Quantities like temperature, velocity, time, viscosity and chemical density are scaled using the relationships to physical properties presented e.g. in [2] to obtain a non-dimensional system of equations. Typically the mantle thickness D is used as length scale, the thermal diffusivity κ as time scale, the temperature drop across the entire mantle ΔT as temperature scale and the contrast in chemical density $\Delta \rho$ as compositional scale. Assuming an infinite Prandtl number, as the mantle of terrestrial bodies is considered highly viscous with negligible inertia, and using the Boussinesq approximation, the non-dimensional equations of thermo-chemical convection are, e.g. [3]:

$$\nabla \cdot \mathbf{u} = 0, \tag{1}$$

$$\nabla \cdot \left[\eta (\nabla \mathbf{u} + (\nabla \mathbf{u})^T) \right] + Ra(T - BC)\mathbf{e}_r - \nabla p = 0, \tag{2}$$

$$\frac{\partial T}{\partial t} + \mathbf{u} \cdot \nabla T - \nabla^2 T - H_{radioactive} = 0, \tag{3}$$

$$\frac{\partial C}{\partial t} + \mathbf{u} \cdot \nabla C = 0, \tag{4}$$

where η is the viscosity, p the dynamic pressure, $\mathbf{u}$ the velocity, t the time, T the temperature, C the chemical component and $\mathbf{e}_r$ is the unit vector in radial direction. The thermal Rayleigh number, Ra, in Eq. (2) is defined as follows:

$$Ra = \frac{\rho g \alpha \Delta T D^3}{\kappa \eta_{ref}} \tag{5}$$

where ρ is the reference density, g the gravitational acceleration, α the thermal expansivity, ΔT the temperature contrast between inner and outer boundaries, D the thickness of the mantle, κ the thermal diffusivity, and η_{ref} the reference viscosity calculated at reference temperature and pressure.

The buoyancy number B in Eq. (2) shows the relative importance of chemical to thermal driven convection and depends on the density contrast ($\Delta \rho$) used:

$$B = \frac{\Delta \rho}{\rho \alpha \Delta T} \tag{6}$$

The factor $H_{radioactive}$ in Eq. (3) represents a non-dimensional source term introduced by the decay of radioactive heat producing elements and is given by:

$$H_{radioactive} = \frac{\rho \alpha Q_m D^2}{\kappa k \Delta T}, \tag{7}$$

where Q_m is the heat production rate in pW/kg and k is the thermal conductivity.

The viscosity varies with temperature and pressure and is calculated according to the Arrhenius law [7]. The non-dimensional formulation is given by [14]:

$$\eta(T,p) = \exp \left(\frac{E + pV}{T + T_0} - \frac{E + p_{ref}V}{T_{ref} + T_0} \right), \tag{8}$$

where E and V are the activation energy and volume, respectively, and T_0 the surface temperature. The variables T_{ref}, p_{ref}, are the reference temperature, pressure, respectively.

3 Technical Realization

The numerical code Gaia is written in C++ and can be used without any additional libraries [4, 6]. If enabled, the MPI libraries allow the usage of thousands of cores in parallel to efficiently solve the system of conservation equations presented in Sect. 2. The governing equations (Eqs. (1)–(3)) are discretized via the finite-volume method on a fixed mesh in arbitrary geometries. In contrast, (4) is modeled via the particle in cell method (PIC) e.g. [12, 17, 18]. Tracer particles are moved to transport various chemical species according to the velocity field obtained on the fixed mesh. In each time-step, the compositional fields are interpolated on the computational grid at the cell centers to be used in Eq. (2). This method has the advantage of being essentially free of numerical diffusion and enables the advection of an arbitrary large number of compositional fields by solving a single transport equation for each particle.

3.1 Numerical Grid

The code Gaia can handle meshes in 2D and 3D geometries as long as they are Voronoi grids [5, 11]. While in a 2D geometry, the mesh can be created during the simulation, for a 3D grid with millions of cells the mesh is build in advance outside the simulation and loaded during the initialization process. Additionally, for large grids, a domain decomposition step, where the mesh is efficiently partitioned into N equal size domains, can be done before the simulation is started on N computational cores. Here we will focus on the description of 3D regional grids. They are particularly interesting since these meshes can reach exceptionally high resolution needed for an accurate solution of the conservation equations.

3.2 File I/O

Gaia periodically writes output files for further post processing of the results as well as snapshot files containing additional information needed to restart a simulation. Executed by N processes, the application writes 1 output file and N snapshot files per defined time period. Tests have shown that the application shows a dramatic performance drop when writing an output file executed by more than 2000 processes.

In order to write a single output file by N processes executing Gaia, the distributed data is first gathered to the master process who writes to file by using POSIX compliant I/O functions. To write N SNAP files, every process opens, writes and closes 1 file with a unique file name calling POSIX compliant I/O functions. Both methods of writing data to files contain time consuming operations and can significantly be improved when using MPI I/O functions.

The procedures for writing output and snapshot files as well as restarting a simulation by reading a snapshot file have been revised and improved [19], and extensively tested in production runs. One key facility used are MPI file views [10]: the MPI data types, with respective offsets and block length for setting the correct file view, are created and committed during the applications initialization procedure. If data has to be written to file, each process sets the respective file view using a precomputed offset and the already committed MPI data type followed by a call to a collective MPI file write function [10]. A special task is given the master process by writing global data such as file header and file footer information [19]. Gaia now produces a single output and snapshot file.

4 Results

4.1 High Resolution Regional Grids

A regional grid describes only a portion of a complete spherical shell. Because the radius ratio is the same and the result is 3D, it is mostly, but not always, sufficient to use regional grids. Because Gaia can handle Voronoi-like grids only, an orthogonal projection was chosen to protect the angles between the faces to remain at 90°. This enables us to use a simple deformed box instead of triangulated Voronoi diagrams. As the portion of the sphere covered increases, the cells at the corners get smaller, so the maximum coverage of the regional grid is a hemisphere.

Figure 1 illustrates this on a low-resolution regional grid. As we can quickly deduce the geometric properties for this grid algebraically, we can create large grids and test the parallel performance on those grids. As it turned out, as long as a single core handles around 40×10^3 degrees of freedom (accounts to 10×10^3 cells per domain), the scaling remains optimal regardless of the total number of cores involved in the computation.

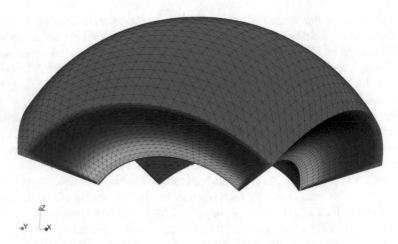

Fig. 1 Generic regional grid that covers a quarter of the complete shell with 15 shells

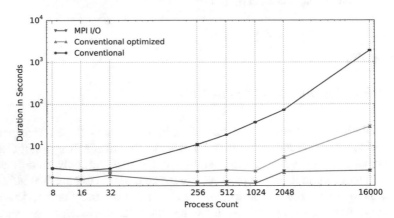

Fig. 2 Writing an OUT file on *Hazel Hen* in Gaia: *MPI I/O* represents the new approach, *Conventional* represents the conventional approach, while *Conventional optimized* represents the conventional approach without collectively calling fopen() in addition to some minor optimizations. From 8 to 2048 processes the file is of the size of approximately 1058 and 2521 MB for 16×10^3 processes

4.2 I/O Performance

We have tested the new writing procedures and compared the results with the older implementation (conventional approach in Fig. 2). The results show a significant improvement when using the MPI I/O interface to write an output file compared to the conventional approach (see Fig. 2).

The values in Fig. 2 show the Median Absolute Deviation of 10 independent measurements defined as:

$$MAD = median(|x_i - median(M)|) \tag{9}$$

4.3 Application to Mantle Convection

A key interest in supercomputers like HazelHen lies in its unique position to evaluate evolution runs of a planetary mantle in 3D. Because the initial phase of a planet is very hot, the simulation demands highly resolved 3D grids that are impossible to use without a supercomputer. As this was always the key interest when using supercomputers before, it is not new to run these kind of simulations.

What we did discover recently is an effect that we did not expect in this kind of scenario. In the early stage of evolution, roughly after 1 billion years (1 Gyr), a weak degree-one formation can be seen in the data if we assume a Mars radius ratio. Compared to Earth, Mars has a smaller radius (or inner-to-outer-core-) ratio and previous studies proposed a degree one convection structure in the interior to be responsible for the observed present-day crustal dichotomy [8, 14, 20].

The underlying weak degree-one obtained with Gaia within an evolution run is highlighted in Fig. 3. This type of simulations require careful preparation: A single evolution run to 1 Gyr requires around 17 days on HazelHen with roughly 2000 cores.

However, further tests are need to confirm our results. As this effect only happens under high-Rayleigh-number and high-viscosity contrast conditions, a setup that requires a large computational effort, only a numerical code highly optimized for parallel computations can address this kind of simulations.

5 Conclusion and Outlook

The discovery of a possible degree-one structure beneath a stagnant-lid needs further investigation as it might provide an answer to the different observations we make between Mars, Earth and the other terrestrial planets. The evolution runs need to be repeated on a full spherical shell and with even higher resolution to prove that this is not a numerical artifact. A next step would provide a regime diagram that connects essential parameters like Rayleigh number and radius ratio to the observed sub-lithospheric degree-one. This would also allow the determination of duration of such a regime and its impact on surface structures that we can observe today.

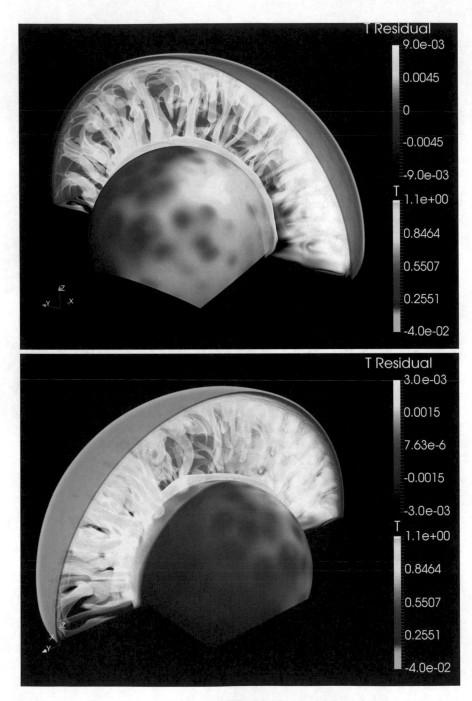

Fig. 3 Convection structure in the interior of Mars after 1 Gyr of thermal evolution. The volume rendering highlights the weak degree-one structure present

References

1. D. Breuer, W.B. Moore, Dynamics and thermal history of the terrestrial planets, the Moon, and Io. Treatise Geophys. **10**, 299–348 (2007)
2. U. Christensen, Convection with pressure- and temperature-dependent non-Newtonian rheology. Geophys. J. R. Astron. Soc. **77**, 343–384 (1984)
3. O. Grasset, E.M. Parmentier, Thermal convection in a volumetrically heated, infinite Prandtl number fluid with strongly temperature-dependent viscosity: implications for planetary thermal evolution. J. Geophys. Res. **103**, 18171–18181 (1998)
4. C. Hüttig, K. Stemmer, Finite volume discretization for dynamic viscosities on Voronoi grids. Phys. Earth Planet. Interiors (2008). https://doi.org/10.1016/j.pepi.2008.07.007
5. C. Hüttig, K. Stemmer, The spiral grid: a new approach to discretize the sphere and its application to mantle convection. Geochem. Geophys. Geosyst. **9** (2008). https://doi.org/10. 1029/2007GC001581. Q02018
6. C. Hüttig, N. Tosi, W.B. Moore, An improved formulation of the incompressible Navier-Stokes equations with variable viscosity. Phys. Earth Planet. Inter. **220**, 11–18 (2013). https://doi.org/ 10.1016/j.pepi.2013.04.002
7. S. Karato, M.S. Paterson, J.D. Fitz Gerald, Rheology of synthetic olivine aggregates: influence of grain size and water. J. Geophys. Res. **91**, 8151–8176 (1986)
8. T. Keller, P.J. Tackley, Towards self-consistent modelling of the Martian dichotomy: the influence of low-degree convection on crustal thickness distribution. Icarus **202**(2), 429–443 (2009)
9. S. Lang, P. Carns, R. Latham, R. Ross, K. Harms, W. Allcock, I/O performance challenges at leadership scale, in *Pages 40:140:12 of: Proceedings of the Conference on High Performance Computing Networking, Storage and Analysis. SC09* (ACM, New York, NY, 2009)
10. Message Passing Interface Forum: MPI: A Message-Passing Interface Standard Version 3.1. Retrieved on March 3, 2016 from http://mpi-forum.org/docs/mpi-3.1/mpi31-report.pdf (2015)
11. A.-C. Plesa, Mantle Convection in a 2D Spherical Shell, in *Proceedings of the First International Conference on Advanced Communications and Computation (INFOCOMP 2011), October 23–29, 2011, Barcelona.* ed. by C.-P. Rückemann, W. Christmann, S. Saini, M. Pankowska (2011), pp. 167–172. ISBN: 978-1-61208-161-8. Retrieved on November 3, 2011, from http://www.thinkmind.org/download.php?articleid=infocomp_2011_2_10_10002
12. A.-C. Plesa, N. Tosi, C. Hüttig, Thermo-chemical convection in planetary mantles: advection methods and magma ocean overturn simulations, in *Integrated Information and Computing Systems for Natural, Spatial, and Social Sciences*, ed. by C.-P. Rückemann (2014)
13. A.-C. Plesa, C. Hüttig, M. Maurice, D. Breuer, N. Tosi, Large scale numerical simulations of planetary interiors, in *High Performance Computing in Science and Engineering '15* ed. by W.E. Nagel, D.H. Kröner, M.M. Resch (2016). https://doi.org/10.1007/978-3-319-24633-8_ 43
14. J.H. Roberts, S. Zhong, Degree-1 convection in the Martian mantle and the origin of the hemispheric dichotomy. J. Geophys. Res. **111**, E06013 (2006). https://doi.org/10.1029/ 2005JE002668
15. G. Schubert, D.L. Turcotte, P. Olson, *Mantle Convection in the Earth and Planets* (Cambridge University Press, Cambridge, 2001)
16. T. Spohn, Editorial: Special issue "Planetary evolution and life". Planet. Space Sci. **98**, 1–4 (2014). https://doi.org/10.1016/j.pss.2014.04.015. Q02018
17. P.J. Tackley, S.D. King, Testing the tracer ratio method for modeling active compositional fields in mantle convection simulations. Geochem. Geophys. Geosyst. **4**(4), 8302 (2003). https://doi. org/10.1029/2001GC000214

18. P.E. van Keken, S.D. King, H. Schmeling, U.R. Christensen, D. Neumeister, M.-P. Doin, A comparison of methods for the modeling of thermochemical convection. J. Geophys. Res. **102**, 22477–22495 (1997)
19. F. Willich, MPI I/O compared to POSIX I/O for Numerical Simulations on Supercomputers: Differences in Performance and Operation Principle (Bachelor Thesis). University of Applied Sciences (HTW), Berlin; German Aerospace Center (DLR), Berlin (2016)
20. S. Zhong, M.T. Zuber, Degree-1 mantle convection and the crustal dichotomy on Mars. Earth Planet. Sci. Lett. **189**, 75–84 (2001)

EXAHD: An Exa-Scalable Two-Level Sparse Grid Approach for Higher-Dimensional Problems in Plasma Physics and Beyond

Mario Heene, Alfredo Parra Hinojosa, Michael Obersteiner,
Hans-Joachim Bungartz, and Dirk Pflüger

Abstract Within the current reporting period (04/2016–04/2017) of our HLRS project we have developed a scalable implementation of the fault-tolerant combination technique. Fault-tolerance is one of the key topics in the ongoing research of algorithms for future exascale systems. Our algorithms enable fault-tolerance for both hard and soft faults, for the efficient and massively parallel computation of high-dimensional PDEs without the need of checkpointing or process replication. The research project EXAHD is part of DFG's priority program "Software for Exascale Computing" (SPPEXA). The project's target application is the large-scale simulation of plasma turbulence with the code GENE. The report combines parts of three publications.

1 Introduction

This report on recent progress of the EXAHD project combines parts of the two publications *Handling Silent Data Corruption with the Sparse Grid Combination Technique* (Proceedings of the SPPEXA Workshop, Springer, 2016) [23] and *A Massively-Parallel, Fault-Tolerant Solver for High-Dimensional PDEs* (Euro-Par 2016: Parallel Processing Workshop, Springer, 2016) [19], and illustrations and excerpts of *Scalable Algorithms for the Solution of Higher-Dimensional PDEs* (Software for Exascale Computing—SPPEXA 2013–2015, Springer, 2016) [18] (all with permission of Springer).

M. Heene • D. Pflüger (✉)
Institute for Parallel and Distributed Systems, University of Stuttgart, Universitätsstr. 38, 70569 Stuttgart, Germany
e-mail: mario.heene@ipvs.uni-stuttgart.de; dirk.pflueger@ipvs.uni-stuttgart.de

A.P. Hinojosa • M. Obersteiner • H.-J. Bungartz
Chair of Scientific Computing, Technical University of Munich, Boltzmannstraße 3, 85748 Garching, Germany
e-mail: hinojosa@in.tum.de; oberstei@in.tum.de; bungartz@in.tum.de

© Springer International Publishing AG 2018   513
W.E. Nagel et al. (eds.), *High Performance Computing in Science and Engineering '17*, https://doi.org/10.1007/978-3-319-68394-2_31

The main motivation behind EXAHD is the development of novel numerical schemes to tackle the exascale-challenges scalability and fault tolerance for the solution of higher-dimensional PDEs. To motivate our work and to introduce the sparse grid combination technique, we first briefly sketch EXAHD's target application, the simulation of plasma turbulence. This additionally illustrates a major strength of our approach: We are able to reuse standard solvers (with only minor adaptations) via a black box approach. For the presentation of results for the fault-tolerant combination technique, however, we will resort to the solution of the d-dimensional advection-diffusion equation. This enabled us to pick arbitrary dimensionalities for the development of our algorithms and software layer and to reduce the computational demand of simulations. Wherever possibly and adequate, we chose $d = 5$ to match the dimensionality of the gyrokinetic equations.

The generation of clean, sustainable energy from plasma fusion reactors is currently limited by the presence of microinstabilities that arise during the fusion process [9]. The numerical simulation of the resulting anomalous transport phenomena plays a decisive role in understanding the mechanisms that prevent the efficient generation of plasma fusion energy. The state-of-the-art code GENE provides the computational tools to achieve this. By means of the *sparse grid* formalism [6], we intend to enable resolutions for such simulations that are out of reach so far. Sparse grids help to alleviate the curse of dimensionality (GENE solves a 5-dimensional PDE) by considerably reducing the number of discretization points with only a small deterioration in the approximation quality. In particular, we construct a sparse grid with the *combination technique* [15] (see Sect. 2). The error bounds of the sparse grid approximation require the solution to fulfill certain regularity conditions, but the combination technique has been successfully used in a wide range of problems.

GENE solves the gyrokinetic equation

$$\frac{\partial f_s}{\partial t} + \left(v_\parallel \mathbf{b}_0 + \frac{B_0}{B_{0\parallel}^*} (\mathbf{v}_{E_\chi} + \mathbf{v}_{\nabla B_0} + \mathbf{v}_c) \right) \cdot \left(\nabla f_s + \frac{1}{m_s v_\parallel} \left(q \bar{\mathbf{E}}_1 - \mu \nabla \left(B_0 + \bar{B}_{1\parallel} \right) \right) \frac{\partial f_s}{\partial v_\parallel} \right) = 0, \tag{1}$$

where $f_s \equiv f_s(\mathbf{x}, v_\parallel, \mu; t)$ is (5+1)-dimensional (see [5, 8] for a thorough description of the model). GENE offers different simulation modes to solve (1). In *local* (or flux-tube) simulations, the x and y coordinates are treated in a pseudo-spectral way, with background quantities like density or temperature (and their gradients) being kept constant in the simulation domain. Since no domain decomposition is used in the radial direction, only a 4D grid is parallelized. In *global* runs, only the y direction is treated in a spectral way, and all five dimensions are parallelized, with background quantities varying radially according to given profiles.

Additionally, GENE distinguishes three main physical scenarios: multiscale problems (local mode, typical grids of size $1024 \times 512 \times 24 \times 48 \times 16$ with two species), stellarator problems (local mode, typical grids of size $128 \times 64 \times 512 \times 64 \times 16$ with two species), and global simulations (expected grid size $8192 \times 64 \times 32 \times 128 \times 64$, up to four species, currently unfeasible due to the curse of dimensionality). The last scenario, for example, would require about one terabyte of data just to store a single snapshot in memory (a complex number per degree of freedom), not to mention the computational and storage demand to solve for a single time step.

With respect to the exascale challenges, the objectives of our project are two-fold:

1. We lift the numerical solution of higher-dimensional PDEs to unprecedented dimensionalities and resolutions, which requires novel numerics implemented in a flexible software framework on massively parallel systems. This lifts parallelization to a new quality, introducing an extra, loosely coupled level of parallelism.
2. We realize a new, hierarchical approach to deal with algorithm-based fault tolerance without the need for checkpoint-restart. Furthermore, the underlying hierarchical scheme provides the means to even detect silent faults, for example due to bit flips in memory.

So far, we have developed new algorithmic tools to implement the combination technique on massively parallel systems, including load balancing strategies, efficient communication schemes among the nodes, and algorithm-based fault tolerance for node failures. In the previous reporting period we have demonstrated that our distributed combination algorithm scales on up to 180,225 cores on *Hazel Hen*.

In this report, we present first results of our massively parallel implementation of the fault-tolerant combination technique. We provide experimental results with hard faults (crashed processes) and soft faults (silent data corruption) obtained on *Hazel Hen* with the help of our home-grown fault simulation software layer.

2 Numerical Methods and Algorithms

2.1 Sparse Grid Combination Technique

The sparse grid combination technique [6, 15] computes the sparse grid approximation of a function u by a linear combination of component solutions u_l. Each u_l is an approximation of u that has been computed on a coarse and anisotropic Cartesian component grid Ω_l. In our case u is the solution of the high-dimensional gyrokinetic equations. The corresponding approximation u_l is the result of a simulation with the application code GENE computed on the grid Ω_l. In general this can be any kind of function which fulfills certain smoothness conditions.

The discretization of each d-dimensional component grid Ω_l is defined by the level vector $\mathbf{l} = (l_1, \cdots, l_d)^T$, which determines the uniform mesh width 2^{-l_i} in dimension i. The number of grid points of a component grid is $|\Omega_l| = \prod_{i=1}^{d}(2^{l_i} \pm 1)$ (+1 if the grid has boundary points in dimension i and −1 if not).

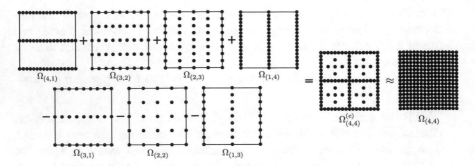

Fig. 1 The classical combination technique with $\mathbf{n} = (4, 4)$ and $\mathbf{l}_{\min} = (1, 1)$. Seven component grids are combined to obtain a sparse grid approximation (on the grid $\Omega_{(4,4)}^{(c)}$) to the full grid solution on the grid $\Omega_{(4,4)}$

In order to retrieve a sparse grid approximation $u_{\mathbf{n}}^{(c)} \approx u$ one can combine the partial solutions $u_{\mathbf{l}}(\mathbf{x})$ as

$$u_{\mathbf{n}}^{(c)}(\mathbf{x}) = \sum_{\mathbf{l} \in \mathscr{I}} c_{\mathbf{l}} u_{\mathbf{l}}(\mathbf{x}), \tag{2}$$

where $c_{\mathbf{l}}$ are the combination coefficients and $\mathscr{I}$ is the set of level vectors used for the combination. $\mathbf{n}$ denotes the maximum discretization level in each dimension. It also defines the discretization of the corresponding full grid solution $u_{\mathbf{n}}$ on $\Omega_{\mathbf{n}}$. Figure 1 shows a two-dimensional example.

There exist different approaches to determine the combination coefficients $c_{\mathbf{l}}$ and the index set $\mathscr{I}$ [20]. Usually,

$$u_{\mathbf{n}}^{(c)}(\mathbf{x}) = \sum_{q=0}^{d-1} (-1)^q \binom{d-1}{q} \sum_{\mathbf{l} \in \mathscr{I}_{\mathbf{n},q}} u_{\mathbf{l}}(\mathbf{x}) \tag{3}$$

is referred to as the classical combination technique with the index set [15]

$$\mathscr{I}_{\mathbf{n},q} = \{\mathbf{l} \in \mathbb{N}^d : |\mathbf{l}|_1 = |\mathbf{l}_{\min}|_1 + c - q : \mathbf{n} \geq \mathbf{l} \geq \mathbf{l}_{\min}\}, \tag{4}$$

where $\mathbf{l}_{\min} = \mathbf{n} - c \cdot \mathbf{e}$, $c \in \mathbb{N}_0$ s.th. $\mathbf{l}_{\min} \geq \mathbf{e}$ specifies a minimal resolution level in each direction, $\mathbf{e} = (1, \ldots, 1)^T$ and $\mathbf{l} \geq \mathbf{j}$ if $l_i \geq j_i \; \forall i$. The computational effort (with respect to the number of unknowns) decreases from $\mathscr{O}(2^{nd})$ for the full grid solution $u_{\mathbf{n}}$ on $\Omega_{\mathbf{n}}$ to $\mathscr{O}(dn-1)$ partial solutions of size $\mathscr{O}(2^n)$. If u fulfills certain smoothness conditions, the approximation quality is only deteriorated from $\mathscr{O}(2^{-2n})$ for $u_{\mathbf{n}}$ to $\mathscr{O}(2^{-2n}n^{d-1})$ for $u_{\mathbf{n}}^{(c)}$. The minimum level $l_{\min}$ has been introduced in order to exclude component grids from the combination. In some cases, if the resolution of a component grid is too coarse this could lead to numerically unstable or even physically meaningless results.

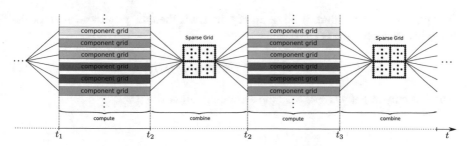

Fig. 2 Concept of recombination

For time-dependent initial value computations, such as they occur in GENE, propagating the combined solution $u_{\mathbf{n}}^{(c)}(t)$ in time requires recombining the component solutions every few time steps. This is necessary to guarantee convergence and stability of the combined solution. Recombination means to combine the component solutions $u_{\mathbf{i}}(t_i)$ according to Eq. (2) and to set the corresponding sparse grid solution $u_{\mathbf{n}}^{(c)}(t_i)$ as the new initial value for each component grid. After that, the independent computations are continued until the next recombination point t_{i+1}, where $t_i = t_0 + i\Delta t$ as illustrated in Fig. 2 (t_0 is the time of a given initial value $u(t_0)$ and the corresponding approximation $u_{\mathbf{i}}(t_0)$).

2.2 Error Splitting Assumption

The classical combination technique is based on the premise that each $u_{\mathbf{i}}$ satisfies the *error splitting assumption* (ESA):

$$u - u_{\mathbf{i}} = \sum_{k=1}^{d} \sum_{\substack{\{e_1,\dots,e_k\} \\ \subset \{1,\dots,d\}}} C_{e_1,\dots,e_k}(\mathbf{x}, h_{i_{e_1}}, \dots, h_{i_{e_k}}) h_{i_{e_1}}^p \cdots h_{i_{e_k}}^p, \tag{5}$$

where $p \in \mathbb{N}$. It is also assumed that each $\{e_1, \dots, e_k\} \subset \{1, \dots, d\}$ is bounded by $|C_{e_1,\dots,e_k}(\mathbf{x}, h_{i_{e_1}}, \dots, h_{i_{e_k}})| \leq \kappa_{e_1,\dots,e_k}(\mathbf{x})$, and that all $\kappa_{e_1,\dots,e_k}$ are bounded by $\kappa_{e_1,\dots,e_k}(\mathbf{x}) \leq \kappa(\mathbf{x})$. It is important to note that Eq. (5) is a *pointwise* relation, which means that it must hold for all points $\mathbf{x}$.

The ESA in one dimension reduces to

$$u - u_i = C_1(x_1, h_i) h_i^p, \quad |C_1(x_1, h_i)| \leq \kappa_1(x_1). \tag{6}$$

In two dimensions it becomes

$$u - u_{\mathbf{i}} = C_1(x_1, x_2, h_{i_1}) h_{i_1}^p + C_2(x_1, x_2, h_{i_2}) h_{i_2}^p + C_{1,2}(x_1, x_2, h_{i_1}, h_{i_2}) h_{i_1}^p h_{i_2}^p. \tag{7}$$

It is possible to show that, if all u_i satisfy the ESA, then the pointwise error of the combination technique is [15]

$$|u - u_n^{(c)}| = \mathcal{O}(h_n^2 (\log(h_n^{-1}))^2),$$ (8)

which is only slightly worse than the error on a full grid, $\mathcal{O}(h_n^2)$.

2.3 Software Framework for Massively Parallel Computations with the Combination Technique

We are developing a general software framework for large-scale computations with the combination technique as part of our sparse grid library SG++ [24]. In order to distribute the component grids over the available compute resources, the framework implements the manager-worker pattern, similarly to [14]. The available compute resources, in form of MPI processes, are arranged in process groups, except for a dedicated manager process. Computation of and access to the component grids is abstracted by so-called *compute tasks*. The manager distributes these tasks to the process groups as illustrated in Fig. 3 (left). The actual application code is then executed by all processes of a group in order to compute the task. Apart from the recombination step, there is no communication between the process groups, so the process groups can compute the tasks independently and asynchronously of each other.

The recombination is the only step that requires global communication between the process groups. Especially for time-dependent simulations, such as the initial value problems in GENE, which require a high number of recombination steps (in the worst case after each time step), an efficient and scalable implementation of this

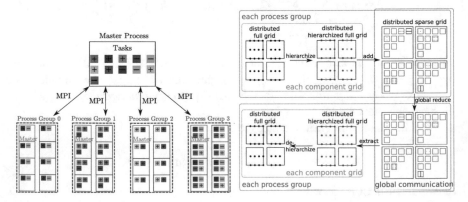

Fig. 3 Left: The manager-worker pattern used by the software framework. Right: The recombination step

step is crucial for the overall performance and the feasibility of the combination technique. Efficient and scalable in this context means that the time for the recombination must not consume more than a reasonable fraction of the overall run time and that this fraction must not scale worse than the run time of the application.

A scalable implementation of the recombination step must avoid gathering the component grid on a single node of the process group, or in a broader sense, on a fixed number of nodes that does not grow with the process group size. Thus, we have developed algorithms which directly work with the distributed data structures of the component grids.

The essential steps of this operation are visualized in Fig. 3 (right) First, all the component grids are transferred into the hierarchical basis representation of the sparse grid by hierarchization. The hierarchization algorithm works directly on the distributed data structures. An efficient and scalable implementation of this algorithm was presented in [17]. Then, the hierarchical coefficients of each component grid are added to a temporary distributed sparse grid data structure. Now, after the component grids within each process group have been reduced, they are reduced globally. The data of the distributed sparse grid is aligned so that the global reduction can be realized by a large number of MPI_Allreduce operations that work on disjoint sets of processes and thus can be executed completely in parallel. After the global reduction the combined solution exists on each process group in the distributed sparse grid. Finally, the combined solution is projected back into the function space of the component grids via dehierarchization (the inverse operation of hierarchization.

For a more detailed discussion of the combination step please refer to [17, 18]. These publications also contain experimental results that demonstrate the scalability of the distributed combination algorithm on up to 180,225 cores of *Hazel Hen*.

2.4 Fault Tolerant Combination Technique

Existing petascale systems can experience faults at various levels, from outright node failure to I/O network malfunctioning or software errors [7]. In this section we focus on hardware failures at node level, which are already quite frequent and will become more common as we approach exascale [7]. As most of the computing time in our application is spent on the actual PDE solver [18], we assume that faults are most likely to occur during the solver call.

If we look back at our parallelization scheme (Fig. 3), we can see what would happen if one or multiple nodes fail: the component grids assigned to the corresponding process groups would get lost. We avoid the need to checkpoint by applying the *Fault Tolerant Combination Technique* (FTCT) [16], an algorithmic approach to recover the combined solution in the presence of faults (see Fig. 4). If faults occur, the component grids that are lost as a result are excluded from the combination and alternative coefficients are computed. Since the resulting grid

EXAHD

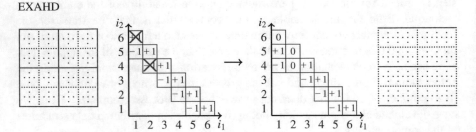

Fig. 4 The Fault-Tolerant Combination Technique in 2D with two faults, and the resulting alternative combination

has less grid points than that corresponding to the original scheme, it is a lossy approach, but as we will show in Sect. 3, the losses are very small. The alternative combination scheme used for the recovery step might need additional component grids that are not included in the original combination scheme (e.g., solution $u_{(1,4)}$ in the figure). In order to have these solutions readily available whenever a fault occurs, the grids below the main diagonals (there are d main diagonals for a d-dimensional combination technique) are computed as well. These extra grids are not used for the combined solution, but only for the recovery. These grids are very coarse and thus relatively cheap to compute (compared to the grids on the main diagonals), especially in higher dimensions. In [22], the overhead to compute these extra grids was predicted to account only for 3–4% of the overall runtime in a large-scale setup with GENE. In rare cases, also some of the coarsest component grids solutions have to be recomputed, but the overhead is very small as well. More details on the FTCT can be found in [16, 22].

2.4.1 Implementation of the FTCT

Our implementation is based on the ULFM specification [3], which is currently the most mature specification of a fault-tolerant MPI. It adds further functionality to the MPI Standard which enables to detect crashed processes and to exclude them from future communication.

After a fault has been detected, the following recovery steps are performed:

1. Create new MPI communicators that exclude the process ranks of the *whole* failed group, using the functions MPI_Comm_revoke and MPI_Comm_ shrink.
2. Compute new combination coefficients to correct the combined solution according to the FTCT algorithm.
3. Redistribute the tasks of the failed group to the living groups. Perform initialization routines if necessary (e.g., set up data structures, etc.).
4. If it is necessary to recompute some of the tasks, initialize these tasks with the last combined solution and compute them (for the required number of time steps).

The last combined solution is still available on all alive process groups from the last combination step (this is basically an in-memory checkpoint of the combined solution that we get "for free" at each combination step).

The corrected combined solution using the alternative coefficients will be computed during the next combination step. Afterwards, the combination coefficients are reset to the original coefficients (for the next combination step). We only use the reduced set of component grids to recover the combined solution at the current time step, but we continue the computation with the full original set of component grids.

2.4.2 Fault Simulation Layer

For our experiments we used our home-grown fault simulation layer [26] as a working implementation of ULFM was not yet available on *Hazel Hen*. This layer extends the MPI specification with some of the functionalities of the ULFM interface.

The functions we extended include the most common point-to-point and collective operations as well as some of the new fault tolerance functions, such as `MPI_Comm_shrink` and `MPI_Comm_revoke`. With the fault simulation layer we can let processes crash virtually by calling a function called `kill_me()`. When this happens, the process goes idle and only handles background operations of the fault simulation layer. It can now be detected as *failed* by other processes.

2.5 Detection of Silent Faults

The previous sections covered sources of faults that can be discovered by the runtime system, e.g. a fault tolerant MPI implementation. In these situations special signals or error codes can be utilized to detect and then recover from faults. However, the occurrence of a fault might not be recognized by the system. This class of errors is known as *silent data corruption* (SDC). In the most general sense, SDC can be any undetected system error, usually in the form of arithmetic computation errors, control errors, or wrong network transfer of data [25]. A well-known type of SDC is an undetected bit flip, which can be caused by cosmic rays interacting with the silicon die or other hardware defects [1, 7]. SDC is still poorly understood, but there is a growing consensus among computational scientists supporting the claim that these errors will affect simulations in future exascale systems [25]. In existing implementations silent faults are resolved by adding redundancy to the MPI implementation [12], using robust algorithms [10, 21] or checking if calculated values are within expected bounds [2].

In our approach we utilize the fact that the different combination solutions should look somewhat similar. This is a direct result of the ESA. Hence, any solutions that deviate too much from the others should be inspected for SDC. We will describe two methods to do this that are based on outlier detection techniques and robust regression. Before going into the details, we emphasize that in both cases we implicitly assume that each u_i is expressed in the hierarchical basis.

2.5.1   Method 1: Comparing Combination Solutions Pairwise via a Maximum Norm

In Sect. 2.2 we argued that each combination solution u_i has to fulfill the ESA (5) pointwise. So suppose we take two arbitrary combination solutions u_t and u_s from the set of all the solutions to be combined. If the two solutions satisfy the ESA, then their difference should satisfy the relation

$$u_t(x_{1,j}) - u_s(x_{1,j}) = C_1(x_{1,j}, h_{t_1})h_{t_1}^p + C_2(x_{1,j}, h_{t_2})h_{t_2}^p + C_{1,2}(x_{1,j}, h_{t_1}, h_{t_2})h_{t_1}^p h_{t_2}^p$$
$$- C_1(x_{1,j}, h_{s_1})h_{s_1}^p - C_2(x_{1,j}, h_{s_2})h_{s_2}^p - C_{1,2}(x_{1,j}, h_{s_1}, h_{s_2})h_{s_1}^p h_{s_2}^p. \tag{9}$$

This equation holds only for the grid points common to both grids Ω_s and Ω_t. Taking the largest value of (9) over all $(\mathbf{l}, \mathbf{j})$ can serve as an indicator of how similar or different two solutions u_t and u_s are. Now assume that SDC has affected one or more function values of either u_t or u_s, causing (9) to be large for the affected grid points. This means we should be suspicious of the grid point $x_{1,j}$ where (9) is largest. Hence, we first determine the point $x_{1,j}^*$ for which the largest difference is observed by comparing all points of each pair of component grids. Next, we compare all component grids at this specific point and try to fit it to the expected error $\tilde{\beta}_{(s,t)}$:

$$\beta_{(s,t)}^* := u_t(x_{1,j}^*) - u_s(x_{1,j}^*) \approx C_1(x_{1,j}^*, h_{t_1})h_{t_1}^p + C_2(x_{1,j}^*, h_{t_2})h_{t_2}^p$$
$$- C_1(x_{1,j}^*, h_{s_1})h_{s_1}^p - C_2(x_{1,j}^*, h_{s_2})h_{s_2}^p \tag{10}$$
$$=: \tilde{\beta}_{(s,t)}$$

We use the robust regression algorithms implemented in the GSL library [13] to fit the errors to the model. Outliers can be detected by large residual values of the robust regression.

2.5.2   Method 2: Comparing Combination Solutions via Their Function Values Directly

We start similar to the first method by identifying the grid point $x_{1,j}^*$ with the largest error. But now we take a look at the function values $u_i(x_{1,j}^*)$ *for all combination*

*solutions u_i containing the grid point $x^*_{1,j}$. Since the different values of $u_i(x^*_{1,j})$ should
be similar across the combination solutions, we try to fit them to a constant $\tilde{u}$:*

$$u_{\min} \leftarrow \min_{\tilde{u}} \sum_{l' \geq l} \rho \left(u_{l'}(x^*_{1,j}) - \tilde{u} \right). \tag{11}$$

Again, outlier detection can be applied to find SDC.

3 Numerical Tests and Results

3.1 Hard Faults

3.1.1 Numerical Setting

Our test problem is the d-dimensional advection-diffusion equation

$$\partial_t u - \Delta u + \mathbf{a} \cdot \nabla u = f \quad \text{in } \Omega \times [0, T) \tag{12}$$

$$u(\cdot, t) = 0 \quad \text{in } \partial\Omega$$

with $\Omega = [0, 1]^d$, $\mathbf{a} = (1, 1, \ldots, 1)^T$ and $u(\cdot, 0) = e^{-100 \sum_{i=1}^{d}(x_i - 0.5)^2}$, implemented
in the PDE-framework DUNE-pdelab [4]. For the spatial discretization we use the
finite volume element (FVE) method on rectangular d-dimensional grids. We use a
simple explicit Euler scheme for the time integration.

For our experiments we use DUNE to compute the component grids. All of the
grids are simulated with the same time step size and we combine after each time
step. We simulate process failures by calling the `kill_me()` function of our fault
simulation layer.

3.1.2 Results

The theoretical convergence of the combination technique (and its fault tolerant
version) has been studied before [15, 16], and our experiments confirm these results.
In all our convergence experiments we vary the number of process groups and let
one random group fail, so the percentage of failed tasks varies accordingly. We
compute the relative l_2-error $e = \frac{\|u_n^{(c)} - u_{\text{ref}}\|_2}{\|u_{\text{ref}}\|_2}$ at the end of the simulation ($t = 0.05$
and $\Delta t = 10^{-3}$), interpolating each combination solution to the resolution of the
reference grid. We combine after every time step.

Figure 5 (left) shows the convergence of the combination technique in 5D
with $\mathbf{i}_{\min} = (3, 3, 3, 3, 3)$ compared to a full grid reference solution of size
$\mathbf{n} = (6, 6, 6, 6, 6)$. The recovered combination technique with faults is only
minimally worse than without faults, even when half of the tasks fail.

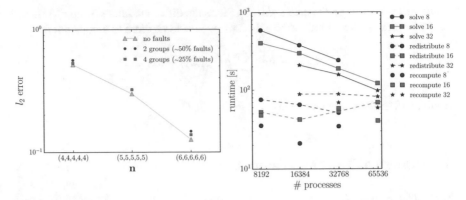

Fig. 5 Left: Convergence of the convection-diffusion equation in 5D using the combination technique, with and without faults. A single process fault causes an entire group to fail. Right: 5D scaling experiments. The number in the legend indicates the number of process groups

We performed scaling experiments on the supercomputer *Hazel Hen* to investigate the overhead of the FTCT in a massively parallel setup. We used a 5D combination technique with $\mathbf{n} = (8, 8, 7, 7, 7)$ and $\mathbf{i}_{min} = (4, 4, 3, 3, 3)$. This resulted in 126 component grids. Note that a computation of the full grid $\Omega_\mathbf{n}$ would not be feasible anymore – not even on the full machine. For the parallelization we used 8192, 16,384, 32,768 and 65,536 processes distributed on 8, 16 or 32 process groups of size 512, 1024, 2048 or 4096. In all cases one process failed in the second iteration and the entire corresponding group is removed. For the same number of groups, always a process in the same group failed.

Figure 5 (right) shows the time to *solve* all the component grids for one time step (using all groups, before the fault occurs), the time to *redistribute* the component grids of the failed group and the time to *recompute* certain tasks if necessary. Our application code has a rather bad node-level performance, but it scales well when the size of the process group is increased. The time for the combination can be neglected in these experiments since it was below 1 s in all cases. Two factors cause our curves to look slightly erratic. First, we show results for only one experiment per configuration. This is due to the long time and large computing resources it takes to run each simulation, which makes statistical studies infeasible. Second, there is some degree of randomness in the assignment of the tasks to the process groups, so even when the same group fails, the tasks to be redistributed or recomputed can vary in each run. The time to redistribute a task essentially is the time its initialization routine takes. For 16 and 32 groups the number of tasks to be redistributed is lower than the number of groups, so the time to redistribute is dominated by the slowest task. Furthermore, in our case the initialization function does not scale with the number of processes. This explains why the time to redistribute did not always decrease.

It is not easy to specify the exact overhead of the FTCT, since it depends on various parameters, such as the expected number of time steps between two failures,

the number of time steps computed in the solve step and the ratio between the
initialization time and the cost of one time step. However, we can easily formulate
upper bounds for the two most costly steps of the recovery process. The redistribute
step can never take longer than to initialize all tasks. The recompute step can
never take longer than the solve step. This means, if the cost of the initialization
is small compared to the total amount of work done between two process failures,
the overhead of the FTCT will be negligible. This is the main conclusion to be drawn
from our experiments.

After the recovery, the time for the solve step increases, since less process groups
are available. In the future we plan to mitigate this problem by not removing the
whole process group, but instead shrinking it to exclude only the failed processes.
The optimal case for our algorithm would be an MPI system that allows to request
new MPI processes at runtime after a node or process failure happened.

3.2 Numerical Tests with Silent Data Corruption

3.2.1 Experimental Setup

To test our algorithm we used once again the d-dimensional advection-diffusion
equation with the DUNE-pdelab framework. We carried out experiments in up to
five dimensions. In all cases we simulated 50 time steps using $\Delta t = 10^{-3}$ and we
combined the solutions every 10 time steps.

Based on the suggestions in [11], three test cases were designed to test our SDC
implementation. In all tests we first choose a combination solution u_i to inject SDC
into. Then we alter one function value in this component grid (in our case in the
middle of the domain) by multiplying it with a constant factor. The test cases only
differ in the size of this constant factor C to investigate very small ($C = 10^{-300}$),
slightly smaller ($C = 10^{-0.5}$) or very large ($C = 10^{+150}$) values of simulated SDC.
In all test cases SDC is only inserted once during the simulation.

We investigated the quality of our two detection methods in three and five
dimensions. We are primarily interested in the percentage of cases where SDC is
detected, as well as the quality of the combination technique after detecting and
removing the wrong solution.

For the simulations in 3D we used $\mathbf{n} = (7, 7, 7)$ and $\mathbf{l}_{min} = (3, 3, 3)$, which results
in 10 combination solutions. For each magnitude of SDC we ran six independent
simulations, injecting SDC at iterations 0, 9, 19, 29, 39 and 49, respectively. We
calculated the error of the solution compared to a full grid solution of level $\mathbf{n} =
(7, 7, 7)$ at the end of each simulation. One can see that Method 1 performs more
poorly when SDC is of moderate magnitude (10^{-300} or $10^{-0.5}$). Method 2 remained
robust, detecting all instances of SDC that would have otherwise led to large errors.
In cases where only small errors were introduced by the SDC the error detection
failed. However, this is not a problem as the errors remain tolerable in these cases.
Figure 6a summarizes our results.

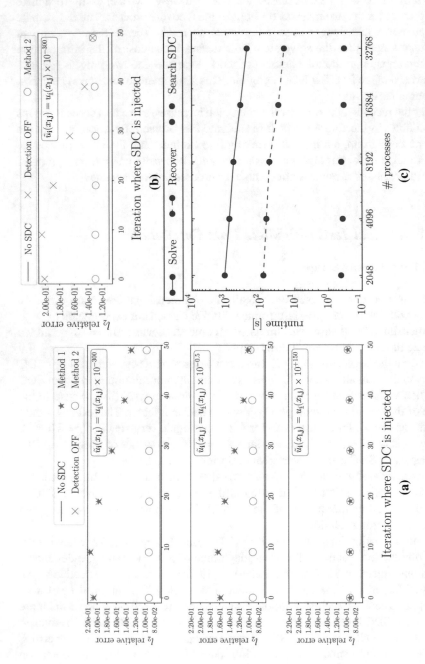

Fig. 6 (a) l_2 relative error of the 3D combination technique with simulated SDC injected in the middle of the domain; (b) l_2 relative error of the 5D combination technique with simulated SDC injected in the middle of the domain; (c) scaling experiments

For our detection tests in 5D we used $\mathbf{n} = (5, 5, 5, 5, 5)$ and a truncation parameter $\mathbf{l}_{min} = (2, 2, 2, 2, 2)$, which resulted in 21 combination solutions. Injection was done as in the 3D case and we compared the combination technique with a full grid solution of level $\mathbf{n} = (5, 5, 5, 5, 5)$. Due to the very high cost of running these tests, we only simulated one scenario, injecting SDC in the middle of the domain and detecting using Method 2 (Fig. 6b). We used two process groups, each with 1024 processes. As in the 3D cases, SDC is not detected when its effect is too small. In all other cases, it is detected and fixed.

3.2.2 Results: Scaling

To test the parallel performance of our algorithm, we measured the time needed to compare all pairs of grids and solve the robust regression problem of Method 2. We used a five dimensional scenario with $\mathbf{n} = (7, 7, 7, 7, 7)$ and $\mathbf{l}_{min} = (3, 3, 3, 3, 3)$, which results in 126 combination solutions. For the parallelization we used eight process groups, doubling the number of processes per group from 256 until 4096. Our time measurements can be seen in Fig. 6c (*Search SDC*). As expected, the time is negligible (3–4 orders of magnitude smaller than the time required for one timestep of the solver, *Solve*). Once the SDC is detected, the wrong solution is removed and the combination technique is adapted. The time to perform these operations is shown in the plot as *Recover*. The main cost of the recovery step comes from reinitializing the wrong task, as we reported in [19], which is still one order of magnitude smaller than one single timestep.

4 Conclusion and Outlook

In the previous reporting period, we have developed new, scalable algorithms to recombine the distributed component grids and demonstrated the scalability on up to 180,225 cores on *Hazel Hen*.

In the current reporting period we extended our algorithms by the capability to detect and correct hard and soft faults. We presented experimental results that demonstrate that our fault-tolerance mechanisms neither do noticeably increase the cost of the computations, nor do they impede the scalability of the distributed combination technique.

Within the upcoming project period we target actual simulations with our application code GENE and the fault-tolerant combination technique. One of our major goals is to demonstrate that with our approach it is possible to achieve new refinement levels for realistic and highly relevant simulation scenarios. Furthermore, we plan to verify our experiments with an actual fault-tolerant MPI implementation, e.g., ULFM, provided that such an implementation will be available on *Hazel Hen* in near future. In future experiments we will also employ state-of-the-art failure models which let the processes crash randomly, in order to study the impact of different fault rates on the accuracy and overhead of the combination technique.

Acknowledgements This work was supported by the German Research Foundation (DFG) through the Priority Programme 1648 Software for Exascale ComputingÌ (SPPEXA) and by the HLRS.

References

1. L. Bautista-Gomez, F. Cappello, Detecting silent data corruption for extreme-scale MPI applications, in *Proceedings of the 22nd European MPI Users' Group Meeting* (ACM, New York, 2015), p. 12
2. E. Berrocal, L. Bautista-Gomez, S. Di, Z. Lan, F. Cappello, Lightweight silent data corruption detection based on runtime data analysis for HPC applications, in *Proceedings of the 24th International Symposium on High-Performance Parallel and Distributed Computing, HPDC '15* (ACM, New York, 2015), pp. 275–278
3. W. Bland et al., A proposal for user-level failure mitigation in the mpi-3 standard. University of Tennessee (2012)
4. M. Blatt, A. Burchardt, A. Dedner, C. Engwer, J. Fahlke, B. Flemisch, C. Gersbacher, C. Gräser, F. Gruber, C. Grüninger et al., The distributed and unified numerics environment, version 2.4. Arch. Numer. Softw. **4**(100), 13–29 (2016)
5. A. Brizard, T. Hahm, Foundations of nonlinear gyrokinetic theory. Rev. Mod. Phys. **79**, 421–468 (2007)
6. H.J. Bungartz, M. Griebel, Sparse Grids. Acta Numer. **13**, 147–269 (2004)
7. F. Cappello et al., Toward exascale resilience: 2014 update. Supercomput. Front. Innov. **1**(1), 5–28 (2014)
8. T. Dannert, Gyrokinetische simulation von plasmaturbulenz mit gefangenen teilchen und elektromagnetischen effekten. Ph.D. thesis, Technische Universität München (2005)
9. E. Doyle, Y. Kamada, T. Osborne et al., Chapter 2: plasma confinement and transport. Nucl. Fusion **47**(6), S18 (2007)
10. J. Elliott, M. Hoemmen, F. Mueller, Evaluating the impact of SDC on the GMRES iterative solver, in *2014 IEEE 28th International Parallel and Distributed Processing Symposium* (IEEE, Piscataway, 2014), pp. 1193–1202
11. J. Elliott, M. Hoemmen, F. Mueller, Resilience in numerical methods: a position on fault models and methodologies (2014). arXiv preprint arXiv:1401.3013
12. D. Fiala, F. Mueller, C. Engelmann, R. Riesen, K. Ferreira, R. Brightwell, Detection and correction of silent data corruption for large-scale High-Performance Computing, in *Proceedings of the International Conference on High Performance Computing, Networking, Storage and Analysis*. (IEEE Computer Society Press, Piscataway, 2012), p. 78
13. M. Galassi, J. Davies, J. Theiler, B. Gough, G. Jungman, P. Alken, M. Booth, F. Rossi, R. Ulerich, GNU scientific library reference manual. Library available online at http://www.gnu.org/software/gsl (2015)
14. M. Griebel, W. Huber, U. Rüde, T. Störtkuhl, The combination technique for parallel sparse-grid-preconditioning or -solution of PDEs on workstation networks, in *Parallel Processing: CONPAR 92 VAPP V.* LNCS, vol. 634 (1992)
15. M. Griebel, M. Schneider, C. Zenger, A combination technique for the solution of sparse grid problems, in *Iterative Methods in Linear Algebra* (IMACS, Elsevier, North Holland, 1992), pp. 263–281
16. B. Harding et al., Fault tolerant computation with the sparse grid combination technique. SIAM J. Sci. Comput. **37**(3), C331–C353 (2015)

17. M. Heene, D. Pflüger, Efficient and scalable distributed-memory hierarchization algorithms for the sparse grid combination technique, in *Parallel Computing: On the Road to Exascale* (2016)
18. M. Heene, D. Pflüger, Scalable algorithms for the solution of higher-dimensional PDEs, in *Software for Exascale Computing - SPPEXA 2013–2015*, ed. by H.-J. Bungartz, P. Neumann, W.E. Nagel (Springer, Berlin, 2016), pp. 165–186
19. M. Heene, A.P. Hinojosa, H.J. Bungartz, D. Pflüger, A massively-parallel, fault-tolerant solver for high-dimensional PDEs, in *Euro-Par 2016: Parallel Processing Workshops*. Lecture Notes in Computer Science, vol. 10104 (Springer, Cham, 2016), pp. 635–647
20. M. Hegland, J. Garcke, V. Challis, The combination technique and some generalisations. Linear Algebra Appl. **420**(2–3), 249–275 (2007)
21. A. Pan, J.W. Tschanz, S. Kundu, A low cost scheme for reducing silent data corruption in large arithmetic circuits, in *IEEE International Symposium on Defect and Fault Tolerance of VLSI Systems, 2008. DFTVS'08* (IEEE, Boston, 2008), pp. 343–351
22. A. Parra Hinojosa et al., Towards a fault-tolerant, scalable implementation of GENE, in *Proceedings of ICCE 2014*. LNCSE (Springer, Berlin, 2015)
23. A. Parra Hinojosa et al., Handling silent data corruption with the sparse grid combination technique, in *Proceedings of the SPPEXA Workshop*. LNCSE (Springer, Berlin, 2016)
24. D. Pflüger et al., SG++ library. http://sgpp.sparsegrids.org/
25. M. Snir, R.W. Wisniewski, J.A. Abraham, S.V. Adve, S. Bagchi, P. Balaji, J. Belak, P. Bose, F. Cappello, B. Carlson et al., Addressing failures in exascale computing. Int. J. High Perform. Comput. Appl. **28**, 129–173 (2014)
26. J. Walter, Design and implementation of a fault simulation layer for the combination technique on HPC systems. Master's thesis, University of Stuttgart, 2016